MACROPOROUS POLYMERS

Production Properties and Biotechnological/Biomedical Applications

MACROPOROUS POLYMERS

Production Properties and Biotechnological/Biomedical Applications

Edited by
Bo Mattiasson
Ashok Kumar
Igor Yu. Galaev

CRC Press
Taylor & Francis Group
Boca Raton   London   New York

CRC Press is an imprint of the
Taylor & Francis Group, an **informa** business

Photographs on the cover are courtesy of Dr. Fatima Plieva.

CRC Press
Taylor & Francis Group
6000 Broken Sound Parkway NW, Suite 300
Boca Raton, FL 33487-2742

First issued in paperback 2017

ISBN 13: 978-1-138-11224-7 (pbk)
ISBN 13: 978-1-4200-8461-0 (hbk)

Library of Congress Cataloging-in-Publication Data

Macroporous polymers : production properties and biotechnological/biomedical
    applications / editors, Bo Mattiasson, Ashok Kumar, Igor Yu. Galaev.
        p. ; cm.
    Includes bibliographical references and index.
    ISBN 978-1-4200-8461-0 (hardcover : alk. paper)
    1. Polymers. 2. Porous materials. I. Mattiasson, Bo, 1945- II. Kumar, Ashok, 1963- III. Galaev, Igor. IV. Title.
    [DNLM: 1. Polymers--chemical synthesis. 2. Polymers--therapeutic use. 3. Cell Separation. 4. Gels--chemical synthesis. 5. Gels--therapeutic use. 6. Regenerative Medicine--methods. 7. Tissue Engineering. QT 37.5.P7 M174 2010]

    QD381.M355 2010
    620.1'92--dc22   2009036995

Visit the Taylor & Francis Web site at
http://www.taylorandfrancis.com

and the CRC Press Web site at
http://www.crcpress.com

Contents

Section I Production of Macroporous Polymers

Section II   Characterization of Macroporous Polymers

Section III   Application of Macroporous Polymers

Introduction

According to the International Union of Pure and Applied Chemistry (IUPAC) definition, macroporous polymers have pores in the range of 50 nm to 1 µm (http://goldbook.iupac.org/MT07177.html). Recently, however, much attention has been brought to the materials with pore sizes between 1 µm and 100 µm, and beyond. Such polymers, sometimes called supermacroporous polymers, are the main target of this book. As the border between these two types of materials is very diffuse, both names *supermacroporous* and *macroporous* are used in the book describing polymer systems with pores of micrometer sizes.

Macroporous hydrogels as a separate class of macroporous polymers are represented by materials composed of three-dimensional hydrophilic polymer networks that in most cases are biocompatible. These macroporous gels offer new and interesting possibilities in biotechnology and biomedicine due to their heterogenic structure, namely the pores filled with solvent and surrounded by relatively thin walls composed of polymer phase. The polymer nature, pore sizes and pore size distribution, pore connectivity, and pore tortuosity are factors that strongly influence the properties and possibilities for success in various applications within the life science area. Access to materials such as macroporous hydrogels has meant that a range of new challenges in life science research can be met. Development in the area is at present intense, both with regard to designing new and better materials and also concerning new applications.

This book gives an up-to-date compilation of modes of production of macroporous hydrogels, characterization of such materials, and applications with regard to both biotechnology and biomedicine. As microbial or mammalian (including human) cells are of the same size as pores in such materials, cells can be handled with macroporous gels and this opens up possibilities with regard to immobilization, separation, and cultivation of cells. The former alternatives are traditionally regarded as biotechnological applications while cultivation of mammalian cells is classified as biomedicine. The porous structure and the compatible properties form a basis to meet challenges in tissue engineering as well as cultivating mammalian cells in bioreactors. With large pores it is possible to modify the surface by grafting, for example, and still have enough pore lumen to allow cells to penetrate or pass.

One can produce macroporous hydrogels using different approaches and from different polymers, biodegradable as well as chemically stable polymers. In biomedicine biodegradability may be desired, while in some industrial applications, such as wastewater treatment processes, it might be more advantageous to use nondegradable robust materials.

The introduction of macroporous hydrogels is one example of polymer chemistry offering interesting solutions to biological problems. By combining the technique of making macroporous gels with other recent advantages in polymer chemistry (stimuli-responsive polymers, polymer brushes, controlled radical polymerization, reversible addition–fragmentation chain transfer (RAFT) polymerization, click chemistry, etc.), new dimensions can be reached.

Under one cover, the editors have done their best to collect chapters written by the scientist most active in the production and study of macroporous polymers and their applications. The intent of the book is to be considered state of the art and will provide further development in the area by attracting fresh recruits, both polymer chemists capable of developing new advanced polymers and biotechnologists ready to use these polymers for new applications. The editors hope that the volume will stimulate the minds of the readers to actively consider this new group of materials as interesting possibilities when addressing different challenges in life science and other areas.

Contributors

Kristina Ambrosch
Department of Pharmaceutical
    Technology
University of Leipzig
Leipzig, Germany

Inmaculada Aranaz
Instituto de Ciencia de Materiales
    de Madrid
Consejo Superior de
    Investigaciones Científicas
Madrid, Spain

Anil K. Bajpai
Department of Chemistry
Bose Memorial Research
    Laboratory
Government Autonomous
    Science College
Jabalpur, India

Maria B. Dainiak
Protista Biotechnology
IDEON
Lund, Sweden

and

Department of Biotechnology
Lund University
Lund, Sweden

Paul A. De Bank
Department of Pharmacy &
    Pharmacology
Centre for Regenerative Medicine
University of Bath
Bath, United Kingdom

Duarte M. de França Prazeres
IBB-Institute for Biotechnology and
    Bioengineering
Centre for Biological and Chemical
    Engineering
Instituto Superior Técnico
Lisbon, Portugal

Francisco del Monte
Instituto de Ciencia de Materiales
    de Madrid
Consejo Superior de
    Investigaciones Científicas
Madrid, Spain

Marianne J. Ellis
Department of Chemical
    Engineering
Centre for Regenerative Medicine
University of Bath
Bath, United Kingdom

Maria L. Ferrer
Instituto de Ciencia de Materiales
    de Madrid
Consejo Superior de
    Investigaciones Científicas
Madrid, Spain

Vida Frankovič
BIA Separations
Ljubljana, Slovenia

Igor Yu. Galaev
Department of Biotechnology
Lund University
Lund, Sweden

Per-Erik Gustavsson
Novo Nordisk A/S
Måløv, Denmark

María C. Gutiérrez
Instituto de Ciencia de
    Materiales de Madrid
Consejo Superior de
    Investigaciones Científicas
Madrid, Spain

Michael C. Hacker
Department of Pharmaceutical
    Technology
University of Leipzig
Leipzig, Germany

Ming-Hua Ho
Department of Chemical
    Engineering
National Taiwan University of
    Science and Technology
Taipei, Taiwan, Republic of China

Hsyue-Jen Hsieh
Department of Chemical
    Engineering
National Taiwan University
Taipei, Taiwan, Republic of China

Era Jain
Department of Biological Sciences
    and Bioengineering
Indian Institute of Technology
    Kanpur
Kanpur, India

Matthew D. Jones
Department of Chemistry
University of Bath
Bath, United Kingdom

Nika Lendero Krajnc
BIA Separations
Ljubljana, Slovenia

Ashok Kumar
Protista Biotechnology
Lund, Sweden

and

Department of Biological Sciences
    and Bioengineering
Indian Institute of Technology
    Kanpur
Kanpur, India

Juin-Yih Lai
Department of Chemical
    Engineering
The Research and Development
    Center for Membrane
    Technology
Chung Yuan University
Chung Li, Taiwan, Republic of
    China

Per-Olof Larsson
Pure and Applied Biochemistry
Lund University
Lund, Sweden

Bo Mattiasson
Department of Biotechnology
Lund University
Lund, Sweden

Sergey V. Mikhalovsky
School of Pharmacy and
    Biomolecular Sciences
Brighton University
Brighton, United Kingdom

Oguz Okay
Istanbul Technical University
Department of Chemistry
Istanbul, Turkey

Hossein Omidian
School of Pharmacy
Purdue University,
West Lafayette, Indiana, USA

Kinam Park
School of Pharmacy
Purdue University,
West Lafayette, Indiana, USA

Fatima M. Plieva
Protista Biotechnology AB
Lund, Sweden

Aleš Podgornik
BIA Separations
Ljubljana, Slovenia

Irina N. Savina
School of Pharmacy and
    Biomolecular Sciences
Brighton University
Brighton, United Kingdom

Michaela Schulz-Siegmund
Department of Pharmaceutical
    Technology
University of Leipzig
Leipzig, Germany

Sandeep K. Shukla
Bose Memorial Research Laboratory
Department of Chemistry
Government Autonomous Science
    College
Jabalpur, India

Franc Smrekar
BIA Separations
Ljubljana, Slovenia

Akshay Srivastava
Department of Biological Sciences
    and Bioengineering
Indian Institute of Technology
    Kanpur
Kanpur, India

Aleš Štrancar
BIA Separations
Ljubljana, Slovenia

Peter Tiainen
PolyPeptide Laboratories AB
Limhamn, Sweden

Paul E. Tomlins
National Physical
    Laboratory
Teddington, Middlesex,
    United Kingdom

Da-Ming Wang
Department of Chemical
    Engineering
National Taiwan University
Taipei, Taiwan, Republic of
    China

and

The Research and Development
    Center for Membrane
    Technology
Chung Yuan University
Chung Li, Taiwan, Republic of
    China

Section I

Production of Macroporous Polymers

1

Production of Macroporous Polymeric Materials by Phase Separation Polymerization

Oguz Okay

CONTENTS

1.1 Introduction

Polymer hydrogels are cross-linked materials absorbing large quantities of water without dissolving. The ability of hydrogels to absorb water arises from their hydrophilic functional groups attached to the polymer backbone while their resistance to dissolution arises from cross-links between network chains (Tanaka 1981; Shibayama and Tanaka 1993). If such a conventional hydrogel is dried by heating, the polymer network obtained has no voids (pores) in its structure. However, as the polymer network swells again in water, the space between the network chains increases so that a type of porosity called *molecular porosity* appears. Thus, molecular porosity in conventional gels depends on the degree of swelling and the distance between the polymer regions is in the range of a few nanometers. In contrast, however, macroporous hydrogels refer to materials having a permanent porous structure that persist even in the dry state (Dusek 1982). According to the IUPAC, material having pores of larger than 50 nm are called macroporous (Sing et al. 1985). Some researchers define macroporous gels as opaque materials with a measurable specific surface area and absorbing nonsolvents in their dry states (Millar et al. 1963; Rabelo and Coutinho 1994).

Macroporous gels contain nanometer to micron size liquid channels separated by cross-linked polymer regions, which provide sufficient mechanical stability. There are two basic techniques to obtain cross-linked polymers with a macroporous structure (Okay 2008). The first technique is the use of inert templates in the hydrogel preparation. By this technique, the polymer formation reactions are carried out in the presence of templates; a macroporous structure in the final hydrogel matrix appears after extraction of template materials. For example, by the cryogelation technique, the polymer formation reactions are carried out below the bulk freezing temperature of the reaction system (Lozinsky 2002). A macroporous structure in the final material appears due to the existence of ice crystals acting as a template for the formation of the pores. Another technique to create a macroporous network structure is the reaction-induced phase separation (i.e., phase separation polymerization). This technique involves the cross-linking copolymerization of the monomer and the cross-linker mixture in the presence of an inert diluent, which is soluble in the monomer mixture (Okay 2000). In order to obtain macroporous structures, a phase separation must occur during the course of the network formation process so that the two-phase structure formed is fixed by the formation of additional cross-links. After the polymerization, the diluent is removed from the network, leaving a porous structure within the highly cross-linked polymer network.

This chapter provides an overview of the formation mechanism of macroporous networks by phase separation polymerization and their characteristics under various experimental conditions. Some examples are also presented to demonstrate the correlation between the preparation conditions and the structure of macroporous hydrogels.

1.2 Formation Mechanism of Macroporous Structures

Macroporous networks are mainly prepared by free-radical, cross-linking copolymerization of vinyl–divinyl monomers in the presence of an inert diluent (Seidl et al. 1967; Guyot and Bartholin 1982; Okay 2000). The diluent, which is a solvent, a nonsolvent, or a linear polymer, is included in the reaction system as a pore forming agent, and plays an important role in the design of the pore structure of cross-linked materials. The diluent is initially soluble in the monomer mixture. At low cross-linker contents and in the presence of solvating diluents, the diluent remains in the gel throughout the reactions so that an expanded network structure is obtained. The expanded gels thus formed collapse during the removal of the diluent after their synthesis and therefore, they are nonporous in the dry state. Heterogeneities in the network structure start to appear and become permanent if the diluent separates out of the gel phase during polymerization.

If the cross-linker content is above a critical level, the reaction mixture of cross-linking copolymerization initially remains homogeneous as long as the growing polymer network is able to absorb all the available monomers and the diluent. As the reactions proceed, that is as the cross-link density of the network increases, a critical point is passed, at which the equilibrium degree of swelling of the network in the diluent becomes equal to its degree of dilution. At this point, since the dilution of a homogeneous network cannot be greater than its equilibrium degree of swelling, the reaction system will separate into two: the network phase and the separated phase. Thus, the condition for incipient phase separation during gelation is given by

$$q_v/q_v^0 \leq 1, \tag{1.1}$$

where q_v is the equilibrium volume swelling ratio of the network in the reaction system and q_v^0 is its dilution degree (reaction volume/volume of polymer network). To test the validity of Equation 1.1, gelation reactions were carried out using the styrene (S), divinylbenzene (DVB) comonomer system and using di-2-ethylhexyl phthalate (DOP) as the diluent (Okay 1986; Okay 1988). Figure 1.1A shows how the q_v/q_v^0 ratio varies with the cross-linker (DVB) concentration used in the network synthesis. The q_v/q_v^0 ratio decreases below unity between 18 and 22 wt % DVB, which is the phase separation condition of Equation 1.1. Indeed, the porosity $P\%$ of the networks starts to increase between the same DVB concentrations (Figure 1B). In Figure 1C, the volume swelling ratio q_v of the networks in toluene is shown as a function of the DVB concentration. The filled and open symbols represent q_v values of the networks prepared with and without using a diluent, respectively. The networks formed below 18 wt % DVB are in a swollen state (i.e., they swell much more than the corresponding, conventional networks prepared without using a diluent). This indicates that the diluent used in the synthesis remains in the gel phase during the reactions so that expanded gels form. However, above 18 wt % DVB, q_v rapidly decreases and approaches to that of homogeneous networks, indicating separation of the diluent out of the network phase. Figure 1.1 supports the relation between equilibrium swelling and the conditions of porosity formation during gelation reactions as expressed in Equation 1.1.

Phase separation during cross-linking can be induced by increasing the cross-linker concentration, (ν-induced syneresis) or by decreasing the solvating power of the diluent (χ-induced syneresis; Dusek 1965; Seidl et al. 1967; Dusek 1982). In both cases, the growing polymer network or the polymer chains cannot absorb all the available solvent in the reaction system. Thus, the reaction de-swells (or collapses) to form reaction particles (microspheres) within the separated continuous liquid phase. As the reaction proceeds, new microspheres are continuously generated due to the successive separation of the growing polymers. The agglomeration of the microspheres

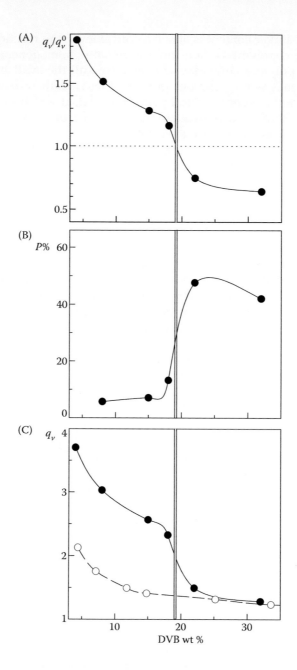

FIGURE 1.1

q_v/q_v^0 ratio (A), the total porosity $P\%$ (B), and the swelling ratio in toluene q_v (C) of S-DVB networks plotted as a function of the cross-linker (DVB) concentration. Diluent = di-2-ethylhexyl phthalate (DOP). Initial monomer concentration = 52 v/v %. The open symbols in (C) represent the swelling ratio of the networks prepared without using a diluent. (Reprinted from Okay, O., *Prog. Polym. Sci.*, 25, 711–79, 2000. With permission from Elsevier.)

results in the formation of a heterogeneous gel consisting of two continuous phases: a gel phase and a diluent phase. Removing of the diluent from the network creates voids (pores) of sizes from 1 nm up to 1 µm when measured in the glassy state of a polymer system. Compared to the χ-induced syneresis, the mechanisms of v-induced syneresis results in a more ordered pore structure consisting of smaller agglomerates rather than formed in the former mechanism.

Almost all macroporous networks formed by phase separation are characterized by a relatively broad pore size distribution ranging from micropores having widths up to 2 nm to macropores having widths greater than 50 nm. Figure 1.2 illustrates typical morphology of a macroporous network formed by phase separation polymerization. The polymer network consists of agglomerates of particles of various sizes that look like cauliflowers. The pores are irregular voids between agglomerates that are typically interconnected. From the scanning electron micrograph (SEM) image in Figure 1.2, one may distinguish microspheres of about 10^2 nm in diameter. The agglomerates of microspheres have sizes between 10^0–10^1 µm. Meso- and macropores having widths in the range 2–200 nm constitute the interstices between the microspheres while larger pores appear between the agglomerates of the microspheres.

As mentioned above, expanded hydrogels collapse during the removal of water and therefore, they are nonporous in the dry state. However, if an expanded hydrogel is freeze-dried after preparation, its expanded network structure is partially preserved so that a porous network can also be generated. For example, Figure 1.3 shows the SEM image of a porous poly(acrylamide) PAAm network prepared by the freeze-drying process. It is seen that the morphology of the network consists of polyhedral large pores

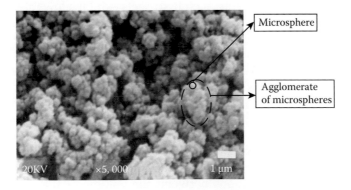

FIGURE 1.2
SEM image of poly(N-isopropylacrylamide; PNIPA) network at ×5000 (1µm bar in the lower right corner). Initial monomer (NIPA) concentration = 20 w/v %. N,N'-methylenebis(acrylamide) content = 30 wt % (with respect to the monomer). Polymerization temperature: 22.5°C. Diluent: Water.

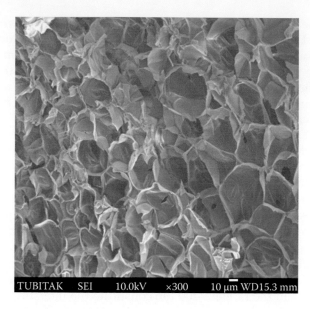

FIGURE 1.3
SEM of porous PAAm network prepared by a freeze-dry process. The network was prepared at 22°C in water. The scaling bar is 10 μm. Magnification = ×300. Initial monomer (AAm) concentration = 15 w/v %.

of about 10^1 μm in size and is completely different from the macroporous gels formed by phase separation (Figure 1.2). However, the porous structure formed by the freeze-dried process is not permanent and collapses during the drying of swollen hydrogel by heating or during its de-swelling in a poor solvent such as acetone (Ozmen, Dinu, and Okay 2008). Further, in contrast to the macroporous hydrogels, such hydrogels exhibit a low rate of response against the external stimuli. Therefore, it is important to note that to obtain macroporous hydrogels with a fast response, separation of water out of the gel phase should take place during the cross-linking process.

Although many experimental studies have dealt with the porosity formation during cross-linking in the presence of several diluents, only a few were concerned with the theory of formation of heterogeneities in such systems. Dusek treated the phase separation during the network formation process under the assumption of thermodynamic equilibrium between the network and separated phases (Dusek 1965; Dusek and Prins 1969). By using Flory's theory of swelling equilibrium and the theory of rubber elasticity, Dusek derived relations between the volume of the network phase and the monomer conversion. Similar thermodynamic relations were also reported by Boots and colleagues to predict the onset of phase separation in cross-linking copolymerization of divinyl monomers (1996). An important assumption involved in these derivations is that all polymer molecules beyond the gel point belong to the gel. Moreover, in these models, different vinyl group reactivities in a gelation system, as well as the variation of the gel cross-link density depending on the reaction condition, are neglected. A kinetic-thermodynamic model developed later predicts the formation conditions of heterogeneities during the gelation reactions by taking into account all the kinetic features of the reaction system and uses conversion-dependent sol and gel properties (Okay 1994; Okay 1999). Thermodynamic equations describe the phase equilibria between the gel and the separated phases and predict the volume of the gel phase as well as the distribution of soluble chains between the phases as a function of the monomer conversion.

1.3 Properties versus Preparation Conditions of Macroporous Materials

Relationships between the formation conditions and the structure of macroporous networks have been the subject of intensive studies during the last four decades (Seidl et al. 1967; Dusek 1982; Okay 2000; Svec 2004; Buchmeiser 2007; Qingquan, Li, and Angu 2007). The main experimental parameters determining the macroporosity of polymer networks are the cross-linker concentration, the type and the amount of the diluent, as well as the polymerization temperature and the type of the initiator.

The effect of the cross-linker concentration on the development of porosity in poly(N-isopropylacrylamide; PNIPA) networks is illustrated in Figure 1.4 (Sayil and Okay 2001). Here, SEM images and integral pore size distributions

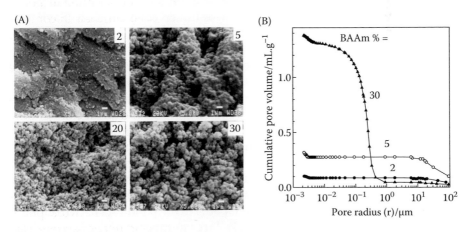

FIGURE 1.4
SEM images at 5000× (left) and integral pore size distributions (right) of PNIPA networks. BAAm contents in wt % (with respect to the monomer) are indicated. Initial monomer concentration = 20 w/v %. Gel preparation temperature = 22.5°C. (Reprinted from Sayil, C., and Okay, O., *Polymer*, 42, 7639–52, 2001. With permission from Elsevier.)

recorded using mercury porosimetry are shown for PNIPA networks formed at various cross-linker contents. PNIPA gels were prepared by free-radical, cross-linking copolymerization of N-isopropylacrylamide (NIPA) monomer and N,N'-methylenebis(acrylamide; BAAm) cross-linker at 22.5°C and at an initial monomer concentration of 20 w/v %. The numbers in the SEM images denote the cross-linker contents. At 2 wt % BAAm (with respect to the total monomer concentration), the network consists of large polymer domains; the discontinuities between the domains are also large. If the cross-linker content increases from 2 to 5 wt %, the morphology changes drastically and a structure consisting of aggregates of spherical domains (microspheres) appears. The microspheres obtained in the reaction medium with 5 wt % cross-linker are more or less fused together to form large aggregates. As the cross-linker content further increases from 5 to 30 wt % BAAm, the morphology changes from a structure with large aggregates of poorly defined microspheres to the structure consisting of aggregates of 1–2 µm dimensions of well-defined microspheres. The microspheres are about 0.1–0.5 µm in diameter. At 30% BAAm, the structure looks like cauliflowers, typical for a macroporous copolymer network. Integral pore size distribution curves also show that, as the cross-linker concentration is increased above 5%, macropores about 0.1–0.2 µm in radius start to appear and these pores contribute about 96% of the total porosity of this sample. The pore size distribution of macropores is very narrow. Considering the SEM picture of the sample obtained with 30 wt % cross-linker, the macropores correspond to the interstices between the microspheres. This indicates that the interstices between the microspheres are accessible to mercury only for the sample obtained with 30 wt % cross-linker.

The following mechanism explains formation and structural changes in PNIPA networks by increasing the cross-linker concentration (Sayil and Okay 2001): When the polymerization is initiated by the decomposition of the initiator molecules, the formed primary radicals start to grow by adding the monomer NIPA and the cross-linker BAAm. Initially, the primary molecules contain NIPA units, BAAm units with one vinyl group unreacted (i.e., with pendant vinyl groups), and BAAm units involved in cycles. As the time goes on, more and more primary molecules are formed so that the intermolecular cross-linking reactions between the primary molecules may also occur during the polymerization. However, previous works indicate the importance of cyclization and multiple cross-linking reactions in free-radical, cross-linking copolymerization (Funke, Okay, and Joos-Muller 1998). For example, in free-radical copolymerization of acrylamide (AAm) or NIPA and BAAm in aqueous solutions, at least 80% of the pendant vinyl groups are consumed by cyclization reactions and a large number of multiple cross-linkages occur per intermolecular cross-link formed (Naghash and Okay 1996; Gundogan, Melekaslan, and Okay 2002; Kizilay and Okay 2003). Thus, cyclization and multiple cross-linking clearly dominate over the intermolecular cross-linking reactions. Since every cycle reduces the coil dimension of the molecules as well as the solvent content inside the coil, the structure of the polymers formed is rather compact and can be considered as clusters. The higher the cross-linker content, the higher the cyclization density of the clusters is, or the lower their solvent content is. When the cyclization density of the clusters exceeds a critical value, the clusters phase separate and form particles called microspheres of about 0.1–0.5 μm in diameter. This point corresponds to a BAAm concentration between 2 and 5%. The agglomeration of the microspheres during cross-linking polymerization, through their peripheral pendant vinyl groups and radical ends, leads to the formation of large, unshaped, discrete agglomerates of 10–100 μm in diameter, which are further agglomerated to form the final network. Macropores of 0.1–0.2 μm in radius constitute the interstices between the microspheres while the voids between the agglomerates build the large pores in PNIPA networks.

It should be noted that, depending on the synthesis parameters, the structural elements, namely the microspheres and their agglomerates are more or less separated by holes. For example, the size and the swellability of the microspheres depend on the cross-linker concentration as well as on the type and the amount of the diluent. At high cross-linker contents, the microspheres are highly intramolecularly cross-linked and largely unswollen. Thus, increasing the cross-linker content in the polymer synthesis creates rigid microspheres so that the number of micropores increases and their size decreases. Since the main part of the specific surface area of a macroporous network comes from the surface of the smallest particles, the factors decreasing the size of the microspheres or increasing their stiffness such as the cross-linker content increase the specific surface area of the networks. Moreover, in the presence of a large amount of a solvating diluent, polymerization is kinetically preferred within the microspheres because the local concentration of the monomers is higher than in the surrounding solution. As a consequence, loose and large microspheres form at high diluent concentrations that tend to be fused after synthesis. On the other hand, poor solvents or linear polymers as a diluent promote phase separation during the polymerization and facilitate formation of bigger clusters and thus, meso- and macropores.

Several experiments show that the pore volume of the network increases as the concentration of the diluent increases, or as the solvating power of the diluent decreases (Seidl et al. 1967; Guyot and Bartholin 1982; Okay 2000; Santa Maria et al. 2004). Thus, the type and the amount of the diluent are also important parameters determining the porosity of polymer networks. Porous structures start to form when the amount of the diluent passes a critical value. The solvating power of the diluent has a critical effect on the pores size as well as on the porous structure of macroporous copolymers. Addition of solvating diluents in the presence of a large amount of cross-linker produces small pores and therefore a considerable specific surface area (50–1000 m²/g) and a relatively low pore volume. In contrast, however, an addition of nonsolvating diluents result in a large pore volume, a relatively large average pore diameter, and a specific surface area varying from 10 to 100 m^2/g. Further, the addition of polymeric diluents such as linear polymers produce very large pores reaching the micrometer range and a small specific surface area of about 0.1 to 10 m^2/g. It should be noted that the values given above only show the general tendencies and, depending on the cross-linker concentration, they change considerably.

Macroporous AAm-based hydrogels were prepared in the presence of methanol, acetic acid, water, dimethylsulfoxide (DMSO), poly(ethylene glycol; PEG) with different molecular weights or dimethylformamide as diluents (Shea et al. 1990; Caykara, Bulut, and Demirci 2007). The results show that more polar solvents, such as water or methanol, produce materials with higher specific surface area and internal pore volume than those prepared with less polar solvents such as DMSO. Also, decreased cross-linker flexibility increased the porosity of the resulting networks. Further, use of DMSO as a diluent in combination with alcohols produces macroporous PAAm networks with pore sizes up to 1000 nm (Xie, Svec, and Frechet 1997). The larger the chain length of the alcohols, the larger the microspheres and thus, the larger the pores in the final material.

Macroporous PAAm hydrogels can also be prepared in DMSO/water mixtures at subzero temperatures (Ozmen, Dinu, and Okay 2008). SEM analysis of the networks revealed the presence of porous morphologies (Figures 1.5 and 1.6). All the network samples formed at or below 0°C have pores with sizes about 10^0 μm that slightly increases with the temperature. It should be mentioned that the DMSO/water mixture exhibit a marked freezing point depression due to the formation of stable DMSO/water complexes. By simply adding DMSO to water, the freezing point of the resulting mixture is lowered and becomes –52°C at a volume ratio of 1:1. Indeed, DSC measurements show that the cross-linking polymerization reactions of AAm in 1:1 DMSO/water proceed in an unfrozen state. Thus, the cryogelation mechanism is not responsible for the formation of porous networks in 1:1 DMSO/water mixture even at –18°C. It was shown that the formation of the macroporous structure at or below 0°C is a result of the cooling-induced phase separation mechanism (Ozmen, Dinu, and Okay 2008).

The temperature sensitivity of PNIPA hydrogels has been used to construct porosity inside the PNIPA network. Poly(N-isopropylacrylamide) hydrogels exhibit a volume transition temperature (VTT) of about 34°C in water (Hirotsu 1993). Thus, PNIPA gel swells in water below the lower critical solution temperature (LCST) but it de-swells or collapses when the temperature is increased above the VTT. Therefore, one may expect that the PNIPA gels formed at low temperatures are homogeneous, whereas those formed at higher temperatures should exhibit heterogeneous structures. Indeed, visual observations show that the gels prepared at 18°C or lower are transparent, whereas those formed at higher temperatures (i.e., at 25°C) are opaque. Thus, at low temperatures, water remains in the gel phase throughout the polymerization, resulting in the formation of expanded homogeneous hydrogels.

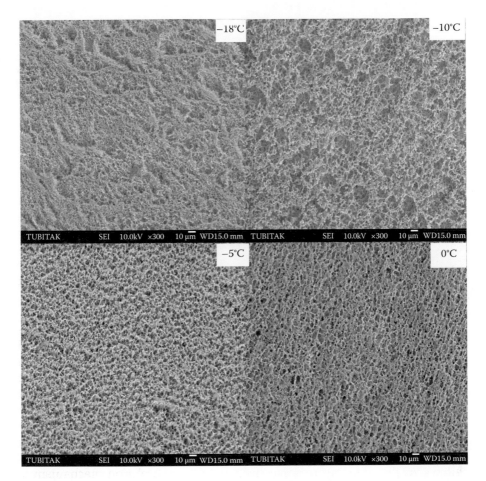

FIGURE 1.5
SEM images of PAAm networks prepared at various temperatures indicated. The scaling bars are 10 μm. Magnification = ×300. Initial monomer (AAm) concentration = 15 w/v %. Polymerization solvent: DMSO:water (1:1 by volume). (Reprinted from Ozmen, M. M., Dinu, M. V., and Okay, O. *Polym. Bull.* 60, 169–80, 2008. With permission from Springer Science + Business Media.)

At high temperatures, water separates out of the network phase due to the polymer-solvent incompatibilities (χ-induced syneresis) resulting in heterogeneous structures. It must be noted that, although a discrete volume transition in PNIPA gels occurs at 34°C, the turbidity in the gel samples appears below this temperature, for example at 25°C (Kayaman et al. 1998). This may be related to a rise in temperature of the polymerization system due to the gel effect (Kara, Okay, and Pekcan 2002).

Figure 1.7 shows the equilibrium swelling ratio of PNIPA hydrogels at 22°C in water (volume of swollen gel/volume of gel after preparation) plotted against the gel preparation temperature (Kayaman et al. 1998; Sayil and

FIGURE 1.6
SEM of PAAm network prepared at −18°C at a larger magnification than in Figure 1.5. The scaling bar is 1 μm. Magnification = ×7500. Initial monomer concentration = 15 w/v %. Polymerization solvent: DMSO:water (3:2 by volume).

Okay 2000). The hydrogel samples were prepared at different temperatures between 1 and 34°C. For the transparent gels (i.e., those formed below 20°C), the equilibrium swelling ratio increases only slightly with increasing gel preparation temperature. However, for heterogeneous gels, V/V_0 strongly depends on the preparation temperature and it drastically increases with increasing temperature. Furthermore, the gels prepared at temperatures higher than 30°C exhibited a loose structure, indicating a very low degree of effective cross-linkages within these gels. These results can be explained as follows. As the temperature is increased, the diluent (water) starts to separate out of the network phase during the reactions. Thus, the polymerization system tends to separate into polymer-rich and polymer-poor regions. Due to the high concentration of pendant vinyl groups in polymer-rich regions, one may expect enhanced rates of cross-linking and multiple cross-linking reactions, which would lead to the formation of highly cross-linked, microgel-like regions in the final gels. On the other hand, the polymer-poor regions become slightly cross-linked due to the high degree of dilution and thus, they constitute the interstices between the microgels in the final networks. According to this picture, the highly cross-linked regions may contribute to the rubber elasticity of the final gels as single junction points. Increasing temperature will increase the extent of phase separation and also, the concentration difference between the gel and diluent phases. As a result, the compactness of the microgels will increase but the connections between the microgels become weaker and weaker as the temperature increases, which would lead to an increased degree of swelling of heterogeneous PNIPA gels during the raising temperature.

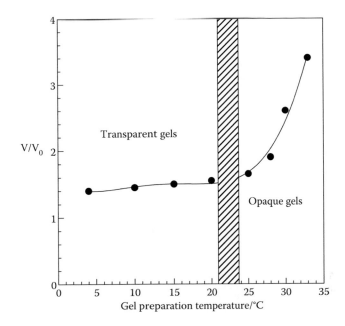

FIGURE 1.7
Variation of the equilibrium swelling ratio of PNIPA gels, V/V_0, with the polymerization temperature. Initial monomer concentration = 8 w/v %. BAAm = 2 wt %. Swelling temperature = 22°C. The shaded area represents the transition region from homogeneous to heterogeneous gels. (Reprinted from Kayaman, N., Kazan, D., Erarslan, A., Okay, O., and Baysal, B. M., *J. Appl. Polym. Sci.*, 67, 805–14, 1998. With permission. Copyright 1998 John Wiley & Sons, Inc.)

The relative values of the weight and the volume swelling ratios of PNIPA networks, q_w and q_v respectively, provide further information about the internal structure of porous networks in their swollen states (Okay 2000). The equilibrium weight swelling ratio q_w includes the amount of solvent taken both by the pores and the polymer region while the volume swelling ratio q_v only includes the amount of solvent taken by the gel portion of the network. Accordingly, the higher the difference between q_w and q_v, the higher the volume of the pores in the network sample is. Figure 1.8 illustrates q_w and q_v of PNIPA hydrogels formed at various temperatures (Sayil and Okay 2002). As the temperature is increased above 20°C, q_w becomes larger than q_v indicating the appearance of liquid channels inside the PNIPA hydrogels. Indeed, calculations indicate that the swollen state porosities in PNIPA hydrogels prepared at or above 22.5°C are around 60% (Sayil and Okay 2002).

In Figure 1.9A, the integral size distributions of the pores in PNIPA networks formed at various temperatures are given (Sayil and Okay 2002). In accord with Figures 1.7 and 1.8, the porosity of the network increases with increasing polymerization temperature from 12°C to 22.5°C. Unexpectedly, however, further increase of the temperature decreases the porosity of the networks. This result is due to the loose pore structure of the hydrogels formed at higher

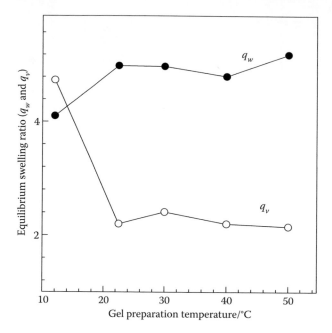

FIGURE 1.8

Equilibrium weight (q_w) and volume swelling ratios (q_v) of PNIPA networks shown as function of the gel preparation temperature. Cross-linker (BAAm) concentration in the initial monomer mixture = 30 wt %. Initial monomer concentration = 20 w/v %. (Reprinted from Sayil, C., and Okay, O., *Polym. Bull.*, 48, 499–506, 2002. With permission from Springer Science + Business Media.)

temperatures that collapses upon drying of the hydrogels. Figure 1.9B shows the effect of the initial monomer concentration on the porosity of the networks. As the initial monomer concentration is decreased from 20 to 10 w/v %, that is, as the amount of the diluent is increased, the porosity decreases due to the loosening of the porous structure that also collapses upon drying. Note that the porosity measurements cannot be carried out on network samples prepared at 5 w/v % monomer concentration due to the weak network structure.

Several inert diluents were used during the cross-linking copolymerization of NIPA and BAAm for the preparation of macroporous PNIPA hydrogels. These include hydroxypropyl cellulose (Wu, Hoffman, and Yager 1992), acetone (Zhang, Zhuo, and Yan 2002), 1,4-dioxane (Zhang, Huang, and Zhuo 2004), sucrose (Zhang et al 2003a), glucose (Zhang et al 2003b), silica particles (Serizawa, Wakita, and Akashi 2002), inorganic salts (Cheng, Zhang, and Zhuo 2003), as well as PEG of various molecular weights (Cicek and Tuncel 1998; Zhang and Zhuo 2000; Zhang et al. 2001; Zhuo and Li 2003; Dogu and Okay 2006; Fanger, Wack, and Ulbricht 2006). It was shown that the rate of de-swelling of swollen PNIPA gels increases with increasing concentration of the diluents due to their action as a pore-forming agent during the gelation process.

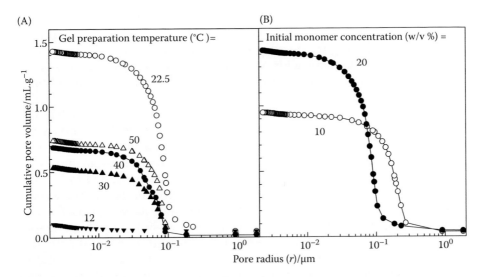

FIGURE 1.9
Integral size distributions of the pores in PNIPA networks. Cross-linker (BAAm) concentration in the initial monomer mixture = 30 wt %. (A) Initial monomer concentration = 20 w/v %. Gel preparation temperatures are indicated in the figure. (B) Gel preparation temperature = 22.5°C. Initial monomer concentrations (w/v %) are indicated in the figure. (Reprinted from Sayil, C., and Okay, O., *Polym. Bull.*, 48, 499–506, 2002. With permission from Springer Science + Business Media.)

Macroporous glycidyl methacrylate (GMA)/ethylene glycol dimethacrylate (EGDM) networks were prepared using lauryl alcohol (LA)/cyclohexanol mixtures as the diluent of the reaction system (Svec et al. 1975; Svec 1986). The materials obtained exhibit a high specific surface area that increased markedly with increasing content of the cross-linker EGDM. An increasing portion of LA in the diluent mixture leads to the formation of large pores and small surface areas. Macroporous GMA/EGDM networks formed in toluene and in the presence of polystyrene as a diluent also consist of large pores (Ferreira, Bigan, and Blondeau 2003). In macroporous GMA/EGDM networks prepared using cyclohexanol/dodecanol mixture as a diluent, although the total volume of the pores does not change much in the temperature range of 60–90°C, the size distribution of pores varies significantly with temperature (Svec and Frechet 1995). Increasing the temperature shifts the pore size distribution toward smaller pores. A similar effect of the polymerization temperature on the pore size has been observed in other vinyl/divinyl monomer copolymerization systems (Xie, Svec, and Frechet 1997; Arrua et al. 2006). The effect of the polymerization temperature on the porous structure is a consequence of the increasing decomposition rate of the initiator on rising the temperature. The higher the reaction temperature, the larger the number of the generated free-radicals per unit time and the larger the number of microspheres formed. Increasing number of microspheres necessarily decreases their size so that smaller voids between them appear in the final copolymer. Similar to the effect of the temperature, the type of the initiator also affects the porous structures. Increasing the decomposition rate of the initiator by replacing benzoyl peroxide initiator with AIBN, the size of the pores decreases at a given polymerization temperature due to the increasing rate of polymerization (Svec and Frechet 1995).

Cross-linked polymer derived from 2-hydroxyethylmethacrylate (HEMA) monomer is an important class of hydrophilic cross-linked polymers, which is widely used as a hydrogel for contact lenses (Wichterle and Lim 1960; Mathur, Moorjani, and Scranton 1996). The maximum equilibrium swelling of poly(HEMA) gels in water is thermodynamically limited to 39 wt %. Thus, if HEMA is polymerized in the presence of greater than 39 wt % water, a phase separation occurs and an opaque heterogeneous gel forms. By varying the amount of water as a diluent in the cross-linking polymerization of HEMA, a large range of porosities can be attained in the final networks. In addition to water, several diluents such as cyclohexanol/LA mixture (Coupek, Krivakova, and Pokorny 1973), benzyl alcohol (Jayakrishnan, Sunny, and Thanoo 1990), cyclohexanol/1-dodecanol mixture (Horak et al. 1986; Horak, Lednicky, and Bleha 1996; Hradil and Horak 2005), toluene (Okay and Gurun 1992), cyclohexane (Gomez, Igarzabal, and Strumia 2004), and 1-octanon (Fang, Pan, and Rempel 2006) were used in the synthesis of macroporous poly(HEMA) networks. In general, increasing solvating power of a diluent mixture produces smaller pores and thus increases the internal surface area although the total volume of the pores decreases. For instance, by varying the composition of cyclohexanol/dodecanol (solvating/nonsolvating) diluent mixture at a given degree of monomer dilution and the cross-linker content, the average diameter of the pores can readily be adjusted in macroporous poly(HEMA) networks. It was shown that macroporous poly(HEMA) hydrogels can also be prepared in the presence of an aqueous NaCl solution (Liu et al. 2000). Various pore morphologies were obtained depending on the concentration of NaCl. The presence of NaCl in the aqueous phase lowers the solubility of both HEMA and poly(HEMA), allowing for the precipitation or salting out, of the polymer phase during gelation. Swelling studies show an increase in equilibrium water content and hydrogel porosity as the NaCl concentration in the polymerization medium increases from 0 to 0.7 M.

1.4 Concluding Remarks

Free-radical cross-linking copolymerization of vinyl/divinyl monomers beyond gelation is a (quasi)ternary system composed of a polymer network, soluble polymers, and low molecular weight compounds such as the monomers and the diluent. All concentrations and properties of the system components change continuously during the reactions. With the onset of phase separation leading to the formation of macroporous structures, the diluent, monomers, and soluble polymers will distribute between the network and separated phases. A clearer understanding of this complicated reaction system will improve structuring hydrogels over all length scales, starting from nanometers to micrometers. Further, formation of porous structures by the phase separation technique requires separation of the diluent out of the gel phase during cross-linking, which necessarily results in gels exhibiting a low degree of volume swelling in good solvents. Although such hydrogels have a short response time, their volume change in external stimuli is fairly small. New strategies are needed to prepare a shorter response time for macroporous hydrogels exhibiting drastic volume changes in response to external stimuli.

References

Arrua, R. D., Serrano, D., Pastrana, G., Strumia, M., and Igarzabal, C. I. A. 2006. Synthesis of macroporous polymer rods based on an acrylamide derivative monomer. *J. Polym. Sci.A: Polym. Chem.* 44:6616–23.

Boots, H. M. J., Kloosterboer, J. G., Serbutoviez, C., and Touwslager, F. J. 1996. Polymerization-induced phase separation. 1. Conversion-phase diagrams. *Macromolecules* 29:7683–89.

Buchmeiser, M. R. 2007. Polymeric monolithic materials: Syntheses, properties, functionalization and applications. *Polymer* 48:2187–98.

Caykara, T., Bulut, M., and Demirci, S. 2007. Preparation of macroporous poly(acrylamide) hydrogels by radiation induced polymerization technique. *Nucl. Instr. Meth. Phys. Res. B* 265:366–69.

Cheng, S., Zhang, J., and Zhuo, R. 2003. Macroporous poly(N-isopropylacrylamide) hydrogels with fast response rates and improved protein release properties. *J. Biomed. Mater. Res.* 67A:96–103.

Cicek, H., and Tuncel, A. 1998. Preparation and characterization of thermoresponsive isopropylacrylamide-hydroxyethylmethacrylate copolymer gels. *J. Polym. Sci. A., Polym. Chem.* 36:527–41.

Coupek, J., Krivakova, M., and Pokorny, S. 1973. Poly[(ethylene glycol dimethacrylate)-co-acrylamide] based hydrogel beads by suspension copolymerization. *J. Polym. Sci.* C42:185–90.

Dogu, Y., and Okay, O. 2006. Swelling-deswelling kinetics of poly(N-isopropylacrylamide) hydrogels formed in PEG solutions. *J. Appl. Polym. Sci.* 99:37–44.

Dusek, K. 1965. Phase separation during the formation of three-dimensional polymers. *J. Polym. Sci.* B3:209–12.

Dusek, K. 1982. Network formation in chain crosslinking (co)polymerization. In *Developments in polymerization 3*, ed. R. N. Haward, 143–206. London: Applied Science.

Dusek, K., and Prins, W. 1969. Structure and elasticity of non-crystalline polymer networks. *Adv. Polym. Sci.* 6:1–102.

Fang, D., Pan, Q., Rempel, G. L. 2006. Preparation and morphology study of microporous poly(HEMA-MMA) particles. *J. Appl. Polym. Sci.* 103:707–15.

Fanger, C., Wack, H., and Ulbricht, M. 2006. Macroporous poly(N-isopropylacrylamide) hydrogels with adjustable size cut-off for the efficient and reversible immobilization of biomacromolecules. *Macromol. Biosci.* 6:393–402.

Ferreira, A., Bigan, M., and Blondeau, D. 2003. Optimization of a polymeric HPLC phase: Poly(glycidyl methacrylate–co-ethylene dimethacrylate) influence of the polymerization conditions on the pore structure of macroporous beads. *React. Func. Polym.* 56:123–36.

Funke, W., Okay, O., and Joos-Muller, B. 1998. Microgels-intramolecularly crosslinked macromolecules with a globular structure. *Adv. Polym. Sci.* 36:139–234.

Gomez, C. G., Igarzabal, C. I. A., and Strumia, M. C. 2004. Effect of the crosslinking agent on porous networks formation of hema-based copolymers. *Polymer* 45:6189–94.

Gundogan, N., Melekaslan, D., and Okay, O. 2002. Rubber elasticity of poly(N-isopropylacrylamide) gels at various charge densities. *Macromolecules* 35:5616–22.

Guyot, A., and Bartholin, M. 1982. Design and properties of polymers as materials for fine chemistry. *Prog. Polym. Sci.* 8:277–332.

Hirotsu, S. 1993. Coexistence of phases and the nature of first-order phase transition in poly-N-isopropylacrylamide gels. *Adv. Polym. Sci.* 110:1–26.

Horak, D., Lednicky, F., and Bleha, M. 1996. Effect of inert components on the porous structure of 2-hydroxyethyl methacrylate-ethylene dimethacrylate copolymers. *Polymer* 37:4243–49.

Horak, D., Svec, F., Kalal, J., Gumargalieva, K., Adamyan, A., Skuba, W., Titova, M., and Trostenyuk, N. 1986. Hydrogels in endovascular embolization. I. Spherical particles of poly(2-hydroxyethyl methacrylate) and their medico-biological properties. *Biomaterials* 7:188–92.

Hradil, J., and Horak, D. 2005. Characterization of pore structure of PHEMA-based slabs. *React. Func. Polym.* 62:1–9.

Jayakrishnan, A., Sunny, M. C., and Thanoo, B. C. 1990. Polymerization of 2-hydroxyethyl methacrylate as large size spherical beads. *Polymer* 31:1339–42.

Kara, S., Okay, O., and Pekcan, O. 2002. Real time temperature and photon transmission measurements for monitoring phase separation during the formation of poly(N-isopropylacrylamide) gels. *J. Appl. Polym. Sci.* 86:3589–95.

Kayaman, N., Kazan, D., Erarslan, A., Okay, O., and Baysal, B. M. 1998. Structure and protein separation efficiency of poly(N-isopropylacrylamide) gels: Effect of synthesis conditions. *J. Appl. Polym. Sci.* 67:805–14.

Kızılay, M. Y., and Okay, O. 2003. Effect of initial monomer concentration on spatial inhomogeneity in poly(acrylamide) gels. *Macromolecules* 36:6856–62.

Liu, Q., Hedberg, E. L., Liu, Z., Bahulekar, R., Meszlenyi, R. K., and Mikos, A. G. 2000. Preparation of macroporous poly(2-hydroxyethyl methacrylate) hydrogels by enhanced phase separation. *Biomaterials* 21:2163–69.

Lozinsky, V. I. 2002. Cryogels on the basis of natural and synthetic polymers: Preparation, properties and application. *Russ. Chem. Rev.* 71:489–511.

Mathur, A. M., Moorjani, S. K., and Scranton, A. B. 1996. Methods for synthesis of hydrogel networks: A review. *J. Macromol. Sci.—Rev. Macromol. Chem. Phys.* C36:405–30.

Millar, J. R., Smith, D. G., Marr, W. E., and Kressman, T. R. E. 1963. Solvent modified polymer networks Part 1. *J. Chem. Soc.* 218:218–25.

Naghash, H. J., and Okay, O. 1996. Formation and structure of polyacrylamide gels. *J. Appl. Polym. Sci.* 60:971–79.

Okay, O. 1986. Heterogeneous styrene-divinylbenzene copolymers. Stability conditions of the porous structures. *J. Appl. Polym. Sci.* 32:5533–42.

Okay, O. 1988. Styrene-divinylbenzene copolymers VI. Porosity formation in the presence of toluene-cyclohexanol mixtures as inert diluents. *Angew. Makromol. Chem.* 157:1–13.

Okay, O. 1994. Kinetic modelling of network formation and properties in free-radical copolymerization. *Polymer* 35:796–807.

Okay, O. 1999. Phase separation in free-radical crosslinking copolymerization: Formation of heterogeneous polymer networks. *Polymer* 40:4117–29.

Okay, O. 2000. Macroporous copolymer networks. *Prog. Polym. Sci.* 25:711–79.

Okay, O. 2008. Macroporous hydrogels from smart polymers. In *Smart polymers: Production, study and application in biotechnology and biomedicine*, ed. B. Mattiason and I. Yu. Galaev, 269–99. Boca Raton, FL: CRC Press, Taylor & Francis.

Okay, O., and Gurun, C. 1992. Synthesis and formation mechanism of porous 2-hydroxyethyl methacrylate–ethylene glycol dimethacrylate copolymer beads. *J. Appl. Polym. Sci.* 46:401–10.

Ozmen, M. M., Dinu, M. V., and Okay, O. 2008. Preparation of macroporous poly(acrylamide) hydrogels in DMSO/water mixture at subzero temperatures. *Polym. Bull.* 60:169–80.

Qingquan, L., Li, W., and Anguo, X. 2007. Research progress in macroporous styrene-divinylbenzene co-polymer microspheres. *Des. Mono. Polym.* 10:405–23.

Rabelo, D., and Coutinho, F. M. B. 1994. Porous structure formation and swelling properties of styrene–divinylbenzene copolymers: Effect of diluent nature. *Macromol. Symp.* 84:341–50.

Santa Maria, L. C., Aguiar, A. P., Aguiar, M. R. M. P., Jandrey, A. C., Guimaraes, P. I. C., and Nascimento, L. G. 2004. Microscopic analysis of porosity of 2-vinylpyridine copolymer networks 1. Influence of diluent. *Materials Letters* 58:563–68.

Sayil, C., and Okay, O. 2000. The effect of preparation temperature on the swelling behavior of poly(N-isopropylacrylamide) gels. *Polym. Bull.* 45:175–82.

Sayil, C., and Okay, O. 2001. Macroporous poly(N-isopropylacrylamide) networks: Formation conditions. *Polymer* 42:7639–52.

Sayil, C., and Okay, O. 2002. Macroporous poly(N-isopropylacrylamide) networks. *Polym. Bull.* 48:499–506.

Seidl, J., Malinsky, J., Dusek, K., and Heitz, W. 1967. Macroporose styrol-divinylbenzol copolymere und ihre anwendung in der chromatographie und zur darstellung von ionenaustauschern. *Adv. Polym. Sci.* 5:113–213.

Serizawa, T., Wakita, K., and Akashi, M. 2002. Rapid deswelling of porous poly(N-isopropylacrylamide) hydrogels prepared by incorporation of silica particles. *Macromolecules* 35:10–12.

Shea, K. J., Stoddard, G. J., Shavelle, D. M., Wakui, F., and Choate, R. M. 1990. Synthesis and characterization of highly crosslinked poly(acrylamides) and poly(methacrylamides): A new class of macroporous polyamides. *Macromolecules* 23:4497–507.

Shibayama, M., and Tanaka, T. 1993. Phase transition and related phenomena of polymer gels. *Adv. Polym. Sci.* 109:1–62.

Sing, K. S. W., Everett, D. H., Haul, R. A. W., Moscou, L., Pierotti, R. A., Rouquerol, J., and Siemieniewska, T. 1985. Reporting physisorption data for gas/solid systems with special reference to the determination of surface area and porosity. *Pure Appl. Chem.* 57:603–19.

Svec, F. 1986. Reactive polymers, 56. Interaction of reactive sites of macroporous copolymers glycidyl methacrylate–ethylene dimethacrylate. *Angew. Makromol. Chem.* 144:39–49.

Svec, F. 2004. Preparation and HPLC applications of rigid macroporous organic polymer monoliths. *J. Sep. Sci.* 27:747–66.

Svec, F., and Frechet, J. M. J. 1995. Temperature, a simple and efficient tool for the control of pore size distribution in macroporous polymer. *Macromolecules* 28:7580–82.

Svec, F., Hradil, J., Coupek, J., and Kalal, J. 1975. Reactive polymers I. Macroporous methacrylate copolymers containing epoxy groups. *Angew. Makromol. Chem.* 48:135–43.

Tanaka, T. 1981. Gels. *Sci. Am.* 244:110–23.

Wichterle, O., and Lim, D. 1960. Hydrophilic gels for biological use. *Nature* (London) 185:117–18.

Wu, X. S., Hoffman, A. S., and Yager, P. 1992. Synthesis and characterization of thermally reversible macroporous poly(N-isopropylacrylamide) hydrogels. *J. Polym. Sci. Part A, Polym. Chem.* 30:2121–29.

Xie, S., Svec, F., and Frechet, J. M. J. 1997. Preparation of porous hydrophilic monoliths: Effect of the polymerization conditions on the porous properties of poly (acrylamide-co-N,N-methylenebisacrylamide) monolithic rods. *J. Polym. Sci. A. Polym. Chem.* 35:1013–21.

Zhang, J. T., Cheng, S. X., Huang, S. W., and Zhuo, R. X. 2003a. Preparation of macroporous poly(N-isopropylacrylamide) hydrogel with improved temperature sensitivity. *J. Polym. Sci. A., Polym. Chem.* 41:2390–92.

Zhang, J. T., Cheng, S. X., Huang, S. W., and Zhuo, R. X. 2003b. Temperature-sensitive poly(N-isopropylacrylamide) hydrogels with macroporous structure and fast response rate. *Macromol. Rapid Commun.* 24:447–51.

Zhang, J., Huang, S., and Zhuo, R. 2004. Preparation and characterization of novel temperature sensitive poly(N-isopropylacrylamide-co-acryloyl beta-cyclodextrin) hydrogels with fast shrinking kinetics. *Macromol. Chem. Phys.* 205:107–13.

Zhang, X., Yang, Y. Y., Chung, T. S., Ma, K. X. 2001. Fabrication and characterization of fast response poly(N-isopropylacrylamide) hydrogels. *Langmuir* 17:6094–99.

Zhang, X., and Zhuo, R. 2000. Preparation of fast responsive, thermally sensitive poly(N-isopropylacrylamide) gel. *Eur. Polym. J.* 36:2301–3.

Zhang, X., Zhuo, R., and Yang, Y. 2002. Using mixed solvent to synthesize temperature sensitive poly(-isopropylacrylamide) gel with rapid dynamics properties. *Biomaterials* 23:1313–18.

Zhuo, R., and Li, W. 2003. Preparation and characterization of macroporous poly(N-isopropylacrylamide) hydrogels for the controlled release of proteins. *J. Polym. Sci. A., Polym. Chem.* 41:152–59.

2

Production and Properties of Cryogels by Radical Polymerization

Fatima M. Plieva, Igor Yu. Galaev, and Bo Mattiasson

CONTENTS

2.1 Introduction

Macroporous polymeric materials are gaining interest in biotechnology and biomedicine [1,2]. There are different approaches to prepare macroporous materials for freeze-drying [3], porogenation [4], microemulsion formation [5], gas-blowing technique [6], and phase separation [7,8]. Macroporous polymeric materials with open porous structures and controlled porosities are produced via cryogelation technology. The cryogelation (or gelation at subzero temperatures) technology renders it possible to prepare the macroporous hydrogels with a wide range of porosities from practically any gel-forming system [9–11]. The macroporous hydrogels (*cryogels*) are synthesized in semifrozen aqueous media where ice crystals perform as porogen and a template for macroporous structure after melting. Contrary to conventional gels (which are homophase systems where solvent is bound to the polymer network), the cryogels are heterophase systems where solvent is presented both inside interconnected macropores and bound to the polymer network. Depending on the gel precursors and chemical reaction used, cryogels with pore sizes from 0.1 to 200 μm can be formed [9–11]. One of the most attractive features of cryogels is their macroporosity, which is sufficient enough for processing the particulate-containing fluids including virus and cell suspensions, crude cell homogenates, wastewater, and so on [10–19].

A wide range of macroporous cryogels were produced via free-radical copolymerization of vinyl and divinyl monomers typically in an aqueous medium. The monomers used are mainly derivatives of acrylic acid such as acrylamide [20–28], N-isopropylacrylamide [29–32], dimethylacrylamide [33,34], and derivatives of methacrylic acid (such as 2-hydroxyethyl methacrylate; HEMA) [35–38]. Among polymeric gel precursors containing polymeriazable double bonds (macromers), the dextran-methacrylate and HEMA-L-lactide-dextran macromers were used for the preparation of macroporous cryogel systems (mainly as scaffolds for tissue engineering) [39,40] or the cryogels based on derivatives of cellulose were prepared via UV irradiation [41,42]. It has been shown that final properties of the macroporous cryogel prepared via free-radical, cross-linking polymerization depend on many parameters including both composition of the reaction mixtures and cryogelation conditions. The preparation of the macroporous cryogels with useful specific moieties (ligands) can be achieved in two ways: functional groups can be added by modification of the already prepared cryogel (i.e., modification after polymerization), or suitable monomers (i.e., bearing the required functionality) can be copolymerized with other monomers during the synthesis of cryogel. This chapter summarizes a recent research on preparation of macroporous cryogels via free-radical polymerization and points to some future perspectives.

2.2 Concept of Cryogel Formation via Free-Radical Polymerization at Subzero Temperatures

The cryogels are produced via a gelation process at subzero temperatures when most of the solvent is frozen and while the dissolved substances

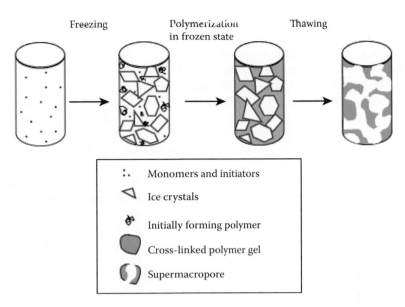

FIGURE 2.1
Scheme of cryogel formation.

(monomers or macromers, cross-linker, initiating system) are concentrated in small nonfrozen regions (or nonfrozen liquid microphase), where the chemical reaction and gel formation proceed with time. While all reagents are concentrated in the nonfrozen liquid microphase, some part of the solvent remains unfrozen and provides the solutes accumulated into the nonfrozen part with sufficient molecular or segmental mobility for reactions to perform. An acceleration of chemical reactions performed in unfrozen liquid microphase compared to the chemical reaction in bulk solution is often observed within a defined range of negative temperatures [43]. After melting solvent crystals (ice in case of aqueous media), a system of large continuous interconnected pores is formed (Figure 2.1). The shape and size of the crystals formed determine the shape and size of the pores formed after defrosting the sample. The pore size depends on the freezing rate, freezing temperature, and initial concentration of monomers/macromers in solution, content of cross-linker (divinyl monomer), thermal prehistory of the reaction mixture, sample size, presence of nucleation agents, and so forth [20,23,26,43,44]. In order to obtain a reproducible freezing pattern, a careful control of all experimental conditions is required.

2.2.1 Freezing Rate and Freezing Temperature

The cooling (freezing) rate is one of the crucial parameters to be controlled during the preparation of cryogels. The slower the freezing rate (or the higher the freezing temperature), the bigger the size of growing ice crystals and, as a result, the cryogels with bigger pore size are prepared [9,21,23,45]. At the same time, the freezing temperature should be low enough to overcome the overcooling problems. Overcooling (or supercooling) is defined as cooling below the initial freezing point of water without forming ice crystals. This is a nonequilibrium, metastable state of water. The nucleation temperature of water or aqueous solution is affected by both the cooling (freezing) rate, volume of the sample, and the addition of a nucleation agent [46,47]. The monitoring of change in temperature during the cooling of a reaction mixture allows for the determination of phase and structural changes in the system. The freezing curves obtained for the solution of monomers (acrylamide [AAm], N,N'-methylenebisacrylamide [MBAAm] and allyl glycidyl ether [AGE]) during the preparation of polyacrylamide cryogels (pAAm-cryogels) at three different final freezing temperatures T_f of −12, −20, and −30°C showed clearly the different phase states of the system under freezing [21] (Figure 2.2). Overcooling down to −11°C for 2 minutes was observed for the reaction mixture frozen at −12°C followed by an abrupt rise of temperature indicating solvent crystallization at a T_c of about −11°C. Once the critical mass of nuclei was reached (2.6 minutes after the reaction mixture was placed in the low temperature thermostat), the system nucleated at a temperature, T_{mc} of about −2.5°C (Figure 2.2). After crystallization was completed, the temperature dropped slowly till the fixed T_f value of −12°C was reached. A freeze concentration process occurred as water was frozen out from the solution. The increase in viscosity of the unfrozen liquid phase slowed down the further crystallization, which was over after about 8 minutes (Figure 2.2). At the lower final freezing temperatures (T_f −20 and −30°C, respectively), the freezing

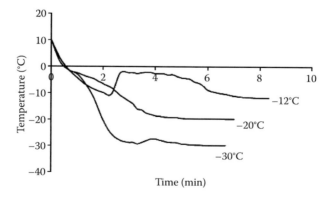

FIGURE 2.2
Thermograms for freezing aqueous solution of acrylamide monomers at −12, −20 and −30°C, respectively, in the presence of 1.2% (w/v) initiating system (ammonium persulfate/N,N,N',N'-tetramethylethylenediamine, APS/TEMED). (Reproduced from F. Plieva, X. Huiting, I.Yu. Galaev, B. Bergenståhl, and B. Mattiasson, *J. Mater. Chem.*, 16, 4065–73, 2006. With permission.)

thermograms revealed practically no overcooling with a small crystallization plateau near 0°C. A small peak observed at about 3.3 minutes for the thermogram obtained at T_f –30°C (a temperature that is below the eutectic point of about –20°C for a water/acrylamide system [21]) indicated the complete crystallization of the liquid microphase that is not frozen at higher temperatures.

The different cooling (freezing) rate of the reaction mixture directly affected the porous properties of cryogels, resulting in decreasing the flow-through properties for the pAAm cryogel monoliths prepared at –18°C as compared to those prepared at –10°C (Figure 2.3) [45]. At lower freezing temperatures, smaller ice crystals are formed, resulting in a smaller cross-section area of each macropore and therefore flow resistance of the cryogel monolith increases. Besides, the gel fraction yield as well as compressive strength were decreased for the pAAm-cryogels prepared at a T_f of –30°C (Table 2.1) [21]. The low gel fraction yield for the pAAm cryogels prepared at T_f of –30°C compared to that prepared at –12°C and was due to the quenching of the polymerization process when the system was completely frozen (no unfrozen liquid microphase remained) at temperatures below the eutectic point (for AAm solution this is about –20°C) [21].

A series of strong polyelectrolyte hydrogels made of 2-acrylamido-2-methylpropane sulfonic acid and MBAAm as the cross-linker were prepared at different temperatures, varied from –22 to 25°C [23]. It was shown that the equilibrium swelling and elasticity properties changed for cryogels compared to hydrogels prepared at temperatures above zero [23]. The volumetric swelling ratio was decreased and the elastic modulus was increased (almost

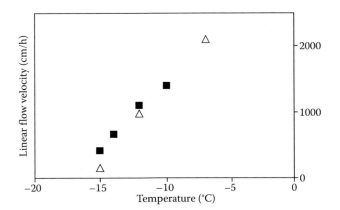

FIGURE 2.3
Flow resistance of the supermacroporous monolithic cryogels prepared at different temperatures: Plain-pAAm (open triangles), epoxy-pAAm (closed squares). The flow rate of water passing through the monolithic columns was measured at the constant hydrostatic pressure equal to 100 cm of a water column corresponding to a pressure of circa 0.01 MPa. For each sample the average of three measurements was taken. (Reproduced from F. M. Plieva, J. Andersson, I.Yu. Galaev, and B. Mattiasson, *J. Sep. Sci.*, 27, 828–36, 2004. With permission.)

TABLE 2.1

Properties of the pAAm Cryogels Prepared at Different Temperatures from the Reaction Mixtures with 1.2% (w/v) Content of Initiating System (Ammonium Persulfate/N,N,N',N'-tetramethylethylenediamine, APS/TEMED)

Temperature/°C	Notation for Prepared MPAAG	Gel Fraction Yield (%)	Compressive Strength/kPa	Water Flow Path/cm h⁻¹
−12	MPAAG-12	79 ± 4	28.3	310
−20	MPAAG-20	83 ± 3	29.2	215
−30	MPAAG-30	68 ± 4	21.0	30

Source: Reproduced from F. Plieva, X. Huiting, I.Yu. Galaev, B. Bergenståhl, and B. Mattiasson, *J. Mater. Chem.*, 16, 4065–73, 2006. With permission.

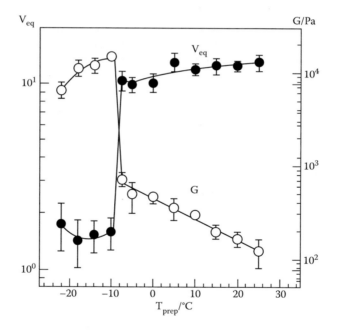

FIGURE 2.4

Volume swelling ratio V_{eq} (filled symbols) and the elastic modulus G (open symbols) of equilibrium swollen poly(2-acrylamido-2-methylpropane sulfonic acid) hydrogels shown as a function of the gel preparation temperature T_{prep}. (Reproduced from M. M. Ozmen, M. V. Dinu, and O. Okay, *Polymer Bull*, 60, 169–80, 2008. With permission.)

10-fold) for the cryogels compared to the conventional hydrogels (Figure 2.4). The much lower swelling degree for cryogels (i.e., hydrogels prepared at subzero temperatures) compared to the conventional gels prepared from the same feeds, but at temperatures above zero, is a typical characteristic of cryogels and was shown for different cryogel systems [11].

2.2.2 Concentration and Composition of Gel Precursors in the Initial Reaction Mixture

An increase in initial comonomer concentration (when other conditions were identical) gives rise to both the polymer concentration in pore walls and the overall strength of macroporous gel material, but often results in the decrease of pore sizes and interconnectivity of macropores [20]. The latter is due to a lower amount of solvent frozen out from the more concentrated initial solution and, as a consequence, a smaller total volume of ice crystals forming the macropores. Pore size and thickness of pore walls in cryogels are controlled in a wide range by simply changing the monomer/macromonomer concentration in the feed and type of a cross-linker used [20,21,29,36]. Increasing the concentration of monomers from 6 to 20% (w/v) for preparation of pAAm cryogels resulted in decreasing pore sizes in the cryogels with simultaneously increasing the thickness of the pore walls. The different states of water in pAAm cryogels prepared from feeds with different monomer concentrations and evaluated by Differential Scanning Calorimetry (DSC) showed that total pore volume in pAAm cryogels decreased at increasing the concentration of monomers in the initial feed (Table 2.2) [20]. The thickness of pore walls in cryogels along with the density of pore walls determine the macroscopic mechanical properties of the pAAm cryogels, while the pore size and pore wall density affect the accessibility of the ligands chemically bound to the polymer backbone [20]. Despite increased density of cryogels prepared from high monomer concentration, the cryogels remained highly elastic. The load displacement curves obtained for pAAm cryogels prepared from the feeds with 6, 10, and 15% (w/v) monomer concentration, respectively, show typical behavior of the highly elastic materials with extensive flexibility and shape recovery property after compression is released [11]. High elasticity of the macroporous cryogels allowed for their extensive deformations without being collapsed [21,30,32,48,49]. Increasing concentration of monomers in the initial reaction mixture resulted in increasing the mechanical strength of pAAm and poly-N-isopropylacrylamide (pNIPA) cryogels [21,32]. Comparative studies on

TABLE 2.2

Water in Different States in pAAm Cryogels Calculated from DSC Measurements

Cryogel Sample	Total Water (%)	Freezable Water[a] (% of Total Water)	Nonfreezable Water (% of Total Water)
6-pAAm	97.3	93.7	6.3
10-pAAm	95.0	96.1	3.9
18-pAAm	92.2	98.1	1.9

Source: Reproduced from F. M. Plieva, M. Karlsson, M.-R. Aguilar, D. Gomez, S. Mikhalovsky, and I.Yu. Galaev, *Soft Matter*, 1, 303–9, 2005. With permission.

[a] Freezable water includes free water and weakly bound water.

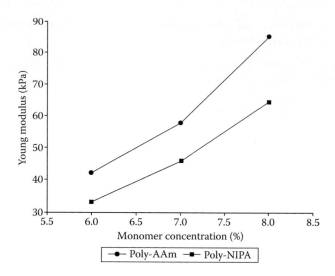

FIGURE 2.5
Comparative study of mechanical strength of pNIPA and pAAm cryogels. The Yung's modulus of 6, 7, and 8% (w/v) for pNIPA (closed squares) and pAAm (closed diamonds) cryogels. The parameters were determined at 80% compression from where the cryogel regains its original shape after swelling in water. (Reproduced from A. Srivastava, E. Jain, and A. Kumar, *Mater. Sci. Eng: A*, 464, 93–100, 2007. With permission.)

pAAm and pNIPA cryogels showed that the mechanical strength of pAAm cryogels were greater than that of pNIPA cryogels at increasing the concentration of monomers in the initial reaction mixture from 6.0 to 8.0% (w/v) (Figure 2.5) [32].

For the cryogels with "smart" properties as in pNIPA cryogels, faster response to stimuli involving a change of environmental conditions compared to that of conventional pNIPA gels was observed [29–32]. It is known that in aqueous solutions, pNIPA undergoes hydrophobic aggregation followed by transition from soluble to insoluble state when the temperature is increased above a critical temperature of 32°C [50]. The swelling degree of pNIPA cryogels prepared at subzero temperatures was much less as compared to that of the conventional pNIPA gels prepared at 22°C (Figure 2.6) [31]. The swelling of pNIPA cryogels depended on the concentration of monomers in the initial reaction mixture and cross-linking degree [29,30,32]. The sharp transition around 30°C was observed for pNIPA cryogels studied in the temperature interval from 4 to 40°C [30]. The stimuli response of pNIPA cryogels to a temperature change from 19°C to 37°C was very fast. The lower the total monomer concentration and the cross-linking degree of pNIPA cryogels, the higher was the change in swelling and the larger amplitude of mechanical deformation occurred when the temperature was changed from 19°C to 37°C and back again (Figure 2.7).

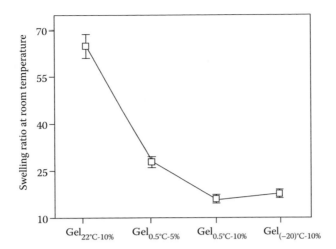

FIGURE 2.6
The time dependence of swelling of pNIPA cryogels produced under semifrozen conditions from polymerization feeds with total monomer concentrations 6 and 10% (w/v) and MBAAm (cross-linker)/NIPA ratios 1/10 and 1/100 mol/mol. (Reproduced from X.-Z. Zhang, and C.-C. Chu, *J. Mater. Chem.*, 13, 2457–64, 2003. With permission.)

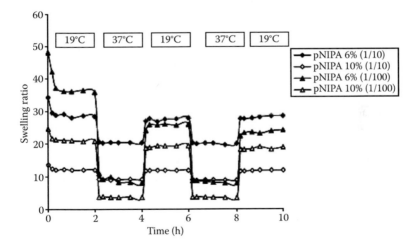

FIGURE 2.7
Swelling ratios of the pNIPA gels in distilled water at room temperature. (Reproduced from I.Yu. Galaev, M. B. Dainiak, F. Plieva, and B. Mattiasson, *Langmuir*, 23, 35–40, 2007. With permission.)

The properties of cryogels depend on the nature of cross-linkers (divinyl monomers) used for the preparation of cryogels and their content in the initial reaction mixture [20,29,30,36]. Thus, HEMA-based cryogel was prepared using both the nondegradable cross-linker as PEG-diacrylate (PEG-DA) and the chemically degradable cross-linker

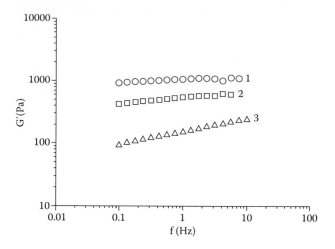

FIGURE 2.8
Elastic moduli of cryogels prepared by UV-vis irradiation of moderately frozen systems based on (1) 3 wt % aqueous hydroxypropylmethylcellulose solutions containing 3 wt % N,N′-methylenebisacrylamide, (2) 3 wt % poly(ethylene glycol) diacrylate, and (3) without cross-linking agent. (Reproduced from P. Petrov, E. Petrova, R. Stamenova, C. B. Tsvetanov, and G. Riess, *Polymer*, 47, 6481–84, 2006. With permission.)

N,N′-bis(methacryloyl)-L-cystine that had different porous structures [36]. Substitution of the cross-linker MBAAm to more hydrophilic cross-linker diallyltartaramide resulted in preparation of pAAm cryogels with two-fold higher swelling degrees [20]. When the macroporous cryogels were prepared as monolithic chromatography columns, the content of the cross-linker in the initial reaction mixture was essential. It was shown that the highly cross-linked pAAm-cryogel monoliths (with cross-linking ratio vinyl(monomers)/divinyl (cross-linker) of 5–10 mol/mol had higher flow-through properties compared to the pAAm-cryogel monoliths formed from the feedstock with low content of the cross-linker (MBAAm) in the reaction mixture (vinyl/divinyl ratio of 20–30 mol/mol) [48,51]. The addition of the cross-linker (MBAAm and PEG-DA) in the initial reaction mixture during preparation of cryogel made of (hydroxypropyl) methyl-cellulose (HPMC) and prepared via UV irradiation of moderately frozen medium resulted in increasing the gel fraction yield from 50 to 67% (when MBAAm was used as cross-linker) and to 63% in the case of PEG-DA as a cross-linker [41]. The preparation of HPMC cryogels in the presence of the cross-linkers also resulted in increased elastic modulus compared to those of the HPMC cryogels prepared without a cross-linker (Figure 2.8).

When preparing the macroporous polymer structures via free-radical polymerization, it is essential to show that the cross-linked matrix is free from nonreacted monomers. The pAAm-cryogels were shown not to contain any nonreacted monomer (AAm) as was confirmed by gas liquid

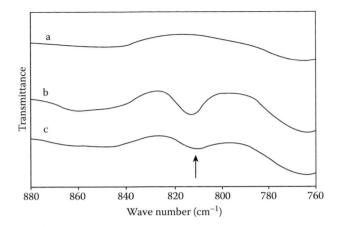

FIGURE 2.9
A fragment of (a) FTIR spectra of dextran, (b) glycidyl metchacrylate derivatized dextran (dextran-MA), and (c) polymerized at –20°C dextran-MA. Arrow indicates a peak at 813 cm⁻¹ originated from a double bond of methacrylate group. (Reproduced from F. Plieva, A. Oknianska, E. Degerman, I.Yu. Galaev, and B. Mattiasson, *J. Biomater. Sci. Polym. Edn.*, 17, 1075–92, 2006. With permission.)

chromatography analysis [45]. When the cryogels were prepared from macromer precursors (as dextran with grafted polymerizable double bonds, dextran-metacrylate, dextran-MA), the content of the remaining nonreacted macromer (dextran-MA) was hardly detectable (Figure 2.9) as was demonstrated by FTIR analysis [39].

2.2.3 The Effect of Initiator Content on Cryogel Porous Structure

In the free-radical polymerization reactions, the concentrations of the initiator ammonium persulfate (APS) and activator, N,N,N',N'-tetra-methylethylenediamine (TEMED) have a great influence on the polymerization rate as well as on the molecular weight of the resulting polymers. The polymerization rate is controlled by the amount of initiating system in the reaction mixture. The polymerization reaction starts with the reaction between APS and TEMED to form free radicals to initiate the polymerization reaction of vinyl and divinyl monomers. Noteworthy is that the APS/TEMED initiating system is a very efficient initiating system to perform the free-radical polymerization at subzero temperatures. Attempts to use other initiating systems to trigger the free-radical polymerization at subzero temperatures were not successful.

When preparing the macroporous cryogels via free-radical polymerization at subzero temperatures, one could differentiate two possible situations: freezing of the solution before the gelation takes place and freezing of the already gelated mixture. The influence of the amount of initiating system was studied for the copolymerization of AAm with the cross-linker MBAAm and functional comonomer AGE in aqueous medium [21], for pNIPA cryogels prepared in aqueous medium (cross-linked with MBAAm), and in dioxane medium (cross-linked with water insoluble dimethacrylate-tyrosine-lysine-tyrosine cross-linker) [29].

The pAAm cryogels were prepared in the presence of a 1.2% (w/v) initiating system in the reaction feed (freezing before gelation) and in the presence of a 5% (w/v) initiating system (freezing after gelation) [21]. Even though the size of pores in the pAAm cryogels prepared according to the possible situations (i.e., freezing before gelation and freezing after gelation) was essentially the same, the flow-through properties for the pAAm cryogels were significantly different [21]. Low flow-through properties of pAAm cryogels (5% (w/v) initiating system) were due to losing the interconnectivity of macropores in these cryogels. It is widely known that the ability for ice crystals to grow and merge with each other results in fabrication of the interconnected channels (pores) after melting of the ice crystals [43]. As ice crystals were formed within the already formed gel (where it was not possible for ice crystals to merge with each other), the porous structure with large (μm in size) closed (not interconnected) macropores were formed resulting in a higher flow resistance of pAAm cryogels prepared from the feed with a 5% (w/v) APS/TEMED system as compared to the pAAm cryogels prepared from the feed with a 1.2% (w/v) APS/TEMED system [21]. The pAAm cryogels prepared with 5% (w/v) APS/TEMED system are not highly elastic and were already broken at compression of 40% compared to the highly elastic

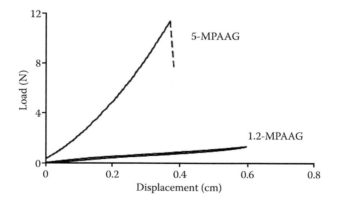

FIGURE 2.10
Load displacement curves for the pAAm cryogels prepared in the presence of 1.2% (w/v) ammonium persulfate/N,N,N',N'-tetramethylethylenediamine (APS/TEMED) system (1.2-MPAAG) and 5% (w/v) APS/TEMED system (5.0-MPAAG) in the reaction feed. (Reproduced from F. Plieva, X. Huiting, I.Yu. Galaev, B. Bergenståhl, and B. Mattiasson, *J. Mater. Chem.*, 16, 4065–73, 2006. With permission.)

pAAm cryogels prepared from the feed with the 1.2% (w/v) APS/TEMED system (Figure 2.10).

2.2.4 Influence of Solvent on Porous Structure of Cryogels

In case of gels prepared at subzero temperatures, the crystals of frozen solvent perform as porogen, thus the presence of other than water solvents has an effect on the porous structure both affecting the solubility of polymer and monomer as well as affecting the size and shape of solvent crystals formed in the semifrozen system and volume of the unfrozen region (unfrozen liquid microphase). The choice of solvents for preparation of cryogels is mainly specified by the melting point of the solvents (i.e., the solvents should be freezable at moderately low temperature and not lower than –20°C). The solvents can be divided in two categories: solvents that are miscible with the resultant polymeric network and solvents that are able to induce phase separation during the polymerization process. Polymerization induced phase separation is a process in which an initially homogeneous solution of monomer and solvent becomes phase separated during the course of the polymerization. The use of organic solvents or the mixture of organic solvent with water for the synthesis of cryogels is necessary, for example in case of water insoluble monomers [40] or cross-linkers [29]. Pure organic solvents such as formamide, 1,4-dioxane and dimethyl sulphoxide (DMSO) or their mixtures with water were used for synthesis of cryogels and their porous properties were evaluated [21,25,29,31].

Solvents like formamide and 1,4-dioxane with melting temperatures at 2 and 11°C, respectively, and different polarities (formamide > water > 1,4-dioxane) were used for preparation of the pAAm cryogels in water-organic media at –20°C [21]. It was shown that due to the association of polar molecules of formamide with the monomer molecules (AAm and MBAAm; i.e., solvent miscible with resultant polymeric network), the porous structure of pAAm cryogels synthesized in formamide was principally different from the structure of pAAm cryogels synthesized in 1,4-dioxane (which, as was expected, induced phase separation during the polymerization process) [21]. The SEM analysis of pAAm cryogel synthesized in 95% formamide indicated a nonporous structure (Figure 2.11a). The association of monomers (AAm and MBAAm) and polymer network with formamide results in better dissolution of hydrophobic MBAAm in the formamide media compared to that in water, and the probability for the formation of highly concentrated MBAAm regions within the network (that were observed for pAAm prepared in water [21]) disappeared when pAAm cryogel was synthesized in formamide (Figure 2.11a). Contrary, the pAAm cryogels synthesized in 95% 1,4-dioxane showed microporous structure due to the segregation of polymeric phase during the polymerization

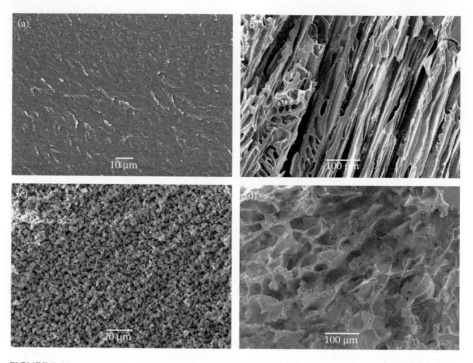

FIGURE 2.11
SEM of conventional pAAm gels prepared at 22°C (a) and (c) and pAAm cryogels (b) and (d) prepared at –20°C in 95/5 (v/v) formamide/water (a,b) and 95/5 (v/v) dioxane/water (c,d), respectively. (Reproduced from F. Plieva, X. Huiting, I.Yu. Galaev, B. Bergenståhl, and B. Mattiasson, *J. Mater. Chem.*, 16, 4065–73, 2006. With permission.)

process (Figure 2.11c). The SEM micrographs of pAAm cryogels prepared at –20°C showed the oriented porous structure with prolonged, aligned pores (Figure 2.11b) as the result of the formation of needle-like formamide crystals. Due to the solvent induced phase separation process, the bimodal porous structure (composed of large macropores with microporous pore walls) was visualized for pAAm cryogels synthesized in dioxane (Figure 2.11d).

2.3 Control over the Free-Radical Polymerization at Subzero Temperatures

The time control over the polymerization reaction can be performed via evaluation of gelation yield or conversion [43]. The gelation yield for the dextran-based cryogels depended on the time during which the reaction

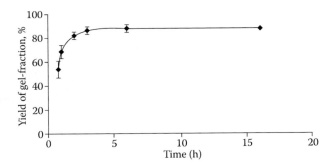

FIGURE 2.12
Kinetics of gelation of supermacroporous dextran cryogels at –20°C. (Reproduced from F. Plieva, A. Oknianska, E. Degerman, I.Yu. Galaev, and B. Mattiasson, *J. Biomater. Sci. Polym. Edn.*, 17, 1075–92, 2006. With permission.)

mixture was kept at –20°C and reached up to 92–96% after 5–6 hours (Figure 2.12) [39].

A very efficient way of controlling the reaction of monomers over time is to carry out *in situ* ¹H-NMR analysis [34]. The behavior of water is of crucial importance for the formation of well-structured cryogels. NMR is an ideal technique to study the behavior of water and the solutes in the nonfrozen water phase. Thus free-radical polymerization of dimethylacrylamide (DMAAm) with the cross-linker polyethylene PEG-DA in a semifrozen aqueous solution was studied using ¹H-NMR, which enables monitoring of the unfrozen water as well as the reaction of monomers over time [34]. The ¹H-NMR was used to monitor the amount of the nonfrozen phase and the progress of the polymerization reaction *in situ* in samples frozen at –12°C directly in the spectrometer. Freezing of the polymerization system results in the formation of a network of ice crystals surrounded by a small amount of unfrozen liquid with accumulated solutes. The amount of unfrozen water in semifrozen samples was quantified as a function of time by measuring the area of the water peak during the reaction, while the polymerization of the monomers was monitored by measuring the decrease in the area of a peak representing the vinyl group of the monomers (Figure 2.13) [34]. Peaks from the vinyl groups of cross-linker are only clearly visible in highly cross-linked samples as they partly overlap the DMAAm peaks. However, two separated peaks of PEG diacrylate are clearly visible at 3.9 and 4.43 ppm. In all cases, the samples were frozen within a few minutes, which can be seen as the sharp decrease in the water signal in Figure 2.13(1). The presence of a liquid phase in the apparently frozen polymerization feed at –12°C is demonstrated by a clearly visible water signal, which was not observed when pure water was frozen at the same temperature. The ¹H-NMR spectra showed complete conversion of

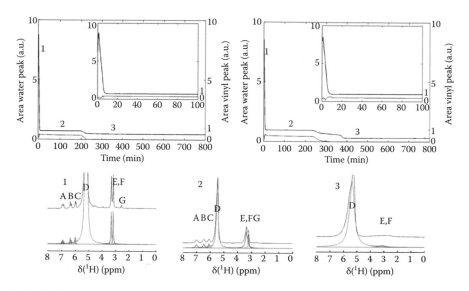

FIGURE 2.13
(See color insert following page 116.) Results of measurements on samples with different
cross-linking degree (6% (w/v) DMAAm:PEG diacrylate) (5:1) top left and (60:1) top right studied at $-12°C \pm 1°C$ in a Bruker DMX 200 spectrometer operating at a ^1H frequency of 200.13
MHz. The ^1H spectra was acquired every 25 seconds for the first 30 minutes, and thereafter
at 10-minute intervals. Top figures show the change in area of the water peak (black) and the
vinyl peak at 6.9 ppm (red) during the reaction. (1) shows the results obtained before freezing
the sample, (2) shows the results from the frozen sample, and (3) shows the spectra after completion of the polymerization reaction. A, B, and C denote the protons in the vinyl group of
the monomer, D the water peak, E and F the methyl peaks, and G the signal from the activator
TEMED. The black lines in (1–3) show the recorded spectra, the red lines show the deconvoluted spectra with Lorentzian functions fitted to each peak, and the blue lines show the sum of
the deconvoluted spectra. (Reproduced from H. Kirsebom, G. Rata, D. Topgaard, B. Mattiasson,
and I.Yu. Galaev, *Polymer*, 381, 3855–58, 2008. With permission.)

the vinyl peaks in the system despite the fact that the average gel yield
of the reactions were approximately 85%, indicating that part of the polymer formed was not incorporated into the gel structure but was washed
out of the macroporous structure. Spectra of both the highly and weakly
cross-linked samples after polymerization showed a characteristic broad
peak in the region of the methyl group, which was more pronounced for
the weakly cross-linked sample indicating that this is a more mobile and
flexible system. It was shown that approximately 5% of the initial water
remained unfrozen even after completion of the polymerization reaction.
This amount of unfrozen water corresponds to approximately five water
molecules per monomer unit of polymer, and the water is most probably
associated with the polymer [34].

2.4 Preparation of Macroporous Cryogels with Controlled Degradability via Free-Radical Polymerization

It has been shown that the incorporation of hydrolysable groups into the cryogel network allows for the preparation of cryogel systems with a controlled rate of degradation. These types of macroporous cryogels are of special interest for biomedical applications. There are three main approaches to produce degradable cryogels via free-radical polymerization: polymerization with comonomers bearing hydrolysable links [40], copolymerization of different monomers with degradable cross-linkers [29,36], or preparation of cryogels from polymers that are biodegradable (e.g., by the action of enzymes) [39,42,52].

The rate of degradation of pHEMA based cryogels cross-linked with degradable water soluble S-S cross-linker N,N′-bis(methacryloyl)-L-cystine (MAS-S) depended on the molar ratio of disulfide cross-linker MAS-S used in the polymerization feed and the concentration of reductive agent like ditiotreitol (DTT) (Figure 2.14). The time frame for degradation of such cryogels varied from a few hours for cryogels with low MAS-S content at 50 mM DTT to about 14 days for cryogels with high MAS-S content in the presence of 1 mM DTT (Figure 2.14) [36]. There was practically no visible degradation of pHEMA cryogels cross-linked with MBAAm and PEG-DA under these conditions [36].

The incorporation of the degradable comonomers (as 2-hydroxyethyl-methacrylate-L-lactide, HEMA-LLA) or macromers (as HEMA-LLA-dextran) into the polymeric backbone allowed for developing the cryogels

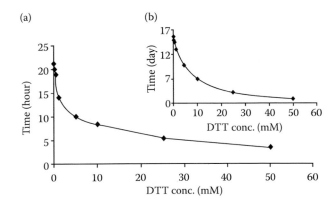

FIGURE 2.14
Dependence of the degradation rate in accordance with the molar ratio of disulfide cross-linker N,N′-bis(methacryloyl)-L-cystine (MAS-S). (a) pHEMA-SS-6 (HEMA/MAS-S molar ratio 6/1) and (b) pHEMA-SS-20 (HEMA/MAS-S molar ratio 20/1). (Reproduced from M. Andac, F. P. Plieva, A. Denizli, I.Yu. Galaev, and B. Mattiasson, *Macromol. Chem. Phys.*, 209, 577–84, 2008. With permission.)

with open porous structure and a desirable rate of degradation level [40]. The dextran-MA based nondegradable cryogels (prepared through the free-radical polymerization of dextran-MA macromer at −20°C) were stable at 37°C in PBS buffer, pH 7.2 for at least four months (Figure 2.15a) while the cryogels with incorporated degradable grafts of HEMA-LLA

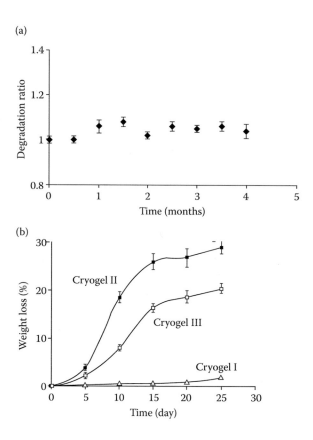

FIGURE 2.15

(a) Degradation ratio of Dextran-MA cryogel at incubation at 37°C and pH 7.4. The dextran-MA cryogels were prepared from 5% (w/v) dextran-MA solution at −20°C. (Reproduced from F. Plieva, A. Oknianska, E. Degerman, I.Yu. Galaev, and B. Mattiasson, *J. Biomater. Sci. Polym. Edn.*, 17, 1075–92, 2006.) (b) Weight loss of nondegradable (cryogel I) and degradable (cryogel II and III) cryogels. The cryogels were dried in air to a constant weight and then placed in vials containing 5 ml saline PBS buffer (pH 7.4) at 37°C. At selected time intervals, the samples were removed from the medium, dried in the air overnight and weighted to determine weight loss. Cryogel I represents the cryogel consisting of 90 wt % HEMA and 10 wt % HEMA-LLA. Cryogel II represents the degradable cryogel consisting of 10 wt % HEMA and 90 wt % HEMA-LLA-dextran. Cryogel III represents the degradable cryogel consisting of 10 wt % HEMA-LLA and 90 wt % HEMA-LLA-dextran. (Reproduced from N. Bölgen, F. M. Plieva, I.Yu. Galaev, B. Mattiasson, and E. Piskin, *J. Biomater. Sci. Polymer Edn.*, 18, 1165–79, 2007. With permission.)

and HEMA-LLA-dextran degraded pronouncedly under these conditions (Figure 2.15b).

2.5 Preparation of Functionalized Cryogels via Free-Radical Polymerization

One of the most straightforward approaches to synthesize the macroporous functionalized cryogels (i.e., cryogels bearing required functionalities) is via direct copolymerization with functional comonomers. Thus, the required functionality is incorporated *in situ* during the synthesis of cryogel in a single freezing–thawing cycle. Typically, the functional comonomer is added to a reaction mixture at a concentration of 5–10 mol % of total monomers [12,24,53,55,58], alternatively the functional comonomer is used as a main monomer for the preparation of cryogel [23]. Both commercially available [12,23,24,53,58] and in-house synthesized [55,58] functional comonomers were used for the preparation of functionalized macroporous cryogels (Table 2.3). The recently published reports on macroporous functionalized cryogels were concerned mainly with synthesis and characterization of the ionic cryogels (i.e., bearing weak or strong ion-exchange functionalities) [12,23,24,53,58]. The final properties of the functionalized ionic cryogels depended on (among other parameters) the content of functional comonomer in a cryogel backbone. Thus,

TABLE 2.3

Macroporous Functionalized Cryogels Prepared Via Free-Radical Polymerization

Main Monomer use for the Preparation of Cryogel	Functional Comonomer	Introduced Functionality	Reference
Acrylamide	2-Dimethylamino ethyl methacrylate	Weak anion-exchange	12, 53
Acrylamide	1-(N,N-bis(carboxymethyl) amino-3-allyl glycerol	Metal-chelate	55
Acrylamide	2-acrylamido-2-methylproanol sulfonic acid sodium salt	Strong cation-exchange	23
2-acrylamido-2-methylpropanol sulfonic acid sodium salt		Strong cation-exchange	24
Acrylamide	Acrylic acid	Week cation-exchange	58

increasing the concentration of ionic comonomer (2-acrylamido-2-methyl-propanol sulfonic acid) in the initial reaction mixture from 0 to 5 mol % resulted in the preparation of ionic cryogels with much faster swelling–de-swelling properties [24].

2.6 Main Applications for Macroporous Cryogels Prepared via Free-Radical Polymerization

Due to the presence of large and interconnected macropores in the cryogels prepared via free-radical polymerization, these macroporous materials were arranged as monolithic devices, (i.e., monolith columns) for processing particulate-containing fluids (virus and cell suspensions). For these purposes, the cryogels were synthesized from available robust monomers as AAm or DMAAm, methyl metacrylic acid (MMA), and cross-linked with MBAAm or PEG-DA cross-linkers [12,13,33,38,53,54]. Specific ligands were introduced *in situ* through incorporation of functional comonomers. The macroporous cryogel monoliths with metal-chelate and ion-exchange functionalities were prepared using this approach [38,53,55] or through grafting of polymer brushes onto the pore surface of polyacrylamide cryogel [15,56,57]. Alternatively, the copolymerization of vinyl monomers with chemically reactive comonomer, AGE, allowed for incorporation of the active oxirane groups into pore walls for further modification of the cryogels with functional ligands. Main requirements for such macroporous monolith matrices are the large size of interconnected pores (for free pass of bioparticles) and proper mechanical strength (to avoid compressibility of the gel matrix under high flow rates) [48]. These applications are discussed in more detail in Chapter 11.

Other important applications for the cryogels prepared via free-radical polymerization are in the biomedical field. The macroporous cryogels with open porous structures and pore sizes up to 200 μm (which are characterized by high elasticity and sponge-like morphology) have an enormous potential as three-dimensional (3D) porous scaffolds for cell culture applications. For these applications, the miniature cryogel 3D scaffolds prepared typically from biocompatible gel precursors and with controlled degree of degradation are required. These cryogel scaffolds were prepared via free-radical polymerization from biocompatible precursors as HEMA, polyethylene glycol (PEG), dextran-methacrylate, pNIPA, and poly(acrylonitrile) [29,35–37,39,52], and the degradation of scaffolds was ensured via copolymerization with monomers bearing hydrolysable ester bonds [40] or via cross-linking with degradable cross-linkers [29,36]. In more detail these applications are discussed in Chapter 14.

2.7 Concluding Remarks

The macroporous cryogels with a broad variety of morphologies are prepared using cryotropic gelation technique meaning gelation at subzero temperatures. These highly elastic hydrophilic materials can be produced from practically any gel-forming system with a broad range of porosities, from elastic and porous gels with pore size up to 1.0 μm to elastic and sponge-like gels with pore size up to 200 μm. The versatility of free-radical, cross-linking reaction at subzero temperatures, especially in an aqueous medium allowed for the development of macroporous cryogels with open porous structures and controlled porosities from a variety of commercially available and in-house synthesized monomers and cross-linkers.

References

1. H.-P. Hentze, and M. Antonietti. 2002. Porous polymers and resins for biotechnological and biomedical applications. *Rev. Mol. Biotechnol.* 90:27–53.
2. N. A. Peppas, Y. Huang, M. Torres-Lugo, J. H. Ward, and J. Zhang. 2000. Physicochemical foundations and structural design of hydrogels in medicine and biology. *Ann. Rev. Biomed. Eng.* 02:9–29.
3. V. R. Patel, and M. M. Amiji. 1996. Preparation and characterization of freeze-dried chitosan-poly (ethylene oxide) hydrogels for site-specific antibiotic delivery in the stomach. *Pharm. Res.* 13:588–93.
4. C. D. Wood, and A. I. Cooper. 2001. Synthesis of macroporous polymer beads by suspension polymerization using supercritical carbon dioxide as a pressure-adjustable porogen. *Macromolecules* 34:5–8.
5. D. J. Bennett, R. P. Burford, T. P. Davis, and H. J. Tilley. 1995. Synthesis of porous hydrogel structure by polymerizing the continuous phase of a microemulsion. *Polymer Int.* 36:219–26.
6. K. Kabiri, H. Omidian, and M. J. Zohuriaan-Mehr. 2003. Novel approach to highly porous superabsorbent hydrogels: Synergistic effect of porogens on porosity and swelling rate. *Polymer Int.* 52:1158–64.
7. S. Hjerten, J.-L. Liao, and R. Zhang. 1989. High-performance liquid chromatography on continuous polymer beds. *J. Chromatogr.* 473:273–75.
8. Y. S. Nam, and T. G. Park. 1999. Porous biodegradable polymeric scaffolds prepared by thermally induced phase separation. *J. Biomed. Mater. Res.* 47:8–17.
9. V. I. Lozinsky, F. M. Plieva, I.Yu. Galaev, and B. Mattiasson. 2002. The potential of polymeric cryogels in bioseparation. *Bioseparation* 10:163–88.
10. V. I. Lozinsky, I.Yu. Galaev, F. M. Plieva, I. N. Savina, H. Jungvid, and B. Mattiasson. 2003. Polymeric cryogels as promising materials of biotechnological interest. *Trends Biotechnol.* 21:445–51.

11. F. M. Plieva, I.Yu. Galaev, and B. Mattiasson. 2007. Macroporous gels prepared at subzero temperatures as novel materials for chromatography of particulate containing fluids and cell culture applications. *J. Sep Sci.* 30:1657–71.

12. P. Arvidsson, F. M. Plieva, I. N. Savina, V. I. Lozinsky, S. Fexby, L. Bülow, I.Yu. Galaev, and B. Mattiasson. 2002. Chromatography of microbial cells using continuous supermacroporous affinity and ion-exchange columns. *J. Chromatogr. A* 977:27–38.

13. M. B. Dainiak, F. M. Plieva, I.Yu. Galaev, R. Hatti-Kaul, and B. Mattiasson. 2005. Cell chromatography: Separation of different microbial cells using IMAC supermacroporous monolithic columns. *Biotechnol. Progr.* 21:644–49.

14. S. L. Williams, M. E. Eccleston, and N. K. H. Slater. 2005. Affinity capture of a biotinylated retrovirus on macroporous monolithic adsorbents: Towards a rapid single-step purification process. *Biotechnol Bioeng.* 89:783–87.

15. J. Yun, S. Shen, F. Chen, and K. Yao. 2007. One-step isolation of adenosine triphosphate from crude fermentation culture of *Sacharomyces cerevisiae* using anion-exchange cryogel chromatography. *J. Chromatogr. B* 860:57–62.

16. A. Hanora, F. M. Plieva, M. Hedström, I.Yu. Galaev, and B. Mattiasson. 2005. Capture of bacterial endotoxins using a supermacroporous monolithic matrix with immobilized polyethyleneimine, lysozyme or polymyxin B. *J. Biotechnol.* 118:421–33.

17. S. Deraz, F. M. Plieva, I.Yu. Galaev, E. Nordberg Karlsson, and B. Mattiasson. 2007. Capture of bacteriocins directly from non-clarified fermentation broth using macroporous monolithic cryogels with phenyl ligands. *Enzyme Microb. Tech.* 40:786–93.

18. M. L. Noir, F. M. Plieva, T. Hey, B. Guiyesse, and B. Mattiasson. 2007. Macroporous molecularly imprinted polymer/cryogel composite systems for the removal of endocrine disrupting trace contaminants. *J. Chromatogr. A* 1154:158–64.

19. W. Noppe, F. M. Plieva, K. Vanhoorelbeke, H. Deckmyn, M. Tuncel, A. Tuncel, I.Yu. Galaev, and B. Mattiasson. 2007. Macroporous monolithic gels, cryogels, with immobilized phages from phage-display library as a new platform for fast development of affinity adsorbent capable of target capture from crude feeds. *J. Biotechnol.* 131:293–99.

20. F. M. Plieva, M. Karlsson, M.-R. Aguilar, D. Gomez, S. Mikhalovsky, and I.Yu. Galaev. 2005. Pore structure in supermacroporous polyacrylamide based cryogels. *Soft Matter* 1:303–9.

21. F. Plieva, X. Huiting, I.Yu. Galaev, B. Bergenståhl, and B. Mattiasson. 2006. Macroporous elastic polyacrylamide gels prepared at subzero temperatures: Control of porous structure. *J. Mater. Chem.* 16:4065–73.

22. F. M. Plieva, D. Pignetti, I.Yu. Galaev, and B. Mattiasson. 2008. Porous structure of supermacroporous composite cryogels.

23. M. M. Ozmen, and O. Okay. 2005. Superfast responsive ionic hydrogels with controllable pore size. *Polymer* 46:8119–27.

24. D. Ceylan, M. M. Ozmen, and O. Okay. 2005. Swelling-deswelling kinetics of ionic poly(acrylamide) hydrogels and cryogels. *J. Appl. Polymer Sci.* 99:319–25.

25. M. M. Ozmen, M. V. Dinu, and O. Okay. 2008. Preparation of macroporous poly(acrylamide) hydrogels in DMSO/water mixture at subzero/temperatures. *Polymer Bull.* 60:169–80.

26. R. V. Ivanov, V. I. Lozinsky, S. K. Noh, Y. R. Lee, S. S. Han, and W. S. Lyoo. 2007. Preparation and characterization of polyacrylamide cryogels produced from a high-molecular-weight precursor. II. The influence of the molecular weight of the polymeric precursor. *J. Appl. Polymer Sci.* 107:382–90.

27. M. M. Ozmen, M. V. Dinu, E. S. Dragan, and O. Okay. 2007. Preparation of macroporous acrylamide-based hydrogels: Cryogelation under isothermal conditions. *J. Macromol. Sci. Part A* 44:1195–1202.

28. K. Yao, S. Shen, J. Yun, L. Wang, X. He, and X. Yu. 2006. Preparation of polyacrylamide-based supermacroporous monolithic cryogel bed under freezing-temperature variation conditions. *Chem. Eng. Sci.* 61:6701–8.

29. P. Perez, F. Plieva, A. Gallardo, J. S. Roman, M. Aguilar, I. Morfin, and F. Ehrburger-Dolle. 2008. Bioresorbable and nonresorbable macroporous thermosensitive hydrogels prepared by cryopolymerization: Role of the cross-linking agent. *Biomacromolecules* 9:66–74.

30. I.Yu. Galaev, M. B. Dainiak, F. Plieva, and B. Mattiasson. 2007. Effect of matrix elasticity on affinity binding and release of bioparticles: Elution of bound cells by temperature-induced shrinkage of the smart macroporous hydrogel. *Langmuir* 23:35–40.

31. X.-Z. Zhang, and C.-C. Chu. 2003. Synthesis of temperature sensitive PNIPAAm cryogels in organic solvent with improved properties. *J. Mater. Chem.* 13:2457–64.

32. A. Srivastava, E. Jain, and A. Kumar. 2007. The physical characterization of supermacroporous poly(N-isopropylacrylamide) cryogel: Mechanical strength and swelling/de-swelling kinetics. *Mater. Sci. Eng: A* 464:93–100.

33. P. Persson, O. Baybak, F. Plieva, I.Yu. Galaev, B. Mattiasson, N. Nilsson, and A. Axelsson. 2004. Characterization of a continuous supermacroporous monolithic matrix for chromatographic separation of large bioparticles. *Biotechnol. Bioeng.* 88:224.

34. H. Kirsebom, G. Rata, D. Topgaard, B. Mattiasson, and I.Yu. Galaev. 2008. *In situ* ¹H-NMR studies of free radical cryopolymerization. *Polymer* 381:3855–58.

35. I. N. Savina, V. Cnudde, S. D'Hollander, L. V. Hoorebeke, B. Mattiasson, I.Yu. Galaev, and F. D. Prez. 2007. Cryogels from poly(2-hydroxyethyl methacrylate): Macroporous, interconnected materials with potential as cell scaffolds. *Soft Matter* 3:1176–84.

36. M. Andac, F. P. Plieva, A. Denizli, I.Yu. Galaev, and B. Mattiasson. 2008. Poly(hydroxyethyl methacrylate)-based macroporous hydrogels with disulfide cross-linker. *Macromol. Chem. Phys.* 209:577–84.

37. F. M. Plieva, P. Ekström, I.Yu. Galaev, and B. Mattiasson. 2008. Monolithic cryogels with open porous structure and unique double-continuous macroporous networks. *Soft Matter* DOI: 10.1039/B804105A.

38. Z. Chen, L. Xu, Y. Liang, J. Wang, M. Zhao, and Y. Li. 2008. Polyethylene glycol diacrylate-based supermacroporous monolithic cryogels as high-performance liquid chromatography stationary phase for protein and polymeric nanoparticle separation. *J. Chromatogr. A* 1182:128–31.

39. F. Plieva, A. Oknianska, E. Degerman, I.Yu. Galaev, and B. Mattiasson. 2006. Novel supermacroporous dextran gels. *J. Biomater. Sci. Polym. Edn.* 17:1075–92.

40. N. Bölgen, F. M. Plieva, I.Yu. Galaev, B. Mattiasson, and E. Piskin. 2007. Cryogelation for preparation of novel biodegradable tissue-engineering scaffolds. *J. Biomater. Sci. Polymer Edn.* 18:1165–79.
41. P. Petrov, E. Petrova, R. Stamenova, C. B. Tsvetanov, and G. Riess. 2006. Cryogels of cellulose derivatives prepared via UV irradiation of moderately frozen systems. *Polymer* 47:6481–84.
42. P. Petrov, E. Petrova, B. Tchorbanov, and C. B. Tsvetanov. 2007. Synthesis of biodegradable hydroxyethylcellulose cryogels by UV irradiation. *Polymer* 48:4943–49.
43. V. I. Lozinsky. 2002. Cryogels on the basis of natural and synthetic polymers: Preparation, properties and applications. *Russ. Chem. Rev.* 71:489–511.
44. F. Plieva, M. Karlsson, M.-R. Aguilar, D. Gomez, S. Mikhalovsky, I.Yu. Galaev, and B. Mattiasson. 2006. Porous structure of macroporous agarose gels prepared at subzero temperatures. *Macromol. Biosci.*
45. F. M. Plieva, J. Andersson, I.Yu. Galaev, and B. Mattiasson. 2004. Characterization of polyacrylamide based monolithic columns. *J. Sep. Sci.* 27:828–36.
46. S.-L. Chen, and T.-S. Lee. 1998. A study of supercooling phenomenon and freezing probability of water inside horizontal cylinders. *Int. J. Heat Mass Trans.* 41:769–83.
47. S.-L. Chen, P.-P. Wang, and T.-S. Lee. 1999. An experimental investigation of nucleation probability of supercooled water inside cylindrical capsules. *Exp. Therm. Fluid Sci.* 18:299–306.
48. F. M. Plieva, E. D. Seta, I.Yu. Galaev, and B. Mattiasson. 2008, Macroporous elastic polyacrylamide monolith columns: Processing under compression and scale-up. *Sep. Purific. Technol.* 65: 110–16.
49. M. B. Dainiak, A. Kumar, I.Yu. Galaev, and B. Mattiasson. 2006. Detachment of affinity-captured bioparticles by elastic deformation of a macroporous hydrogel. *Proc. Natl. Acad. Sci, USA* 103:849–54.
50. H. G. Schild. 1992. Poly(N-isopropylacrylamide): Experiment, theory and application. *Prog. Polym. Sci.* 17:163–249.
51. P. Arvidsson, F. M. Plieva, V. I. Lozinsky, I.Yu. Galaev, and B. Mattiasson. 2003. Direct chromatographic capture of enzyme from crude homogenate using immobilized metal affinity chromatography on a continuous supermacroporous adsorbent. *J. Chromatogr. A* 986:275–90.
52. E. Jain, A. Srivastava, and A. Kumar. 2008, July 3. Macroporous interpenetrating cryogel network of poly(acrylonitrile) and gelatin for biomedical applications. *J. Mater. Sci: Mater. Med.*
53. F. M. Plieva, I. N. Savina, S. Deraz, J. Andersson, I.Yu. Galaev, and B. Mattiasson. 2004. Characterization of supermacroporous monolithic polyacrylamide based matrices designed for chromatography of bioparticles. *J. Chromatogr. B* 807:129–37.
54. A. Kumar, F. M. Plieva, I.Yu. Galaev, and B. Mattiasson. 2003. Affinity fractionation of lymphocytes using supermacroporous monolithic cryogel. *J. Immunol. Methods* 283:185–94.
55. F. Plieva, B. Bober, M. Dainiak, I.Yu. Galaev, and B. Mattiasson. 2006. Macroporous polyacrylamide monolithic gels with immobilized metal affinity ligands. The effect of porous structure and ligand coupling chemistry on protein binding. *J. Mol. Recogn.* 19:305–12.

56. I. N. Savina, I. Yu. Galaev, and B. Mattiasson. 2005. Anion-exchange super-macroporous monolithic matrices with grafted polymer brushes of *N,N*-dimethylaminoethyl-methacrylate. *J. Chromatogr. A* 1092:199–205.
57. K. Yao, J. Yun, S. Shen, and F. Chen. 2007. *In situ* graft-polymerization preparation of cation-exchange supermacroporous cryogel with sulfo groups in glass columns. *J. Chromatogr. A* 1157:246–51.
58. F. M. Plieva, M. Hedström, I. Yu. Galaev, and B. Mattiasson. 2008. Cryogel monolith columns with double-continuous macroporous networks and mixed functionalities.

3

Macroporous Polymer Scaffolds through Leaching Processes

Michael C. Hacker, Kristina Ambrosch, and Michaela Schulz-Siegmund

CONTENTS

3.1 Introduction

Porosity was often regarded as a defect to be avoided in the early years of engineered materials, but now it has become an important parameter for the processing of metals [1], ceramics [2], and polymeric materials [3]. Porous polymer matrices, which are the focus of this book, have found numerous applications in various fields, especially chemistry and the biomedical sciences where they have been explored as supports for catalysts, artificial extracellular matrices for *in vitro* tissue culture, drug delivery devices, and more [3–5].

In general, such three-dimensional (3-D) highly porous matrices, typically addressed as scaffolds, can resemble woven or nonwoven fiber meshes or porous monoliths with inherently different mechanical properties and permeation characteristics for liquid media and cell dispersions. While fiber meshes are extremely well permeable due to high porosities, higher mechanical stabilities can generally be achieved for porous solids of the same material. The permeability of porous solids in turn is critically influenced by parameters such as pore interconnectivity, pore throat size, and tortuosity [6,7]. Whenever porous materials are described, average pore diameter, pore geometry, and pore interconnectivity are important parameters. However, the nomenclature for describing porous structures found in diverse fields is not consistent [1]. Rigid porous monoliths are typically referred to as porous solids, cellular solids, or foams. The term sponge usually refers to flexible foams. The IUPAC recommendations for the characterization of porous solids, which were initially defined for catalysts, absorbents, and membranes, describe pore diameters of less than 2 nm as micro. Mesopores range between 2 and 50 nm and macroporous structures exhibit pores greater than 50 nm. The same range of sizes (micro, meso, and macro) has been used to define the diameter of the connecting channels (interconnects) in matrices with an open pore structure. For porous polymers used in biomedical applications, which will be the focus of this chapter, these definitions are typically not maintained [8].

Considering that in these applications the porous structures are typically designed to support the ingrowth of cells or provide support for cells to attach, proliferate, differentiate, and lay down extracellular matrices within the pores, pore sizes of interest characteristically exceed the size of single cells. As most eukaryotic animal cells range between 10 and 50 μm in size, pores or interconnects that are within this size range are described as micro in tissue engineering literature (microporous: <100 μm). Foams with pore sizes above 100 μm are typically described as macroporous [9]. Ideally, interconnect diameters larger than the dimensions of cells are preferred to enable infiltration of the cells into the scaffold, smaller pores and interconnects may only be sufficient to positively influence the exchange of nutrients and cellular waste products [10].

Various fabrication techniques have been developed to process different kinds of biomaterials, especially synthetic biodegradable polymers into porous scaffolds with pore sizes ranging from the nanometer scale up to the millimeter range [9,11,12]. A large number of studies have aimed to identify optimum pore size ranges for different kinds of cells or tissues [9,13], for example, average pore sizes around 5 µm for neovascularization, 5–15 µm for fibroblast ingrowth, 20–125 µm for skin regeneration, 70–120 µm for chondrocyte ingrowth, 100–400 µm for bone regeneration, and 200–350 µm for osteoconduction. These numbers further depend on scaffold porosity and materials used but nevertheless indicate that macroporous solids are warranted for many biomedical applications. Several techniques have been developed and advanced to fabricate porous polymer scaffolds with controlled 3-D pore structure [1]. The average pore sizes that have been achieved using the most common of these processes are summarized in Figure 3.1.

Techniques to generate fibrous scaffolds include electrospinning [14,15], bonding of nonwoven meshes [16], textile processing of extruded polymer fibers [17], and melt spinning [18,19]. Average pore sizes in fibrous scaffolds typically do not significantly exceed 100 µm. Especially for electrospun scaffolds, which ideally resemble nonbonded, nonwoven meshes in which pore size is determined by fiber diameter, macroporous structures can hardly

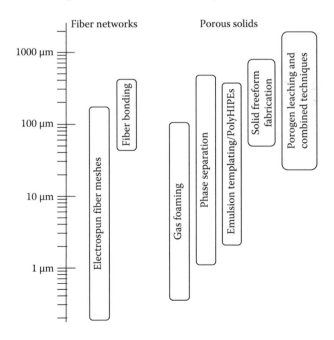

FIGURE 3.1
Pore sizes typically obtained by different techniques used for the fabrication of porous solids. (Reproduced from Weigel, T., Schinkel, G., and Lendlein, A., *Expert Rev. Med. Devices*, 3, 835–51, 2006; Karageorgiou, V., and Kaplan, D., *Biomaterials*, 26, 5474–91, 2005; Karande, T. S., Ong, J. L., and Agrawal, C. M., *Ann. Biomed. Eng.*, 32, 1728–43, 2004. With permission.)

be achieved [20]. The fiber diameter of such fibrous scaffolds can be easily reduced to the nanometer range and matrices that structurally resemble natural extracellular matrix (ECM) can be generated and have found wide application as synthetic ECM mimetics [21,22]. Some of the processes used to fabricate porous solids, such as gas foaming [23], emulsion templating [24,25], microsphere sintering [26], and phase separation processes [27] yield average pore sizes that are immanently limited to a few hundred micrometers (Figure 3.1).

To generate average pore sizes above 300 µm the methods of choice include solid freeform fabrication (SFF) [5,7,28] and leaching techniques that employ pore forming templates from various materials that can be extracted or leached from a polymer–porogen composite under conditions that do not affect the polymer phase. The SFF techniques allow for defect-tailored scaffold fabrication and generation of defined microstructures but, in addition to being time-consuming, the need for sophisticated equipment, and high material demands, some techniques lack sufficient resolution or require toxic solvents or binders [7,28]. The maximum porosities obtained by SFF are often limited to approximately 80%. Porogen leaching techniques, however, can be performed in any well-equipped chemical laboratory and are very versatile with regard to the material to be processed. These advantages in combination with the ability to template macroporous structures with porosities up to 90% may likely explain the popularity of porogen leaching techniques and their widespread application in medical and engineering laboratories.

This chapter will describe different leaching techniques that have been used to fabricate porous polymer solids for biomedical applications. In addition, advancements of the basic techniques to enhance pore interconnectivity and versatility are being discussed. A short overview on different polymers that have been processed by leaching techniques is given and trends in macroporous solid fabrication by leaching processes are highlighted.

3.2 Leaching Processes for the Fabrication of Macroporous Solids

The term leaching is commonly defined as the process by which soluble constituents (e.g., salts, nutrients, or chemicals) are washed into a lower layer of soil or are dissolved and carried away by a percolating fluid, typically water. Based on this definition, this chapter discusses scaffold fabrication techniques that employ solid or semisolid particles as pore templates, which are dispersed into a polymer matrix and later dissolved or extracted from the matrix by a solvent inert to the polymer. Leaching processes are one major concept to fabricate macroporous solids and commonly involve the following steps: (1) inclusion of a dispersed porogen within a discontinuous polymer matrix (polymer solution or dry polymer powder); (2) a technique leading to the creation of a continuous polymer matrix around the dispersed porogen; and (3) leaching of the porogen from the polymer matrix leaving a macroporous solid [29]. Natural and synthetic polymers have been processed into macroporous scaffolds by leaching techniques, with the majority of studies focusing on synthetic, and especially biodegradable polymers due to these polymers' attractiveness as temporary matrices for 3-D cell cultivation and tissue engineering attempts. Commonly used polymers will be discussed in the third section of this chapter.

In foams fabricated by leaching processes, key structural differences are obtained depending on whether the processes of polymer matrix formation and porogen leaching occur sequentially or almost simultaneously. In the latter case the continuous polymer phase is not yet fully formed or hardened when porogen leaching is initiated, which leaves the polymer matrix moldable during the initial stages of porogen extraction and allows for the formation of more or larger interconnects through additional processes. Such processes are described by the keyword combination porogen dispersion–composite leaching and solidification (PD/CLS) in this chapter. A temporal separation of the processes of continuous polymer matrix formation and porogen extraction, however, increases the robustness of the method and provides better control over pore size and pore structure. During such processes, which are addressed as porogen embedding–composite leaching (PE/CL) methods in this chapter, a fully hardened composite of polymer and porogen is formed prior to the leaching step and the formation of interconnects depends on direct contacts between neighboring porogen particles. Table 3.1 summarizes different porogen materials, sizes, and shapes that have been utilized in different PE/CL, PD/CLS, and combined techniques as outlined in the following paragraphs.

3.2.1 Porogen Embedding–Composite Leaching (PE/CL) Techniques

Leaching techniques have become popular with the increasing demand for macroporous foams from various biodegradable polymers as synthetic ECM substitutes for *in vitro* tissue engineering [30]. Therefore, key leaching techniques have been developed and advanced for the processing of synthetic and especially biodegradable polymers.

3.2.1.1 Solvent Casting–Particulate Leaching (SC/PL) and Derived Techniques

Classically, porous solids from lipophilic synthetic polymers have been fabricated by dispersing water soluble particles into an organic solution of the polymer. The mixture is then poured into a mold or dish and allowed to solidify by evaporation of the polymer solvent. The polymer–porogen matrix is finally immersed in water to leach out the water soluble particles and generate

TABLE 3.1

Porogen Materials, Geometrical Properties (Dimensions and Shape) and Details on Porogen Fusion and Applied Technique as Used for the Fabrication of Macroporous Solids by Leaching Processes

Porogen Material	Porogen Size [µm]	Particle Shape (Spherical, Angular, Fibrillar)	Porogen Fusion	Combination Technique	PD	References
Hydrophilic porogens						
Ice	Ø 237	s	–	freeze drying	+	125
Sugar	250–380	s	–	–	–	38
	590–2000	a	–/+	freezing/polymer precipitation	+	51, 126
	106–710	a	–/+	freeze drying/phase separation	–	64
NaCl	up to 500	a	–	–	–	32, 89, 127
	100–500	s	+	–	–	39
	106–710	a	–/+	freeze drying/phase separation	–	64
	250–470	a	–	solvent merging	–	36
	125–200	a	–	phase separation	–	62
	250–425	a	+	–	–	29
	250–425	a	–/+	gas foaming (CO_2)	+	68, 69
NaCl + Urea	106–500	a	*	molding (p↑, ΔT↑)	–	50
NaCl + PEO	250–500	a	*	molding (p↑, ΔT↑)	–	49
NaCl + PEO	not specified	a	*	extrusion/phase separation	–	48
NaCl + NH_4HCO_3	150–710	a	–	gas foaming (effervescent salt)	+	67
Na-tartrate	up to 150	a	–	–	–	32
Na-citrate	up to 150	a	–	–	–	32
NH_4Cl	150–250	a	–	–	–	40
$Na_2HPO_4 \bullet 7H_2O$	90–400	a	–	phase inversion	–	86

Porogen	Size	Shape		Processing		Ref
$(NH_4)_2SO_4$	100 × 600	needles	–	–	–	128
NH_4HCO_3	300–400	a	–	freeze drying/phase separation	–	129
NH_4HCO_3	100–500	a	–	gas foaming (effervescent salt)	+	65
Gelatin	106–710	s	–	–	–	45
Gelatin	280–450	a	+	freezing/freeze drying	–	63
Lipophilic porogens						
Camphor	not specified	not specified	*	molding ($\Delta T\uparrow$)	–	47
Hydrocarbons	106–300	s	–	–	–	40
	300–500	s	*	–	+	55
	100–500	s	+	–	–	42
	150–700	s	–	freeze drying	–	130
	355–450	s	–	molding (s, $\Delta T\uparrow$)	–	41
Lipids	100–710	s	*	–	+	56, 57
Wax	300–500	s	*	–	+	55
Poly (methyl methacrylate)	120–500	s	–	–	–	101
Poly (ethyl methacrylate)/ Poly (methyl methacrylate)	~200	s	+	–	–	43
Poly (L-lactic acid)	Ø 12	f	–	–	–	131
Polycaprolactone	not specified	f	–	–	–	75

Note: PD indicates a porogen dispersion technique characterized by a semisolid polymer/porogen intermediate that is subjected to the leaching step.

*Indicates that the applied processing technique resulted in the formation of a continuous porogen phase in absence of a classical porogen fusion process.

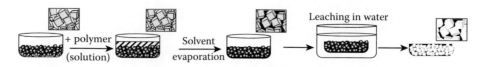

FIGURE 3.2
Schematic illustration of the classical solvent casting/salt leaching technique as described by Mikos and coworkers. Inserted schemes are magnified illustrations of the polymer/porogen composite. ▨ polymer solution, ■ solid polymer matrix, ◢ angular salt particle. (Reproduced from Mikos, A. G., Thorsen, A. J., Czerwonka, L. A., Bao, Y., Langer, R., Winslow, D. N., and Vacanti, J. P., *Polymer*, 35, 1068–77, 1994. With permission.)

a network of macropores. Along with the concept of tissue engineering [30] a leaching technique was introduced by the laboratory of Robert Langer at the Massachusetts Institute of Technology that employed salt particles for the processing of biodegradable poly(L-lactic acid) (PLLA) and poly(D,L-lactic-*co*-glycolic acid) (PLGA) [31,32] (Figure 3.2). Salt particles, namely sodium chloride, sodium tartrate, and sodium citrate crystals were classified by sieving. Salt crystals of a defined particle size range were added to a polymer solution in chloroform or dichloromethane (DCM) (both nonsolvents for the salts) and the vortexed dispersion was cast in a Petri dish [32]. The solvent was allowed to evaporate from the covered dish and residual solvent amounts were removed by vacuum drying. Different thermal treatments were applied to the polymer–salt composite membranes to modulate polymer crystallinity. To leach out the salt, the composites were immersed in distilled water on a shaker table at room temperature for 48 h with water changes every 6 h. The salt-free, porous membranes were first air and subsequently vacuum dried and stored under desiccation. Using 70–90 wt % salt yielded homogenous membranes with interconnected pores. The membrane properties porosity and median pore diameter were directly related to the salt weight fraction and particle size, respectively. Foams fabricated utilizing this technique have been tested in several tissue engineering attempts with various cell types *in vitro* and have shown no adverse effects on cell proliferation and new tissue formation [12]. Porous membranes with improved pliability could be fabricated using this SC/PL process by blending poly(ethylene glycol) (PEG) with the matrix polymer [33]. The leached substrates could be rolled over to form tubular scaffolds without macroscopic damage to the scaffold. It is believed that micropores, which resulted from dissolution of PEG during the leaching step, altered the structure of the pore walls, and changed the membrane's mechanical properties accordingly. Pliability is considered as a critical parameter of scaffolds for soft tissue regeneration and tubular scaffolds are required as matrices for artificial intestines or vascular grafts.

Certain limitations of the classic SC/PL, however, motivated the development of a modified technique in which a compression molding step is added. In particular, membrane thickness was limited to 3 mm in SC/PL because particle sedimentation accompanied with the formation of asymmetric membranes topped by a solid skin layer was observed for substrates exceeding this limit. In addition, morphological differences between the two surfaces, the air–composite interface and the dish–composite interface of scaffolds fabricated by SC/PL were observed. In the combined SC/compression molding/PL technique, pieces of the polymer–salt composite fabricated by SC are shaped into cylindrical form at temperatures just above the melting or glass transition temperature of the matrix polymer under elevated pressures [34]. The resulting composite can then be cut into the desired thickness or shape. Subsequent leaching of the salt leaves a scaffold with open pores and a more uniform pore distribution than typically obtained by SC/PL for membranes with increased thickness.

Another strategy to assemble 3-D constructs of increased thickness and complex shape from porous membranes fabricated by SC/PL involves membrane lamination and follows three steps [35]. First, a contour plot is drawn from the particular 3-D shape. Second, porous membranes are fabricated by SC/PL. Third, the constituent membrane sections are laminated in proper order by wetting the surfaces with a suitable solvent to build up a structure with the desirable shape. Implants with specific shapes were prepared from poly (D, L-latic acid) (PLA) and PLGA in a proof-of-concept study. The scaffolds produced had pore morphologies similar to those of the constituent membranes and the pores of adjacent layers were interconnected resulting in a continuous pore network.

A more recent technique, which is addressed as a solvent merging/PL and can be classified as a PE/CL technique, claims to provide a quicker and easier process for the preparation of macroporous polymeric scaffolds than conventional PE/CL technologies [36]. Matrices with porosities exceeding 85% and a well-interconnected porous structure were achieved. Unlike in previously described techniques, the biodegradable polymer was not mixed with the salt particles in organic solution, but was directly mixed with the salt particles in the solid state. Ground and sieved biodegradable polymer granules (250–470 µm) were mixed with sifted sodium chloride particles in different weight proportions and cast into a circular-cylindrical mold with a stainless-steel mesh bottom to hold the mixture. A defined volume of polymer solvent was added for a short time to partially dissolve the polymer particles and quickly removed by applying vacuum to a collection chamber attached beyond the mesh. Swollen polymer particles were coagulated during this suction step and solidified by adding a nonsolvent. Finally, a large volume of distilled water was passed through the material to wash out the embedded porogen particles, residual solvent, and nonsolvent. As expected for a porogen leaching process, scaffold porosity and pore size distribution were controlled by porogen weight ratio and average particle size, respectively.

Lately, an ambient temperature injection molding and PL technique was presented with the objective to avoid thermal degradation of polymers due to processing at elevated temperatures and the use of high-pressure equipment typical of classical compression molding [37]. A special mold facilitated shaping and composite discharging. Polymer solutions in chloroform were mixed with sieved sodium chloride particles and the resulting semisolid paste-like dispersion was charged into the cavity of the injection molder and quickly injected into the mold under relatively low pressure at room temperature. Following vacuum drying, the composite was removed from the mold and placed in deionized water to leach out salt particulates. Scaffolds in complex shape and high porosities (up to 94%) were fabricated with reasonable control of porosity and pore size of the resulting scaffolds by ratio of porogen to polymer and particulate size, respectively.

3.2.1.2 PE/CL with Controlled Porogen Assembly or Porogen Fusion

With the objective to control interconnect formation and to increase pore throat size, different strategies were developed. In general, two different concepts were applied. One focuses on predefining particle assembly and interparticle contact areas physically by minimizing disparity, controlling particle shape, or particle size distribution. The other aims at bonding the porogen particles to different degrees thus controlling interconnect size.

A recent variant of the classical SC/PL technique focused on the formation of spherical and interconnected pore structures [38]. Sugar particles (200–500 µm) were selected as porogens because this material can be easily spheronized by pouring the particles through a horizontal Meker burner. Upon flame passage, the particles partially melted and adopted a spherical geometry when falling onto the collector plate. A defined amount of spheronized porogen was poured into a cylindrical mold and compressed manually to ensure dense packing. The PLLA was dissolved in chloroform and cast in the sugar assembly. After trapped air has been removed by the vacuum treatment and solvent was allowed to evaporate, the composites were leached in distilled water and scaffolds with highly spherical and interconnecting pores were obtained (Figure 3.3A). The interconnects ranged from 10 to 100 µm in diameter, which is respectable considering that the interconnect-templating interparticle contacts were generated just by densely packing spherical porogen particles. A similar modified SC/PL technique that utilizes spheronized salt particles has also been described for the fabrication of interconnected scaffolds [39]. Milled, sifted, and dried rock salt was spheronized by external injection into a flame spray torch. Compared to sugar as a porogen material, the spheronization of the salt requires a larger amount of energy due to the higher melting point. Additional sintering of the salt particles in a furnace resulted in scaffolds with interconnecting pores. In comparison, the process involving the tightly packed spherical sugar particles seemed more efficient with regard to energy consumption and processing steps.

Improved interconnectivity by tightly packing can also be obtained by a technique named pressure differential/PL [40]. Solid polymer–porogen composites were formed by infiltrating the prepacked porogen bed (angular ammonium chloride particles or spherical paraffin microspheres) with

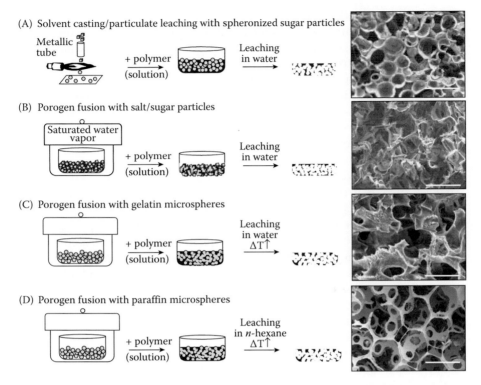

FIGURE 3.3
Schematic illustration of porogen embedding/composite leaching (PE/CL) techniques employing different porogen materials: (A) spheronized sugar particles (Micrograph reprinted from Vaquette, C., Frochot, C., Rahouadj, R., and Wang, X., *J Biomed Mater Res* B, 86, 9–17, 2008, with permission of John Wiley & Sons, Inc.) (B) fused salt/sugar particles (Micrograph reprinted from Murphy, W. L., Dennis, R. G., Kileny, J. L., and Mooney, D. J., *Tissue Eng.*, 8, 43–52, 2002, with permission of Mary Ann Liebert, Inc. publishers.) (C) fused gelatin microspheres (Micrograph reprinted from Zhou, Q., Gong, Y., and Gao, C., *J Appl Polym Sci.*, 98, 1373–79, 2005, with permission of John Wiley & Sons, Inc.) (D) fused paraffin microspheres (Micrograph reprinted from Ma, P. X., and Choi, J. W., *Tissue Eng.*, 7, 23–33, 2001, with permission of Mary Ann Liebert, Inc. publishers.) ■ solid polymer matrix, ◪ angular porogen particle, ○ spherical porogen particle. Scale bars represent 500 μm.

polymer solution. After drying and leaching with water or hexane, macroporous scaffolds with high interconnectivity were generated. The sphericity of the particulates was not altered during compression and macropore size of the scaffolds fabricated correlated with the sizes of the presieved particles. Pore shape was affected by the porogen type and scaffolds made with angular ammonium chloride particles showed an irregular pore shape. On the other hand, spherical pores and thin struts characterized the macroporous structures when the paraffin porogen was used.

Well-interconnected macroporous structures were also generated with a lipophilic porogen material by taking advantage of the assembly characteristics of densely packed spherical particles [41]. Paraffin microspheres were mixed with solutions of biodegradable polymers in acetone to yield paste-like dispersions that were compressed into molds and allowed to dry creating a solid composite. The porogen microspheres were leached out using a Soxhlet extractor with *n*-pentane for 24 h. Porous scaffolds with ordered internal spherical macropores of various pore sizes and porosities (80–97%) were obtained.

These studies indicate that enhancing control over pore interconnectivity and interconnect diameter with PE/CL techniques is a major subject in method development. Different strategies have been investigated to produce, control, and enlarge interparticle connections that determine interconnect and pore throat size and homogeneity. Spherical particles have the inherent advantage to exhibit a better packing efficiency than angular porogen particles such as ground salt. As a result of the high packing density and resulting interparticle contact areas regular interconnects can be formed [38]. These effects can be even more pronounced when spherical porogen particles are sintered to form a coherent template [39]. Similar strategies have been applied for angular porogen particles, which can assemble and form contacts in different configurations [39]. The largest throat size results in the comparably rare case that the faces of two adjacent angular particles are in direct contact. More likely configurations include a particle corner touching the corner, side, or face of a neighboring particle.

A different approach to impart improved pore interconnectivity to polymer scaffolds fabricated by PE/CL with angular particles is to partially fuse the solid porogen together prior to the creation of a continuous polymer matrix. In the pioneering study, molds were filled with sodium chloride particles of a predefined size range [29] (Figure 3.3B). The salt-filled molds were subjected to 95% humidity at 37°C for periods up to 24 h in a cell culture incubator. Polymer solution was cast over the dried salt templates and solvent was allowed to evaporate. Porous solids with interconnecting pores were obtained after leaching out the salt in water. Microscopically, salt fusion prior to SC resulted in enhanced pore interconnectivity within PLGA scaffolds. As expected for a PE/CL technique the scaffolds pore structure appeared similar to the structure of the fused salt template. It was also shown that fusion time significantly affected pore and interconnect structure. Pores within cross sections of scaffolds processed with salt fused for 1 h exhibited a less organized pore structure and a large density of small interconnects. The interconnect diameter doubled after a fusion time of 24 h. A porogen fusion method involving spheronized salt particles has also been described [39].

Most techniques for the fabrication of polymer scaffolds by leaching utilize water-soluble porogen materials and a hydrophilic solvent, typically water, for their extraction. An aqueous leaching process, which may last up to several days, however, may not be compatible with matrix polymers that are prone to hydrolysis or with strategies where water-soluble bioactive molecules are coembedded in the polymer matrix for sustained release. With these limitations in mind a class of techniques was described, which employs hydrophobic porogen materials in combination with organic solvents for porogen leaching that are nonsolvents for the scaffold material. One of these techniques can be classified as a PE/CL technique and includes a porogen bonding step utilizing paraffin microspheres as porogen [42] (Figure 3.3D). Paraffin microspheres were fabricated by melt dispersion in an aqueous poly(vinyl alcohol) solution. After washing and drying, the spheres were sifted into different size ranges and filled in a Teflon mold. The leveled mold was placed into an oven at 37°C for 20 min. After cooling to room temperature, a polymer (PLLA or PLGA) solution was added by dropping onto the paraffin microsphere assembly. Pyridine was selected as the polymer solvent due to its compatibility with paraffin. Trapped air was removed from the composite through the application of a low vacuum. Thereafter, solvent was removed and a solid polymer–paraffin composite was formed under a high vacuum. After removal from the mold, the CL step was performed in hexane at room temperature for 2 days with repeated changes. Following solvent exchange with the higher freezing hydrocarbon cyclohexane, the leached matrices were frozen, lyophilized, and vacuum dried to ensure complete solvent removal. Microscopically, the generated pores were well interconnected and retained the shape of the original paraffin spheres. Correspondingly, longer heat treatment time of the paraffin spheres resulted in larger interconnect diameters. With regard to scaffold mechanics, it was shown that compressive modulus of the scaffolds decreased with increasing porosity and increased with pore size.

In a similar approach that was also motivated by the advantages of spherical interconnecting pores in which the pore size and the diameter of the pore throats can be well controlled, a sintered template from acrylic resin— poly(ethyl methacrylate) and poly(methyl methacrylate)—microspheres was employed [43]. In a mold, the microspheres were heated to 140°C for 30 min to form the sintered template. Through the application of pressure during the sintering step, denser templates that yielded scaffolds with higher porosities were obtained. Following a solvent-free approach in order not to interfere with the acrylic template, a polycaprolactone (PCL) melt was injected in the template by pressurized nitrogen gas. After cooling, the composites were immersed in ethanol for several days at room temperature to dissolve the acrylic template and finally vacuum dried.

Depending on the extent of porogen fusion and shaping of the resulting template, similar processes have been addressed as replication techniques and used to process ceramics and polymeric materials [44]. This procedure involves the formation of a replica of the desired porous structure from wax or another material that can easily be removed by melting or dissolution. This replica is then used as a negative casting mold meaning that the interstices are filled with the matrix polymer in liquid phase. After composite formation by curing, cooling, or precipitation of the liquid polymer phase, the pore template is removed. Several methods including leaching in water or another applicable solvent may be suitable for template removal depending on the properties of the template material.

3.2.1.3 Compression–Melt Molding: Porogen Embedding without Organic Solvents

Most techniques discussed previously, utilize organic solvents to dissolve the polymer and disperse the porogen particles. The combination of compression molding and PL, however, results in a technique that employs particulate porogens but avoids the use of organic solvents. It should be noted that the phrases compression molding, melt molding, and injection molding are sometimes used to describe similar processes depending on the focus of the authors. In the context of porogen leaching technology, molding processes involve plastic polymer–porogen mixtures that are shaped using a negative mold of the target 3-D object. Plasticity of the polymer–porogen mixture can be achieved by applying high pressures, adding a solvent, or heating the polymer above its melting temperature or glass transition temperature. In typical molding steps combinations of added solvents, increased temperatures, and pressures are used. Compression molding typically involves increased pressure in combination with increased temperature or added solvents. In this section, which refers to solvent-free processing, compression molding refers to molding steps involving pressure and elevated temperatures. As the polymers are handled above their transition temperatures or melting points, such processes are sometimes referred to as melt molding. In injection molding, the plastic dispersion is typically transferred into the mold by elevated pressure but plasticity can either be achieved through heating or the addition of solvent. In Table 3.1, molding processes that have been applied in scaffold fabrication are specified by naming the specific parameters used to plastify the polymer (solvent, temperature, pressure) instead of relying on the terms compression, melt, or injection molding.

The PLGA scaffolds have been fabricated by a compression molding/PL technique with gelatin microspheres as porogens [45]. Following a single emulsion protocol, gelatin microspheres were manufactured. After drying and sieving, microparticles of defined size ranges were mixed with a finely ground PLGA powder and filled into Teflon molds. While a constant compressive force was applied, the molds were heated to a temperature above the glass transition temperature of the polymer in a convection oven, maintained at that temperature for 60 min and removed from the oven to cool down. During the thermal and compressive treatment a coherent polymer–porogen composite is formed from the mixture of polymer powder and porogen particles. Subsequently, the PLGA–gelatin microsphere composites were removed from the molds and placed in distilled water at 37°C for 28 h with repeated water changes. The leached cylinders were air and vacuum dried before they were cut with a low speed diamond saw. Foam porosity and pore size were controlled by the initial relative amount of gelatin utilized and size range of the gelatin microspheres, respectively. Both the yield strength and the compressive modulus of PLGA foams decreased with increasing porosity but were unaffected by pore size. The combination of compression molding and CL allows for the fabrication of porous scaffolds in the shape of the mold utilized. Porosity and pore size can be independently controlled by varying the amount and size of porogen used, respectively. With this technique it should be possible to incorporate bioactive molecules in either polymer or porogen for controlled delivery, because this process does not utilize organic solvents. However, the amount of bioactive molecules that might be released from the matrix during porogen leaching or inactivated during the heating–compression step has to be considered.

The PE/CL technique that is referred to as melt molding–PL applies heat and pressure to form a solid polymer–porogen composite. Leaching the composite in water leads to macroporous solids fabricated in a solvent-free process. The PLGA blends have been processed by such a technique using presieved sodium chloride particles [46].

Molding under an elevated temperature in combination with a thermoplastic or meltable porogen material is a concept that can be applied to generate a continuous porogen template during polymer–porogen composite formation in order to improve pore interconnectivity within the leached scaffold. Recently, a leaching technique with camphor as the porogen material has been established [47]. Porous matrices of poly(vinylidene fluoride) were fabricated by mixing the polymer with camphor and sealed in a glass tube. The tubes were placed in an oven and the polymer–porogen mixture was melted at 210°C for 30 min under constant shaking. Quenching to room temperature yielded a solid mass that was removed from the tube and leached with cyclohexane for 10 days with regular solvent changes. The resulting porous gels were washed in methanol and dried. Pore size diameters ranging between 4 nm and 400 μm were determined by mercury intrusion porosimetry. Better control of pore structure was achieved by combining sodium chloride and poly(ethylene oxide) (PEO) as porogen materials. Molding under an elevated temperature resulted in melting of the PEO particles that formed a coherent porogen phase while the salt particles remained unaffected templating macropores of controlled diameter [48,49]. A similar method has also been described for a porogen combination of sodium chloride and urea [50] (Table 3.1).

3.2.2 Porogen Dispersion–Composite Leaching and Solidification (PD/CLS) Techniques

This class of techniques shares the common characteristic of a sufficient plastic polymer–porogen composite intermediate during processing that allows for the formation of interconnects or micropores by additional processes while the porogen is leached out.

The methodology of a representative technique utilizing fractionated glucose crystals is only briefly described in the peer-reviewed literature [51], but more detailed information can be found in the corresponding patent literature [52,53]. The PLGA scaffolds with an extensively interconnected network of macropores (diameters range between 0.5–3.5 mm) were achieved by dispersing glucose particles in a solution of the polymer in dimethylsulfoxide (DMSO). The dispersion was transferred into aluminum molds and chilled at –15°C for 30 min [54]. For CLS, the molds were immersed in a water bath where DMSO was extracted into the water causing the polymer to precipitate and the glucose crystals were leached by the water from the precipitating polymer matrix. Subsequent to a vacuum drying step the matrices were cut into the desired shape.

Alternate PD/CLS techniques exclusively employ nonaqueous media for polymer processing in order to overcome problems attributed to extended exposure times to water such as in typical leaching steps for hydrophilic porogens. As mentioned in the previous section on PE/CL techniques, anhydrous leaching techniques can be used for the processing of hydrolytically sensitive polymers and allow for the incorporation of water soluble bioactive molecules during scaffold fabrication. In addition, the specific processing steps applied in these techniques cause porogen melting during the leaching step. The melted porogen material, which is hardly miscible with the solidifying polymer phase, forms a continuous porogen phase generating a well-interconnecting pore structure [55,56]. The ability to generate microstructures resembling the microstructure of trabecular bone makes these techniques attractive for the processing of all kinds of biodegradables and not just polymers that are particularly prone to hydrolysis. A technique called hydrocarbon templating utilizes microspheres made from waxy hydrocarbons, such as paraffin, as porogens [55]. Paraffin microspheres were dispersed in solutions of biodegradable polymers, such as PLLA and PLGA, in DCM or chloroform. The paste-like dispersions were packed in Teflon molds, which were immersed in a warm hydrocarbon solvent (e.g., pentane or hexane), to remove the porogen without dissolving or significantly swelling the matrix polymer. Hydrocarbon templating has been shown to yield large volume foams with homogenous microstructure that can be carved or cut into scaffolds of a desired size and shape.

Solid lipid templating is also an anhydrous processing technique that, as compared to hydrocarbon templating, utilizes a biocompatible porogen material and nonhalogenated solvents for the processing of a variety of polymers, including low molecular weight block copolymers suitable for bone tissue engineering [56–58] (Figure 3.4A). Different size fractions of solid lipid microparticles—fabricated by melt dispersion—serve as porogen and are mixed with solutions of different polymers in ethylacetate or methyl ethyl ketone-tetrahydrofuran mixtures. The viscous dispersion is transferred to Teflon molds and immersed in warm hexane for CLS. During the CLS step, which only lasts 30 min, pore interconnectivity is obtained by precipitating the polymer from its solution at the phase boundary with

(A) Solid lipid templating with solid lipid microspheres

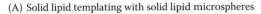

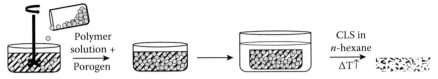

(B) Gas foaming with ammonium bicarbonate/sodium chloride particles

(C) Compression molding/gas foaming (CO_2) with salt particles

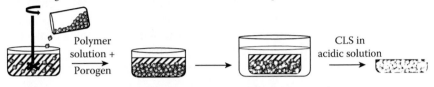

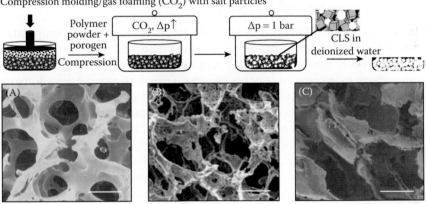

FIGURE 3.4

Schematic illustration of different particulate dispersion/composite leaching and solidification (PD/CLS) techniques. (A) solid lipid templating (Reproduced from Hacker, M., Ringhofer, M., Appel, B., Neubauer, M., Vogel, T., Young, S., Mikos, A. G., Blunk, T., Gopferich, A., and Schulz, M. B., *Biomaterials*, 28, 3497–3507, 2007; Hacker, M., Tessmar, J., Neubauer, M., Blaimer, A., Blunk, T., Gopferich, A., and Schulz, M. B., *Biomaterials*, 24, 4459–73, 2003. With permission.) (B) gas foaming/salt leaching (Reproduced from Lin, H. R., Kuo, C. J., Yang, C. Y., Shaw, S. Y., and Wu, Y. J., *J Biomed Mater Res.*, 63, 271–79, 2002. With permission.) (C) compression molding/carbon dioxide foaming/salt leaching (Reproduced from Sheridan, M. H., Shea, L. D., Peters, M. C., and Mooney, D. J., *J Control Release.*, 64, 91–102, 2000. With permission.) ▧ dry polymer powder, ■ solid polymer matrix, ▨ semisolid polymer matrix (e.g., polymer solution), ▰ angular porogen particle, ○ spherical porogen particle. Scale bars represent 200 μm. (Copyright permissions of inserted micrographs B and C are still pending.)

a continuous phase of melted lipid that is formed within the solidifying matrix. After complete extraction of the porogen, the solidified porous matrices are retrieved from the mold, vacuum dried, and cut to the desired size and shape. Macroporous structures with good interconnectivity that closely resemble the structure of spongy bone can be fabricated with this method [56]. Despite porogen melting during the CLS step, scaffold porosity and pore size were controlled by initial porogen content and pore size, respectively.

3.2.3 Combined Techniques Involving Leaching Processes

Leaching processes of both the PE/CL and the PD/CLS type have been used in combination with other pore forming techniques with the objective to improve major scaffold characteristics including scaffold porosity, pore interconnectivity, and interconnect size. Important examples of such combined processes are highlighted in the following section.

3.2.3.1 Combinations of Leaching with Freeze-Drying or Phase Separation Processes

Phase separation, which can be induced by changing solvent composition or system temperature, and freeze-drying processes are widely accepted methods to generate foams with high porosities and open pore networks [59,60]. A drawback of such techniques is that pore diameters exceeding 300 μm can hardly be obtained. Modifications of these techniques that include a particulate porogen have been described to overcome these limitations in pore size. A straightforward example combined ice particles with a freeze-drying process for the fabrication of polycaprolactone (PCL) scaffolds [61]. A solution of PCL in chloroform was mixed with ice particulates at 0°C and the mixture was frozen in liquid nitrogen. Freeze-drying of the composite yielded porous scaffolds in which pore size and porosity were determined by the size and mass fraction of ice particulates indicating the functionality of ice particles as typical porogen. Although, the ice particles are not really leached out in the true sense of the word but sublimated in vacuum, the process is referred to as porogen leaching.

A method that can be classified as a PE/CL technique has been described for a PLA-dextran blend [62]. Instead of one of the classical polymer solvents DCM or chloroform, a DCM-benzene mixture was used. Following a PE/CL routine, presieved NaCl particles (125–200 μm) were mixed into the solution of PLA and chemically protected dextran and the dispersion was cast into a Teflon mold. Solvents were removed by subsequent air and vacuum drying steps during which phase separation occurred. The subsequent evaporation of benzene under a vacuum resulted in a microporous structure with pore diameters of approximately 5–10 μm. In the CL step, the salt particles were finally leached out of the composite and macroscopic pores were formed. In comparison to scaffolds fabricated using the solvent DCM, the additional phase separation generated scaffolds having higher porosity and an open porous structure.

Other processes combining porogen leaching with freeze drying or phase separation processes can be found in Table 3.1.

3.2.3.2 Combinations of Porogen Bonding, Freeze-Drying and Porogen Leaching

A PE/PL technique that employs bonded gelatin microspheres as a pore forming template and a freeze-drying step to form the polymer–porogen composite has been shown to yield well-controlled interconnected PLLA scaffolds [63] (Figure 3.3C). Gelatin particles (280–450 μm) were treated in a saturated water vapor at 70°C to form a bonded 3-D assembly in a mold. A solution of PLLA in dioxane was cast onto the gelatin assembly. The mixtures were then freeze-dried or dried at room temperature to generate a microporous solid composite. The leaching step in water that removed the template of gelatin particles yielded interconnected, macroporous foams.

A combined PD/CLS technique has been described that utilizes the concepts of porogen fusion–replication and freeze-drying for the processing of biodegradable polymers [64]. Fused templates of sugar or salt were predesigned and prepared by two methods, one involving a humid atmosphere and the other a partial solvent mixture. For the first method, sugar–salt particles with the desired size range were filled into a plastic container and exposed to a controlled atmosphere (relative humidity of 75%, 81%, or 97% at 25°C) in a desiccator for two days. Thereafter, the fused templates were dried and removed from the container. For the second method, sugar or salt particles of predefined size fractions were soaked in an acetone-water mixture and gently packed into a plastic mold. Upon evaporation of the solvent mix, the dissolved fraction of the porogen materials precipitates on the undissolved porogen grains causing fusion of the grains and generation of a template in the size and shape of the mold.

3.2.3.3 Combinations of Gas Foaming and Particulate Leaching

A set of strategies centers around the generation of gas bubbles during the CLS step to generate interconnects and additional micropores in scaffolds. One concept of gas foaming employs effervescent salt particles or organic compounds, which are dispersed in the polymer matrix and decompose under formation of a gas upon contact with the leaching media triggered by moisture, heat or changes in pH. Different effervescent salts have been employed as porogen and gas forming agent or have been used in combination with inert salts as additional porogens (Table 3.1; Figure 3.4B). A representative method used ammonium bicarbonate as an efficient gas foaming agent as well as a particulate porogen salt to produce highly open porous PLLA foams [65]. Ammonium bicarbonate was dispersed in organic PLLA solution to form a viscous paste, which was cast in a mold and subsequently immersed in hot water to initiate CLS. Upon contact with water, the salt was not only extracted but also degraded under the formation of ammonia and carbon dioxide within the solidifying polymer matrix. The gas formation expanded the templated pores to a significant extent and led to a well-interconnected network of macropores. Several variations of such gas foaming–salt leaching processes have been described. In an improved method, citric acid was added to the leaching media and macroporous PLGA foams with porosities exceeding 90% and pore size of about 200 μm were obtained [66]. The porosity and mechanical strength could be controlled by adjusting the extent of acid-base gas evolving reaction between the effervescent salt and the leaching media. Figure 3.4B shows the pore structure of a foam fabricated by gas foaming–PL using ammonium bicarbonate and sodium chloride as mixed porogen [67].

Another gas foaming concept makes use of the specific properties of carbon dioxide (CO_2). When pressurized, CO_2 serves as a solvent for several biodegradable polymers and polymer matrices can be saturated with CO_2 under such conditions. Upon a quick pressure release, a thermodynamic instability in the polymer gas solution is generated and dissolved CO_2 nucleates and forms pores within the polymer matrix. Depending on nucleation type and speed, pore size and distribution can be varied in certain limits, but pore sizes far exceeding 100 μm can hardly be obtained. Consequently, a technique combining CO_2 foaming and PL has been proposed [68,69] (Figure 3.4C). In a solvent free, cold process, disks comprised of PLGA and presieved sodium chloride were compression molded at room temperature and subsequently allowed to equilibrate with high pressure CO_2 gas. Upon pressure relief, the nucleation and growth of gas pores started and a continuous polymer matrix with entrapped salt particles and gas bubbles formed. After leaching in water, interconnected macroporous foams were obtained in which porosity and the level of pore connectivity were regulated by the ratio of polymer–salt particles and the size of salt particles.

3.2.4 Anisotropic Macroporous Solids through Leaching Processes

In general, PL techniques are not known for their ability to generate anisotropic pore structures. In applications where controlled pore gradients or oriented pores are desired, the methods of choice include SFF techniques [70], certain electrospinning techniques [14,71], laser excimer ablation [72], and thermally induced phase separation [13,73,74]. Technically, PL techniques could be used to generate scaffolds with gradients in pore size or porosity. Using compression molding, polymer–salt composite disks that vary in average porogen size or porogen content can be fused into cylindrical composites, which—after the leaching step—provide a predefined zonal organization of the scaffold microstructure regarding pore sizes and porosities.

PL techniques that employ waxy porogen materials, such as hydrocarbon templating or solid lipid templating, suffer far less from particle sedimentation than solvent casting techniques and can be used for the fabrication of large volume scaffolds [55,56]. One can imagine that dispersion of waxy porogen particles in polymer solution that vary in porogen particle size or porogen content are used to fill a mold to finally obtain scaffolds with gradients in pore size and porosity.

With the objective to obtain poly(2-hydroxyethyl methacrylate) (pHEMA) scaffolds with longitudinally oriented channels for guided nerve regeneration, a fiber templating technology was developed [75]. Grouped bundles of extruded PCL fibers were fused by melting the ends of the strands and inserted into glass tubes with the fused ends first. Relevant quantities of HEMA, water, and excipients were degassed, combined with initiator, and accelerating agent and added to the fiber-filled molds in which polymerization was initiated. After removal from the glass molds, the pHEMA/PCL composites were cut into sections and placed in vials filled with acetone to dissolve the PCL fibers. To complete extraction, the etched scaffolds were Soxhlet extracted in water to remove residual acetone. By varying the diameter of the extruded fibers, scaffolds of differing channel diameters were formed. In comparison with PLA or PLGA, PCL was particularly suited for this method because PCL was insoluble in HEMA monomers yet soluble in acetone. Acetone in turn was appropriate as it did not significantly swell the pHEMA gel allowing the cross-linked polymer scaffold to maintain its structural integrity during the PCL extraction process.

3.3 Polymeric Materials Processed by Leaching Techniques

3.3.1 Synthetic Linear Polymers

Biodegradable synthetic polymers are the most prominent materials for the fabrication of leached macroporous foams for regenerative medicine and tissue engineering [76]. Poly(α-hydroxy esters), such as PLA and PLGA, are the main representatives of this group of polymers and have been most frequently used for the fabrication of macroporous biodegradable foams. Most techniques described in the second section of this chapter have been established using these polymers. Several reviews provide a thorough overview on the different polymers that have been processed and the characteristics and possible applications of the resulting foams and may be recommended to the reader for further study [77–80]. A combined polymer and salt PL technique [48] and a novel technique that integrates soft lithography, melt molding, and PL of PLGA microspheres [81] have been described for the fabrication of PCL foams. Classical PE/CL has been used for the fabrication of scaffolds from polyhydroxyalkanoates to be tested in tissue engineering applications [82,83]. Biodegradable polyurethanes have been processed by SC/PL and combined phase separation–salt leaching techniques to yield elastomeric foams [84–87]. Other biodegradable polymers that have been processed using leaching techniques include poly(amino acids) [88] and polyphosphazenes [89].

3.3.2 Synthetic Cross-Linked Polymer Networks

Various cross-linkable biodegradable macromonomers have been developed and characterized for applications in hard tissue reconstruction or regeneration due to the superior mechanical properties of cross-linked networks over linear polymers. One prominent representative is poly(propylene fumarate) (PPF), which forms mechanically strong biodegradable matrices upon cross-linking or cross-copolymerization [90]. Highly porous PPF scaffolds have been produced by a PE/CL technique [91]. In a solvent-free process, presieved salt particles were dispersed in viscous PPF macromer and PPF was cross-linked around the porogen particles in a mold. Leaching of the PPF–salt composites yielded foams with pore sizes ranging from 300 to 800 μm and porosities of 60–70%. A more recent study on salt-leached PPF scaffolds involved microcomputed tomography and provided detailed information on pore structure and pore interconnectivity [92]. Porogen leaching techniques have also been applied to yield macroporous foams from cross-linked polyanhydrides [93] and photopolymerized lactic acid based oligomers [94]. Porous cross-linked solids of water-soluble poly(propylene fumarate-*co*-ethylene glycol) were fabricated by a gas forming process with sodium bicarbonate and ascorbic acid upon macromer cross-linking in an aqueous solution [95].

Photopolymerized macroporous foams from a model dimethacrylate macromer were prepared by salt leaching and a systematic investigation of parameters controlling the scaffold structure was performed [96]. Architectural properties, such as porosity, pore size, strut thickness, strut size distributions, and pore interconnectivity were analyzed by combined techniques of microcomputed tomography, mercury porosimetry, and gravimetric analysis.

3.3.3 Natural Polymers

Silks have gained importance as scaffold materials for different applications toward the regeneration of bone, cartilage, ligament, and skin [97]. Macroporous 3-D scaffolds have been derived from silk fibroin solutions with salt leaching, gas foaming, and freeze-drying processes. Silk scaffolds obtained by salt leaching exhibited a very high porosity (>95%), uniform, interconnecting pores with controllable pore size and size distribution, and high compressive strength. Silk fibroin has been introduced into the salt leaching process in aqueous or organic, typically hexafluoroisopropanol (HFIP), solution [98]. The use of HFIP as a solvent enabled scaffold fabrication from a larger range of silk concentrations as the processing of aqueous solutions [97]. In addition, scaffolds from regenerated silk fibroin have been fabricated using a gas foaming technique with ammonium bicarbonate granules as a pore forming and a gas foaming agent [99].

Polymeric blends of cornstarch, which have shown promising degradative and mechanical properties, have been processed to yield starch-based foams for regenerative applications [100]. Several processing methodologies were originally developed and optimized. Two processes, namely compression molding/PL and classical SC/PL, involve porogen leaching steps.

Cellulose scaffolds with and without dispersed nano-sized hydroxyapatite were fabricated by a PE/CL technique. Following the paradigm of PE/PL techniques to utilize porogen materials with solubility properties opposite to those of the matrix material, hydrophobic poly(methyl methacrylate) particles (150–250 μm and 250–500 μm) were used for the processing of hydrophilic cellulose [101]. The acrylic particles were dispersed into the cellulose solutions and the suspensions were allowed to gel, immersed in water and methanol for washing, and finally dried at 40°C. From the dry composite, the acrylic particles were leached with dichloromethane. This process combines phase separation and PL utilizing a hydrophobic porogen for the processing of a hydrophilic polymer.

A special foaming process known from soft ice fabrication is called overrun process and has been used for the fabrication of porous foams from water-soluble polymers [102]. Overrun-processed scaffolds are typically characterized by a higher porosity than scaffolds prepared by a conventional freeze-drying method and showed a more uniform pore structure resulting from the injection of gas bubbles and ice recrystallization during the process [103]. Porous gelatin scaffolds were prepared using a modified overrun process because the traditional process of injecting air into a polymer solution in a refrigerated barrel under constant stirring yielded closed pore foams [104]. In the modified process sodium chloride or sucrose particles were added into the gelatin solution and homogeneously dispersed. The dispersion was poured into the overrun process apparatus, which was maintained at the comparably high temperature of 20°C to avoid gelation of the solution. Nitrogen was injected into the dispersion and the gas bubbles were homogeneously dispersed into the dispersion by a rotating impeller. The resulting gel-like dispersion was deep frozen and lyophilized. To initiate cross-linking, the dried composites were immersed in an acetone-water mixture containing the carbodiimide cross-linking agent EDC. Thereafter, the foams are extensively rinsed to remove any remaining porogen, frozen and again lyophilized. The combined overrun–particle leaching process resulted in an uniformly distributed and interconnected open structure.

3.4 Trends in Macroporous Solid Fabrication by Leaching Processes

Biomedical materials research has identified various needs that should be addressed by macroporous solids used as mechanical support, to deliver cells and bioactive molecules, and in other regenerative strategies. Depending on the application type and location different architectural properties, such as scaffold size, pore size, pore structure, interconnect size, and porosity, and mechanical properties may be appropriate [9,13,105]. Regarding material chemistry, control of degradation kinetics, and chemical microenvironment within the degrading foam is another important criterion [106]. In addition, scaffold design for tissue regeneration has the principal objective to recapitulate the structural organization and cell-matrix interactions of natural ECM [107,108]. In such attempts the incorporation of bioactive molecules, such as adhesion peptides or proteins and growth factors, play a key role. In general, scaffold fabrication should be focused on achieving an optimized balance between architectural features, chemical, physical, and biological properties.

Degradative, Mechanical and Architectural Properties. As outlined in the previous sections, porogen leaching techniques allow for the reliable processing of different polymeric materials with versatile properties and provide excellent control over architectural features. Therefore, leaching processes and their combination with other techniques allow for the generation of macroporous solids tailored to a specific application. Polymer chemistry can be varied to adjust degradative and mechanical properties of the resulting foams. Composite scaffolds with improved mechanical properties have also been fabricated by leaching techniques [90]. Especially, nano-composite scaffolds from PPF and functionalized carbon nanostructures [92] or PPF and acrylate-derivatized alumoxane [90], which can be covalently incorporated in the cross-linked composite, represent promising matrices for hard tissue engineering. The dispersion of phosphate ceramics, such as β-tricalcium phosphate or nano-sized hydroxyapatite, within macroporous biodegradable matrices is a strategy that focuses on improving both the mechanical properties and the biological properties of the composite [109–113]. Biomimetic concepts that focus on improving the architectural properties of scaffolds center on the generation of nanostructures within a macroporous foam (e.g., by combining phase separation and porogen leaching processes) [22,109,114].

Drug-Loaded Foams and Matrices with Improved Biological Properties. The incorporation of bioactive molecules into macroporous matrices or the modification of the foam surfaces with said molecules are strategies to induce specific biological responses and improve interactions of the matrix with cells and tissues [115–119]. Such strategies to generate biomimetic matrices have typically been applied to hydrogel-based scaffolds, but few studies have involved macroporous foams that were fabricated by leaching processes. Hydrocarbon templating, an anhydrous leaching process, has been used to incorporate and release the model protein alkaline phosphatase with minimal loss of activity from biodegradable foams [55]. Matrices produced by a solvent-free gas foaming-salt leaching technique were loaded with vascular endothelial growth factor [69]. The protein retained 90% of its bioactivity and was released in a controlled manner. Another interesting strategy to obtain controlled release of proteins from macroporous polymeric scaffolds involves emulsion coating of porous foams prefabricated by a molding/PL technique [120,121]. In this process, a prefabricated structure is infiltrated by a water-in-oil emulsion that is fabricated from an aqueous protein solution and an organic polymer solution. After solvent evaporation, a polymer film containing the protein is deposited on the surface of the prefabricated structure. By adjusting the emulsion parameters the release kinetics of lysozyme could be controlled.

Anhydrous solid lipid templating has successfully been employed to process amine reactive biodegradable copolymers to form macroporous solids for instant surface modification [57]. The material processed in this study has been shown effective to control osteoblastic cell adhesion to surfaces covalently modified with cyclic adhesion peptides [122]. A PE/CL strategy has been employed to fabricated scaffolds from a novel biodegradable graft copolymer of PLLA and the ECM component chondroitin sulphate [123].

In addition to the release of bioactive molecules and the design of biomimetic surface, gene delivery is another concept that has been realized with leached macroporous solids. Using a template of fused methacrylic microspheres, macroporous fibrin scaffolds suitable for soft tissue engineering applications were fabricated and loaded with nonviral gene carriers [124]. Different spatial distribution allowed for temporal control of maximal transgene expression and the expression profile.

3.5 Conclusion

A variety of leaching processes is available for the robust and comparably facile processing of different materials, especially polymers, into macroporous solids with defined pore architectures and due to the improved processing methods well-interconnected porosity. Initial problems of skin formation, dimensional limitations, and closed pore structures have been overcome and techniques are available that allow for the processing of both hydrophilic and lipophilic materials as well as macromonomers that require cross-linking during processing and specialty polymers with chemically sensitive functionalities that need to be preserved during processing. Leaching processes provide adequate control of pore size, especially in the high micrometer range, almost independent of the material properties. This makes such processes the first choice in many science and engineering laboratories, especially in those where new materials are being developed. The development of designed polymers with tailored properties and functionalities progresses and leaching processes enable a robust lab scale processing of such materials into macroporous scaffolds for material testing in 3-D cell culture.

3.6 Acknowledgments

Financial support by the Deutsche Forschungsgemeinschaft (TRR 67 TP A1) is kindly acknowledged. The publisher for the copyrighted material in Figure 3.3 B and D is Mary Ann Liebert, Inc. publishers.

References

1. Kelly, A. 2006. Why engineer porous materials? *Philos Transact A Math Phys Eng Sci* 364:5–14.
2. Chevalier, E., Chulia, D., Pouget, C., and Viana, M. 2008. Fabrication of porous substrates: A review of processes using pore forming agents in the biomaterial field. *J Pharm Sci* 97:1135–54.
3. Hentze, H.-P., and Antonietti, M. 2002. Porous polymers and resins for biotechnological and biomedical applications. *Rev Mol Biotechnol* 90:27–53.
4. Zhang, H., and Cooper, A. I. 2005. Synthesis and applications of emulsion-templated porous materials. *Soft Matter* 1:107–13.
5. Sachlos, E., and Czernuszka, J. T. 2003. Making tissue engineering scaffolds work. Review on the application of solid freeform fabrication technology to the production of tissue engineering scaffolds. *Eur Cell Mater* 5:29–40.
6. Otsuki, B., Takemoto, M., Fujibayashi, S., Neo, M., Kokubo, T., and Nakamura, T. 2006. Pore throat size and connectivity determine bone and tissue ingrowth into porous implants: Three-dimensional micro-CT based structural analyses of porous bioactive titanium implants. *Biomaterials* 27:5892–900.
7. Weigel, T., Schinkel, G., and Lendlein, A. 2006. Design and preparation of polymeric scaffolds for tissue engineering. *Expert Rev Med Devices* 3:835–51.
8. Christenson, E. M., Soofi, W., Holm, J. L., Cameron, N. R., and Mikos, A. G. 2007. Biodegradable fumarate-based polyHIPEs as tissue engineering scaffolds. *Biomacromolecules* 8:3806–14.
9. Karageorgiou, V., and Kaplan, D. 2005. Porosity of 3D biomaterial scaffolds and osteogenesis. *Biomaterials* 26:5474–91.
10. Ruiyun Zhang, P. X. M. 2000. Synthetic nano-fibrillar extracellular matrices with predesigned macroporous architectures. *J Biomed Mater Res* 52:430–38.
11. Karande, T. S., Ong, J. L., and Agrawal, C. M. 2004. Diffusion in musculoskeletal tissue engineering scaffolds: Design issues related to porosity, permeability, architecture, and nutrient mixing. *Ann Biomed Eng* 32:1728–43.
12. Mikos, A. G., and Temenoff, J. S. 2000. Formation of highly porous biodegradable scaffolds for tissue engineering. *Electron J Biotechnol* 3:114–19.
13. Oh, S. H., Park, I. K., Kim, J. M., and Lee, J. H. 2007. *In vitro* and *in vivo* characteristics of PCL scaffolds with pore size gradient fabricated by a centrifugation method. *Biomaterials* 28:1664–71.
14. Pham, Q. P., Sharma, U., and Mikos, A. G. 2006. Electrospinning of polymeric nanofibers for tissue engineering applications: A review. *Tissue Eng* 12:1197–1211.

15. Sill, T. J., and von Recum, H. A. 2008. Electrospinning: Applications in drug delivery and tissue engineering. *Biomaterials* 29:1989–2006.

16. Mikos, A. G., Bao, Y., Cima, L. G., Ingber, D. E., Vacanti, J. P., and Langer, R. 1993. Preparation of poly(glycolic acid) bonded fiber structures for cell attachment and transplantation. *J Biomed Mater Res* 27:183–89.

17. Freed, L. E., Vunjak, N. G., Biron, R. J., Eagles, D. B., Lesnoy, D. C., Barlow, S. K., and Langer, R. 1994. Biodegradable polymer scaffolds for tissue engineering. *Biotechnology* 12:689–93.

18. Sittinger, M., Reitzel, D., Dauner, M., Hierlemann, H., Hammer, C., Kastenbauer, E., Planck, H., Burmester, and G. R., Bujia, J. 1996. Resorbable polyesters in cartilage engineering: Affinity and biocompatibility of polymer fiber structures to chondrocytes. *J Biomed Mater Res* 33:57–63.

19. Fambri, L., Pegoretti, A., Fenner, R., Incardona, S. D., and Migliaresi, C. 1997. Biodegradable fibres of poly(-lactic acid) produced by melt spinning. *Polymer* 38:79–85.

20. Pham, Q. P., Sharma, U., and Mikos, A. G. 2006. Electrospun poly(ε-caprolactone) microfiber and multilayer nanofiber/microfiber scaffolds: Characterization of scaffolds and measurement of cellular infiltration. *Biomacromolecules* 7:2796–805.

21. Teo, W. E., He, W., and Ramakrishna, S. 2006. Electrospun scaffold tailored for tissue-specific extracellular matrix. *Biotechnol J* 1:918–29.

22. Ma, P. X. 2008. Biomimetic materials for tissue engineering. *Adv Drug Deliv Rev* 60:184–98.

23. Mooney, D. J., Baldwin, D. F., Suh, N. P., Vacanti, J. P., and Langer, R. 1996. Novel approach to fabricate porous sponges of poly(D,L-lactic-co-glycolic acid) without the use of organic solvents. *Biomaterials* 17:1417–22.

24. Barbetta, A., Dentini, M., De Vecchis, M. S., Filippini, P., Formisano, G., and Caiazza, S. 2005. Scaffolds based on biopolymeric foams. *Adv Funct Mater* 15:118–24.

25. Cameron, N. R. 2005. High internal phase emulsion templating as a route to well-defined porous polymers. *Polymer* 46:1439–49.

26. Borden, M., Attawia, M., Khan, Y., and Laurencin, C. T. 2002. Tissue engineered microsphere-based matrices for bone repair: Design and evaluation. *Biomaterials* 23:551–59.

27. Nam, Y. S., and Park, T. G. 1999. Porous biodegradable polymeric scaffolds prepared by thermally induced phase separation. *J Biomed Mater Res* 47:8–17.

28. Liu Tsang, V., and Bhatia, S. N. 2004. Three-dimensional tissue fabrication. *Adv Drug Deliv Rev* 56:1635–47.

29. Murphy, W. L., Dennis, R. G., Kileny, J. L., and Mooney, D. J. 2002. Salt fusion: An approach to improve pore interconnectivity within tissue engineering scaffolds. *Tissue Eng* 8:43–52.

30. Langer, R., and Vacanti, J. P. 1993. Tissue engineering. *Science* 260:920–26.

31. Mikos, A. G., Sarakinos, G., Leite, S. M., Vacanti, J. P., and Langer, R. 1993. Laminated three-dimensional biodegradable foams for use in tissue engineering. *Biomaterials* 14:323–30.

32. Mikos, A. G., Thorsen, A. J., Czerwonka, L. A., Bao, Y., Langer, R., Winslow, D. N., and Vacanti, J. P. 1994. Preparation and characterization of poly(L-lactic acid) foams. *Polymer* 35:1068–77.

33. Wake, M. C., Gupta, P. K., and Mikos, A. G. 1996. Fabrication of pliable bio-degradable polymer foams to engineer soft tissues. *Cell Transplant* 5:465–73.
34. Widmer, M. S., Gupta, P. K., Lu, L., Meszlenyi, R. K., Evans, G. R., Brandt, K., Savel, T., Gurlek, A., Patrick-CW, J., and Mikos, A. G. 1998. Manufacture of porous biodegradable polymer conduits by an extrusion process for guided tissue regeneration. *Biomaterials* 19:1945–55.
35. Mikos, A. G., Sarakinos, G., Leite, S. M., Vacant, J. P., and Langer, R. 1993. Laminated three-dimensional biodegradable foams for use in tissue engineering. *Biomaterials* 14:323–30.
36. Liao, C. J., Chen, C. F., Chen, J. H., Chiang, S. F., Lin, Y. J., and Chang, K. Y. 2002. Fabrication of porous biodegradable polymer scaffolds using a solvent merging/particulate leaching method. *J Biomed Mater Res* 59:676–81.
37. Wu, L., Jing, D., and Ding, J. 2006. A "room-temperature" injection molding/particulate leaching approach for fabrication of biodegradable three-dimensional porous scaffolds. *Biomaterials* 27:185–91.
38. Vaquette, C., Frochot, C., Rahouadj, R., and Wang, X. 2008. An innovative method to obtain porous PLLA scaffolds with highly spherical and interconnected pores. *J Biomed Mater Res B* 86:9–17.
39. Gross, K. A., and Rodriguez-Lorenzo, L. M. L. 2004. Biodegradable composite scaffolds with an interconnected spherical network for bone tissue engineering. *Biomaterials* 25:4955–62.
40. Grenier, S., Sandig, M., and Mequanint, K. 2007. Polyurethane biomaterials for fabricating 3D porous scaffolds and supporting vascular cells. *J Biomed Mater Res A* 82:802–9.
41. Zhang, J., Zhang, H., Wu, L., and Ding, J. 2006. Fabrication of three dimensional polymeric scaffolds with spherical pores. *J Mater Sci* 41:1725–31.
42. Ma, P. X., and Choi, J. W. 2001. Biodegradable polymer scaffolds with well-defined interconnected spherical pore network. *Tissue Eng* 7:23–33.
43. Lebourg, M., Sabater Serra, R., Más Estellés, J., Hernández Sánchez, F., Gómez Ribelles, J. L., and Suay Antón, J. 2008. Biodegradable polycaprolactone scaffold with controlled porosity obtained by modified particle-leaching technique. *J Mater Sci Mater Med* 19:2047–53.
44. White, R. A., Weber, J. N., and White, E. W. 1972. Replamineform: A new process for preparing porous ceramic, metal, and polymer prosthetic materials. *Science* 176:922–24.
45. Thomson, R. C., Yaszemski, M. J., Powers, J. M., and Mikos, A. G. 1995. Fabrication of biodegradable polymer scaffolds to engineer trabecular bone. *J Biomater Sci Polym Ed* 7:23–38.
46. Oh, S. H., Kang, S. G., Kim, E. S., Cho, S. H., and Lee, J. H. 2003. Fabrication and characterization of hydrophilic poly(lactic-*co*-glycolic acid)/poly(vinyl alcohol) blend cell scaffolds by melt-molding particulate-leaching method. *Biomaterials* 24:4011–21.
47. Dasgupta, D., and Nandi, A. K. 2005. Multiporous polymeric materials from thermoreversible poly(vinylidene fluoride) gels. *Macromolecules* 38:6504–12.
48. Reignier, J., and Huneault, M. A. 2006. Preparation of interconnected poly([epsilon]-caprolactone) porous scaffolds by a combination of polymer and salt particulate leaching. *Polymer* 47:4703–17.
49. Ghosh, S., Viana, J. C., Reis, R. L., and Mano, J. F. 2007. The double porogen approach as a new technique for the fabrication of interconnected poly(L-

lactic acid) and starch based biodegradable scaffolds. *J Mater Sci Mater Med* 18:185–93.

50. Katoh, K., Tanabe, T., and Yamauchi, K. 2004. Novel approach to fabricate keratin sponge scaffolds with controlled pore size and porosity. *Biomaterials* 25:4255–62.

51. Holy, C. E., Dang, S. M., Davies, J. E., and Shoichet, M. S. 1999. *In vitro* degradation of a novel poly(lactide-*co*-glycolide) 75/25 foam. *Biomaterials* 20:1177–85.

52. Holy, C. E., Shoichet, M. S., and Davies, J. E. 1998. Biodegradable polymer scaffold. Patent #WO9925391.

53. Holy, C. E., Shoichet, M. S., and Davies, J. E. 2000. Polymer scaffold having microporous polymer struts defining interconnected macropores. Patent #US6472210.

54. Karp, J. M., Shoichet, M. S., and Davies, J. E. 2003. Bone formation on two-dimensional poly(DL-lactide-co-glycolide) (PLGA) films and three-dimensional PLGA tissue engineering scaffolds *in vitro*. *J Biomed Mater Res A* 64:388–96.

55. Shastri, V. P., Martin, I., and Langer, R. 2000. Macroporous polymer foams by hydrocarbon templating. *Proc Natl Acad Sci USA* 97:1970–75.

56. Hacker, M., Ringhofer, M., Appel, B., Neubauer, M., Vogel, T., Young, S., Mikos, A. G., Blunk, T., Gopferich, A., and Schulz, M. B. 2007. Solid lipid templating of macroporous tissue engineering scaffolds. *Biomaterials* 28:3497–507.

57. Hacker, M., Tessmar, J., Neubauer, M., Blaimer, A., Blunk, T., Gopferich, A., and Schulz, M. B. 2003. Towards biomimetic scaffolds: Anhydrous scaffold fabrication from biodegradable amine-reactive diblock copolymers. *Biomaterials* 24:4459–73.

58. Lieb, E., Tessmar, J., Hacker, M., Fischbach, C., Rose, D., Blunk, T., Mikos, A. G., Goepferich, A., and Schulz, M. B. 2003. Poly(D,L-lactic acid)-poly(ethylene glycol)-monomethyl ether diblock copolymers control adhesion and osteoblastic differentiation of marrow stromal cells. *Tissue Eng* 9:71–84.

59. Nam, Y. S., and Park, T. G. 1999. Biodegradable polymeric microcellular foams by modified thermally induced phase separation method. *Biomaterials* 20:1783–90.

60. Whang, K., Healy, K. E., Elenz, D. R., Nam, E. K., Tsai, D. C., Thomas, C. H., Nuber, G. W., Glorieux, F. H., Travers, R., and Sprague, S. M. 1999. Engineering bone regeneration with bioabsorbable scaffolds with novel microarchitecture. *Tissue Eng* 5:35–51.

61. Ning, C., Cheng, H., Yin, Z., Zhu, W., Chen, H., Lei, S., Yin, S., and Liu, H. 2008. Preparation of porous polycaprolactone scaffolds by using freeze-drying combined porogen-leaching methods. *Key Eng Mater* 368–72 Part 2: 1224–26.

62. Cai, Q., Yang, J., Bei, J., and Wang, S. 2002. A novel porous cells scaffold made of polylactide-dextran blend by combining phase-separation and particle-leaching techniques. *Biomaterials* 23:4483–92.

63. Zhou, Q., Gong, Y., and Gao, C. 2005. Microstructure and mechanical properties of poly(L-lactide) scaffolds fabricated by gelatin particle leaching method. *J Appl Polym Sci* 98:1373–79.

64. Hou, Q., Grijpma, D. W., and Feijen, J. 2003. Preparation of interconnected highly porous polymeric structures by a replication and freeze-drying process. *J Biomed Mater Res* 67B:732–40.

65. Nam, Y. S., Yoon, J. J., and Park, T. G. 2000. A novel fabrication method of macroporous biodegradable polymer scaffolds using gas foaming salt as a porogen additive. *J Biomed Mater Res* 53:1–7.

66. Yoon, J. J., and Park, T. G. 2001. Degradation behaviors of biodegradable macroporous scaffolds prepared by gas foaming of effervescent salts. *J Biomed Mater Res* 55:401–8.
67. Lin, H. R., Kuo, C. J., Yang, C. Y., Shaw, S. Y., and Wu, Y. J. 2002. Preparation of macroporous biodegradable PLGA scaffolds for cell attachment with the use of mixed salts as porogen additives. *J Biomed Mater Res* 63:271–79.
68. Harris, L. D., Kim, B. S., and Mooney, D. J. 1998. Open pore biodegradable matrices formed with gas foaming. *J Biomed Mater Res* 42:396–402.
69. Sheridan, M. H., Shea, L. D., Peters, M. C., and Mooney, D. J. 2000. Bioabsorbable polymer scaffolds for tissue engineering capable of sustained growth factor delivery. *J Control Release* 64:91–102.
70. Woodfield, T. B. F., Van Blitterswijk, C. A., De Wijn, J., Sims, T. J., Hollander, A. P., and Riesle, J. 2005. Polymer scaffolds fabricated with pore-size gradients as a model for studying the zonal organization within tissue-engineered cartilage constructs. *Tissue Eng* 11:1297–1311.
71. Courtney, T., Sacks, M. S., Stankus, J., Guan, J., and Wagner, W. R. 2006. Design and analysis of tissue engineering scaffolds that mimic soft tissue mechanical anisotropy. *Biomaterials* 27:3631–38.
72. Doi, K., Nakayama, Y., and Matsuda, T. 1996. Novel compliant and tissue-permeable microporous polyurethane vascular prosthesis fabricated using an excimer laser ablation technique. *J Biomed Mater Res* 31:27–33.
73. Guan, J., Fujimoto, K. L., Sacks, M. S., and Wagner, W. R. 2005. Preparation and characterization of highly porous, biodegradable polyurethane scaffolds for soft tissue applications. *Biomaterials* 26:3961–71.
74. Yang, F., Qu, X., Cui, W., Bei, J., Yu, F., Lu, S., and Wang, S. 2006. Manufacturing and morphology structure of polylactide-type microtubules orientation-structured scaffolds. *Biomaterials* 27:4923–33.
75. Flynn, L., Dalton, P. D., and Shoichet, M. S. 2003. Fiber templating of poly(2-hydroxyethyl methacrylate) for neural tissue engineering. *Biomaterials* 24:4265–72.
76. Gunatillake, P. A., and Adhikari, R. 2003. Biodegradable synthetic polymers for tissue engineering. *Eur Cell Mater* 5:1–16.
77. Athanasiou, K. A., Agrawal, C. M., Barber, F. A., and Burkhart, S. S. 1998. Orthopaedic applications for PLA-PGA biodegradable polymers. *Arthroscopy* 14:726–37.
78. Hutmacher, D. W. 2000. Scaffolds in tissue engineering bone and cartilage. *Biomaterials* 21:2529–43.
79. Agrawal, C. M., and Ray, R. B. 2001. Biodegradable polymeric scaffolds for musculoskeletal tissue engineering. *J Biomed Mater Res* 55:141–50.
80. Yang, S., Leong, K. F., Du, Z., and Chua, C. K. 2001. The design of scaffolds for use in tissue engineering. Part I. Traditional factors. *Tissue Eng* 7:679–89.
81. Sarkar, S., Lee, G. Y., Wong, J. Y., and Desai, T. A. 2006. Development and characterization of a porous micro-patterned scaffold for vascular tissue engineering applications. *Biomaterials* 27:4775–82.
82. Zhao, K., Deng, Y., Chun Chen, J., and Chen, G. Q. 2003. Polyhydroxyalkanoate (PHA) scaffolds with good mechanical properties and biocompatibility. *Biomaterials* 24:1041–45.
83. Chen, G. Q., and Wu, Q. 2005. The application of polyhydroxyalkanoates as tissue engineering materials. *Biomaterials* 26:6565–78.

84. Fromstein, J. D., and Woodhouse, K. A. 2002. Elastomeric biodegradable polyurethane blends for soft tissue applications. *J Biomater Sci Polym Ed* 13:391–406.
85. Heijkants, R. G. J. C., Van Calck, R. V., De Groot, J. H., Pennings, A. J., Schouten, A. J., van Tienen, T. G., Ramrattan, N., Buma, P., and Veth, R. P. H. 2004. Design, synthesis and properties of a degradable polyurethane scaffold for meniscus regeneration. *J Mater Sci Mater Med* 15:423–27.
86. Gorna, K., and Gogolewski, S. 2006. Biodegradable porous polyurethane scaffolds for tissue repair and regeneration. *J Biomed Mater Res A* 79:128–38.
87. Guelcher, S. A. 2008. Biodegradable polyurethanes: Synthesis and applications in regenerative medicine. *Tissue Eng Part B Rev* 14:3–17.
88. Bourke, S. L., and Kohn, J. 2003. Polymers derived from the amino acid -tyrosine: Polycarbonates, polyarylates and copolymers with poly(ethylene glycol). *Adv Drug Deliv Rev* 55:447–66.
89. Laurencin, C. T., El-Amin, S. F., Ibim, S. E., Willoughby, D. A., Attawia, M., Allcock, H. R., and Ambrosio, A. A. 1996. A highly porous 3-dimensional polyphosphazene polymer matrix for skeletal tissue regeneration. *J Biomed Mater Res* 30:133–38.
90. Mistry, A. S., and Mikos, A. G. 2005. Tissue engineering strategies for bone regeneration. *Adv Biochem Eng Biotechnol* 94:1–22.
91. Fisher, J. P., Holland, T. A., Dean, D., and Mikos, A. G. 2003. Photo initiated cross-linking of the biodegradable polyester poly(propylene fumarate). Part II. *In vitro* degradation. *Biomacromolecules* 4:1335–42.
92. Shi, X., Sitharaman, B., Pham, Q. P., Liang, F., Wu, K., Edward Billups, W., Wilson, L. J., and Mikos, A. G. 2007. Fabrication of porous ultra-short single-walled carbon nanotube nanocomposite scaffolds for bone tissue engineering. *Biomaterials* 28:4078–90.
93. Burkoth, A. K., Burdick, J., and Anseth, K. S. 2000. Surface and bulk modifications to photo crosslinked polyanhydrides to control degradation behavior. *J Biomed Mater Res* 51:352–59.
94. Burdick, J. A., Frankel, D., Dernell, W. S., and Anseth, K. S. 2003. An initial investigation of photo curable three-dimensional lactic acid based scaffolds in a critical-sized cranial defect. *Biomaterials* 24:1613–20.
95. Behravesh, E., Timmer, M. D., Lemoine, J. J., Liebschner, M. A. K., and Mikos, A. G. 2002. Evaluation of the *in vitro* degradation of macroporous hydrogels using gravimetry, confined compression testing, and microcomputed tomography. *Biomacromolecules* 3:1263–70.
96. Lin-Gibson, S., Cooper, J. A., Landis, F. A., and Cicerone, M. T. 2007. Systematic investigation of porogen size and content on scaffold morphometric parameters and properties. *Biomacromolecules* 8:1511–18.
97. Wang, Y., Kim, H. J., Vunjak-Novakovic, G., and Kaplan, D. L. 2006. Stem cell-based tissue engineering with silk biomaterials. *Biomaterials* 27:6064–82.
98. Kim, U. J., Park, J., Joo Kim, H., Wada, M., and Kaplan, D. L. 2005. Three-dimensional aqueous-derived biomaterial scaffolds from silk fibroin. *Biomaterials* 26:2775–85.
99. Nazarov, R., Jin, H. J., and Kaplan, D. L. 2004. Porous 3-D scaffolds from regenerated silk fibroin. *Biomacromolecules* 5:718–26.
100. Gomes, M. E., Godinho, J. S., Tchalamov, D., Cunha, A. M., and Reis, R. L. 2002. Alternative tissue engineering scaffolds based on starch: Processing methodologies, morphology, degradation and mechanical properties. *Mater Sci Eng , C* 20:19–26.

101. Tsioptsias, C., and Panayiotou, C. 2008. Preparation of cellulose-nanohydroxyapatite composite scaffolds from ionic liquid solutions. *Carbohydr Polym* 74:99–105.
102. Kang, H. G., Lee, S. B., and Lee, Y. M. 2005. Novel preparative method for porous hydrogels using overrun process. *Polym Int* 54:537–43.
103. Su, J. L., So, Y. K., and Lee, Y. M. 2007. Preparation of porous collagen/hyaluronic acid hybrid scaffolds for biomimetic functionalization through biochemical binding affinity. *J Biomed Mater Res B* 82:506–18.
104. Kang, H. G., Kim, S. Y., and Lee, Y. M. 2006. Novel porous gelatin scaffolds by overrun/particle leaching process for tissue engineering applications. *J Biomed Mater Res B* 79:388–97.
105. Guarino, V., Causa, F., and Ambrosio, L. 2007. Porosity and mechanical properties relationship in PCL porous scaffolds. *J Appl Biomater Biomech* 5:149–57.
106. Liu, W., and Cao, Y. 2007. Application of scaffold materials in tissue reconstruction in immunocompetent mammals: Our experience and future requirements. *Biomaterials* 28:5078–86.
107. Causa, F., Netti, P. A., and Ambrosio, L. 2007. A multi-functional scaffold for tissue regeneration: The need to engineer a tissue analogue. *Biomaterials* 28:5093–99.
108. Shin, H., Jo, S., and Mikos, A. G. 2003. Biomimetic materials for tissue engineering. *Biomaterials* 24:4353–64.
109. Ma, P. X. 2004. Scaffolds for tissue fabrication. *Mater Today* 7:30–40.
110. Yaszemski, M. J., Payne, R. G., Hayes, W. C., Langer, R., and Mikos, A. G. 1996. *In vitro* degradation of a poly(propylene fumarate)-based composite material. *Biomaterials* 17:2127–30.
111. Wolfe, M. S., Dean, D., Chen, J. E., Fisher, J. P., Han, S., Rimnac, C. M., and Mikos, A. G. 2002. *In vitro* degradation and fracture toughness of multilayered porous poly(propylene fumarate)/β-tricalcium phosphate scaffolds. *J Biomed Mater Res* 61:159–64.
112. Chen, P. P., Zhang, S. M., Cheng, L., Huang, S. L., Liu, J., Zhou, W., Gong, H., and Luo, Q. M. 2006. Adhesion and proliferation of chondrocytes on nano-HA/PDLLA scaffolds for bone tissue engineering. *Key Eng Mater* 309–11 part II:919–22.
113. Zhang, L., Li, Y., Wang, X. J., Wei, J., and Peng, X. 2005. Studies on the porous scaffold made of the nano-HA/PA66 composite. *J Mater Sci* 40:107–10.
114. Chen, V. J., and Ma, P. X. 2004. Nano-fibrous poly(-lactic acid) scaffolds with interconnected spherical macropores. *Biomaterials* 25:2065–73.
115. Sohier, J., Moroni, L., Blitterswijk, C. V., Groot, K. D., and Bezemer, J. M. 2008. Critical factors in the design of growth factor releasing scaffolds for cartilage tissue engineering. *Expert Opin Drug Deliv* 5:543–66.
116. Chung, H. J., and Park, T. G. 2007. Surface engineered and drug releasing pre-fabricated scaffolds for tissue engineering. *Adv Drug Deliv Rev* 59:249–62.
117. Sokolsky-Papkov, M., Agashi, K., Olaye, A., Shakesheff, K., and Domb, A. J. 2007. Polymer carriers for drug delivery in tissue engineering. *Adv Drug Deliv Rev* 59:187–206.
118. Tessmar, J. K., and Pferich, A. M. 2007. Matrices and scaffolds for protein delivery in tissue engineering. *Adv Drug Deliv Rev* 59:274–91.

119. Drotleff, S., Lungwitz, U., Breunig, M., Dennis, A., Blunk, T., Tessmar, J., and Gopferich, A. 2004. Biomimetic polymers in pharmaceutical and biomedical sciences. *Eur J Pharm Biopharm* 58:385–407.
120. Sohier, J., Haan, R. E., de Groot, K., and Bezemer, J. M. 2003. A novel method to obtain protein release from porous polymer scaffolds: Emulsion coating. *J Control Release* 87:57–68.
121. Sohier, J., Vlugt, T. J. H., Cabrol, N., Van Blitterswijk, C., de Groot, K., and Bezemer, J. M. 2006. Dual release of proteins from porous polymeric scaffolds. *J Control Release* 111:95–106.
122. Lieb, E., Hacker, M., Tessmar, J., Kunz-Schughart, L. A., Fiedler, J., Dahmen, C., Hersel, U., Kessler, H., Schulz, M. B., and Gopferich, A. 2005. Mediating specific cell adhesion to low-adhesive diblock copolymers by instant modification with cyclic RGD peptides. *Biomaterials* 26:2333–41.
123. Lee, C. T., Huang, C. P., and Lee, Y. D. 2006. Biomimetic porous scaffolds made from poly(L-lactide)-*g*-chondroitin sulfate blend with poly(L-lactide) for cartilage tissue engineering. *Biomacromolecules* 7:2200–9.
124. Saul, J. M., Linnes, M. P., Ratner, B. D., Giachelli, C. M., and Pun, S. H. 2007. Delivery of non-viral gene carriers from sphere-templated fibrin scaffolds for sustained transgene expression. *Biomaterials* 28:4705–16.
125. Chen, G., Ushida, T., and Tateishi, T. 2001. Preparation of poly(L-lactic acid) and poly(D, L-lactic-*co*-glycolic acid) foams by use of ice microparticulates. *Biomaterials* 22:2563–67.
126. Guan, L., and Davies, J. E. 2004. Preparation and characterization of a highly macroporous biodegradable composite tissue engineering scaffold. *J Biomed Mater Res A* 71:480–87.
127. Agrawal, C. M., McKinney, J. S., Lanctot, D., and Athanasiou, K. A. 2000. Effects of fluid flow on the *in vitro* degradation kinetics of biodegradable scaffolds for tissue engineering. *Biomaterials* 21:2443–52.
128. Horák, D., Hlídkováa, H., Hradila, J., Lapèíkováa, M., and Šloufa, M. 2008. Superporous poly(2-hydroxyethyl methacrylate) based scaffolds: Preparation and characterization. *Polymer* 49:2046–54.
129. Zhang, L. F., Yang, D. J., Chen, H. C., Sun, R., Xu, L., Xiong, Z. C., Govender, T., and Xiong, C. D. 2008. An ionically crosslinked hydrogel containing vancomycin coating on a porous scaffold for drug delivery and cell culture. *Int J Pharm* 353:74–87.
130. Ma, Z., Gao, C., Gong, Y., and Shen, J. 2003. Paraffin spheres as porogen to fabricate poly(L-lactic acid) scaffolds with improved cytocompatibility for cartilage tissue engineering. *J Biomed Mater Res* 67:610–17.
131. Studenovská, H., Šlouf, M., and Rypáèek, F. 2008. Poly(HEMA) hydrogels with controlled pore architecture for tissue regeneration applications. *J Mater Sci Mater Med* 19:615–21.

4

Production and Properties of Poly(Vinyl Alcohol) Cryogels: Recent Developments

María C. Gutiérrez, Inmaculada Aranaz,
María L. Ferrer, and Francisco del Monte

CONTENTS

4.1 Introduction

The preparation of polyvinyl alcohol (PVA) cryogels has been the subject of research for numerous works since their discovery in 1970 [1]. The cryogenic treatment basically consists of freezing an initially homogeneous polymer solution at low temperatures, storing it in the frozen state for a definite time, and finally defrosting it (Figure 4.1) [2]. Freezing is generally performed at a relatively high cooling rate, and, although solid physical gels are already obtained by imposing a single freeze-thaw cycle, freezing and thawing can

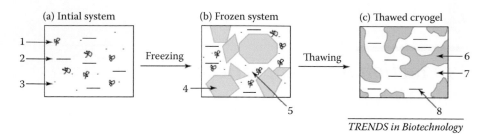

FIGURE 4.1
Schematic representation of PVA cryogels formation by freeze-thawing process. (1) macromolecules in a solution, (2) solvent, (3) low-molecular solutes, (4) polycrystals of frozen solvent, (5) unfrozen liquid microphase, (6) polymeric framework of a cryogel, (7) macropores, and (8) solvent. (Reprinted from Lozinsky, V. I., Galaev, I. Yu., Plieva, F. M., Savina, I. N., Jungvid, H., and Mattiasson, B. *Trends Biotechnol.*, 21, 445–51, 2003. With permission.)

be repeated a number of times to further increase the degree of crystallinity and the size of the crystals and improve the gel strength [3]. The PVA structure obtained by cryotropic treatments induces outstanding properties to the resulting hydrogels such as high modulus, rubber elasticity, and mechanical stability over a large range of deformation and temperatures. The PVA cryogels exhibit improved physical properties with respect to PVA hydrogels obtained with other techniques [2,4] also.

From a structural point of view, the freeze-thawing treatment on PVA solutions having high water content (85–90% by weight) produce heterogeneous materials with different morphology compared to the more conventional gels obtained from nonfrozen systems. The main characteristic of PVA cryogels is a network including interconnected micro- and macropores [5], where PVA chains and water molecules are organized over different hierarchical length scales. Ultimately, the outstanding properties of these systems depend on their complex architecture [6]. This attractive combination of properties (e.g., mechanical and structural), along with a lack of toxicity, biocompatibility, and high water content makes PVA cryogels very attractive matrices for applications in biotechnology and biomedicine [7].

In this chapter, we will focus on the most recent achievements on PVA cryogels. During these last five years, most of the work has been devoted to the incorporation of different organic and inorganic additives into the PVA formulation, aiming to extend the range of compositions and properties of the resulting PVA cryogels. Of special interest are those works remarking the biocompatible character of the freeze-thawing process by the *in situ* immobilization of bacteria and microorganisms within the resulting PVA cryogel, or by making use of these cryogels for biomedical purposes (for instance, as scaffolds for tissue engineering purposes). The other important field, where numerous groups have focused their research interest, has been the understanding of the mechanisms governing the structural and morphological features of the cryogels resulting after freeze-thawing processes. We do not claim that this review includes all of the published work within these last five years, so we apologize to the authors of the excellent work we may have unintentionally left out.

4.2 Extending the Compositional Nature of PVA Cryogels

4.2.1 Cryogels from PVA Solutions Containing Organic Additives

The incorporation of surfactants into the PVA cryogel formulations has been attempted for both anionic (e.g., sodium decylsulfate, SDS) and cationic (e.g., dodecyltrimethylammonium bromide, DTAB) ones [8]. In the former case, the addition of SDS micelles inside the gel provided hydrophobic and hydrophilic regions of interest for drug delivery purposes. In the later case, it was demonstrated that both the micelle intradiffusion coefficients (measured by pulsed gradient spin-echo [PGSE] NMR) and the spin-probe correlation times (measured by EPR) were affected by the cryogel polymeric scaffold, but not the structure of the micellar aggregates. These results indicated that dynamic processes in the hydrogel were much slower than in bulk water. Interestingly, this work also demonstrated that surfactant molecules were mostly solubilized in the polymer poor phase that, eventually, also contained more polymer than expected. These results suggested that the spinodal decomposition, which occurs during the freezing step of cryogel preparation, was not complete or was prevented by ice formation, and a simple consequence of the increase was the overall solid content in the solution due to the presence of surfactant aggregates. Similar effects are observed for PVA solutions of increased concentration. Surfactants (either ionic or nonionic) have also been used to obtain foamed PVA cryogels. Gas-filled PVA cryogels formed as a result of freezing-thawing of such foams contained large pores of up to ≅180 μm due to coalescence of air bubbles incorporated into the heterogeneous gel matrix. As mentioned above, mechanical and thermal properties of cryogels depended on the nature and concentration of surfactants, as well as on the regime of cryogenic treatment. The morphology of the frozen foam also depended on the type of surfactant added to the initial system [9].

The incorporation of cyclodextrines (γ-CD, also capable of self-assembling) in PVA hydrogels prepared by freeze-thawing cycles corroborated that the presence and the concentration of additives in the hydrogel strongly influence its swelling and reological responses. Interestingly, crystalline inclusion compound was observed between PVA and γ-CD in hydrogels with low concentration of γ-CD [10].

Nonetheless, one of the most interesting properties of the freeze-thawing processes for preparation of hydrogels meant for biotechnology and biomedicine uses is their biocompatible character. Such a biocompatibility has

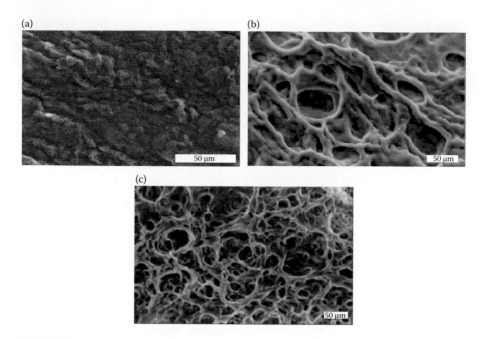

FIGURE 4.2
Scanning electron micrographs of (a) PVA (23 kV, ×500), (b) PVA-DNA (25 kV, ×350), and (c) PVA-DNA-NaBr (25 kV, ×200) gel membranes. Bars are 50 µm. (Reprinted from Papancea, A., Valente, A. J. M., Patachia, S., Miguel, M. G., and Lindman, B. *Langmuir*, 24, 273–79, 2008. With permission.)

very recently been demonstrated in the encapsulation of DNA into PVA hydrogels (Figure 4.2) that, depending on the crystallinity of the PVA matrix, was in a globular (amorphous PVA) or an extended conformation (crystalline PVA). The strong interactions established between PVA and DNA were also reflected in the cryogel surface, the bulk morphology, and the swelling properties, which differed from those characteristics of plain PVA hydrogels. Studies of DNA retention and delivery kinetics demonstrated the utility of these hydrogels for drug delivery purposes [11].

4.2.2 Cryogels from PVA-Based Blend Solutions

The application of the freezing and thawing cycle processes to polymer blends has also been recently explored for PVA solutions containing alginate with the morphology of the resulting cryogels being highly dependent on the PVA/alginate ratio. Applications of PVA/alginate cryogels (in this case, processed in the form of spheres) have been explored in the field of biotechnology, using their macroporous structure for immobilization of different microorganisms such as bacteria and fungi [12]. In this case, freeze-thawing cycles were somehow deleterious for microorganism viability, but it was recovered with a proper time of incubation in glucose solutions. A different behavior was observed for bacteria and fungi. The former proliferated inside the hydrogel, attached on top of the fibril network formed by the polymers, while fungi was recovered predominantly on the surface of the spheres. Porous PVA/alginate (with different PVA compositions up to 50 wt %) cryogels have also been utilized as bioartificial cell scaffolds to improve cell compatibility as well as flexibility of the scaffolds. The scaffolds exhibited highly open-cellular pore structures with almost the same surface and cross-sectional porosities (total porosities about 85%, regardless of PVA composition) and the pore sizes approximately span from 290 μm to 190 μm with increasing PVA content. The PVA/alginate hybrid scaffolds were softer and more elastic than the control alginate scaffolds without significant changes of mechanical strength. It was observed that the PVA/alginate scaffolds had better cell adhesion and faster growth than the control scaffold. PVA/alginate blends with a 30/70 ratio (in wt %) were found to be the most desirable composition to improve performance in terms of both flexibility and cell compatibility.

Blends of PVA and water soluble chitosan (ws-chitosan) have also been used for cryogel preparation by freeze-thawing [13] and also by freeze-thawing combined with γ-irradiation (both freeze-thawing followed by irradiation and irradiation followed by freeze-thawing) [14]. Among the three types of resulting cryogels, the later (i.e., γ-irradiation followed by freeze-thawing) showed the largest swelling capacity and mechanical strength, the highest thermal stability, and the lowest water evaporation rate and turbidity for wound dressing. Interestingly, swelling and mechanical strength of cryogels resulting from this combined technique could even be further improved by tuning the proper PVA/ws-chitosan weight ratio. The presence of ws-chitosan was of special interest given its antibacterial activity.

Of special interest, in terms of mechanical properties, are those blends of PVA and silk fibroin (SF) reported by Li et al [15]. The authors observed that both the relation of PVA/SF blending ratio and the freezing temperature determine the final morphology, the fine structure, and the properties of the resulting porous membranes. Thus, for blend ratios of PVA/SF ranging from 25/75 to 50/50 wt %, the resulting membranes exhibited smaller pore diameter, larger pore density, higher crystallinity, and higher strength and elongation. A similar behavior was observed by decreasing the freezing temperature (e.g., from –10 to –80°C).

Wu et al. have recently demonstrated that blends of PVA and water soluble dendrimers were also quite interesting for applications in biomedical fields. The authors have prepared novel, physically cross-linked hydrogels by a cyclic freeze-thawing treatment of aqueous solutions containing PVA and polyamidoamine (PAMAM), an amine terminated PAMAM dendrimer G6-NH$_2$. In these hydrogels, physical cross-link occurred via hydrogen bonds among hydroxyl groups, amide groups, and amino groups in PVA and PAMAM dendrimer. The high hydrophobicity of PAMAM dendrimer G6-NH$_2$ provided hydrogels exhibiting larger swelling ratios and faster re-swelling than plain PVA hydrogels. Combining the special host property of PAMAM dendrimer, there were certain expectations about the use of these novel, physically cross-linked hydrogels in drug delivery, due to their potentially improved drug-loading capacity and prolonged drug release time [16].

4.2.3 Cryogels from PVA-Based Hybrids and Composites Colloidal Suspensions

The validity of freeze-thawing processes for preparation of cryogels from colloidal suspensions is another interesting peculiarity of this process. It is worthy to note that the resulting cryogels are homogeneous throughout the whole macrostructure provided the suspension prior to freezing is also homogeneous. Thus, the number of entities suitable to form homogeneous suspensions has been studied, some providing mechanical reinforcement and others providing functionality to the resulting hydrogels. Among the former, the use of cellulose fibers provided biocompatible nanocomposites exhibiting a broad range of mechanical properties. Of particular interest were the PVA/cellulose nanocomposite stress-strain properties fitting to those of the porcine aorta in both the circumferential and the axial tissue directions [17]. This feature made PVA/cellulose composites very promising materials for cardiovascular soft tissue replacement applications such as aorta and heart valve leaflets.

The entrapment of particles of the strong anion-exchange resin Amberlite into a PVA cryogel by the freeze-thawing process has provided another composite material, which properties depend strongly on whether the resin was used in OH$^-$ form or Cl$^-$ form. The ion-exchange filler in OH$^-$ form caused both a significant reinforcement of the composite material and an increase in the gel fusion temperature. These effects were thought to be associated with the additional ionic bonding between the continuous and disperse phases. Beads of 200–600 μm in size were prepared from the composite material and used in expanded-bed, ion-exchange chromatography for the capture of the negatively charged solutes benzoate and lactate from the suspension of negatively charged cells. The plausibility of the approach has been demonstrated on model systems composed of yeast cells and benzoate and with a real fermentation broth produced after lactic acid fermentation [18].

Fibers of inorganic nature (e.g., clays) have also been widely used for mechanical reinforcement in a number of hydrogels prepared by different methods. Cryogel structures of PVA/clay nanocomposites were also prepared through freeze-thawing of PVA/clay suspensions (clay content ranges 0.5–4 wt %) [19]. The morphology of the resulting cryogels resembled a house of cards structure (Figure 4.3) of reduced density (from that characteristic of clays, 2.35 to 0.05 g/cm^3). The relative changes in T_g were most likely the result of two competing effects: (i) surface interaction that strengthens the interface (decreasing chain mobility), and (ii) enhanced interfacial-free volume due to the lower bulk crystallinity of polymer chains that increases chain mobility.

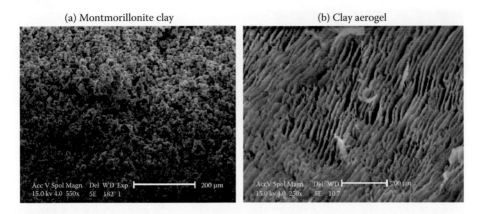

(a) Montmorillonite clay   (b) Clay aerogel

FIGURE 4.3
Scanning electron micrographs of particulate structure and platelets of clay and clay aerogel. (Reprinted from Bandi, S., and Schiraldi, D. A. *Macromolecules*, 39 (19), 6537–45, 2006. With permission.)

Following a similar approach, the use of colloidal silica formed *in situ* by hydrolysis and condensation (sol-gel process) of a silicon alkoxyde precursor (e.g., tetramethoxysilane, TMOS) has also been studied [20]. The continuous phase of such gel materials represented the supramolecular PVA network, which was supposed to be additionally cured with the silicon-containing oligomeric cross agents formed from TMOS in the course of hydrolytic polycondensation. The incorporated silica components influenced the morphology of cryogels, with a significant increase in gel strength and heat resistance with increasing TMOS content.

The use of nanoparticles may also provide nanocomposite materials with enhanced properties. Thus, an intelligent magnetic hydrogel (ferrogel) was fabricated by freeze-thawing a PVA solution containing Fe_3O_4 magnetic nanoparticles and a drug [21]. The operation mechanism was the following: the drug was accumulated around the ferrogel in the presence of an external direct current magnetic field (e.g., on), but the accumulated drug was released to the environment as soon as the magnetic field was switched off (Figure 4.4). The drug release behavior from the ferrogel was strongly dominated by the particle size of Fe_3O_4 under a given magnetic field. The best magnetic-sensitive effects were observed for the ferrogels with larger Fe_3O_4 particles due to its stronger saturation magnetization and smaller coercive force. Furthermore, the amount of drug release could be controlled by fine-tuning the switching duration time (SDT) through an externally controllable on-off operation in a given magnetic field. It was demonstrated that the highest burst drug amounts and best close configuration of the ferrogel were observed for the SDT of 10 and 5 min, respectively. These results were highly promising for practical clinical needs where controllable or programmable drug release profiles are required.

Polymer blends of PVA and poly(N-vinyl pyrrolidone) in a 70/30 ratio (in wt %) containing silver nanoparticles (Ag-NP, 0.1–1.0 wt %) have also been

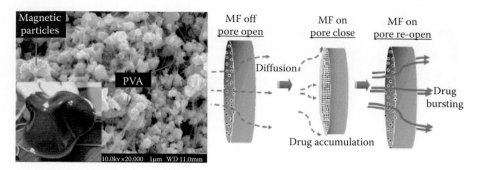

FIGURE 4.4

(Left) Cross-sectional SEM image of magnetic particles dispersed in PVA hydrogels and OM photos of PVA5-LM17 ferrogels; **(Right)** Mechanism of "close" configuration of the ferrogels due to the aggregation of Fe_3O_4 nanoparticles under "on" magnetic fields causes the porosity of the ferrogels to decrease. (Reprinted from Liu, T.-Y., Hu, S.-H., Liu, T.-Y., Liu, D.-M., and Chen, S.-Y. *Langmuir*, 22, 5974–78, 2006. With permission.)

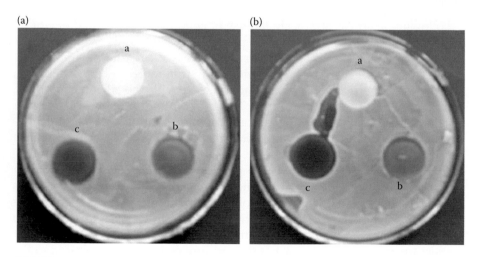

FIGURE 4.5

Antibacterial test results for (a) *E. coli* and (b) *S. aureus* after 24 hours incubation. (a) PVA–PVP hydrogel; (b) 0.2 wt % Ag/PVA–PVP hydrogel; (c) 1.0 wt % Ag/PVA–PVP hydrogel. (Reprinted from Yu, H., Xu, X., Chen, X., Lu, T., Zhang, P., and Jing, X. *J. Appl. Polym. Sci.*, 103, 125–33, 2007. With permission.)

recently used for preparation of cryogels by repeated freeze-thawing treatment [22]. The SEM and TEM studies allowed the observation of typical freeze-dried morphologies of the resulting cryogels, with excellent dispersion of the Ag-NP (ranging from 20 to 100 nm) within the cryogel matrix. The resulting cryogels exhibited excellent antibacterial properties against *Escherichia coli* and *Staphylococcus aureus* by simple water absorption, swelling, and release of silver ions (Figure 4.5).

4.2.4 Cryogels from PVA Solutions Containing Biological Entities (Proteins, Enzymes and Microorganisms)

As mentioned in the introduction, one of the most important features of freeze-thawing processes for hydrogel preparation is its biocompatible character, which should favor *in situ* immobilization of different biological species, from enzymes to living microorganisms.

Moreover, cryogels can be easily processed in the form of monoliths, which lately are emerging in chromatography as a solid alternative to classical packed columns [23]. Early studies with monolithic columns clearly demonstrated that this configuration enables column permeability and mass transfer to be maximized simultaneously so that extremely fast chromatographic separations at high flow rates and at reasonably low back pressure can be obtained. The combination of these interesting features (monoliths capable of immobilizing large biological entities within their large macroporous structure) could make cryogels highly suitable for immunoaffinity chromatography, a powerful analytical tool for high throughput resolution and characterization of complex protein mixtures.

Thus, a new concept for the preparation of selective sorbents with high flow path properties was presented by embedding molecularly imprinted polymers (MIPs) into various macroporous gels (MGs) [24]. A MIP was first synthesized with 17β-estradiol (E2) as a template for the selective adsorption of this endocrine disrupter. The composite macroporous gel/MIP (MG/MIP) monoliths were then prepared at subzero temperatures. Complete recovery of E2 from a 2 µg/L aqueous solution was achieved using the PVA MG/MIP monoliths whereas only 49–74% was removed with nonimprinted polymers (when no template was used). The PVA MG/MIP monolith columns were operated at almost 10 times a higher flow rate (50 mL/min) compared to the MIP columns with an operation flow rate of 1–5 mL/min. The possibility for processing the particulate containing wastewater effluents at high flow rates with selectivity for E2 removal, as well as the easy preparation of the monoliths, make the macroporous MG/MIP systems attractive as robust sorbents for the clean up of water from endocrine disrupting trace contaminants.

Combinatorial chemistry is another interesting research field where PVA cryogels have been applied lately. Thus, a sophisticated protocol that comprises the colyophilization of α-chymotrypsin and different mixtures of polystyrene sodium sulfonate, PVA, dextran, Ludox HS30 silica nanoparticles, polyethylene glycol, and sodium dodecyl sulfate has recently been published by Cooper and coworkers for the preparation of uniform spherical beads [25]. In this case, the hierarchical structure of the chymotrypsin spheres (with up to five levels of spatial organization obtained using PVA solutions containing many of the above mentioned compositional cases, from organic additives through blends with inorganic colloidal additives) provided particular functional and structural properties. Thus, the composites were assayed for catalytic activity by monitoring a nonaqueous *trans*-esterification reaction, while their mechanical strength was measured using a compression assay. Initial screening identified a set of six support materials that contributed favorably to either the enzyme activity or to the mechanical strength of the composite. A design-of-experiments methodology was employed to screen 80 combinations of these six base materials. A model representing this formulation space was constructed that could be used to predict both the catalytic activity and mechanical strength with reasonable accuracy for any combination of the six-base component materials. The model was used to predict optimized materials with an enzyme activity that was 50 times greater than that of the free enzyme and to set a minimum acceptable level of mechanical stability for these composites, which would ultimately allow reusability for at least 10 reaction cycles. Somewhat disappointingly, it was found that the composites that produced the strongest beads were also associated with the lowest catalytic activity.

The immobilization of whole microorganisms (e.g., bacteria) within the macroporous structure of a cryogel has certainly been one of the most interesting examples recently reported in biotechnology, given the intrinsic difficulty that implies the preservation of cells viability (even for the simplest ones, such as bacteria) during the preparation process of the cryogel [12]. Interest in the immobilization of microbial cells in membranes and bioreactors resides in the achievement of enhanced catalytic activity and stability (as a result of enhanced protection of the microorganism from mechanical degradation and deactivation), which ultimately generates an overall intensification of the biochemical reactions [26]. In particular, bacteria and other microbial cells can be used for bioremediation and biocatalysis, offering the possibility of decontaminating polluted environmental media and of implementing environment-friendly synthetic chemoenzymatically catalyzed processes [27]. The study of bacterial growth and proliferation in three-dimensional (3D) scaffolds has been typically achieved by soaking the scaffold in a cell suspension. However, cell proliferation throughout the entire scaffold structure is impeded using this procedure and colonization is limited to just a few layers at the periphery of the scaffold [28]. It should be noted that scaffold colonization occurs initially at the periphery and that nutrients and oxygen tend to be consumed by such outer bacteria before they can diffuse to the inner reaches of the scaffold. This problem would be overcome to a large extent if cells could be grown from the inner to the outer parts of the scaffold. The application of freeze-thawing processes has recently allowed for the new design of robust matrices for cell immobilization. Thus, MGs with immobilized microbial cells were prepared at subzero temperatures and were formed inside a plastic core (so-called, protective housing). Highly retained activity of immobilized yeast (77–92%) and *E. coli* (50–91%) cells was due to the high structural stability of the heterogeneous porous PVA cryogels. The authors demonstrated that due to the protective housing the macroporous gel particles with immobilized cells can be used in well-stirred bioreactors [29].

The immobilization of hydrocarbon-oxidizing bacteria in PVA cryogels (processed in the form of granules) has allowed for the *n*-hexadecane oxidation with efficiencies of 51% after the 10-day incubation. The achievement of such a high reaction yield was ascribed to the ability of partially hydrophobized PVA cryogel granules to, first, a sufficient number (ca. 6.5×10^3) of viable bacterial cells and, second (but principal), an allocation of cells at the interface in water-hydrocarbon system, favoring high contact of immobilized biocatalyst with hydrophobic substrate and water phase. The PVA cryogels with increased hydrophobicity could even be used for immobilization of bacterial cultures (e.g., *Rhodococcus* bacteria) performing oxidative transformations of water-immiscible organic compounds. Interestingly, PVA cryogel granules also provided high viability (up to 10 months at room temperature) to immobilized bacteria (Figure 4.6) [30].

Spores of the filamentous fungus *Rhizopus oryzae* have also been entrapped in macroporous PVA cryogels. In this case, the fungus cells were cultivated inside the carrier beads to prepare immobilized biocatalyst capable of

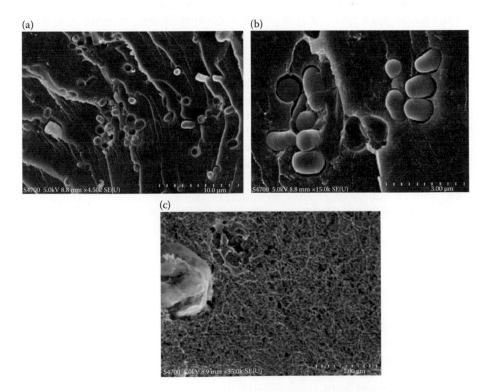

FIGURE 4.6
SEM images of PVA cryogel-immobilized *R. ruber* cells at (a, b) low and (c) high magnification. (Reprinted from Kuyukina, M. S., Ivshina, I. B., Gavrin, A. Yu., Podorozhko, E. A., Lozinsky, V. I., Jeffree, C. E., and Philp, J. C., *J. Microbiol. Meth.*, 65, 596–603, 2006. With permission.)

producing L(+)-lactic acid (LA). The immobilized cells possessed increased resistance to high concentrations of accumulated product and gave much higher yields of LA, which makes immobilized cells more attractive for biotechnological applications [31].

A similar approach has also been recently reported for immobilization of alkaliphilic *Bacillus agaradhaerens* LS-3C microorganism in PVA cryogel beads as a convenient source of cyclodextrin glycosyltransferased (CGTase) with an productivity of 2–3 mg β-cyclodextrin (β-CD) h^{-1} g^{-1} beads. The immobilized biocatalyst exhibited maximum activity at pH 9 and 50°C and formed cyclodextrins comprising of 92–94% β-CD and remaining α-CD. The cyclodextrin product from the immobilized cell bioreactor was continuously recovered by adsorption to Amberlite XAD-4 in a recycle batch mode [32].

The PVA cryogels have also been used for immobilization of hydrocarbon-degrading microorganisms for the clean up of diesel-contaminated soil [33]. The authors prepared biopiles to compare immobilized bioaugmentation with liquid culture bioaugmentation and biostimulation. In terms of percentage removal of diesel after 32 days, the immobilized systems were found to be the most successful (more than commercial liquid bioaugmentation agents), with greatest removal in a coimmobilized system containing PVA-entrapped microorganisms and a synthetic oil absorbent.

4.2.5 Functionalized PVA Cryogels

The composition of PVA cryogels can be extended not only prior, but also after the freeze-thawing process that is, by functionalization of the resulting PVA cryogel. This approach has mostly been applied for PVA cryogels working as supports to promote cell growth, where the biocompatible character of freeze-thawing processes is of great interest. One of the most interesting works on PVA cryogels-based biomaterials has been published by Kaneto et al., who through an elegant approach obtained uniaxially oriented PVA cryogels by mechanical drawing. The oriented hydrogels were later functionalized (in this particular case, mineralized) with hydroxyapatite (HAp) by repeating alternate immersion into solutions of phosphate and calcium ions. *In vitro* mouse fibroblast (L929) culture growth onto HAp-PVA cryogels resembled the oriented structure (Figure 4.7), which may be of interest for yielding oriented tissues [34].

One of the most recent works published by Mattiasson and coworkers implies one step forward in the use of PVA cryogels as support for cell cultures [35]. The capture of human acute myeloid leukemia KG-1 cells expressing the CD34 surface antigen and the fractionation of human blood lymphocytes were evaluated on PVA cryogel beads and dimethyl acrylamide (DMAAm) monolithic cryogel with immobilized protein A (Figure 4.8). The affinity ligand (protein A) was chemically coupled to the reactive PVA cryogel beads and epoxy-derivatized monolithic cryogels through different immobilization

Not drawn   Drawn

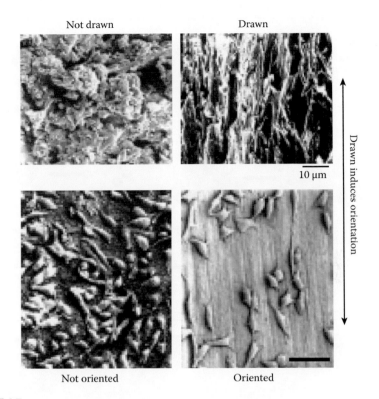

10 μm

Drawn induces orientation

Not oriented   Oriented

FIGURE 4.7
Top row: Scanning electron microscopic photographs of a freeze-fractured hybrid of hydroxy-apatite with nondrawn hydrogels (left) and with hydrogels drawn five times as long as its original length and then chemically cross-linked with 20 mol % glutaraldehyde (right). Bottom row: Scanning electron microscopic photographs of mouse fibroblast cells extending onto the above mentioned nondrawn (left) and drawn (right) hydrogels. (Reprinted from Kaneko, T., Ogomi, D., Mitsugi, R., Serizawa, T., and Akashi, M., *Chem. Mater.*, 16 (26), 5596–5601, 2004. With permission.)

techniques and the binding efficiency of the cell surface receptors specific antibody-labeled cells to the gels and beads was determined. The binding of cells to monolithic cryogel was higher (90–95%) compared with cryogel beads (76%). The B-lymphocytes, which bound to the protein A-cryogel beads, were separated from T-lymphocytes with yields of 74 and 85% for the two cell types, respectively. About 91% of the bound B-cells could be recovered without significantly impairing their viability. This work revealed significant differences in the percentage of cell binding to the immunosorbents caused by ligand density, flow shear forces, and bond strength between the cells and the affinity surface once distinct chemical coupling of protein A, size of beads, sequence of antibody binding to protein A adsorbents, morphology, and geometry of surface matrices were compared.

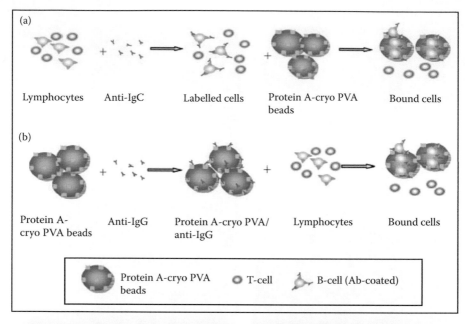

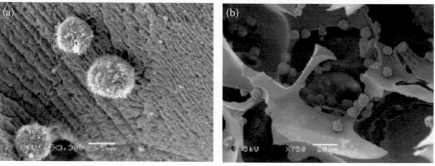

FIGURE 4.8

Top row: Schematic representation of the binding of human blood B-lymphocytes on cryogel adsorbents. (a) The cells are treated with goat antihuman IgG and applied to the protein A-adsorbent, (b) the antibodies are first applied to the protein A adsorbent and then cells allowed to bind. Bottom row: Scanning electron micrograph of the bound CD34+ human acute myeloid leukemia KG-1 cells on the surface of protein A–PVA cryogel bead (a) and in the inner part of supermacroporous protein A–cryogel monolithic matrix (b). The cells are affinity-bound on the surface of the bead and on the pore walls of the monolith through the interaction of protein A and anti-CD34 antibodies labeled on the cell surface. Magnification (×3,300) for (a) and (×750) for (b). (Reprinted from Kumar, A., Rodríguez-Caballero, A., Plieva, F. M., Galaev, I. Yu., Nandakumar, K. S., Kamihira, M., Holmdahl, R., Orfao, A., and Mattiasson, B., *J. Molec. Recog.*, 18 (1), 84–93, 2005. With permission.)

4.3 Morphology Control in Ice-Templating Processes

The above mentioned examples provide an overview of the versatility of freeze-thawing processes in compositional terms, with PVA cryogels having organic additives, polymers, clays, inorganic particles, and nanoparticles and even biological species incorporated within the porous structure. However, PVA cryogels obtained by freeze-thawing processes tend to exhibit a heterogeneous structure resulting from the occurrence of at least three concomitant and, at the same time, conflicting processes during freezing, namely, crystallization of the solvent, liquid-liquid phase separation, and crystallization of PVA. At subzero temperatures, the solvent, water, is frozen, while the solute is concentrated in small regions constituting the so-called unfrozen liquid microphase of eutectic composition [2,3]. Crystallization of PVA takes place inside these regions. Although the temperature of the system is likely below the eutectic temperature, the crystallization of PVA may not be complete, because the unfrozen liquid microphase jellifies and prevents the attainment of thermodynamic equilibrium. Therefore, formation of the gels upon defrosting is the result of two processes, namely, incomplete water crystallization and incomplete PVA crystallization. This is why although strong physical gels are obtained already by imposing a single freeze/thawing cycle, repetitive cycles are required to improve the strength of these gels. Improvement of mechanical properties is related to the increased degree of crystallinity of PVA, resulting from the more efficient segregation of PVA and ice phases after successive cycles [2,3].

Decrease of the freezing temperature or slow down of the freezing rate should also favor the formation of crystalline ice and, hence, phase segregation. Studies on the influence of the freezing temperature and freezing rate on PVA cryogel properties obtained by freeze-thawing processes are basically restricted to those performed by Lozinsky and coworkers. One of their latest studies states the relevance of the freezing and thawing regime and the temperature of freezing storage on the ice crystals size and the PVA crystallinity [6], which ultimately determine the cryogel porosity and mechanical strength, respectively. This interesting study reveals that the slower the freezing rate, the smaller the amount of both crystals, being their sizes larger. The occurrence of this mechanism for ice crystals is supported by SEM micrographs while for PVA crystals it is supported by the increase in ΔH values (indicative of the increased degree of cooperative hydrogen binding). Interestingly, the decrease of the freezing temperature seems to provide large number of PVA crystals in a reduced size, accordingly to the observed moderate increase in the fusion temperature.

Moreover, the formation of small or large ice crystals upon freezing is related to both the solute concentration in the solution-dispersion or gel and the solute size [36]. The ability for ice crystal growth in the presence of a given solute depends on the adsorption–desorption balance of the particular solute at the ice crystal surface. Irreversible adsorption would completely abrogate ice-crystal growth, while complete desorption would allow free crystal growth. In practice, adsorption and desorption at the ice surface are in dynamic equilibrium. For a particular solute, the effectiveness in inhibiting the growth of ice crystals simply depends on the extent of the surface area of ice that is covered by such solute. Thus, large solutes favor the formation of small ice crystals, while small solutes favor the formation of large ones. Under the same principle (i.e., inhibition of ice crystal growth determined by the extent to which the solute covers the ice surface), solute concentration also plays a role in tailoring the final morphology of the microchanneled structure; ice crystals formation is favored at low concentrations while it tends to be inhibited at high concentrations. Eventually, very high solute concentrations would fully inhibit ice crystal formation and, thus, the formation of any microchanneled structures at all. On the other side though, very low concentrations may yield mechanically weak microchanneled structures as a consequence of the resulting small amount of polymer supporting the whole structure.

One of the most successful techniques used for observation of ice crystals formation upon freezing of water solutions and even hydrogels of different nature is cryo-etch-SEM. For cryo-etch-SEM experiments, samples are plunge-frozen by immersion in subcooled liquid nitrogen (e.g., liquid nitrogen at vacuum pressure). Under these circumstances, the freezing rate is about 10^4 °C s^{-1} [37]. This freezing rate is far from that needed to obtain supercooled water from pure aqueous solutions (ca. 10^6 °C min^{-1}, by quenching tiny sprayed droplets on a cryoplate), [38] but it is fast enough to impede the formation of crystalline ice if water contains impurities (e.g., aqueous solutions). Then, the sample temperature was subsequently raised to −90°C. Such temperature (ca. 47°C above T_g) favors the formation of crystalline ice, which readily frees itself of any dissolved solute. Sublimation of ice under vacuum conditions allows the visualization of solutes concentrated at the surrounding empty areas where ice originally resided [39]. The micrographs obtained by cryo-etch-SEM of PVA solutions with different concentrations show regular patterned structures (Figure 4.9) [40], resembling the original formation of ice crystals of homogeneous sizes and shapes. There are two obvious differences between cryo-etch-SEM and freeze-thawing: the freezing temperature and the freezing rate. However, we think the procedure followed for ice crystal formation (i.e., a 2D surface homogenously exposed to etching) can also play an important role in the homogeneity of the structures found by cryo-etch-SEM. It is noteworthy that the plunge-freezing of a 3D monolith tends to produce thermal gradients across the sample (due to the Leidenfrost effect), which ultimately results in the achievement of ice crystals with heterogeneous sizes and shapes.

On the basis of these experiments, it is evident that, besides efficient ice crystallization and phase segregation, the design of a thermal route for freezing a homogeneous sample is mandatory for the achievement of controlled

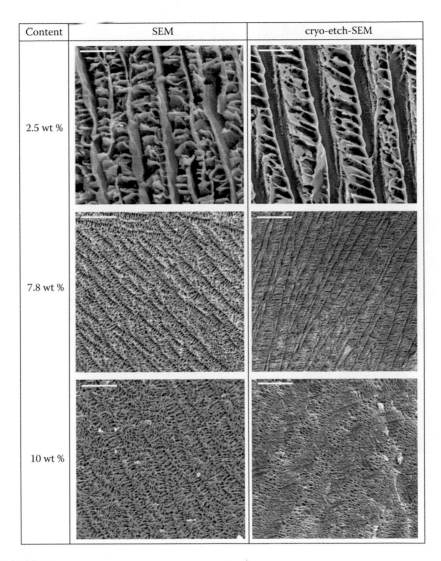

Content	SEM	cryo-etch-SEM
2.5 wt %		
7.8 wt %		
10 wt %		

FIGURE 4.9
SEM (left column) and cryo-etch-SEM (right column) micrographs of cross-sectioned (perpendicular to the direction of freezing) monolithic PVA scaffolds. The average molecular weight of the PVA was 72000 for every ISISA (ice segregration induced self assembly) processed sample. Scale bar is 20 μm for every picture. (Reprinted from Gutierrez, M. C., Garcia-Carvajal, Z. Y., Jobbagy, M., Rubio, F., Ferrer, M. L., and del Monte, F., *Adv. Funct. Mater.*, 17, 3505–13, 2007. With permission.)

structures. An interesting experiment (e.g., the partial immersion of aqueous solution-suspension or gel into a nitrogen liquid bath) allows exemplifying this issue [41]. Under these circumstances, three distinct zones can be clearly distinguished in the frozen sample, each characterized by a particular pore shape and dimension (Figure 4.10). In Zone 1, the closest to the initial cold

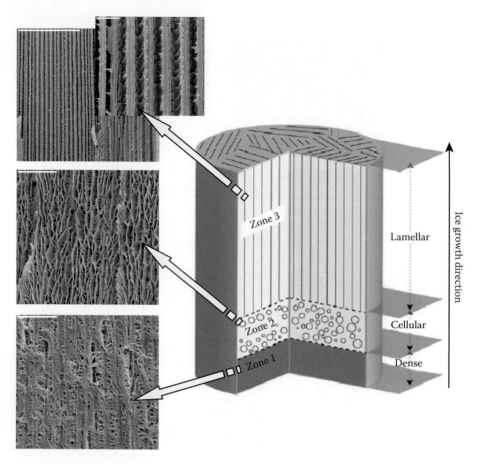

FIGURE 4.10
Schematic representation and SEM micrographs of the three distinctive zones found in the PVA cryogel after unidirectional freezing and freeze-drying. It is worthy of note that their dimensions depending largely upon processing conditions. (Reprinted from Deville, S., Saiz, E., and Tomsia, A. P. *Biomaterials*, 27, 5480–89, 2006. With permission.)

finger, no porosity is observed and the material is dense. In Zone 2, the material is characterized by a cellular morphology. Finally, in the upper zone (Zone 3), the structure is a lamellar type, with long parallel pores aligned in the direction of movement of the ice front. When the probe is immersed in liquid nitrogen (i.e., at −196°C), water supercools and amorphous rather than crystalline ice is formed. The formation of amorphous ice does not occur when pure water is frozen at this temperature (even though it is well below the glass transition temperature of water, $T_g \sim −137$°C), since T_g is not an equilibrium transition temperature and, hence, being only quasithermodynamic, may vary with experimental conditions [42]. Moreover, ice crystallization depends on the presence of impurities in the water, being fully inhibited for pseudoconcentrated solutions (threshold values of concentration bringing about this inhibition vary widely and depend on the nature of the solute). Thus, the presence of solutes determines the formation of amorphous ice in the immersed portion, so that neither matter segregation nor formation of any porous structure occur (Zone 1). Meanwhile, an ice front running along the nonimmersed portion of the probe can be visualized. The temperature at the ice front is obviously higher than that of liquid nitrogen, because of differential diffusion of heat coming in from the environment surrounding the probe (air and liquid nitrogen, respectively). The further the ice front moves away from the liquid nitrogen immersion level the more its temperature increases. Eventually, it is not low enough for supercooling to occur and, therefore, crystalline rather than amorphous ice is formed. As pointed out above, the formation of crystalline ice determines matter segregation and results in the formation of porous structures after freeze-drying. The pore size depends on the distance from the liquid nitrogen immersion level (i.e., small in Zone 2, where temperatures are still close to that of liquid nitrogen and large at Zone 3 where temperatures are higher) so that the resulting scaffold may exhibit a long range longitudinal pore-size heterogeneity. However it is worth noting that, independently of the PVA content (within the 2.5–10 wt % range), and despite the heterogeneity of the longitudinal structure at Zone 3 being heterogeneous, SEM micrographs of cross-sectioned monoliths already reveal a homogeneous and well-patterned structure.

4.3.1 PVA Cryogels Obtained by Unidirectional Freezing and Freeze-Drying

On the basis of the above mentioned observations, one wonders if there may be any procedure suitable to provide homogeneous structures. Thus, homogeneous structures have indeed been obtained by not just partial, but continuous immersion (at a constant rate) into liquid nitrogen. Under these circumstances, the ice front was forced to be closer to and, eventually (at high freezing rates during rapid immersion), to coincide with the liquid nitrogen immersion level. When keeping the distance between the ice front and the immersion level constant, the ice crystals size remained unmodified along the sample, which allowed the achievement of a homogeneous and well-patterned microchanneled structure. Thus, the size of the microchannels formed at Zone 3 is determined by the distance from the ice front to the liquid nitrogen immersion level; for example, the closer to, then the narrower the channels as a consequence of the smaller size of the ice crystals that formed (Figure 4.11) [39,40,43].

The above SEM micrographs revealed the remarkable capability of a unidirectional freezing process to tailor the cryogels morphology. The main difference between unidirectional freezing and freeze-thawing processes is the combination of low freezing temperatures and homogeneous freezing (typically by using relatively low freezing rates and thus avoiding thermal gradients across the sample during freezing) used for the former, which favor

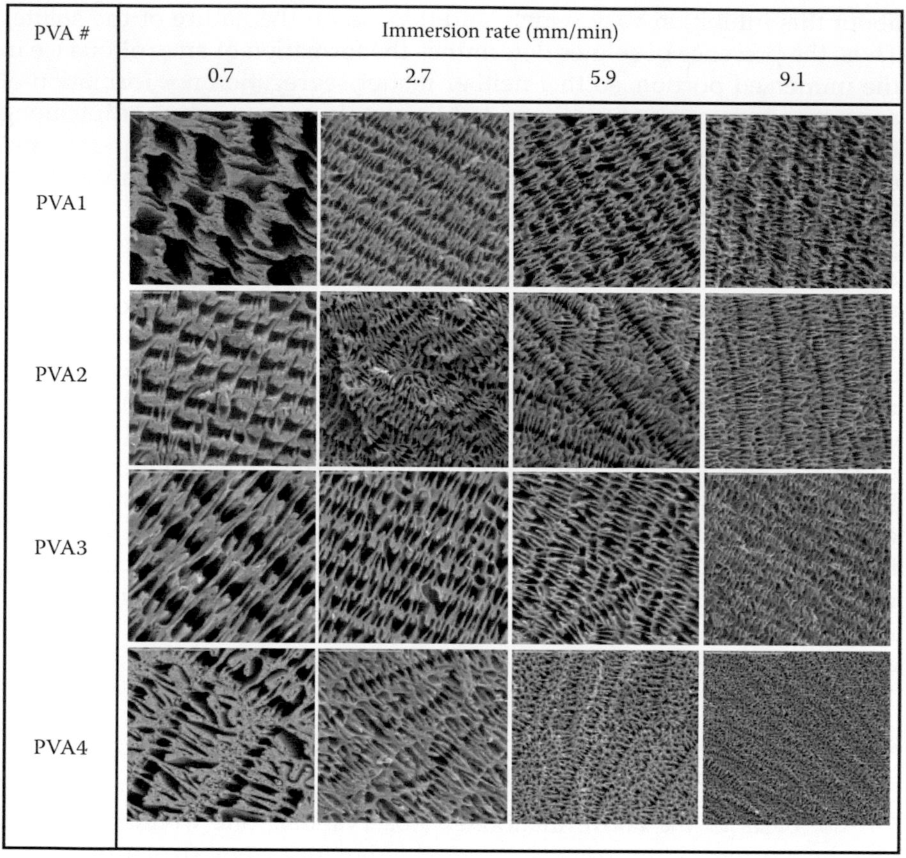

PVA #	Immersion rate (mm/min)			
	0.7	2.7	5.9	9.1
PVA1				
PVA2				
PVA3				
PVA4				

FIGURE 4.11

SEM micrographs of cross-sectioned (perpendicular to the direction of freezing) monolithic PVA scaffolds. These tailored morphologies were obtained by using PVA with different average molecular weights (13,000–23,000 for PVA1; 72,000 for PVA2; 89,000–98,000 for PVA3; 130,000 for PVA4), and by processing the PVA solution at different freezing rates. The bar is 20 μm in length in every picture. Note that, in some cases, the actual magnification (indicated by the length of the bar scale) is varied for better visualization of the scaffold macrostructure. The PVA content of all of the samples was 7.8 wt %. (Reprinted from Gutierrez, M. C., Garcia-Carvajal, Z. Y., Jobbagy, M., Rubio, F., Ferrer, M. L., and del Monte, F., *Adv. Funct. Mater.*, 17, 3505–13, 2007. With permission.)

the formation of crystalline ice with homogeneous sizes and shapes (at least, in the upper zones of the monolith, Zone 3 in Figure 4.10) and hence, phase segregation. Cryogenic processes based on unidirectional freezing were so-called ice segregation induced self-assembly (ISISA).

The effectiveness of phase segregation was also reflected in the degree of crystallinity of PVA cryogels. It is worthy to note that unidirectional freezing processes at liquid nitrogen temperatures promoted further polymer depletion from the depleted polymer phase and, as a consequence, polymer accumulation between adjacent ice crystals (e.g., the enriched polymer phase). Accumulation of polymer at the enriched polymer phase increased the crystallinity degree of the cryogel, reaching values of 40%. Interestingly, the achievement of a homogeneous morphology (e.g., unidirectional micro-channeled structure), besides a high crystalline character, resulted in an enhancement of the mechanical properties of the scaffolds. Thus, the increase in tensile strength was noticeable (reaching values close to 4 MPa), mostly for samples with wider pore channels and a large accumulation of polymer at the scaffold structure. High values of tensile strength are indicative of bulky character enhancement for those samples with thicker structures. It is worthy to note that the tensile strength for bulk PVA without macrostructural defects range from 20 to 80 MPa (depending on processing, crystalline or amorphous polymer character and polymer molecular weight, among other variables) [44], but values above 1 MPa are rare for macroporous PVA cryogels obtained by freeze-thawing processes [45]. Nonetheless, SEM micrographs revealed small pores in the enriched polymer phase, indicating the remaining presence of water in the polymer enriched phase even after unidirectional freezing. The skeletal density measured by mercury porosimetry was well below that of amorphous PVA, which corroborates this issue.

4.3.2 Applications in Biotechnology and Drug Delivery

The versatility of the ISISA process for the preparation of scaffolds is tremendous, not only allowing a large variety of compositions (inorganic, organic, hybrid, biohybrid, composites, and biocomposites) mentioned above for simple freeze-thawing processes, but also allowing an excellent control of the macroporous structure of the resulting cryogels. Thus, the use of simple PVA solutions and of PVA in combination with nanoparticles, has been recently reported for preparation of a diverse array of complex structures such as polymer-inorganic nanocomposites, aligned gold microwires and microwire networks, porous composite microfibers, and biaxially aligned composite networks, with excellent control of structural features, both at the micro- and the nano-scale thanks to the use of the unidirectional freezing and freeze-drying processes (Figure 4.12) [46].

Cooper and his group have also recently reported a very elegant work where aligned porous poly(ε-caprolactone) microparticles were formed on an aligned PVA cryogel (Figure 4.13). The method can also be extended to different compositions (e.g., polystyrene microparticles) and tailored for the achievement of different morphologies (i.e., shape and size of pores). The aligned porous microparticles were released by dissolving the supporting composite in water. The authors demonstrated the validity of poly(ε-caprolactone) microparticles as suitable supports for the growth of mouse embryo stem cells [47].

Our group has also recently demonstrated that the application of the ISISA process to bacteria suspended in an aqueous PVA solution could provide a PVA scaffold with immobilized bacteria throughout its 3D structure [48].

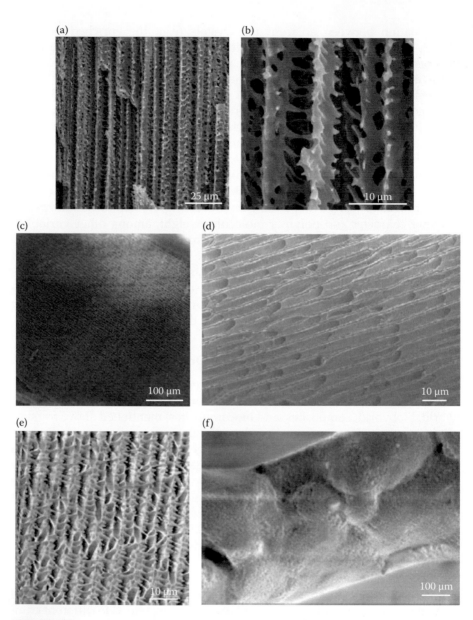

FIGURE 4.12
SEM micrographs of: (a,b) aligned porous PVA showing pore alignment (a) and the PVA fish-
bone morphology (b); (c,d) aligned porous PVA–silica composite showing broad alignment of
the pore structure (c), and at a higher magnification the parallel channels, each a few micrometers in diameter (d); (e) cross-section of the porous PVA–silica composite perpendicular to
the direction of alignment; (f) nano-particulate structure of the composite—the silica particles
that make up the structure have an average diameter of 15 nm. (Reprinted from Zhang, H.,
Hussain, I., Brust, M., Butler, M. F., Rannard, S. P., and Cooper, A. I., *Nat. Mater.*, 4, 787, 2005.
With permission.)

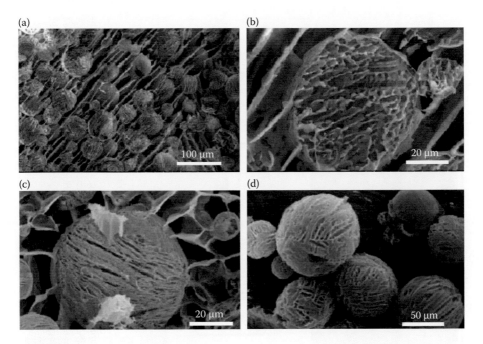

FIGURE 4.13
(a) Aligned porous poly(ε-caprolactone) (PCL) particles entrapped in an aligned PVA-DS matrix. (b) Internal aligned porous structure of a single sectioned PCL particle. (c) Aligned porous surface of a single PCL particle in the matrix. (d) The supporting matrix of PVA and SDS was dissolved and the aligned porous PCL particles were released. (Reprinted from Zhang, H., Edgar, D., Murray, P., Rak-Raszewska, A., Glennon-Alty, L., and Cooper A. I., *Adv. Funct. Mater.*, 18, 222–28, 2008. With permission.)

However, preservation of the membrane structure of bacteria occurred only in those cases when the thickness of the walls that support the scaffold macrostructure is able to fully embed bacteria cells [47,49]. Bacteria are typically much larger than scaffold wall thicknesses so that they are damaged upon simple ice-templating immobilization. Our group entrapped bacteria within beads composed of a natural calcium-alginate polymer that contains glucose for further cryo-protection (Figure 4.14). A close inspection of the confocal fluorescence micrograph revealed high levels of bacteria comparable to those observed when the bacteria were immobilized by simple soaking of a preformed scaffold.

It is worthy to emphasize that the use of suspensions and emulsions for preparation of ice-templated materials is possible for almost any composition, as far as one is able to ensure the homogeneity of such suspension-emulsion. Our group extended this approach to organic crystalline needles (e.g., ciprofloxacin, CFX) suspended in an aqueous solution of PVA (Figure 4.15) [40]. The CFX is a synthetic fluoroquinolone antimicrobial agent whose primary mechanism of action against bacteria (both gram-positive and gram-negative) involves inhibition of topoisomerase IV and DNA gyrase

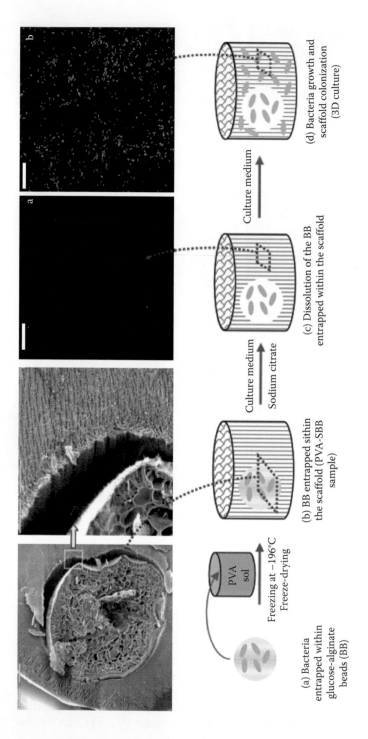

FIGURE 4.14

Scheme presenting the process followed for preparation of a PVA scaffold with immobilized bacteria-containing alginate beads. SEM micrographs show a bead of ca. 1 mm entrapped within the microchanneled structure of the PVA scaffold. Confocal fluorescence microscopic images of the PVA scaffold entrapping bacteria containing alginate beads soaked in culture medium: (a) before, and (b) after, incubation at 37°C for 24 hours. (Reprinted from Gutierrez, M. C., Garcia-Carvajal, Z. Y., Jobbagy, M., Yuste, L., Rojo, F., Abrusci, C., Catalina, F., del Monte, F., and Ferrer, M. L., *Chem. Mater.*, 19, 1968, 2007. With permission.)

[50]. The CFX crystalline needles were prepared *in situ* by incorporation of doubly lyophilized CFX into a warm (90°C) solution of PVA. Cooling down to room temperature promoted CFX crystallization in the form of needles homogenously suspended in the PVA solution. The ISISA processing of this suspension resulted in the formation of PVA scaffolds with crystalline CFX needles homogeneously entrapped within the PVA fences that maintain the microchannel structure. The resulting PVA-CFX scaffolds were used for drug delivery and controlled release, with special attention being paid to the influence of the scaffold morphology (mainly tailored by the use of different freezing rates, see Figure 4.11) on the kinetics of the drug release. The extremely different morphologies of the achieved scaffolds (one with narrow channels separated by thin polymer walls, another with wide channels separated by thick polymer walls) determine different rates of controlled release. Thus, release from PVA cryogels characterized by wide channels was mostly based on scaffold hydration, swelling and dissolution of the entrapped drug, and can be extended for longer periods of time (complete release required more than 30 hours), while that from PVA scaffolds characterized by narrow channels was mostly based on scaffold erosion and dissolution of the entrapped drug, and the kinetic release was much faster than in the previous case (Figure 4.15). The interest in introducing CFX as crystalline needles resides in the possibility of loading the PVA scaffold with different drug concentrations with minimum modification of the scaffold morphology (which ultimately determines the kinetic of the release). Thus, we can release the proper drug dose at a higher than minimum inhibitory concentration, but without any risk of toxic overdoses [51]. *In vitro* experiments have also demonstrated the ability of PVA-CFX to inhibit the growth of *E. coli* in different culture media (Figure 4.16).

4.4 Conclusions

Summarizing, this review has exemplified the versatility of ice templating processes for the preparation of PVA cryogels with different compositions. During the last five years, the homogeneous incorporation of numerous additives of any size and shape (ranging from the nano to the microscale and from particles to fibers up to more complex morphologies like biological entities, respectively) and different nature (ranging from organic to inorganic) into the PVA solution has allowed the achievement of PVA cryogels with extended compositions and, hence, properties. In this review, we have paid special attention to PVA cryogels applied in biotechnology and biomedicine. It is worthy to note that one of its most important features of this process is its biocompatibility, which makes the resulting PVA cryogels of great interest for this sort of application.

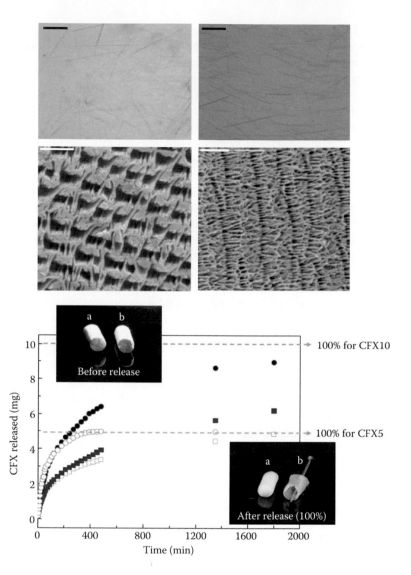

FIGURE 4.15
CFX needles dispersed in an aqueous PVA solution at (a) 5 mg/mL, and (b) 10 mg/mL (bars are 50 μm in length); SEM micrographs of PVA scaffolds (average molecular weight 72,000) prepared at immersion rates in (c) 0.7 and in (d) 9.1 mm/min (bars are 20 μm). (e) Kinetics of CFX release from scaffolds prepared at immersion rates of 0.7 (red squares) and 9.1 mm/min (blue circles). Experiments were conducted on scaffolds containing 5 (open symbols) and 10 mg (solid symbols) of CFX, respectively. Dashed orange lines indicate 100% CFX release. Photographs of monolithic PVA-CFX5 scaffolds prepared at immersion rates of (a) 0.7, and (b) 9.1 mm/min are also shown. Before release: lyophilized gels. After release: wet gels. In the latter case, erosion can be seen in the PVA-CFX5 scaffold prepared at an immersion rate of 9.1 mm/min. (Reprinted from Gutierrez, M. C., Garcia-Carvajal, Z. Y., Jobbagy, M., Rubio, F., Ferrer, M. L., and del Monte, F., *Adv. Funct. Mater.*, 17, 3505–13, 2007. With permission.)

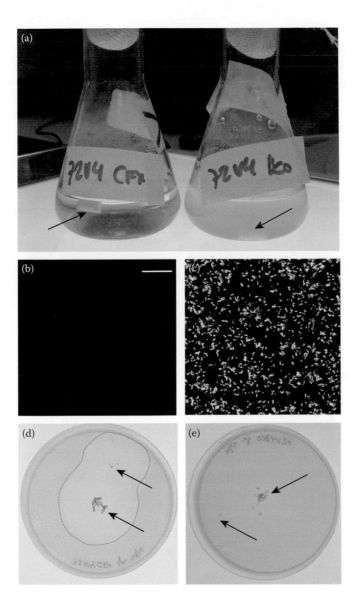

FIGURE 4.16
Picture (a, left) shows a culture medium in which bacteria growth was inhibited (no turbidity) by release of CFX from the PVA-CFX scaffold. In the presence of a PVA scaffold lacking CFX, bacterial growth was efficient and the culture medium became turbid (a, right). Red arrows point to PVA-CFX and PVA monoliths immersed in the non-turbid and turbid media, respectively. Confocal fluorescence microscopy images of PVA-CFX (b) and PVA (c) monoliths were obtained after soaking for 24 h in the bacteria culture medium. Depth of focus is 4 μm. Scale bar is 20 μm. Picture of finely ground PVA-CFX (d) and PVA (e) scaffolds deposited on top of an agar bacteria culture medium. Blue line in (d) delimits the zone where bacteria growth is inhibited by CFX while no inhibition is observed in (e) for PVA scaffolds. Red arrows point to powder of scaffolds on the agar plate. Scaffolds were prepared at freezing rate of 0.7 mm min⁻¹.

This review has also remarked the possibilities of unidirectional freezing for the achievement of PVA cryogels with controlled microchanneled porous structures. We have discussed how cryogel microstructure is always a replica of the ice crystals, accordingly control over ice crystal formation is, ultimately, the key to obtaining scaffolds with tailored morphologies and with freezing might be achieved by different thermal treatments (e.g., freezing of the aqueous medium by immersion in cryogenic liquids of different nature or by the application of some other kind of freezing device). Thus, we have determined the three main process control variables available to tailor the final morphology of cryogels obtained by unidirectional freezing, those are the PVA content, the molecular weight of the PVA, and the freezing rate. The hydrolysis degree of PVA can also play a role in the polymer behavior upon freezing (e.g., derivatization of hydroxyl groups of the PVA cryogel can eventually allow incorporation of any required functionality). However, except for a few cases [52], most of the work has used PVAs with degrees of hydrolysis from 98 to 99%, for that reason, we consider there are not enough data currently to open a serious discussion about the relevance of this parameter on the morphology and properties of the resulting cryogels.

Acknowledgments

The authors thank MEC, CSIC, and Comunidad de Madrid for financial support (MAT2006-02394, 200660F01, 200760I009, and S-0505/PPQ-0316 Projects). TPA Inc. is also acknowledged for valuable support. María C. Gutierrez acknowledges CSIC for an I3P research contract.

References

1. (a) Inoue, T., Japan Patent 47-012854, 1972. (b) Inoue, T., U.S. Patent 3875302, 1972. (c) Inoue, T., Japan Patent 48-030462, 1973. (d) Inoue, T., Japan Patent 48-030463, 1973. (e) Kukharchik, M. M., and Baramboim, N. K. 1972. *Vysokomol. Soedin., Ser. B* 14:843. (f) Peppas, N. A. 1975. Turbidimetric studies of aqueous polyvinyl-alcohol) solutions. *Makromol. Chem.* 176:3433.
2. (a) Lozinsky, V. I., and Plieva, F. M. 1998. Poly(vinyl alcohol) cryogels employed as matrices for cell immobilization: 3. Overview of recent research and developments. *Enzyme Microb. Technol.* 23:227–42. (b) Arvidsson, P., Plieva, F. M., Savina, I. N., Lozinsky, V. I., Fexby, S., Bülow, L., Galaev, I.Yu., and Mattiasson, B. 2002. Chromatography of microbial cells using continuous supermacroporous affinity

and ion-exchange columns. *J. Chromatog. A* 977 (1):27–38. (c) Hassan, C. M., and Peppas, N. A. 2000. Structure and morphology of freeze/thawed PVA hydrogels. *Macromolecules* 33 (7):2472–79. (d) Hassan, C. M., Ward, J. H., and Peppas, N. A. 2000. Modeling of crystal dissolution of poly(vinyl alcohol) gels produced by freezing/thawing processes. *Polymer* 41 (18):6729–39. (e) Ricciardi, R., Auriemma, F., Gaillet, C., De Rosa, C., and Laupretre, F. 2004. Investigation of the crystallinity of freeze/thaw poly(vinyl alcohol) hydrogels by different techniques. *Macromolecules* 37:9510–16. (f) Auriemma, F., De Rosa, C., Ricciardi, R., Lo Celso, F., Triolo, R., and Pipich V. 2008. Time-resolving analysis of cryotropic gelation of water/poly(vinyl alcohol) solutions via small-angle neutron scattering. *J. Phys. Chem. B* 112:816–23.

3. (a) Lozinsky, V. I., Galaev, I.Yu., Plieva, F. M., Savina, I. N., Jungvid, H., and Mattiasson, B. 2003. Polymeric cryogels as promising materials of biotechnological interest. *Trends Biotechnol.* 21:445–51. (b) Kobayashi, M., Ando, I., Ishii, T., and Amiya, S. 1998. Structural and dynamical studies of poly(vinyl alcohol) gels by high-resolution solid-state C-13 NMR spectroscopy. *J. Mol. Struct.* 440:155–64. (c) Lozinsky, V. I., Damshkaln, L. G., Kurochkin, I. N., and Kurochkin, I. I. 2008. Study of cryostructuring of polymer systems: 28. Physicochemical properties and morphology of poly(vinyl alcohol) cryogels formed by multiple freezing-thawing. *Colloid J.* 70 (2):189–98.

4. Watase, M., and Nishinari, K. 1989. Effect of the degree of saponification on the rheological and thermal-properties of polyvinyl-alcohol) gels. *Makromol. Chem.* 190:155–63.

5. Lozinsky, V. I. 1998. Cryotropic gelation of poly(vinyl alcohol) solutions. *Russ. Chem. Rev.* 67:573.

6. Lozinsky, V. I., Damshkaln, L. G., Shaskol'skii, B. L., Babushkina, T. A., Kurochkin, I. N., and Kurochkin, I. I. 2007. Study of cryostructuring of polymer systems: 27. Physicochemical properties of poly(vinyl alcohol) cryogels and specific features of their macroporous morphology. *Colloid. J.* 69 (6):747–64.

7. (a) Hassan, C. M., and Peppas, N. A. 2000. Structure and applications of poly(vinyl alcohol) hydrogels produced by conventional crosslinking or by freezing/thawing methods. *Adv. Polym. Sci.* 153:37–65. (b) Stauffer, S., and Peppas, N. A. 1992. Poly(vinyl alcohol) hydrogels prepared by freezing-thawing cyclic processing. *Polymer* 33:3932–36.

8. Tedeschi, A., Auriemma, F., Ricciardi, R., Mangiapia, G., Trifuoggi, M., Franco, L., De Rosa, C., Heenan, R. K., Paduano, L., and D'Errico, G. 2006. A study of the microstructural and diffusion properties of poly(vinyl alcohol) cryogels containing surfactant supramolecular aggregates. *J. Phys. Chem. B* 110:23031–40.

9. Lozinsky, V. I., Damshkaln, L. G., Kurochkin, I. N., and Kurochkin, I. I. 2005. Study of cryostructuring of polymer systems: 25. The influence of surfactants on the properties and structure of gas-filled (foamed) poly(vinyl alcohol) cryogels. *Colloid J.* 67 (5):589–601.

10. Hernández, R., Rusa, M., Rusa, C. C., López, D., Mijangos, C., and Tonelli, A. E. 2004. Controlling PVA hydrogels with gamma-cyclodextrin. *Macromolecules* 37:9620–25.

11. Papancea, A., Valente, A. J. M., Patachia, S., Miguel, M. G., and Lindman, B. 2008. PVA-DNA cryogel membranes: Characterization, swelling, and transport studies. *Langmuir* 24:273–79.

12. Cho, S. H., Oh, S. H., and Lee, J. H. 2005. Fabrication and characterization of porous alginate/polyvinyl alcohol hybrid scaffolds for 3D cell culture. *J. Biomater. Sci., Polym. Ed.* 16 (8):933–47.
13. Park, H., and Kim, D. 2006. Swelling and mechanical properties of glycol chitosan/poly(vinyl alcohol) IPN-type superporous hydrogels. *J. Biomed. Mater. Res. A* 78 (4):662–67.
14. Yang, X., Liu, Q., Chen, X., Yu, F., and Zhu, Z. 2008. Investigation of PVA/ws-chitosan hydrogels prepared by combined γ-irradiation and freeze-thawing. *Carbohyd. Polym.* 73 (3):401–8.
15. Li, M., Minoura, N., Dai, L., and Zhang, L. 2001. Preparation of porous poly(vinyl alcohol)-silk fibroin (PVA/SF) blend membranes. *Macromol. Mater. Eng.* 286 (9):529–33.
16. Wu, X.-Y., Huang, S.-W., Zhang, J.-T., and Zhuo, R.-X. 2004. Preparation and characterization of novel physically cross-linked hydrogels composed of poly(vinyl alcohol) and amine-terminated polyamidoamine dendrimer. *Macromol. Biosci.* 4 (2):71–75.
17. Millon, L. E., and Wan, W. K. 2006. The polyvinyl alcohol-bacterial cellulose system as a new nanocomposite for biomedical applications. *J. Biomed. Mater. Res. B: Appl. Biomater.* 79B (2):245–53.
18. Savina, I. N., Hanora, A., Plieva, F. M., Galaev, I. Yu., Mattiasson, B., and Lozinsky, V. I. 2005. Cryostructuration of polymer systems: 24. Poly(vinyl alcohol) cryogels filled with particles of a strong anion exchanger: Properties of the composite materials and potential applications. *J. Appl. Polym. Sci.* 95 (3):529–38.
19. Bandi, S., and Schiraldi, D. A. 2006. Glass transition behavior of clay/poly(vinyl alcohol) composites. *Macromolecules* 39 (19):6537–45.
20. Lozinsky, V. I., Bakeeva, I. V., Presnyak, E. P., Damshkaln, L. G., and Zubov, V. P. 2007. Cryostructuring of polymer systems: 26. Heterophase organic-inorganic cryogels prepared via freezing-thawing of aqueous solutions of poly(vinyl alcohol) with added tetramethoxysilane. *J. Appl. Polym. Sci.* 105 (5):2689–2702.
21. Liu, T.-Y., Hu, S.-H., Liu, T.-Y., Liu, D.-M., and Chen, S.-Y. 2006. Magnetic-sensitive behavior of intelligent ferrogels for controlled release of drug. *Langmuir* 22:5974–78.
22. Yu, H., Xu, X., Chen, X., Lu, T., Zhang, P., and Jing, X. 2007. Preparation and antibacterial effects of PVA-PVP hydrogels containing silver nanoparticles. *J. Appl. Polym. Sci.* 103:125–33.
23. Plieva, F. M., Galaev, I.Yu., and Mattiasson, B. 2007. Macroporous gels prepared at subzero temperatures as novel materials for chromatography of particulate-containing fluids and cell culture applications. *J. Sep. Sci.* 30 (11):1657–71.
24. Le Noir, M., Plieva, F., Hey, T., Guieysse, B., and Mattiasson, B. 2007. Macroporous molecularly imprinted polymer/cryogel composite systems for the removal of endocrine disrupting trace contaminants. *J. Chromatogr. A* 1154 (1–2):158–64.
25. Long, J., Hutcheon, G. A., and Cooper, A. I. 2007. Combinatorial discovery of reusable noncovalent supports for enzyme immobilization and nonaqueous catalysis. *J. Comb. Chem.* 9:399–406.
26. (a) Hecht, V., Langer, O., and Deckwer W. D. 2000. Degradation of phenol and benzoic acid in a three-phase fluidized-bed reactor. *Biotechnol. Bioeng.* 70:391–99. (b) Pekdemir, T., Keskinler, B., Yildiz, E., and Akay, G. 2003. Process

intensification in wastewater treatment: Ferrous iron removal by a sustainable membrane bioreactor system. *J. Chem. Technol. Biotechnol.* 78:773–80. (c) Giorno, L., and Drioli, E. 2000. Biocatalytic membrane reactors: Applications and perspectives. Trends Biotechnol. 18:339–49. (d) Erhan, E. Keskinler, B. Akay, G., and Algur, O. F. J. 2002. Removal of phenol from water by membrane-immobilized enzymes: Part I. Dead-end filtration. *Membr. Sci.* 206:361–73. (e) Kwak, M. Y., and Rhee, J. S. 1992. Cultivation characteristics of immobilized *Aspergillus oryzae* for kojic acid production. *Biotechnol. Bioeng.* 39:903–06.

27. (a) Ishige, T., Honda, K., and Shimizu, S. 2005. Whole organism biocatalysis. *Curr. Opin. Chem. Biol.* 9:174–80. (b) White, C., Sharman, A. K., and Gadd, G. M. 1998. An integrated microbial process for the bioremediation of soil contaminated with toxic metals. *Nat. Biotechnol.* 16:572–75. (c) Schmid, A.; Dordick, J. S., Hauer, B., Kiener, A., Wubbolts, M., and Witholt, B. 2001. Industrial biocatalysis today and tomorrow. *Nature* 409:258–68.

28. (a) Wolffberg, A., and Sheintuch, M. 1993. Density distribution of growing immobilized cells. *Chem. Eng. Sci.* 48:3937–44. (b) Akay, G., Erhan, E., and Keskinler, B. 2005. Bioprocess intensification in flow-through monolithic microbioreactors with immobilized bacteria. *Biotechnol. Bioeng.* 90:180–90.

29. Plieva, F. M., Oknianska, A., Degerman, E., and Mattiasson, B. 2008. Macroporous gel particles as robust macroporous matrices for cell immobilization. *Biotechnol. J.* 3:410–17.

30. Kuyukina, M. S., Ivshina, I. B., Gavrin, A. Yu., Podorozhko, E. A., Lozinsky, V. I., Jeffree, C. E., and Philp, J. C. 2006. Immobilization of hydrocarbon-oxidizing bacteria in poly(vinyl alcohol) cryogels hydrophobized using a biosurfactant. *J. Microbiol. Meth.* 65:596–603.

31. Efremenko, E., Spiricheva, O., Varfolomeyev, S., and Lozinsky, V. 2006. *Rhizopus oryzae* fungus cells producing L(+)-lactic acid: Kinetic and metabolic parameters of free and PVA-cryogel-entrapped mycelium. *Appl. Microbiol. Biotechnol.* 72 (3):480–85.

32. Martins, R. F., Plieva, F. M., Santos, A., and Hatti-Kaul, R. 2005. Integrated immobilized cell reactor-adsorption system for β-cyclodextrin production: A model study using PVA-cryogel entrapped *Bacillus agaradhaerens* cells. *Biotechnol. Lett.* 25 (18):1537–43.

33. Cunningham, C. J., Ivshina, I. B., Lozinsky, V. I., Kuyukina, M. S., and Philp, J. C. 2004. Bioremediation of diesel-contaminated soil by microorganisms immobilised in polyvinyl alcohol. *Int. Biodeter. Biodegr.* 54:167–74.

34. Kaneko, T., Ogomi, D., Mitsugi, R., Serizawa, T., and Akashi, M. 2004. Mechanically drawn hydrogels uniaxially orient hydroxyapatite crystals and cell extension. *Chem. Mater.* 16 (26):5596–601.

35. Kumar, A., Rodríguez-Caballero, A., Plieva, F. M., Galaev, I. Yu., Nandakumar, K. S., Kamihira, M., Holmdahl, R., Orfao, A., and Mattiasson, B. 2005. Affinity binding of cells to cryogel adsorbents with immobilized specific ligands: Effect of ligand coupling and matrix architecture. *J. Molec. Recog.* 18 (1):84–93.

36. Inada, T., and Lu, S.-S. 2003. Inhibition of recrystallization of ice grains by adsorption of poly(vinyl alcohol) onto ice surfaces. *Cryst. Growth Des.* 3:747–52.

37. Kirsop, B. E., and Doyle, A. 1991. *Maintenance of microorganism and cultured cells.* London: Academic Press.

38. Velikov, V., Borick, S., and Angell, C. A. 2001. The glass transition of water, based on hyperquenching experiments. *Science* 294:2335–38.
39. (a) Menger, F. M., Zhang, H., Caran, K. L., Seredyuk, V. A., and Apkarian, R. P. 2002. Gemini-induced columnar jointing in vitreous ice. Cryo-HRSEM as a tool for discovering new colloidal morphologies. *J. Am. Chem. Soc.* 124:1140–41. (b) Menger, F. M., Galloway, A. L., Chlebowski, M. E., and Apkarian, R. P. 2004. Ultrastructure in frozen/etched saline solutions: On the internal cleansing of ice. *J. Am. Chem. Soc.* 126:5987–89. (c) Liang, J., Ma, Y., Zheng, Y., Davis, H. T., Chang, H.-T., Binder, D., Abbas, S., and Hsu, F.-L. 2001. Solvent-induced crystal morphology transformation in a ternary soap system: Sodium stearate crystalline fibers and platelets. *Langmuir* 17:6447–54.
40. Gutierrez, M. C., Garcia-Carvajal, Z. Y., Jobbagy, M., Rubio, F., Ferrer, M. L., and del Monte, F. 2007. Poly(vinyl alcohol) scaffolds with tailored morphologies for drug delivery and controlled release. *Adv. Funct. Mater.* 17:3505–13.
41. (a) Deville, S., Saiz, E., and Tomsia, A. P. 2006. Freeze casting of hydroxyapatite scaffolds for bone tissue engineering. *Biomaterials* 27:5480–89. (b) Deville, S., Saiz, E., and Tomsia, A. P. 2007. Ice-templated porous alumina structures. *Acta Materialia* 55:1965–74.
42. Williams, E., and Angell, C. A. 1977. Pressure-dependence of glass-transition temperature in ionic liquids and solutions: Evidence against free volume theories. *J. Phys. Chem.* 81:232–37.
43. Gutierrez, M. C., Ferrer, M. L., and del Monte, F. 2008. Ice-templated materials: Sophisticated structures exhibiting enhanced functionalities obtained after unidirectional freezing and ice-segregation-induced self-assembly. *Chem. Mater.* 20:634–48.
44. (a) Strawhecker, K. E., and Manias, E. 2000. Structure and properties of poly (vinyl alcohol)/Na+ montmorillonite nanocomposites. *Chem. Mater.* 12:2943–49. (b) Mbhele, Z. H., Salemane, M. G., van Sittert, C. G. C. E., Nedeljkovic, J. M., Djokovic, V., and Luyt, A. S. 2003. Fabrication and characterization of silver-polyvinyl alcohol nanocomposites. *Chem. Mater.* 15:5019–24. (c) Coleman, J. N., Cadek, M., Blake, R., Nicolosi, V., Ryan, K. P., Belton, C., Fonseca, A., Nagy, J. B., Gunko, Y. K., and Blau, W. J. 2004. High-performance nanotube-reinforced plastics: Understanding the mechanism of strength increase. *Adv. Funct. Mater.* 14:791–98.
45. (a) Wan, W. K., Campbell, G., Zhang, Z. F., Hui, A. J., and Boughner, D. R. 2002. Optimizing the tensile properties of polyvinyl alcohol hydrogel for the construction of a bioprosthetic heart valve stent. *J. Biomed. Mater. Res. B* 63:854–61. (b) Ricciardi, R., D'Errico, G., Auriemma, F., Ducouret, G., Tedeschi, A. M., De Rosa, C., Lauprêtre, F., and Lafuma, F. 2005. Short time dynamics of solvent molecules and supramolecular organization of poly(vinyl alcohol) hydrogels obtained by freeze/thaw techniques. *Macromolecules* 38:6629–39.
46. Zhang, H., Hussain, I., Brust, M., Butler, M. F., Rannard, S. P., and Cooper, A. I. 2005. Aligned two- and three-dimensional structures by directional freezing of polymers and nanoparticles. *Nat. Mater.* 4:787–93.
47. Zhang, H., Edgar, D., Murray, P., Rak-Raszewska, A., Glennon-Alty, L., and Cooper A. I. 2008. Synthesis of porous microparticles with aligned porosity. *Adv. Funct. Mater.* 18:222–28.

48. Gutierrez, M C., Garcia-Carvajal, Z. Y., Jobbagy, M., Yuste, L., Rojo, F., Abrusci, C., Catalina, F., del Monte, F., and Ferrer, M. L. 2007. Hydrogel scaffolds with immobilized bacteria for 3D cultures. *Chem. Mater.* 19:1968–73.
49. Soltmann, U., Böttcher, H., Koch, D., and Grathwohl, G. 2003. Freeze gelation: A new option for the production of biological ceramic composites (biocers). *Mater. Lett.* 57:2861–65.
50. Schulz, M., and Schmoldt, A. 2003. Therapeutic and toxic blood concentrations of more than 800 drugs and other xenobiotics. *Pharmazie* 58:447–74.
51. Choi, S. H., Kwon, J.-H., and Kim, C.-W. 2004. Microencapsulation of insulin microcrystals. *Biosci. Biotechnol. Biochem.* 68:749–53.
52. Plieva, F. M., Karlsson, M., Aguilar, M.-R., Gomez, D., Mikhalovsky, S., Galaev, I.Yu., and Mattiasson, B. 2006. Pore structure of macroporous monolithic cryogels prepared from poly(vinyl alcohol). *J. Appl. Polym. Sci.* 100:1057–66.

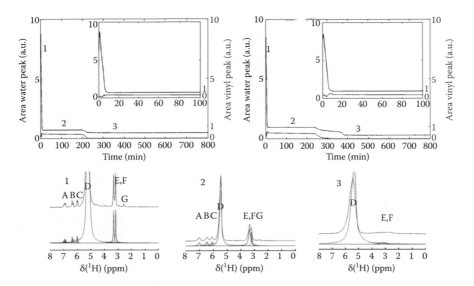

FIGURE 2.13
Results of measurements on samples with different cross-linking degree (6% (w/v) DMAAm:PEG diacrylate) (5:1) top left and (60:1) top right studied at –12°C ± 1°C in a Bruker DMX 200 spectrometer operating at a ¹H frequency of 200.13 MHz. The ¹H spectra was acquired every 25 seconds for the first 30 minutes, and thereafter at 10-minute intervals. Top figures show the change in area of the water peak (black) and the vinyl peak at 6.9 ppm (red) during the reaction. (1) shows the results obtained before freezing the sample, (2) shows the results from the frozen sample, and (3) shows the spectra after completion of the polymerization reaction. A, B, and C denote the protons in the vinyl group of the monomer, D the water peak, E and F the methyl peaks, and G the signal from the activator TEMED. The black lines in (1–3) show the recorded spectra, the red lines show the deconvoluted spectra with Lorentzian functions fitted to each peak, and the blue lines show the sum of the deconvoluted spectra. (Reproduced from H. Kirsebom, G. Rata, D. Topgaard, B. Mattiasson, and I.Yu. Galaev, *Polymer*, 381, 3855–58, 2008. With permission.)

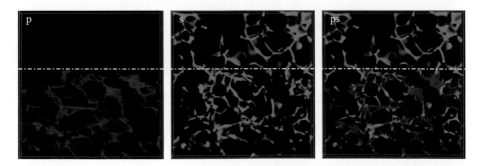

FIGURE 6.12
Confocal laser scanning microscopy (CSLM) 3D reconstituted images of the secondary polyacrylamide (pAAm) cryogel formed inside the interconnected pores of the primary dextranmethacrylate (dextran-MA) cryogel. The dextran-MA was stained with Rhodamine B (red colored) and pAAm cryogel was stained with FITC (green colored). Image (p) for the dextran-MA cryogel stained with Rhodamine B, image (s) for the pAAm stained with FITC and image (ps) for the stained with both dyes dextran-MA with pAAm cryogels with gradient porosity. (Reproduced from F. M. Plieva, P. Ekström, I. Yu. Galaev, and B. Mattiasson, *Soft Matter*, 2008. With permission.)

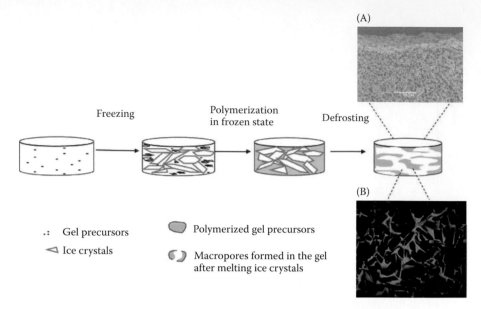

FIGURE 14.1

Formation of cryogels. Insert graph: scanning electron micrograph of p(2-hydroxyethyl methacrylate cryogel, ×50 (A); porous structure of laminin-containing protein cryogel labeled with FITC and viewed by confocal laser scanning microscopy. (Reproduced from Dainiak, M. B., B. Mattiasson, and I. Yu. Galaev, *BioForum Europe*, 11, 28–29, 2007.)

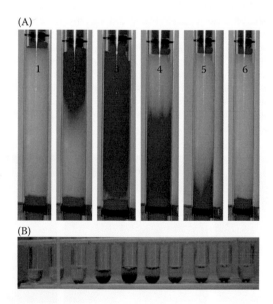

FIGURE 14.3

Flow of the pulse of whole blood through an underivatized cryogel column. (A) One ml blood was applied to the naked cryogel column at a flow rate of 0.5 ml/min in isotonic buffer solution. Column 1: column before application; column 2–5: column during the run; column 6: column after the flow of blood sample. (B) Flow through fractions of the column were collected and left to stand. No red blood cells lysis was observed in the fractions. (Reproduced from Noppe, W., F. M. Plieva, K. Vanhoorelbeke, H. Deckmyn, M. Tuncel, A. Tuncel, I.Yu. Galaev, and B. Mattiasson, *J. Biotechnol.*, 131, 293–99, 2007.)

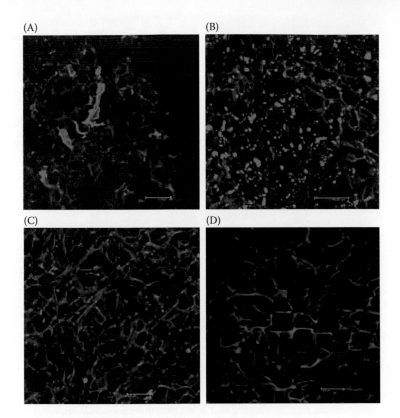

FIGURE 14.15

Confocal laser scanning microscopy images of (A,B) spatial distribution of collagen and fibrinogen in bulk-modified and (C,D) surface-modified pHEMA cryogel scaffolds. (Reproduced from Savina, I. N., M. B. Dainiak, H. Jungvid, B. Mattiasson, and I.Yu. Galaev, *J. Biomater. Sci.*, 20, 1781–95, 2009, DOI:10.1163/156856208X386390.)

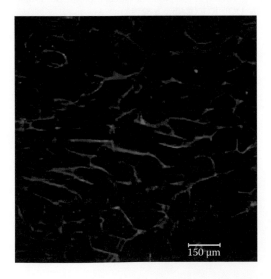

FIGURE 14.17

Distribution of fibrinogen in fibrinogen-containing protein-based cryogel scaffolds viewed by confocal laser scanning microscopy. Samples were immunostained with monoclonal antifibrinogen mouse antibodies and with secondary FITC-labeled antimouse antibodies.

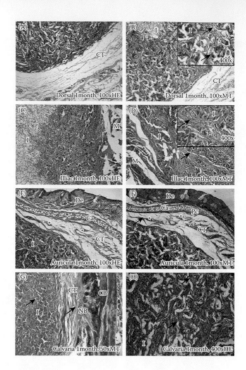

FIGURE 14.18

Representative histological data at one month: the particles of cryogel scaffold are surrounded by a moderate capsule characterized by infiltration of mononuclear phagocytic cells, macrophages, lymphocytes, fibroblasts, and some polymorphonuclear leukocytes with foreign body giant cells in hard and soft tissues at the implantation sites. Collagen fibers and the blood vessels are in close relation with the implant. Auricular cartilage appear healthy. HE: haematoxylin eosin, MT: Masson's trichrome, I: implant, CT: connective tissue, FC: fibrous callus, CB: compact bone, NB: new trabecular bone, Ca: cartilage, Pc: pericondrium, De: dermis, Sk: skin, M: muscles, (*): blood vessels, Arrow: collagen fibers. (Reproduced from Bolgen, N., I. Vargel, P. Korkusuz, E. Guzel, F. Plieva, I.Yu. Galaev, B. Mattiasson, and E. Piskin, *J. Biomed. Mater. Res. Part A*, 2009, DOI: 10.1002/jbm.a.32193.)

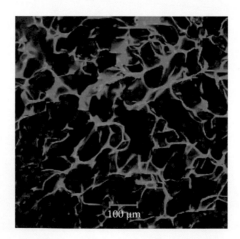

FIGURE 14.19

Confocal laser scanning microscopy of primary dermal fibroblasts SKF375 cultured on protein-based cryogel dermal regeneration template for 6 days. Snapshot of the bottom of the scaffold. Cells were stained with 4′,6-diamidino-2-phenylindole (DAPI), cryogel was stained with fluorescein isothiocyanate (FITC).

5

Preparation of Polylactide Scaffolds

Ming-Hua Ho, Da-Ming Wang, Hsyue-Jen Hsieh, and Juin-Yih Lai

CONTENTS

Porous polymers with suitable porous structures and chemical properties can serve as templates to guide cells to differentiate and grow into tissues or even organs. Such biomedical applications have drawn a lot of research attention and forms a research field called tissue engineering. For an application in tissue engineering, the polymer should have appropriate biocompatibility and biodegradability. In addition, high porosity, a large surface area, a suitable pore size, and a highly interconnected pore structure are also needed. In this chapter, we concentrate on giving an overview of the methods for preparing a three-dimensional porous polymer structure. Polylactide (PLA), an Food and Drug Administration (FDA)–approved biodegradable polymer for clinical use, is used as an example polymer to demonstrate the pore-forming techniques. First, brief introductions to tissue engineering and PLA are given. An overview of the techniques to fabricate porous PLA then follows.

5.1 Tissue Engineering and Biodegradable Polymer Scaffolds

Organ or tissue failure remains a frequent, costly, and serious problem in health care despite recent advances in biomedical and biochemical technologies. Millions of surgical procedures are performed every year to treat patients who suffer from the loss or failure of organs and tissues, resulting from accidents or diseases (Langer and Vacanti 1993). A general therapy is to perform organ transplantation to replace the damaged parts; however, the number of donors is usually limited. Tissue engineering, a technique to create new tissue from cultured cells, has now been considered as a potential alternative to organ or tissue transplantation (Langer and Vacanti 1993).

Tissue engineering is a research field relating to the development and manipulation of *in vitro* grown cells, tissues, or organs and to replace or support the function of defective or injured body parts. It combines biological science with materials and engineering sciences to quantify structure–function relationships in normal and pathological tissues, to develop new approaches to repair tissues, and to develop replacements for tissues (Nerem 1991; Langer and Vacanti 1993). Tissue engineering involves a combination of disciplines to achieve new therapies.

Three general strategies have been utilized in the creation of new tissues (Langer and Vacanti 1993). These include (1) the replacement of only those isolated cells or cell substitutes needed for function; (2) production and delivery of tissue-inducing substances such as growth factor and signal molecules; and (3) cells placed on or within a scaffold fabricated from synthetic polymers or natural substances. Scaffolds, cells, and signals are the three key elements in tissue engineering and are usually called "the triangle of tissue engineering," which is illustrated in Figure 5.1. Combining these three vital essentials to form a tissue-like complex that is then transplanted back into the human body may be a feasible method for tissue regeneration.

Scaffolds, serving as transplanting vehicles for cultured cells and templates to guide tissue regeneration, play an important role in transforming the cultured cells to a new tissue (Langer and Vacanti 1993; Freed et al. 1993; Cao

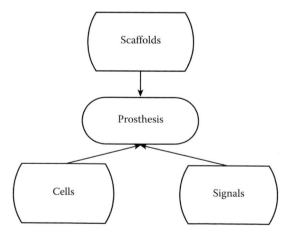

FIGURE 5.1
The tissue engineering triad.

et al. 1997). Isolated cells injected into the body at random cannot form new tissue on their own without scaffolds (Folkman and Haudenshild 1980). Most mammalian cells are anchorage dependent and cannot perform their normal function if not provided with an adhesion substrate. Traditional culturing of nontransformed animal cells involves the attachment and growth of the cells on flat sheets in containers, such as petri dishes, flasks, and roller bottles. The cultures on these two-dimensional surfaces do not reflect the normal *in vivo* environment; moreover, they may lead to artifacts such as de-differentiation (Lucas-Clerc, Massart, and Campion 1993). Preparation of scaffolds that can serve as three-dimensional templates to guide cells to construct tissues has become an essential research topic in tissue engineering.

To serve as cell vehicles, scaffolds should be biocompatible; biodegradable, inducing no tissue response to the host; and completely resorbable, leaving a totally natural tissue replacement after degradation. They should also be easily and reliably reproducible into a variety of shapes and structures that retain their shape when being implanted. In addition, to allow a high density of seeded cells and to be able to promote neovascularization (the formation of a blood-vessel system within tissue) after being implanted, the scaffolds should have a high porosity, a large surface area, a suitable pore size, and a highly interconnected pore structure (Mikos et al. 1993; Mooney et al. 1994, 1995; Schugens et al. 1995). It has been reported that the pore size of scaffolds should be close to 20 μm for the ingrowth of hepatocytes, about 20–150 μm for skin regeneration, and in the range of 90–150 μm for bone regeneration (Maquet and Jerome 1997). The interaction between scaffolds and cells should support the function and growth of differentiated cells (Cima et al. 1991), and in certain situations it should induce ingrowth of desirable cell types from surrounding tissues.

5.2 Introduction to Polylactide

An important class of synthetic biodegradable polymers is polylactides (PLAs). It is among the few synthetic materials approved for human clinical use by the FDA in the United States. Polylactide has been used to fabricate biodegradable bone nails (Rokkanen, Bostman, and Hirvensalo 1994) and is extensively used to prepare scaffolds for tissue engineering (Vacanti et al. 1988). By far, the family of polylactides is the most commonly used synthetic biomaterial.

A polylactide is a polymer of lactic acid, usually synthesized by a ring-opening polymerization of cyclic diesters lactide (Dubois et al. 1991). Lactic acid has D- and L-stereoisomers (see Figure 5.2). The polymer of the D-stereoisomer is called PDLA and that of L-stereoisomers is called PLLA. Whereas PLLA possesses about 37% crystallinity, the optically inactive poly

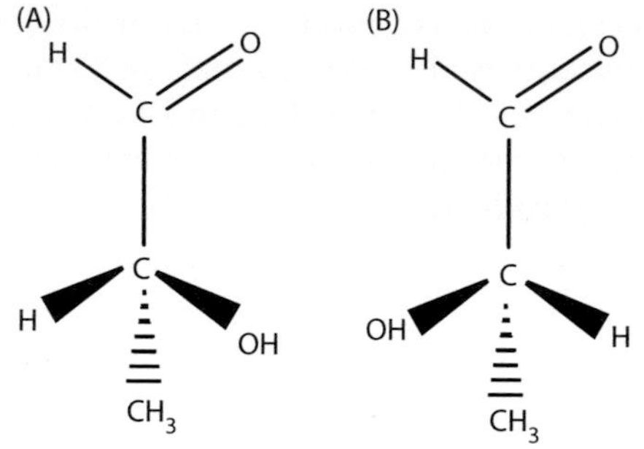

FIGURE 5.2
Chemical structures of lactic acids: (A) L-lactic acid and (B) D-lactic acid.

(d,l-lactic acid; PDLLA) is amorphous. Both types of PLA are biodegradable via the hydrolysis mechanism. PLLA is employed more often than PDLA, because the hydrolysis of PLLA yields L-lactic acid, which is the naturally occurring stereoisomer of lactic acid, a by-product of anaerobic metabolism in the human body. After the degradation of PLLA, the resulting lactic acid can be incorporated into the tricarboxylic acid (TCA) cycle to be finally excreted by the body as carbon dioxide and water (Bostman 1991). Therefore, PLLA is not only biodegradable but also completely resorbable in the human body.

For tissue engineering, different target systems require different scaffold degradation times. Compared to other biodegradable polyesters, PLLA degrades more slowly because the ester bond in PLLA is less labile to hydrolysis due to the steric hindrance of the methyl group (Reed and Gilding 1981). The crystallinity of PLLA also slows down its degradation rate. The longer degradation time of PLLA makes PLLA scaffolds suitable for those applications that need to sustain mechanical integrity longer in the human body. But, note that the scaffold degradation rate depends not only on polymer degradability but also on its surface morphology, pore size, and porosity (Boss 2000). Therefore, the scaffolds should be designed in accordance with their application targets (Langer and Vacanti 1993; Freed et al. 1993; Mikos et al. 1993; Mooney et al. 1994, 1995; Schugens et al. 1995; Cao et al. 1997).

5.3 The Preparation of PLA Scaffolds

To design suitable PLA scaffolds for tissue engineering applications, both the material chemistry and the scaffold structure need to be tailored. In this chapter, we only concentrate on the technologies that are currently used to generate porous scaffolds. Several preparation methods are introduced in the following sections, including porogen leaching (Mikos et al. 1995; Holy et al. 1999; Chen, Ushida, and Tateishi 2001), saturation and release of CO_2 (Mooney et al. 1996; Harris, Baldwin, and Mooney 1998), three-dimensional printing (Park, Nam, and Ha 1997), and phase separation techniques (Schugens et al. 1996a,b; Nam and Park 1999; Han 2000).

5.3.1 Porogen Leaching

Among all the methods for scaffold preparation, porogen leaching is the most convenient. This method provides easy control on pore structure, pore size, and porosity. Figure 5.3 is a schematic description of the porogen leaching process. First, water-soluble particles, functioning as porogens, are mixed with a solution of PLA and organic solvent. The porogens must be well suspended in the PLA solution by agitation or an other method. After the suspension, the solution is gelled by quenching, and the solvent contained in the gel is removed. Next, the porogens are leached out by immersing the dried gel in water. The space occupied by the porogens becomes pores after they are leached out. Any water-soluble particles with suitable size can serve as porogens, such as salts (Mikos et al. 1995), glucose (Holy et al. 1999), and gelatin microspheres (Thomson et al. 1998). The pore size and porosity of

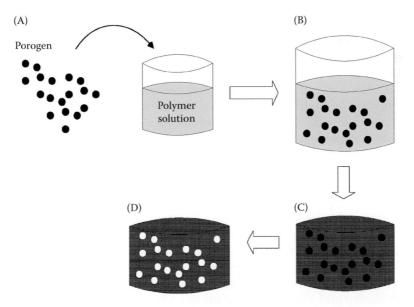

FIGURE 5.3
Preparation of scaffolds by porogen leaching: (A) mixing of porogens and PLA solution; (B) well suspension of porogens in the PLA solution; (C) solidification of the PLA solution containing suspended porogens; and (D) leaching out the porogens with water to form porous PLA scaffolds.

the prepared scaffolds can be tailored by varying the size, morphology, and concentration of the porogens.

Scaffolds that are fabricated by the porogen-leaching method usually exhibit a porous structure with closed pores. However, scaffolds with interconnected pores are preferred in tissue engineering, to allow seeded cells to move freely in the scaffolds and to enhance the transport of nutrients, growth factors, and metabolic wastes. Obviously, closed pores are not a suitable structure for applications in tissue engineering. The disadvantages of closed pores were investigated and reported in the literature (Yang et al. 2002). To overcome this shortcoming, modified porogen-leaching processes have been proposed. A combination of porogen-leaching and gas-foaming methods (Harris, Baldwin, and Mooney 1998; Nam, Yoon, and Park 2000) was shown to be able to produce biodegradable PLA matrices with open pore structures. Oh and colleagues (2003) proposed a method for scaffold preparation by combining polymer melting and porogen leaching to avoid closed pores.

5.3.2 Textile Technologies

Textile technologies have also been employed to prepare biodegradable polymer scaffolds. The PLA, PGA, and other semicrystalline polymers can be processed into fibers using textile technologies. Woven or nonwoven scaffolds can then be formed by using the fibers. The fibrous scaffolds possess interconnected porous structures and high porosity, which are highly desirable for applications in tissue engineering. The woven scaffolds have good mechanical strength, but the pores are often too large for cell seeding. The preparation of nonwoven scaffolds is easier than that of the woven ones. The porosity and pore size can be controlled by the degree of compression of the fibers. However, there are several shortcomings of PLA nonwoven scaffolds, such as low mechanical strength, difficulty in controlling pore morphology, and limited fiber diameter variations.

Electrospinning, first introduced in the early 1930s (Formhals 1934) to fabricate industrial or household nonwoven fabric products to resolve some of the difficulties discussed above, is now widely used to spin biodegradable polymer fibers. In the electrospinning procedure, electrical fields provide the driving force for fiber spinning. The setup for the electrospinning process is depicted schematically in Figure 5.4. First, the PLA solution is forced through a capillary, forming a drop of polymer solution at the tip. A high voltage is applied between the tip and a grounded collection target. When the electric field strength overcomes the surface tension of the droplet, a polymer solution jet is initiated and accelerated toward the collection target. As the jet travels through the air, the solvent evaporates and a nonwoven polymer fabric is formed on the target collector. To generate preferential orientation or a tubular structure, an electrically grounded rotating drum is usually used as the collector (Katta et al. 2004; Kim et al. 2006). The diameter of the fibers prepared

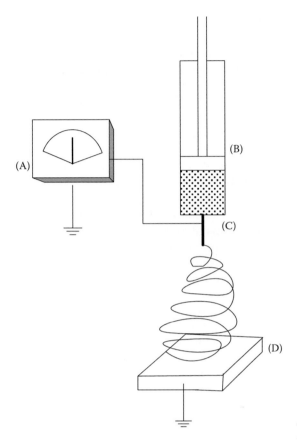

FIGURE 5.4
Setup for electrospinning: (A) power supply; (B) syringe pump; (C) needle; and (D) collector.

by electrospinning typically ranges from 50 nm to a few micrometers. The nonwoven fabrics of the electrospun fibers are an interconnected network with an ultra-high surface-to-volume ratio. The high surface-to-volume ratio was shown to be able to enhance cell adhesion, which in turn can influence cell migration, proliferation, and differentiation (Elsdale and Bard 1972; Ma and Langer 1999). It should be mentioned that the mechanical strength of electrospun scaffolds is usually too low for engineering hard tissues.

5.3.3 Phase Separation Methods

Phase separation is a widely used method to prepare commercial porous membranes (Mulder 1991). The structure of the membrane can be tailored by adjusting the phase separation conditions. Phase separation techniques are also often used to prepare porous scaffolds for tissue engineering (Whang, Thomas, and Healy 1995; Schugens et al., 1996a,b; Nam and Park 1999; Han,

2000; Ma et al. 2001), including thermally induced phase separation (TIPS) and nonsolvent induced phase separation (NIPS). In the phase separation processes, homogeneous polymer solution separates into polymer-rich and poor phases as the solubility of polymers decrease due to the change in temperature or composition. After the removal of the liquid contained in the phase separated solution, the polymer-rich phase forms the solid matrix of the scaffold, while the polymer-poor phase becomes pores (Mulder 1991).

On one hand, the phase separation mechanism could be liquid–liquid demixing (Han 2000), which generates polymer-poor and polymer-rich liquid phases. The subsequent growth and coalescence of the polymer-poor phase would then develop to form the pores in scaffolds. On the other hand, when the temperature is low enough to allow the freeze of the polymer solution, the phase separation mechanism could be solid–liquid demixing, which forms phases of frozen solvent and concentrated solution. After the removal of the frozen solvent, the remaining space becomes pores. By adjusting the polymer concentration, using a different solvent, and varying the cooling rate, phase separation could occur via a different mechanism and result in scaffolds with various morphologies.

In the stage of solvent removal, the porous structure contained in the solution needs to be carefully retained. Therefore, freeze-drying is usually used for solvent removal (Zhang and Ma 1999; Shen et al. 2000; Zhang and Zhang 2001). Without freeze-drying, a rise in temperature during the drying stage could result in remixing of the phase separated solution or remelting of the frozen solution, leading to the destruction of the porous structure. Although freeze-drying can prevent the disintegration of the porous structure, it would be so time- and energy-consuming that the whole scaffold fabrication process would become inefficient and economically uncompetitive. To improve the efficiency of the fabrication process, a freeze-extraction process was developed for the preparation of polyester porous scaffolds, including PLA scaffolds. The freeze-extraction method fixes the porous structure under freezing conditions so that in the subsequent drying stage the freeze-drying process is not needed. The process of freeze extraction is described in Figure 5.5. In the freeze-extraction process, after the PLA solution is frozen, the frozen solution is immersed in a nonsolvent bath to allow for the exchange of solvent and nonsolvent at a temperature lower than the freezing point of the polymer solution. Such a freeze-extraction technique can extract the solvent out of the polymer matrix before the drying stage. After that, the polymer matrix is surrounded with nonsolvent, and redissolution of the polymer would not occur in the drying stage. The freeze-extraction method for the preparation of PLA scaffolds has the advantage of being more efficient and economical. Another advantage of freeze-extraction is discussed next. For the freeze-drying process, the solvent vapor pressure at the drying temperature (usually very low) needs to be high enough to allow for its removal. Hence, the choice of solvent is quite limited. The limitation of choosing a solvent can be lifted when scaffolds are prepared by the freeze-extraction method, because solvent is

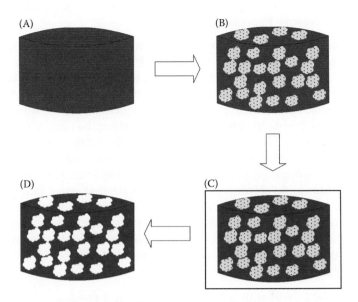

FIGURE 5.5
Procedures of the freeze-extraction method: (A) PLA solution is placed in a mold; (B) polymer-rich and polymer-poor phases are generated after the freezing of the PLA solution; (C) the frozen PLA solution is immersed into a nonsolvent bath, where the exchange of solvent and nonsolvent takes place; and (D) after the solvent is extracted out, the structure contained in the frozen solution can be retained with a drying stage.

removed by extraction but not by vaporization. Furthermore, due to the low solvent-removal efficiency of freeze-drying, the solvent might not be able to be completely removed and the residual solvent might limit the application of the scaffolds. For the freeze-extraction process, the solvent is not removed by volatilization but by extraction with the nonsolvent. Once the amount of non-solvent is large enough, the solution can be almost completely replaced with the nonsolvent. By using a nonsolvent, which is safe for biomedical use (such as ethanol aqueous solution), the problem of residual solvent can be resolved.

Nonsolvent induced phase separation (NIPS) is another method to generate a porous structure. In NIPS, a PLA solution is cast on a substrates and immersed in a nonsolvent bath, where the exchange of solvent and non-solvent occurs. With the replacement of the solvent by the nonsolvent, the homogeneous solution phase separates into polymer-rich and-poor phases, and a porous structure can form. After a certain period of time, the solvent in the scaffold is extracted out completely by the nonsolvent and the formed structure is then fixed. The PLA–dioxane–water and PLA–chloroform–alcohol are two common systems for the preparation of PLA scaffolds using NIPS (van de Witte et al. 1996). However, by using the NIPS method, a dense skin often appears on the scaffold surface, which may obstruct cell spreading and nutrient transport when applied to tissue engineering. Therefore, NIPS is used less to prepare PLA scaffolds when compared with TIPS.

5.3.4 High-Pressure Processing

Supercritical fluid (SCF) processing methods have been developed for scaffold preparation. By using the SCF technique, PLA scaffolds can be prepared without using any organic solvent, which is highly beneficial for biomedical applications (Montjovent et al. 2005). In such a process, the PLA is first saturated with CO_2 at a high pressure. The system is then quenched to the supersaturated state either by reducing the pressure or by increasing the temperature to result in phase separation. The use of SCF allows for the tuning of the solvency of CO_2 to PLA simply by regulating the pressure or temperature. Therefore, CO_2 can serve as a solvent or a nonsolvent to PLA, depending on the operation condition. Because no solvent or nonsolvent is needed other than CO_2, there is definitely no remaining solvent or nonsolvent in the prepared scaffolds. However, the control of pore size and porosity is difficult to achieve and the dense skin layers often appear on the surface of the scaffolds prepared by SCF. Further improvement on this process is still needed.

5.3.5 Scaffold Fabrication with Computer-Aided Design

The fabrication techniques discussed above cannot build up scaffolds with complicated and predefined shapes. Such scaffolds can be prepared with computer-aided design; for example, rapid prototyping (RP) (Chua 1994; Kochan, Chua, and Du 1999) and solid free-form (SFF) (Taboas et al. 2003). RP techniques include shape deposition manufacturing (Zhang et al. 2005), fused deposition modeling (Woodfield et al. 2004; Ang, Leong, and Chua 2006), nonfused liquid deposition modeling (Ruecker, Laschke, and Junker 2006), three-dimensional printing (Curodeau, Sachs, and Caldarise 2000), selective laser sintering (Tan, Chua, and Leong 2005), and so on. These RP techniques can create specific pore shapes by the selective addition of material, layer-by-layer, and guided by the computer program. The step-by-step development of scaffolds improves the reproducibility and accuracy, including the control of pore size, porosity, and interconnectivity. Nevertheless, how to enhance the resolution and to reduce the cost still remains a challenge.

5.4 Conclusions

Tissue engineering is one of the most exciting interdisciplinary and multi-disciplinary research areas today. Scaffold materials and fabrication technologies play a pivotal role in tissue engineering. This brief review is intended to illustrate how to fabricate PLA, an FDA-approved biodegradable polymer for clinical use, into scaffolds for applications in tissue engineering. Several currently used fabrication methods are briefly discussed, including porogen leaching, electrospinning, phase separation methods, SCF processes, and computer-aided design.

References

Ang, K. C., Leong, K. F., Chua, C. K., and Chandrasekaran, M. 2006. Investigation of the mechanical properties and porosity relationships in fused deposition modeling-fabricated porous structures. *Rapid Prototyping J* 12:100–105.

Boss, J. H. 2000. Biocompatibility: Review of the concept and its relevance to clinical practice. In *Biomaterials and bioengineering handbook* 1–86, ed. D. L. Wise et al. New York: Marcel Dekker.

Bostman, O. M., 1991. Osteolytic changes accompanying degradation of absorbable fracture fixation *Bone Joint Surg A* 73:679–82.

Cao, Y., Vancanti, J. P., Paige, K. T., Upton, J., and Vacanti, C. A. 1997. Transplantation of chondrocytes utilizing a polymer-cell construct to produce tissue-engineering cartilage in the shape of a human ear. *Plast Reconstr Surg* 100:297–302.

Chen, G., Ushida, T., and Tateishi, T. 2001. Preparation of poly (L-latide) and poly (D, L-lactic-co-glycolic acid) by use of ice microparticulates. *Biomaterials* 22:2563–67.

Chua, C. K. 1994. Three-dimensional rapid prototyping technologies and key development area. *Comp Control Eng J* 5:200–206.

Curodeau, A., Sachs, E., and Caldarise, S. 2000. Design and fabrication of cast orthopedic implants with freeform surface textures from 3-D printed ceramic shell. *J Biomed Mater Res* 53:525–35.

Dubois, P. H., Jacobs, C., Jérôme, R., and Teyssié, P. H. 1991. Macromolecular engineering of polylactone and polylactide. 4 mechanism and kinetics of lactide homopolymerization by aluminum isopropoxide. *Macromol* 24:2266–70.

Elsdale, T., and Bard, J. 1972. Collagen substrata for studies on cell behavior. *Cell Biol* 54:626–37.

Folkman, J., and Haudenshild, C. 1980. Angiogenesis *in vitro*. *Nature* 288:551–56.

Formhals, A. 1934. Process and apparatus for preparing artificial threads. US Patent #1,975,504.

Freed, L. E., Marquis, J. C., Nohria, A., Emmanual, J., Mikos, A. C., and Langer, R. 1993. Neocartilage formation *in vitro* and *in vivo* using cells cultured on synthetic biodegradable polymers. *J Biomed Mater Res* 27:11–23.

Han, M. J. 2000. Biodegradable membranes for the controlled release of progesterone. 1. Characterization of membrane morphologies coagulated from PLGA/Progesterone/DMF solutions. *J Appl Polym Sci* 75:60–67.

Harris, L. D., Baldwin, D. F., and Mooney, D. J. 1998. Open pore biodegradable matrices formed with gas foaming. *J Biomed Mater Res* 42:396–402.

Holy, C. E., Dang, S. M., Davies, J. E., and Shiochet, M. S. 1999. *In vitro* degradation of a novel poly (lactide-*co*-glycolide) 75/25 foam. *Biomaterials* 20:1177–85.

Katta, P., Alessandro, M., Ramsier, R. D., and Chase, G. G. 2004. Continuous electrospinning of aligned polymer nanofibers onto a wire drum collector. *Nano Lett* 4:2215–18.

Kim, C. H., Jung, Y. H., Kim, H. Y., Lee, D. R., Dharmaraj, N., and Choi, K. E. 2006. Effect of collector temperature on the porous structure of electrospun fibers. *Macromol Res* 14:59–65.

Kochan, D., Chua, C. K., and Du, Z. H. 1999. Rapid prototyping issues in the 21st century. *Comput Ind* 39:3–10.

Langer, R., and Vacanti, J. P. 1993. Tissue engineering. *Science* 260:920–26.

Lucas-Clerc, C., Massart, C., and Campion, J. P. 1993. Long-term culture of human pancreatic islets in an extracellular matrix: Morphological and metabolic effects. *Mol Cell Endocrinol* 94:9–20.

Ma, J., Wang, H., He, B., and Chen, J. T. 2001. A preliminary *in vitro* study on the fabrication and tissue engineering applications of a novel chitosan bilayer materials as a scaffold of human neofetal fibroblast. *Biomaterials* 22:331–36.

Ma, P. X., Langer, R. 1999. *Tissue engineering method and protocols.* Ed. M. Yarmush and J. Morgan, pp. 47–56. Totowa, NJ: Human Press, Inc.

Mikos, A. G., Sarakinos, G., Lyman, M. D., Ingber, D. E., Vacanti, J. P., and Langer, R. 1993. Prevascularization of porous biodegradable sponges. *Biotechnol Bioeng* 42:716–23.

Mikos, A. G., Thorsen, A. J., Czerwonka, L. A., Bao, Y., Langer, R., Winslow, D. N., Vacanti, J. P. 1995. Preparation and characterization of poly (L-lactic acid) foams. *Polymer* 35:1067–77.

Montjovent, M. O., Mathieu, L., Hinz, B., Applegate, L. L., Bourban, P. E., Zambelli, P. Y., Manson, J. A., and Pioletti, D. P. 2005. Biocompatibility of bioresorbable poly(L-lactic acid) composite scaffolds obtained by supercritical gas foaming with human fetal bone cells. *Tissue Eng* 11:1640–49.

Mooney, D. J., Baldwin, D. F., Suh, N. P., Vacanti, L. P., and Langer, R. 1996. Novel approach to fabricate porous sponges of poly (D, L-lactic-co-glycolic acid) without the use of organic solvents. *Biomaterials* 17:1417–22.

Mooney, D. J., Kaufmann, P. M., Sano, K., Mcnamara, K. M., Vacanti, J. P., and Langer, R. 1994. Transplantation of hepatocytes using porous biodegradable sponges. *Transplant Proc* 26:3425–26.

Mooney, D. J., Park, S., Kaufmann, P. M., Sano, K., Mcnamara, K., Vacanti, J. P., and Langer, R. 1995. Biodegradable sponges for hepatocyte transplantation. *J Biomed Mater Res* 29:959–66.

Mulder, M. 1991. *Basic principles of membrane technology.* Dordrecht: Kluwer Academic Publishers.

Nam, Y. S., and Park, T. G. 1999. Porous biodegradable polymeric scaffolds prepared by thermally induced phase separation. *J Biomed Mater Res* 47:8–17.

Nam, Y. S., Yoon, J. J., and Park, T. G. 2000. A novel fabrication method of macroporous biodegradable polymer scaffolds using gas foaming salt as a porogens additive. *J Biomed Mater Res* 53:1–7.

Nerem, R. M. 1991. Cellular engineering. *Ann Biomed Eng* 19:529–45.

Oh, S. H., Kang, S. G., Kim, E. S., Cho, S. H., and Lee, J. H. 2003. Fabrication and characterization of hydrophilic poly(lactic-*co*-glycolic acid)/poly(vinyl alcohol) blend cell scaffolds by melt-molding particulate-leaching method. *Biomaterials* 24:4011–21.

Park, Y. J., Nam, K. H., and Ha, S. J. 1997. Porous poly (L-latide) membranes for guided tissue regeneration and controlled drug delivery: Membrane fabrication and characterization. *J Control Release* 43:151–60.

Reed, A. M., and Gilding, D. K. 1981. Biodegradable polymers for use in surgery—poly(glycolic)/poly(lactic acid) homo and copolymers: II. *In vitro* degradation. *Polymer* 22:494–98.

Rokkanen, P., Bostman, O., and Hirvensalo, E. 1994. Adsorbable implants in the fixation of fractures, osteomies, arthrodeses and ligaments. *Acta Orthopaedica Scandinavica Suppl* 260:19–20.

Ruecker, M., Laschke, M. W., and Junker, D. 2006. Angiogenic and inflammatory response to biodegradable scaffolds in dorsal skinfold chambers of mice. *Biomaterials* 27:5027–29.

Schugens, C. H., Grandfils, C. H., Jerome, R., Teyssie, P., Delree, P., Martin, D., Malgrange, B., and Moonen, G. 1995. Preparation of a macroporous biodegradable polylactide implant for neuronal transplantation. *J Biomed Mater Res* 29:1349–62.

Schugens, C., Maquet, V., Grandfils, C., Jerome, R., and Teyssie, P. 1996a. Biodegradable and macroporous polylactide implants for cell transplantation. I. Preparation of macroporous polylactide supports by solid-liquid phase separation. *Polymer* 37:1027–38.

Schugens, C., Maquet, V., Grandfils, C., Jerome, R., and Teyssie, P. 1996b. Biodegradable and macroporous polylactide implants for cell transplantation. II. Preparation of polylactide foams by liquid-liquid phase separation. *J Biomed Mater Res* 30:449–61.

Shen, F., Cui, Y. L., Yang, L. F., Yao, K. D., Dong, X. H., Jia, W. Y., and Shi, H. D. 2000. A study on the fabrication of porous chitosan/gelatin network scaffold for tissue engineering. *Polym Int* 49:1596–99.

Taboas, J. M., Maddox, R. D., Krebsbach, P. H., and Hollister, S. J. 2003. Indirect solid free form fabrication of local and global porous, biomimetic and composite 3D polymer-ceramic scaffolds. *Biomaterials* 24:181–94.

Tan, K. H., Chua, C. K., and Leong, K. F. 2005. Selective laser sintering of biocompatible polymers for applications in tissue engineering. *Bio-Med Mater Eng* 15:113–24.

Thomson, R. C., Yaszemski, M. J., Powers, J. M., Mikos, and A. G. 1998. Hydroxyapatite fiber reinforced poly(α-hydroxy ester) foams for bone regeneration. *Biomaterials* 19:1935–43.

Vacanti, J. P., Morse, M. A., Saltzman, W. M., Domb, A. J., Perez-Atayde, A., and Langer, R. 1988. Selective cell transplantation using bioabsorbable artificial polymers as matrices. *J Pediatr Surg* 23:3–9.

van de Witte, P., Esselbrugge, H., Dijkstra, P. J., van der Berg, J. W. A., and Feijen, J. 1996. Phase transitions during membrane formation of polylactides. I. A morphological study of membranes obtained from system polytide-chloroform-methanol. *J Member Sci* 113:223–36.

Whang, K., Thomas, C. H., and Healy, K. E. 1995. A novel method to fabricate biodegradable scaffolds. *Polymer* 36:837–42.

Woodfield, T. B. F., Malda, J., de Wijn, J., Peters, F., Riesle, J., and van Blitterswijk, C.A. 2004. Design of porous scaffolds for cartilage tissue engineering using a three dimensional fiber-deposition technique. *Biomaterials* 25:4149–61.

Yang, J., Shi, G. X., Bei, J. Z., Wang, S. G., Cao, Y. L., Shang, Q. X., Yang, G. H., Wang, W. Q., and Li, J. 2002. Fabrication and surface modification of macroporous poly(L-lactic acid) and poly(L-lactic-*co*-glycolic acid) (70/30) cells scaffold for human skin fibroblast cells culture. *J Biomed Mater Res* 62:438–46.

Zhang, H., Hutmacher, D. W., Chollet, F., Poo, A. N., and Burdet, E. 2005. Microrobotics and MEMS-based fabrication techniques for scaffold-based tissue engineering. *Macromol Biosci* 5:477–89.

Zhang, R. Y., and Ma, P. X. 1999. Poly(alpha-hydroxyl acids) hydroxyapatite porous composites for bone-tissue engineering. I. Preparation and morphology. *J Biomed Mater Res* 44:446–55.

Zhang, Y., and Zhang, M. Q. 2001. Synthesis and characterization of macroporous chitosan/calcium phosphate composite scaffolds for tissue engineering. *J Biomed Mater Res* 55:304–12.

6

Macroporous Polysaccharide Gels

Fatima M. Plieva, Igor Yu. Galaev, and Bo Mattiasson

CONTENTS

6.1 Introduction

Polysaccharides, stereoregular polymers of monosaccharides (sugars), are abundant natural polymers (biopolymers), which are widely available, renewable, and inexpensive. Polysaccharides are hydrophilic biopolymers that are characterized by nontoxicity, biodegradability, and high chemical reactivity that allow for their easy modification. Due to their properties and availability, the polysaccharide-based materials find a wide range of applications in biotechnology and biomedicine [1–4].

There is considerable interest in the preparation of macroporous polysaccharide gels from hydrophilic polysaccharide polymers. These polymers, typically used for the preparation of gels, can be divided into cationic, anionic, and neutral polymers (Table 6.1). Among abundant polysaccharides polymers, cellulose is an attractive raw material that can be converted to different derivatives of significant industrial potential and has found a wide application, for example, as matrices for chromatography [5,6]. Starch, the most abundant storage polysaccharide in plants [7], is used as chromatographic support [8] and as promising biodegradable polymeric material in tissue engineering [9,10]. Many starch derivatives (e.g., cyclodextrins), which are a family of cyclic oligosaccharides, have found wide application in biotechnology as useful molecular complexation agents [11]. Alginate and chitosan have been widely used due to their low toxicity and good biocompatibility and biodegradability [12–17]. Hyaluronic acid, a naturally occurring polysaccharide composed of N-acetyl-D-glucosoamine and D-glucuronic acid, is a major component of extracellular matrix (ECH) in connective tissues [18]. Dextran is used in various applications due to its bioavailability and biocompatibility. Soluble dextrans are used as blood plasma expanders

TABLE 6.1

Precursors Used for the Preparation of Polysaccharide Gels

Type of Polysaccharide	Precursors	Methods Used for the Preparation of Macroporous Gels	References
Anionic polysaccharide	Alginic acid Hyaluronic acid Pectin Carrageenan Gellan Cellulose acetate Dextran sulphate Cellulose phosphate	Physical and chemical cross-linking, ionotropic cross-linking	18, 70, 71 2, 12, 47, 48, 56, 72–74
Cationic polysaccharide	Chitosan	Physical and chemical cross-linking, ionotropic cross-linking	28, 36, 49, 66, 75
Neutral polysaccharides	Cellulose Dextran Agarose Pullulan Starch Cyclodextrins	Physical and chemical cross-linking	8, 20–22, 41–44, 63, 64

[19] and dextran hydrogels have potential in drug delivery [20] or as scaffolds for tissue engineering [21–23]. Dextran-based adsorbents prepared via cross-linking with epichlorohydrin (e.g., Sephadex™) adsorbents are well-known media for chromatography. Agarose has been typically used as an electro-phoresis media for separation of DNA fragments [24] and recently macropo-rous agarose scaffolds were exploited for tissue engineering [25,26]. Beads of agarose (prepared from agarose of 2, 4, or 6% concentration, respectively, and cross-linked with epichlorohydrin) were commercialized under the trade-mark Sepharose and are widely used for protein chromatography.

The polysaccharide precursors most often used for the preparation of poly-saccharide gels are listed in Table 6.1. Properties of porous polysaccharide gels depend on the nature and physical–chemical characteristics of the poly-saccharide polymer. These gels can be cross-linked physically, chemically, or via polyelectrolyte complexation (ionotropic links; see Table 6.1). Some poly-saccharides (like agarose, karrageenan, and gellan) form gels when cooling the solution; while polysaccharides like curdlan, some cellulose derivatives, and starch form gels when heating the solution of biopolymers.

Polysaccharide gels, which are characterized by macroporous structure with pores up to 200 μm in size, have found wide applications in the bio-medical area, especially as cell culture scaffolds for tissue engineering. The main approaches used for the fabrication of macropores in polysaccharide systems are listed in Table 6.2. Most of the macroporous matrices made from polysaccharides are prepared via freeze-drying [27–30], freeze-extraction (or solvent-exchange-phase separation SEPS) [16,28], thermally induced phase separation (TIPS) [28], mold compression/salt leaching [9,10,31] and double-emulsification techniques [32]. The cryogelation technique (or gelation at

TABLE 6.2

Main Approaches for the Preparation of Macroporous Polysaccharide Gels

Approach Used	Polysaccharide	Pore Size, μm	References
Freeze-drying (FD)	Alginate, chitosan, HA, starch, cellulose	Up to 200	27–30, 43, 49
Solvent induced phase separation (SIPS) and thermally-induced phase separation (TIPS)	Alginate, chitosan	Up to 400 (for SIPS) Up to 2 (for TIPS)	15, 16, 28
Double emulsification	Agarose	Up to 30	32, 55
Compression molding/ solvent casting–particle leaching approaches	Starch, chitosan	Up to 100	9, 10, 31
Cryogelation (gelation at subzero temperatures mainly in aqueous medium)	Agarose, alginate, cellulose, dextran, chitosan	Up to 200	22, 23, 26, 36, 66, 69

subzero temperatures) that has been widely exploited during the last 10 years allows for the preparation of macroporous polysaccharide systems with well-defined structures and a wide range of pore sizes [25,26,33–38]. The main approaches for the preparation of macroporous gels from polysaccharide precursors are briefly described in this chapter with a focus on the polysaccharide gels prepared via cryogelation technique (polysaccharide cryogels). The term "macroporous polysaccharide gels" will be used throughout the text. This term means polysaccharide matrices with pore sizes from hundreds of nanometers to hundreds of micrometers.

6.2 Preparation of Macroporous Polysaccharide Gels via Chemical Cross-Linking

Polysaccharide based materials (especially those intended for use as sorbents) are prepared via cross-linking reactions between the hydroxyl or amino groups present in polysaccharide molecules with appropriate cross-linkers to form water-insoluble, cross-linked networks [4,8,39–42]. The pore sizes of the cross-linked polysaccharide gels range in the hundreds of nanometers and are controlled to a large extent by the degree of cross-linking the polymer chains, the concentration of polymer in bulk solution, and the solvent medium used. Thus, thermally sensitive gels were prepared by cross-linking hydroxypropylcellulose (HPC) with poly-ethyleneglycol diglycidyl ether (PEGDE) at alkaline conditions in water/dimethylsulfoxide media [42]. The cross-linking reaction was performed both in homogeneous solutions and heterogeneous suspensions. The HPC gels prepared at heterogeneous conditions had a porous structure with interconnected channels and a higher swelling rate as compared to compact and nonporous structure of HPC gels obtained from homogeneous solutions [42].

6.3 Preparation of Macroporous Polysaccharide Gels via Freeze-Drying

The most widely used approach to form macroporous polymeric materials involve the freeze-drying technique when the polymer is cryoconcentrated between growing crystals during freezing of the polymer solution. Thin walls are formed between the macropores that are created during the sublimation of frozen solvent (most frequently, water or dioxane). Thus, HPC gels that have a fast response to thermal stimuli were prepared via freeze-drying and subsequent rehydration of the gels [43]. Water content was the key point to control the porosity and shrinking range in the HPC gels. Pores of different size ranges were produced by freezing the gels with different amounts of water at −20°C (Figure 6.1) [43]. A much faster shrinking process was observed for HPC gels prepared in the presence of higher amounts of water. Freeze-drying was also used for the preparation of gels prepared from cellulose fibers [44]. The fibers were suspended in a salt hydrate melt (hydrated calcium thiocyanate melt) and after gelation and washing it with water the system was freeze-dried (or supercritical drying was applied after washing the cellulose gel with ethanol) [44]. These drying methods were specially used in the production process aiming to create porosity in the materials prepared from fibrillar cellulose.

Freeze-drying is one of the most commonly used techniques for fabrication of porous polysaccharide scaffolds for tissue engineering with pore sizes up to 200 μm [27–30; Table 6.2]. Thus, the most traditional way to prepare macroporous alginate sponges includes three steps: cross-linking of the alginate with calcium ions, freezing the cross-linked alginate solutions, and removal of ice crystals by sublimation [12]. The resulting alginate scaffolds displayed 90% porosity with sizes of macropores sufficient for efficient cell seeding. The mechanical properties of the alginate scaffolds depended on

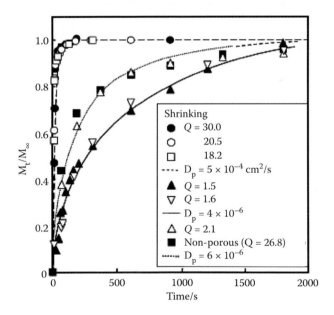

FIGURE 6.1
Kinetics of weight changes for the freeze-dried hydroxypropyl cellulose (HPC) gels. The gels contained different amounts of water ($Q = 1.5$–30; Q is the equilibrium swelling degree) were frozen at −20°C and then were freeze-dried. The gels were re-swollen in water and were equilibrated at 22°C. Time evolution of the gel weight was determined when the gel was rapidly transferred to a water bath at 60°C. (Reproduced from N. Kato, and S. H. Gehrke, *Colloids Surf. B: Biointerfaces*, 38, 191–96, 2004. With permission.)

the type of alginate, that is the ratio of guluronic (G) to mannuronic (M), residues. The effect of G content in alginate on porous properties of alginate sponges is in accordance with the theory of alginate gelation by ionic cross-linking. According to the egg-box model proposed by Grant et al. [45], the bivalent cations bridge the negatively charged guluronic acid residues on the alginate, and the mannuronic residues play only a subordinate role in establishing the gel framework. Thus, increasing the G content generally produces a matrix with increased pore size [46]. It was shown that the freezing regime during the fabrication of alginate scaffolds via freeze-drying technology was influenced significantly on porous properties of the alginate scaffolds [47]. Thus, the scaffolds prepared by slowly cooling an alginate solution in a freezer (–20°C) had spherically shaped interconnected pores, while the scaffolds prepared at –35°C or by freezing in liquid nitrogen (–196°C) that is with a steep temperature gradient across the solution, displayed complex directional pore structure [47]. The scaffolds with uniform interconnected pores prepared in a freezer had higher elasticity modulus as compared to gels with directional pore structure (Figure 6.2) [47]. The pore structure had strong influence on compressibility of the alginate scaffolds showing higher elasticity modulus for the scaffolds with uniform, interconnected pores prepared in a freezer (Figure 6.2) [47]. Incorporation of inorganic particles as hydroxyapatite (HAP) into the alginate allowed for the preparation of composite alginate/HAP scaffolds of increased mechanical strength [48]. A freeze-drying technique was used for the fabrication of superporous chitosan sponges [49] and agarose gels with uniaxial channels [30,50] intended for tissue engineering scaffolds.

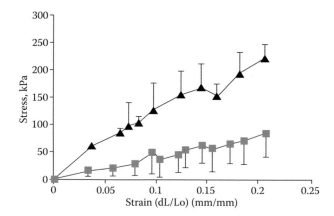

FIGURE 6.2
Strain-stress behavior of alginate scaffolds fabricated by freezing the alginate solution in a freezer (closed triangles) or in liquid nitrogen (closed squares). (Reproduced from S. Zmora, R. Glicklis, and S. Cohen, *Biomaterials*, 23, 4087–94, 2002. With permission.)

6.4 Preparation of Macroporous Polysaccharide Gels via Freeze-Extraction (Solvent-Exchange Phase Separation, SEPS)

During freeze-extraction (or cryo-extraction), a frozen polymeric solution is immersed into a nonsolvent with a low freezing temperature (hence liquid while the solvent is already frozen) to allow the exchange of frozen solvent for a liquid nonsolvent at a temperature lower than the freezing point of the polymer solution. This method is often referred to as *solvent-exchange phase separation* (SEPS) [15,16,28]. Such a freeze-extraction technique can extract the solvent from the polymeric structure. After that, the polymeric matrix is surrounded with nonsolvent and the simple air-drying or freeze-drying can be used to remove the nonsolvent. Alginate, chitosan, and starch-based macroporous matrices have been prepared via this approach [16,28]. Chitosan-based macroporous gels with pore sizes from 100 to 400 µm were prepared via freeze-extraction when water in 2.4% chitosan solution frozen at –15°C was exchanged for methanol containing aqueous ammonia (Figure 6.3a) [28]. However, if a chitosan solution of the same concentration was first precipitated from an alkaline medium and subsequently freeze-dried, chitosan gels with pores in the range of 0.1–1 µm were formed (Figure 6.3b) [28]. Thus, the pore size of chitosan macroporous gels varied in wide ranges by changing the mechanism of the formation of porous structures. The formation of ice crystals in the diluted chitosan solution induced the formation of a macroporous structure (Figure 6.3a), while the precipitation of the polymer matrix was the driving force in the formation of microporous chitosan structure (Figure 6.3b).

6.5 Macroporous Polysaccharide Gels Prepared via Compression Molding/Solvent Casting—Particle Leaching

Starch-based macroporous gels were prepared via compression molding and particle leaching methods. The method includes blending powders of starch-based polymer and salt (leachable particles of sizes from 50 to 1000 µm), mold-compression of the blend followed by salt leaching (typically by immersion of the samples into distilled water to dissolve the salt particles) [9,10,31]. Alternatively, starch polymers were first dissolved in an appropriate organic solvent and mixed with salt particles of different sizes, followed by solvent evaporation, and eventual salt leaching [31]. The prepared starch-based macroporous gels were designed for the tissue engineering application. By using these techniques it was possible to develop a wide range of

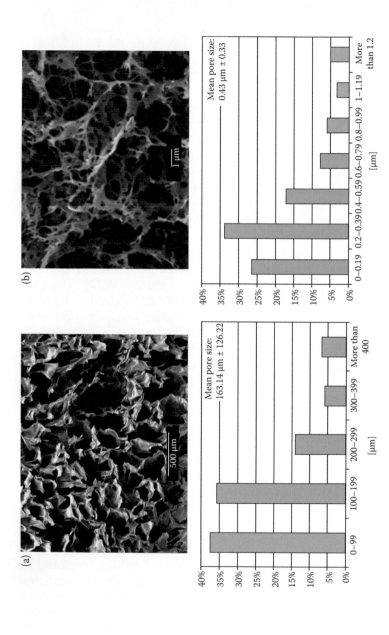

FIGURE 6.3

(a) SEM micrograph and pore size distribution of a chitosan scaffold frozen at −15°C and produced by solvent exchange phase separation (SEPS) showing macroporous anisotropic properties. (Reproduced from J. Nakamatsu, F. G. Torres, O. P. Troncoso, Y. Min-Lin, and A. R. Boccaccini, *Biomacromolecules*, 7, 3345–55, 2006. With permission. (b) SEM micrograph and pore size distribution of a chitosan scaffold produced by thermally induced phase separation (TIPS) showing micro- and nanoporous isotropic structure. (Reproduced from J. Nakamatsu, F. G. Torres, O. P. Troncoso, Y. Min-Lin, and A. R. Boccaccini, *Biomacromolecules*, 7, 3345–55, 2006. With permission.)

starch-based tissue engineering scaffolds. Porosity was controlled by the size and amount of the used leachable salt particles [31].

6.6 Preparation of Macroporous Polysaccharides via Double Emulsification Procedure

Superporous agarose matrices were prepared by a double emulsification procedure and had two sets of pores: normal diffusion pores (tens of nanometers in size) and flow through pores (several micrometers in size) [51]. The superporous agarose as a new chromatography medium was introduced by Gustavsson et al. [51,52] and was prepared in beaded form [32,53,54] or as continuous rods (monolith) [55]. The presence of flow through pores (superpores, which can be up to 30 μm in size) in the agarose improved mass transfer via these adsorbents compared to the conventional agarose-based sorbents. The preparation of macroporous agarose via the double emulsification procedure is discussed in detail in Chapter 7.

6.7 *In Situ* Forming Gels of Polysaccharides

One interesting approach to form macroporous polysaccharide gels is the self-cross-linking of polysaccharide chains (i.e., without using special cross-linking agents). Such *in situ* cross-linking gels are formed when mixing two polymers and the technology holds great promise in biomedicine. For this purpose, polysaccharide chains were modified accordingly usually by partial oxidation with sodium periodate, which resulted in the appearance of aldehyde groups in the macromolecule. Thus, *in situ* forming polymer scaffolds were prepared by cross-linking oxidized alginate with gelatin in the presence of small concentrations of borax [56] or by mixing hydrazide-modified carboxymethyldextran (CMDX-ADH) with aldehyde-modified dextran (DX-CHO) or aldehyde-modified carboxymethylcellulose (CMC-CHO) [57]. For all hydrogel formulations, higher concentrations of gel precursors gelled more rapidly (Figure 6.4). The gelation times for gels of carboxymethyldextran cross-linked with dextran (CMDX–DX) were shorter than for gels of carboxymethyldextran cross-linked with carboxymethylcellulose (CMDX–CMC) at the same concentration of polymers. These dextran-based in-situ formed (or injectable) gels had minimal cytotoxicity and demonstrated a good efficacy in preventing peritoneal adhesions [57].

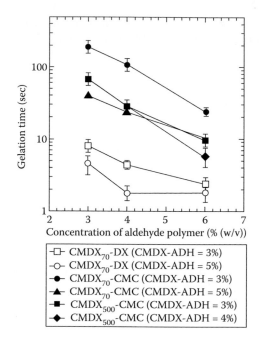

FIGURE 6.4
Gelation time for the gels prepared from different compositions of hydrazide-modified carboxymethyl dextran and aldehyde-modified carboxymethylcellulose. (Reproduced from T. Ito, Y. Yeo, C. B. Highleya, E. Bellasa, and D. S. Kohaneb, *Biomaterials*, 28, 3418–26, 2007. With permission.)

6.8 Preparation of Polysaccharide Macroporous Gels via Cryogelation (Polysaccharide Cryogels)

The cryogelation technique was extensively exploited for the investigations of the behavior of different polysaccharide systems at subzero temperatures [22,23,25,26,33–38,60,63–67,69]. Cryotropic gelation (or cryostructuration) is a specific type of gel formation that takes place as a result of the cryogenic treatment of systems that potentially are capable of gelation. The formation of a cryogel occurs in the partly frozen aqueous medium, when growing ice crystals perform as porogen which after melting creates macropores. The cryogels are produced via single or repeated freezing-thawing cycle(s). The freezing-thawing regime is used to obtain the desired porous and mechanical properties of cryogels [38,58,59,60,61]. Polysaccharide gel precursors are more complex as compared to low molecular weight precursors as monomers, the latter often used for the synthesis of cryogels. The conditions for the cryostructuration are strongly dependent on the physical-chemical properties of the polysaccharide polymer used. The decrease of apparent critical concentration of gelation for cryogels as compared to that for

(a)

(b)

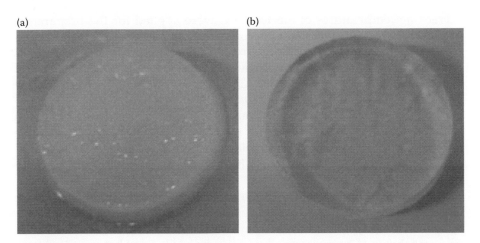

FIGURE 6.5
Typical appearance of (a) agarose cryogel sponge and (b) agarose conventional gel prepared from the same feed (2 wt % agarose solution).

conventional gels prepared from the same feedstock is a common phenomenon in cryotropic gelation [62].

Two main types of polysaccharide cryogel systems can be identified. The first type of polysaccharide cryogels is essentially formed during the thawing step of the frozen systems (thawing-induced cryogelation) and the second type is formed during the storage in the frozen state. The thawing-induced cryogelation is mainly typical for the physically cross-linked polysaccharide cryogels formed via the hydrogen bonding, for example, amylopectin, amylose, maldodextrin, xantan, and locust bean gum (LBG) [33,35,38,63,64].

The other type of cryogelation (when the cryogelation occurs during the storage in the frozen state) is mainly typical for the chemically cross-linked systems (when the polysaccharide gels are prepared via cross-linking with appropriate cross-linkers or free-radical polymerization) and for some physically cross-linked systems (e.g., agarose). These cryogels are often highly elastic sponges with well-defined macroporous structures. The typical appearance of a cryogel as compared to a conventional gel (prepared from a solution of the same concentration) is presented in Figure 6.5. The cryogels are opaque, elastic, often have a sponge-like morphology, and differ from the traditional gels, which are often transparent and rigid (Figure 6.5b).

6.8.1 Physically Cross-Linked Polysaccharide Cryogels

Aqueous solutions of many polysaccharides such as xantan, LBG, amylopectin, amylose, and maltodextrins undergo physical changes as a result of their freezing, frozen storage, and subsequent thawing of their aqueous dispersions resulting in the formation of physically (noncovalently) cross-linked networks [33,35,38,63,64].

Freezing temperatures of −18 to −20°C were applied for the preparation of polysaccharide cryogels. It was confirmed for a range of synthetic and natural polymers that lowering the freezing temperature from −10 to −20°C resulted in the increase in gelation yield and gel strength. However, further decrease of the temperature to −30°C (a temperature much lower than the glass transition temperature of the system [T_g']), resulted in losing mechanical stability of the cryogels. There is always a competition between the high polymer concentration (favorable for gelation) in the unfrozen microphase and the low mobility of the macromolecular chains and their segments arising from the low temperatures and high viscosity of the unfrozen microphase [35,38,62].

Noncovalent thermoreversible polymeric cryogels made of LBG, one of the important galactomannans, were prepared as result of gelation at subzero temperatures [33]. The nature of physical cross-linking in the LBG cryogels was shown to be numerous hydrogen bonds formed between hydroxyl groups of the polygalactomannan chains. Chaotropic agents like urea, when added to the level commonly used for strong suppression of hydrogen bonding in water-biopolymer systems (e.g., up to 4 M) decreased the strength of cryostructurates formed [33].

LBG cryogels of different morphologies have been prepared depending on the concentration of polymer in the initial solution. The soft sponges were produced from dilute (0.05–0.25 g/dL) LBG solutions, whereas cellular materials and nonspongy cryogels were formed in semiconcentrated (0.5–2 g/dL) and concentrated (> 2.5 g/dL) LBG solutions, respectively [33].

One of the inherent features of cryostructuration of the physically cross-linked polysaccharide gels is the dependence of their properties (mechanical and thermal stability of the final cryogels) on the thawing rate during defrosting of the frozen systems. A thawing rate of 0.03°C/min has been found optimal with regard to gelation yield, mechanical strength, and fusion temperature [33,35,59,63,64,65]. The LBG cryogels thawed at a lower rate showed higher elasticity modulus and fusion temperature compared to those cryogels produced when faster thawing of the frozen systems was exploited (Figure 6.6). The addition of sucrose, glucose, fructose, or sorbitol at concentrations of 40–60 wt % resulted in initial increases in the cryogel strength but further increasing the concentration of sugars decreased the cryogel strength [35]. The initial increase in cryogel strength was ascribed to the reduction in water content with increasing concentration of sugar; the subsequent decrease was ascribed to the inhibition of polymer–polymer association due to the binding of sugar molecules to the polymer chains, with differences in gel strength arising from the differences in binding strength [35].

A prolonged incubation of the frozen systems at temperatures in the range of −5 to 0°C (at which the system is above T_g' and several degrees below the ice melting point) provided conditions favoring the interactions between polymeric chains due to the increased mobility of the segments of polymeric chains [33,63–65]. A similar effect was shown for the system of

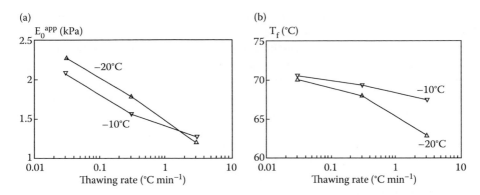

FIGURE 6.6
The dependencies of (a) rheological (apparent elastic modules, E_0^{app}) and (b) thermal (gel fusion temperature, T_f) properties of cryogels prepared from 20 g/L LBG aqueous solutions, which were frozen at –10 or –20°C, respectively and thawed at various rates. (Reproduced from V. I. Lozinsky, L. G. Damashkaln, R. Brown, and I. T. Norton, *J. Appl. Polym. Sci.*, 78, 371–81, 2000. With permission.)

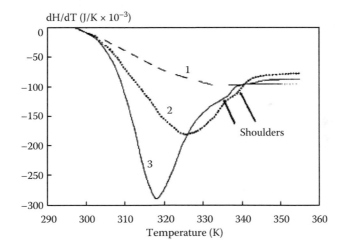

FIGURE 6.7
DSC profiles for the fusion of maltodextrin cryogels produced from 100 g/L maltodextrin solutions by freezing at –12°C for 18 h and thawing at rates of (1) 3.0°C/min, (2) 0.3°C/min, and (3) 0.03°C/min. (Reproduced from V. I. Lozinsky, L. G. Damshkaln, R. Brown, and I. T. Norton, *J. Appl. Polym. Sci.*, 83, 1658–67, 2002. With permission.)

maltodextrin-water, where the highest fusion enthalpy was found for the samples at the lowest thawing rate of 0.03°C (Figure 6.7). Parameters affecting the final properties of such systems include the composition of the starting material (concentration of polymer, molecular size) and the cryostructuration conditions such as the temperatures of freezing-thawing and the number of freeze-thaw cycles.

With an increasing number of freeze-thaw cycles, the strength of physically cross-linked cryogels was also increasing. Thus it was shown that after four freeze–thaw cycles applied during cryostructuration of LBG in the presence of 30% sucrose, the viscosity fluctuation becomes large, indicating that the elasticity became significant [58]. It was confirmed by many studies that the strength of the final cryogels was increased with an increasing number of freeze-thawing cycles [38,58–60]. During the freezing process, the polymers are excluded from the growing ice crystals, producing a polymer-rich region in the unfrozen phase, thus promoting the associations between LBG chains. Successive freeze–thaw cycles probably pushed the already-formed network together and that eventually lead to enhancement of the elasticity and hardness of cryogel matrices.

6.8.2 Chemically Cross-Linked Polysaccharide Cryogels

Basically, the same chemical routes, which are used for cross-linking of polysaccharide gels at ambient temperatures, can be exploited for the preparation of polysaccharide cryogels at subzero temperatures. Chemically cross-linked polysaccharide cryogels were prepared via cross-linking of agarose with epichlorohydrin in alkaline media [36], free-radical polymerization of methacrylate derivatives of dextran [22], cross-linking of chitosan with glutaraldehyde [36,66], and via UV-irradiation of the frozen solution of hydroxyethylcellulose (HEC) [37,67]. The pores in such cryogels varied from μm-sized pores [37,67] to hundreds of μm-sized pores [22,36,66]. Mechanical strength and elasticity depended on the used biopolymer system and production procedure. The HEC cryogels, prepared via UV irradiation of the frozen HEC solutions at –30°C, had higher values for the storage and loss moduli (G' and G'', respectively) as compared to those for the gels obtained from the same HEC solutions but irradiated at room temperature (Figure 6.8). The reason for that was the known cryoconcentration effect. Due to the concentration of HEC chains in the nonfrozen liquid microphase, the polymeric network of HEC cryogels was formed from the solution with higher HEC concentration, thus ensuring higher mechanical strength for the HEC cryogels compared to traditional gels.

6.8.3 Macroporous Polysaccharide Cryogels
with Pores up to 200 μm in Size

Polysaccharide-based macroporous cryogels with hundreds of μm-sized pores were prepared from polysaccharide precursors such as agarose, alginate, cellulose, dextran, and chitosan. Scanning electron microscopy (SEM) studies showed well-defined macroporous structures of large pores surrounded with thin and dense walls—all features typical for the macroporous material (Figure 6.9). The polysaccharide cryogels were formed via different routes (physical or chemical cross-linking, ionotropic complexation) depending on the properties of biopolymer.

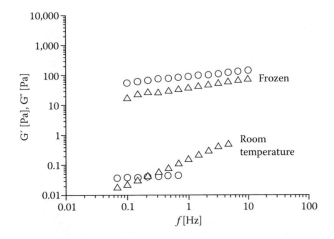

FIGURE 6.8
Variation of elasticity (open circles) and loss moduli (open triangles) in the 0.1–10 Hz frequency range for the gels prepared by UV-irradiation of 3 wt % aqueous hydroxyethyl cellulose at room temperature and in the frozen state. (Reproduced from P. Petrov, E. Petrova, R. Stamenova, C. B. Tsvetanov, and G. Riess, *Polymer*, 47, 6481–84, 2006. With permission.)

6.8.3.1 Agarose Macroporous Cryogels

The properties of agarose cryogels are, to a large extent, dependent on the thermal and chemical prehistory of the agarose polymer. Agarose is a gel forming polymer that is gelling at temperatures in the range of 36–40°C (depending on the type of agarose) due to the formation of numerous hydrogen bonds. Modification of agarose is known as a way to hinder the hydrogen bond formation to some extent. Derivatization of agarose with alkyl-derivatives as amphiphilic compounds with cationic functionalities (as 3-chloro-2-hydroxypropyldimethyl-dodecylammonium chloride, known as Quab 342) affected the formation and final properties (electroosmotic mobility and hydrophobicity) of agarose gel [68].

The physically cross-linked agarose cryogels prepared without any modification via cryogelation of aqueous solution of agarose at −35°C were shown to have irregular porous structure with channel-like pores in the direction of the cooling front from the outer surface to the center (Figure 6.10b) followed by spherically shaped macropores in the center of the cryogel (Figure 6.10a). The reason for such irregularities of the porous structure of agarose cryogels, most probably, was the formation of ice crystals in the already gelled agarose. Different approaches were used for decreasing the rate of hydrogen bond formation during the preparation of agarose cryogels [26,36]. One of the approaches was to use a pH shift in order to deprotonate agarose hydroxyl groups and thus influence the rate of the agarose gelation [26]. Time was needed then to reach the gelling point of 20°C, which was increased from 3.5 min (when the pH of the agarose solution was 7.0) to 30 min (when the pH

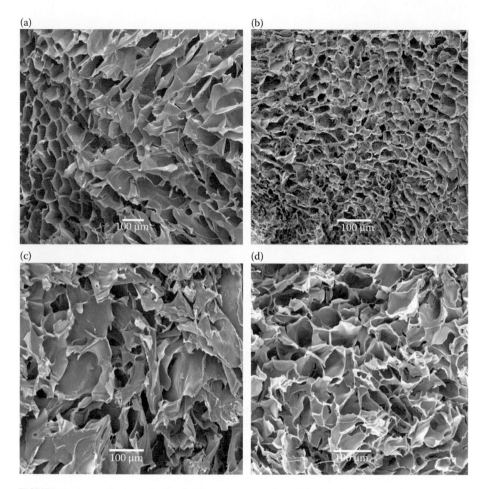

FIGURE 6.9
SEM images of the polysaccharide cryogels prepared from (a) agarose, (b) dextran, (c) alginate, and (d) chitosan. (From F. M. Plieva, I. Y. Galaev, and B. Mattiasson, Unpublished manuscript, 2008.)

of agarose solution was 12.6) [26]. The increase in polymer concentration in the gel phase of agarose cryogels compared to the concentration of agarose in bulk solution was studied for all cryostructuration conditions [26]. The agarose sponges, with more uniformly distributed macropores, were prepared in the presence of chaotropic agents (like urea) added to the reaction mixture at 3.5 M concentration required for suppressing the hydrogen bond formation [36]. Optical microscopy and SEM showed uniformly distributed pores of different sizes in the agarose scaffolds prepared in the presence of urea [36]. The regular macroporous structure formed in this case was probably due to the simultaneous formation of urea crystals, which affected the overall pattern of ice crystal formation.

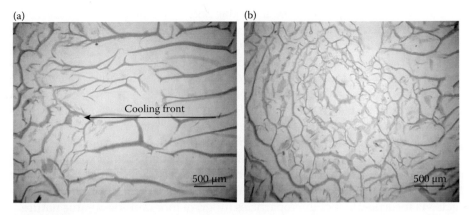

FIGURE 6.10
Optical microscopy images of a diametrical cross section of agarose cryogel sponges prepared at –35°C: (a) section from outer to intermediate part and (b) center part of the cryogel. Experimental conditions: agarose solution (3 wt %) was frozen inside a plastic tube (inner diameter 12 mm) at –35°C for 1 h followed by storage in the frozen state at –5°C for 10 h. The agarose sponges were stained with the dye Cibacron Blue (3.5 mg/ml in 0.05 M NaOH containing 4 M NaCl for 24 h) followed by sectioning into thin slices (10 μm) using a microtome. Samples were studied using a Nikon Labophot-2 light microscope.

Another approach to form macroporous agarose cryogels was via derivatization of agarose with allyl-glycidyl ether (AGE) in order to incorporate the polymerizable double bonds onto the agarose chain [36]. The derivatized AGE-agarose was used for the preparation of agarose sponges via free radical polymerization [36]. The typical structure of the agarose sponge prepared via AGE-derivatized agarose is shown in Figure 6.9a. The agarose sponges prepared via free-radical polymerization and chemical cross-linking with epichlorohydrin were more thermostable compared to the thermoreversible physically cross-linked agarose cryogel sponges [36]. The properties of agarose cryogels and gel surface chemistry can be modulated by the incorporation of another biopolymer, for example, gelatin and formation of cryogels with interpenetrating networks [69]. So far the macroporous agarose cryogels were used as 3D scaffolds for cell culture [25,26,69], potentially, however, they could be used as monolithic adsorption media.

6.8.3.2 Alginate Macroporous Cryogels

Alginate is a well-known example of a polymer that can be cross-linked by ionic interactions. Alginate cryogels were prepared via cross-linking with bivalent metal ions like Ca^{2+} [36], or they were prepared via cross-linking with amino-containing cross-linkers (ethylenediamine) in the presence of water-soluble carbodiimides (not published). The porous properties of alginate gels depended on the concentration of the alginate and the procedure of casting cryogels. Thus, for the preparation of alginate scaffolds via cross-linking with bivalent metal ions two approaches were used [36]. Soft and sponge-like alginate cryo-gels were prepared after freezing a 1 wt % solution of sodium alginate in the presence of $CaCl_2$ at $-20°C$ (Figure 6.9C). Alternatively, alginate cryogels were prepared via freeze-extraction as follows: the frozen solution of alginate was immersed at $-12°C$ into 95% ethanol containing 1% $CaCl_2$. Thus, the exchange of water to nonsolvent and simultaneous cross-linking of polymer chains with Ca^{2+} ions occurred with time. Such alginate cryogel sponges were character-ized with open porous structure, pore volume 91%, and oriented porosity [36]. It was shown that macroporosity, elasticity, and mechanical strength of the alg-inate scaffolds were effectively controlled via concentration of alginate in the bulk solution, formation of interpenetrating networks with other biopolymers (e.g., gelatin, chitosan, and agarose), and formation of composite alginate cryo-gels via incorporation of microparticles (filler) of inorganic or organic nature.

6.8.3.3 Chitosan Macroporous Cryogels

The aminopolysaccharide, chitosan, undergoes reactions typical for amines like N-acylation and Schiff base formation. Essentially all the methods used for the preparation of chitosan macroporous gels (cross-linking with glutaraldehyde, SEPS, and ionotropic gelation) can be exploited for the preparation of chitosan cryogels at subzero temperatures. The macroporous chitosan cryogels were pre-pared via cross-linking with glutaraldehyde at $-18°C$ [22,66] (Figure 6.9D). The porous and swelling properties of the chitosan cryogels depended on the degree of cross-linking with glutaraldehyde and the concentration of biopolymer.

6.8.3.4 Dextran Macroporous Cryogels

Dextran-based cryogels with well-defined macroporous structures were prepared via free-radical polymerization at subzero temperatures after modification of dextran with polymerizable double bonds [22]. The dextran-methacrylate (dextran-MA) macromer was obtained after the modification of dextran with glycidyl methacrylate in DMSO as a polar aprotic solvent in the presence of dimethylaminopyridine as catalyst (Figure 6.11). The polymer-ization of the dextran-MA was mainly dependent on the degree of substitu-tion (DS) measured as an average number of the polymerizable double bonds introduced into a dextran macromolecule. Dense and sponge-like cryogels were prepared from dextran-MA with DS 42 whereas using dextran-MA with DS 4 resulted in very weak cryogels [22]. Due to a well-defined macrop-orous structure of dextran cryogels (Figure 6.9b) they exhibit low flow resis-tance when arranged as monolithic columns [22]. The porous properties of the dextran cryogels were dependent on the molecular weight of the dextran macromer, DS with polymerizable double bonds, concentration of the mac-romer in the reaction mixture, and content of the cross-linker [22,23].

The dextran-MA can be easily copolymerized with different biocompatible monomers, for example, hydroxyethyl methacrylate or poly(ethylene glycol)

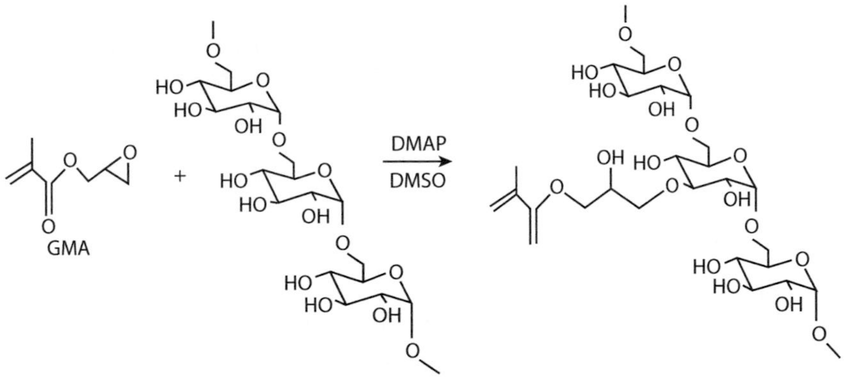

FIGURE 6.11
Modification of dextran with glycidyl methacrylate. (Reproduced from F. Plieva, A. Oknianska, E. Degerman, I. Y. Galaev, and B. Mattiasson, *J. Biomater. Sci. Polym. Edn.*, 17, 1075–92, 2006. With permission.)

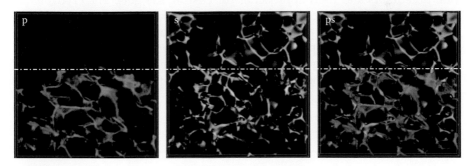

FIGURE 6.12
(See color insert following page 116.) Confocal laser scanning microscopy (CSLM) 3D reconstituted images of the secondary polyacrylamide (pAAm) cryogel formed inside the interconnected pores of the primary dextranmethacrylate (dextran-MA) cryogel. The dextran-MA was stained with Rhodamine B (red colored) and pAAm cryogel was stained with FITC (green colored). Image (p) for the dextran- MA cryogel stained with Rhodamine B, image (s) for the pAAm stained with FITC and image (ps) for the stained with both dyes dextran-MA with pAAm cryogels with gradient porosity. (Reproduced from F. M. Plieva, P. Ekström, I. Yu. Galaev, and B. Mattiasson, *Soft Matter*, 2008. With permission.)

acrylate or it can be modified with ECM components to prepare the dextran-based cryogel scaffolds for tissue engineering. The sequential freezing-thawing allowed for the preparation of polysaccharide cryogels consisting of two continuously interlacing cryogel networks with gradient porosities [61]. A new macroporous polyacrylamide cryogel (secondary cryogel) was formed inside the interconnected macropores of the already preformed dextran cryogel (primary cryogel) [61]. Confocal Laser Scanning Microscopy (CLSM) clearly demonstrated the macroporous structure of the dextran cryogel (red colored in the region below the dashed line on the image [ps], Figure 6.12) showing the inside interconnected macroporous structure of a secondary cryogel. This approach was shown to be efficient for the formation of macroporous polysaccharide cryogels with required gel surface chemistry.

6.9 Conclusion and Future Perspectives

Polysaccharides are abundant biopolymers with great potential in different areas. The versatility of chemical structures with known properties and chemistries, allowed for the development of advanced functionalized polysaccharide materials with required properties. Due to the unique properties (biodegradability, nontoxicity, hydrophilicity, and ease of chemical modifications), the polysaccharide materials have been widely exploited in the biomedical applications. The rational design of polysaccharide macroporous gels depends on the physical–chemical properties of the used polysaccharide precursors and used approaches for casting the gels. The cryogelation was shown to be an efficient technique for the formation of polysaccharide gels with well-defined macroporous structures. Compared to the other techniques used for the preparation of macroporous polysaccharide gels (e.g., freeze-drying), the cryogelation renders it possible to produce macroporous polysaccharide gels (cryogels) with wide ranges of porosities and in different geometrical formats.

References

1. A. S. Hoffman. 2002. Hydrogels for biomedical applications. *Adv. Drug Del. Rev.* 43:3–12.
2. J. L. Drury, and D. J. Mooney. 2003. Hydrogels for tissue engineering: Scaffolds design variables and applications. *Biomaterials* 24:4337–51.
3. M. A. Barbosa, P. L. Granja, C. C. Barrias, and I. F. Amaral. 2005. Polysaccharides as scaffolds for bone regeneration. *ITBM RBM* 26:212–17.
4. G. Crini. 2005. Recent development in polysaccharide-based materials used as adsorbents in wastewater treatment. *Progr. Polym. Sci.* 30:38–70.
5. S. Kuga. 1980. New cellulose gel for chromatography. *J. Chromatogr. A* 195:221–30.
6. G. Grznarova, S. Yu, V. Stefuca, and M. Polakovic. 2005. Quantitative characterization of pore structure of cellulose gels with or without bound protein ligand. *J. Chromatogr. A* 1092:107–13.
7. S. Jobling. 2004. Improving starch for food and industrial applications. *Cur. Opin. Plant Biol.* 7:210–18.
8. F. Delval, G. Crini, S. Bertini, C. Filiatre, and G. Torri. 2005. Preparation, characterization and sorption properties of crosslinked starch-based exchangers. *Carbohydr. Polym.* 60:67–75.

9   S. C. Mendes, R. L. Reis, Y. P. Bovell, A. M. Cunha, C. A. V. Blitterswijk, and J. D. D. Bruijn. 2001. Biocompatability testing of novel starch-based materials with potential application in orthopaedic surgery: A preliminary study. *Biomaterials* 22:2057–64.

10. M. E. Gomes, A. S. Ribeiro, P. B. Malafaya, R. L. Reis, and A. M. Cunha. 2001. A new approach based on injection moulding to produce biodegradable starch-based polymeric scaffolds: Morphology, mechanical and degradation behaviour. *Biomaterials* 22:883–89.

11. M. Singh, R. Sharma, and U. C. Banerjee. 2002. Biotechnological applications of cyclodextrins. *Biotechnol. Adv.* 20:341–59.

12. L. Shapiro, and S. Cohen. 1997. Novel alginate sponges for cell culture and transplantation. *Biomaterials* 18:583–90.

13. G. Miralles, R. Baudoin, D. Dumas, D. Baptiste, P. Hubert, J. F. Stoltz, E. Dellacherie, D. Mainard, P. Netter, and E. Payan. 2001. Sodium alginate sponges with or without sodium hyaluronate: *In vitro* engineering of cartilage. *J. Biomed. Mat. Res.* 57:268–78.

14. H. L. Lai, A. Abu'khalil, and D. Q. M. Craig. 2003. The preparation and characterisation of drug-loaded alginate and chitosan sponges. *Int. J. Pharm.* 251:175–81.

15. F.-L. Mi, S.-S. Shyu, Y.-B. Wu, S.-T. Lee, J.-Y. Shyong, and R.-N. Huang. 2001. Fabrication and characterization of a sponge-like asymmetric chitosan membrane as a wound dressing. *Biomaterials* 22:165–73.

16. M.-H. Ho, P.-Y. Kuo, H.-J. Hsein, T.-Y. Hsien, L.-T. Hou, J.-Y. Lai, and D.-M. Wang. 2004. Preparation of porous scaffolds by freeze-extraction and freeze-gelation methods. *Biomaterials* 25:129–38.

17. E. Khor, and L. Y. Lim. 2003. Implantable applications of chitin and chitosan. *Biomaterials* 24:2339–49.

18. S.-N. Park, J.-C. Park, H. O. Kim, M. J. Song, and H. Suh. 2002. Characterization of porous collagen/hyaluronic acid scaffold modified by 1-ethyl-3-(3-dimethylaminopropyl)carbodiimide cross-linking. *Biomaterials* 23:1205–12.

19. J. Neyts, D. Reymen, D. Letourneur, J. Jozefonvicz, D. Schols, J. Este, G. Andrei, et al. 1995. Differential antiviral activity of derivitized dextrans. *Biochem. Pharmacol.* 50:743–51.

20. W. E. Hennik, H. Talsma, J. C. H. Borchert, S. C. D. Smedt, and J. Demeester. 1996. Controlled release of proteins from dextran hydrogels. *J. Control. Release* 39:47–55.

21. Q. Cai, J. Yang, J. Bei, and S. Wang. 2002. A novel porous cells scaffold made of polylactide-dextran blend by combining phase-separation and particle-leaching techniques. *Biomaterials* 23:4483–92.

22. F. Plieva, A. Oknianska, E. Degerman, I. Y. Galaev, and B. Mattiasson. 2006. Novel supermacroporous dextran gels. *J. Biomater. Sci. Polym. Edn.* 17:1075–92.

23. N. Bölgen, F. M. Plieva, I. Y. Galaev, B. Mattiasson, and E. Piskin. 2007. Cryogelation for preparation of novel biodegradable tissue-engineering scaffolds. *J. Biomater. Sci. Polym. Edn.* 18:1165–79.

24. P. G. Righetti, and C. Gelfi. 1996. Electrophoresis gel media: The state of the art. *J. Chromatogr. B* 699:63–75.

25. K. Bloch, V. I. Lozinsky, I. Y. Galaev, K. Yavriyanz, M. Vorobeychik, D. Azarov, L. G. Damashkaln, B. Mattiasson, and P. Vardi. 2005. Functional activity of insulinoma cells (INS-1E) and pancreatic islets cultured in agarose cryogel sponges. *J. Biomed. Mater. Res.* 75A:802–9.

26. V. I. Lozinsky, L. G. Damshkaln, K. O. Bloch, P. Vardi, N. V. Grinberg, T. V. Burova, and V. Y. Grinberg. 2008. Cryostructuring of polymer systems. XXIX. Preparation and characterization of supermacroporous (spongy) agarose-based cryogels used as three-dimensional scaffolds for culturing insulin-producing cell aggregates. *J Appl. Polym. Sci.* 108:3046–62.

27. Y. Wan, Y. Fang, H. Wu, and X. Cao. 2007. Porous polylactide/chitosan scaffolds for tissue engineering *J. Biomed. Mater. Res.* 80A:776–89.

28. J. Nakamatsu, F. G. Torres, O. P. Troncoso, Y. Min-Lin, and A. R. Boccaccini. 2006. Processing and characterization of porous structures from chitosan and starch for tissue engineering scaffolds. *Biomacromolecules* 7:3345–55.

29. A. Nussinovitch, M. G. Corradini, M. D. Normand, and M. Peleg. 2001. Effect of starch, sucrose and their combinations on the mechanical and acoustic properties of freeze-dried alginate gels. *Food Res. Int.* 34:871–78.

30. S. Stocols, and M. H. Tuszynski. 2006. Freeze-dried agarose scaffolds with uniaxial channels stimulate and guide linear axonal growth following spinal cord injury. *Biomaterials* 27:443–51.

31. M. E. Gomes, J. S. Godinho, D. Tchalamov, A. M. Cunha, and R. L. Reis. 2002. Alternative tissue engineering scaffolds based on starch: Processing methodologies, morphology, degradation and mechanical properties. *Mater. Sci. Eng. C* 20:19–26.

32. P.-E. Gustavsson, and P.-O. Larsson. 1996. Superporous agarose, a new material for chromatography. *J. Chromatogr. A* 734:231–40.

33. V. I. Lozinsky, L. G. Damashkaln, R. Brown, and I. T. Norton. 2000. Study of cryostructuring of polymer systems. XIX. On the nature of intermolecular links in the cryogels of locust bean gum. *Polym. Int.* 49:1434–43.

34. A. Lazaridou, C. G. Biliaderis, and M. S. Izydorczyk. 2003. Molecular size effects on rheological properties of oat β-glucans in solution and gels. *Food Hydrocoll.* 17:693–712.

35. J. P. Doyle, P. Giannouli, E. J. Martin, M. Brooks, and E. R. Morris. 2006. Effect of sugars, galactose content and chain length on freeze-thaw gelation of galactomannans. *Carbohydr. Polym.* 64:391–401.

36. F. M. Plieva, I. Y. Galaev, and B. Mattiasson. 2008. Porous structure of macroporous polysaccharide gels prepared at subzero temperatures. Unpublished manuscript.

37. P. Petrov, E. Petrova, R. Stamenova, C. B. Tsvetanov, and G. Riess. 2006. Cryogels of cellulose derivatives prepared via UV irradiation of moderately frozen systems. *Polymer* 47:6481–84.

38. A. Lazaridou, and C. G. Biliaderis. 2004. Cryogelation of cereal β–glucans: Structure and molecular size effects. *Food Hydrocoll.* 18:933–47.

39. X. Zeng, and E. Ruckenstein. 1998. Cross-linked macroporous chitosan anion-exchange membranes for protein separations. *J. Membr. Sci.* 148:195–205.

40. B. P. Adrados, I. Y. Galaev, K. Nilsson, and B. Mattiasson. 2001. Size exclusion behavior of hydroxypropylcellulose beads with temperature-dependent porosity. *J. Chromatogr. A* 930:73–78.

41. J. C. Yu, Z.-T. Jiang, H.-Y. Liu, J. Yu, and L. Zhang. 2003. β-Cyclodextrin epichlorohydrin copolymer as a solid-phase extraction adsorbent for aromatic compounds in water samples. *Anal. Chim. Acta* 477:93–101.

42. E. Marsano, E. Bianchi, and A. Viscardi. 2003. Microporous thermally sensitive hydrogels based on hydroxypropyl cellulose cross-linked with poly-ethyleneglycol ether. *Polymer* 44:6835–41.

43. N. Kato, and S. H. Gehrke. 2004. Microporous, fast response cellulose ether hydrogel prepared by freeze-drying. *Colloids. Surf. B: Biointerfaces,* 38:191–96.
44. S. Hoepfner, L. Ratke, and B. Milow. 2008. Synthesis and characterization of nanofibrillar cellulose aerogels. *Cellulose* 15:121–29.
45. G. T. Grant, E. R. Morris, D. A. Rees, P. J. C. Smith, and D. Thom. 1973. Biological interactions between polysaccharides and divalent cations: the egg box model. *FEBS Lett.* 32:195–98.
46. J. P. Halle, D. Landry, A. Fourbier, M. Beaudry, and A. Leblond. 1993. Method for the quantification of alginate in microcapsules. *Cell Transplant.* 2:429–36.
47. S. Zmora, R. Glicklis, and S. Cohen. 2002. Tailoring the pore architecture in 3-D alginate scaffolds by controlling the freezing regime during fabrication. *Biomaterials* 23:4087–94.
48. H.-R. Lin, and Y.-J. Yeh. 2004. Porous alginate/Hydroxyapatite composite scaffolds for bone tissue engineering: preparation, characterization and *in vitro* studies. *Mater. Res. Part B: Appl. Biomater.* 71B:52–65.
49. Y.-J. Seol, J.-Y. Lee, Y.-J. Park, Y.-M. Lee, Y. Ku, I.-C. Rhyu, S.-J. Lee, S.-B. Han, and C.-P. Chung. 2004. Chitosan sponges as tissue engineering scaffolds for bone formation. *Biotechnol. Lett.* 26:1037–41.
50. S. Stokols, and M. H. Tuszynski. 2004. The fabrication and characterization of linearly oriented nerve guidance scaffolds for spinal cord injury. *Biomaterials* 25:5839–46.
51. P.-E. Gustavsson, A. Axelsson, and P.-O. Larsson. 1998. Direct measurement of convective fluid velocities in superporous agarose beads. *J. Chromatogr. A* 795:199–210.
52. P.-E. Gustavsson, A. Axelsson, and P.-O. Larsson. 1999. Superporous agarose as a hydrophobic interaction chromatography support. *J. Chromatogr. A* 830:275–84.
53. P.-E. Gustavsson, K. Mosbach, K. Nilsson, and P.-O. Larsson. 1997. Superporous agarose as an affinity chromatography support. *J. Chromatogr. A* 776:197–203.
54. Q.-H. Shi, X. Zhou, and Y. Sun. 2005. A novel superporous agarose medium for high-speed protein chromatography. *Biotechnol. Bioeng.* 92:643–51.
55. P.-E. Gustavsson, and P.-O. Larsson. 1999. Continuous superporous agarose beds for chromatography and electrophoresis. *J. Chromatogr. A* 832:29–39.
56. B. Balakrishnan, and A. Jayakrishnan. 2005. Self-cross-linking biopolymers as injectable *in situ* forming biodegradable scaffolds. *Biomaterials* 26:3941–51.
57. T. Ito, Y. Yeo, C. B. Highleya, E. Bellasa, and D. S. Kohaneb. 2007. Dextran-based *in situ* cross-linked injectable hydrogels to prevent peritoneal adhesions. *Biomaterials* 28:3418–26.
58. C.-F. Mao, and J.-C. Chen. 2006. Interaction association of locust bean gum in sucrose solutions: an interpretation based on thixotropic behavior. *Food Hydrocoll.* 20:730–39.
59. H. Vaikousi, and C. G. Biliaderis. 2005. Processing and formulation effects on rheological behavior of barley β-glucan aqueous dispersions. *Food Chem.* 91:505–16.
60. A. Lazaridou, H. Vaikousi, and C. G. Biliaderis. 2008. Effects of polyols on cryo-structurization of barley β-glucans. *Food Hydrocoll.* 22:263–77.
61. F. M. Plieva, P. Ekström, I.Yu. Galaev, and B. Mattiasson. 2008. Monolithic cryo-gels with open porous structure and unique double-continuous macroporous networks. *Soft Matter* 4:2418–28.

62. V. I. Lozinsky. 2002. Cryogels on the basis of natural and synthetic polymers: preparation, properties and applications. *Russ. Chem. Rev.* 71:489–511.
63. V. I. Lozinsky, L. G. Damashkaln, R. Brown, and I. T. Norton. 2000. Study of cryostructuration of polymer systems. XVIII. Freeze-thaw influence on water-solubilized artificial mixtures of amylopectin. *J. Appl. Polym. Sci.* 78:371–81.
64. V. I. Lozinsky, L. G. Damashkaln, R. Brown, and I. T. Norton. 2000. Study of cryostructuration of polymer systems. XVI. Freeze-thaw-induced effects in the low concentration systems amylopectin-water. *J. Appl. Polym. Sci.* 75:1740–48.
65. V. I. Lozinsky, L. G. Damshkaln, R. Brown, and I. T. Norton. 2002. Study of cryo-structuration of polymer systems. XXI. Cryotropic gel formation of the water-maltodextrin systems. *J. Appl. Polym. Sci.* 83:1658–67.
66. H. Kirsebom, M. R. Aguilar, and J. S. Roman. 2007. Macroporous scaffolds based on chitosan and bioactive molecules. *J. Bioact. Comp. Polym.* 22:621–36.
67. P. Petrov, E. Petrova, B. Tchorbanov, and C. B. Tsvetanov. 2007. Synthesis of biodegradable hydroxyethylcellulose cryogels by UV irradiation. *Polymer* 48:4943–49.
68. A. Maruska, and O. Kornysova. 2004. Homogeneous reversed-phase agarose thermogels for electrochromatography. *J. Chromatogr. A* 1044:223–27.
69. A. Tripathi, N. Kathuria, and A. Kumar. 2009. Elastic and macroporous agarose-gelatin cryogels with isotropic and anisotropic porosity for tissue engineering. *J. Biomed. Mat. Res. Part A* 90A:680–94.
70. K. Lindenhayan, C. Perka, R. S. Spitzer, N. H. Heilmann, K. Pommerening, J. Menniske, and M. Sit. 1999. Retention of hyaluronic acid in alginate beads: Aspects for *in vitro* cartilage engineering. *J. Biomed. Mater. Res.* 44:149–55.
71. P. Gacesa. 1988. Alginates. *Carbohydr. Polym.* 8:161–82.
72. C.-T. Chiu, J.-S. Lee, C.-S. Chu, Y.-P. Chang, and Y.-J. Wang. 2008. Development of two alginate-based wound dressings. *J. Mater. Sci: Mater. Med.* 19:2503–13.
73. E. Amici, A. H. Clark, V. Normand, and N. B. Johnson. 2000. Interpenetrating net-work formation in gellan-agarose gel composites. *Biomacromolecules* 1:721–29.
74. M. Boissiere, A. Tourrette, J. M. Devoisselle, F. D. Renzo, and F. Quignard. 2006. Pillaring effects in macroporous carrageenan–silica composite microspheres. *J. Colloid. Interface Sci.* 294:109–16.
75. C. C. Leffler, and B. W. Muller. 2000. Influence of the acid type on the physi-cal and drug liberation properties of chitosan-gelatin sponges. *Int. J. Pharm.* 194:229–37.

7

Superporous Agarose Gels: Production, Properties, and Applications

Per-Erik Gustavsson, Peter Tiainen, and Per-Olof Larsson

CONTENTS

7.1 Introduction

Agarose is a well-established material in many applications of biochemistry and biotechnology. Agarose gels are used for electrophoresis of very large molecules such as DNA and agarose-based particles are used as chromatographic media for separation of all sorts of biomolecules, particularly proteins. Prominent commercial examples of agarose-based separation particles are exemplified by the brand names Sepharose, Uniflow, Affi-gel, and Superose. The chromatography particles are available for many modes of chromatography, for example, gel filtration, ion exchange chromatography, hydrophobic interaction chromatography, and affinity chromatography.

Chemically, agarose is a polysaccharide material prepared from naturally occurring agar that is obtained from seaweed. It consists of long polymer chains with a repeating unit made up of D-galactose and 3,6-anhydro-L-galactose (Medin 1995). The pore size of agarose gels depends on the agarose content. Beads with a 6% agarose content are frequently used in chromatography applications and have an average pore size of approximately 30 nm, whereas 4 and 2% agarose beads have a pore size of 70 and 150 nm, respectively (Sofer and Hagel 1997). Agarose with even larger diffusion pores are used in gel electrophoresis to allow the passage of very large molecules such as DNA.

The attractiveness of agarose as a separation medium and as a support for biomolecules is based on several factors. It is easy to obtain agarose gels with large pores suitable for biochromatography and an appreciated property is their hydrophilicity and inertness toward proteins and other biomolecules (low unspecific protein adsorption, etc.). Nevertheless, it is easy to functionalize the agarose structures and to introduce ion exchange groups, affinity ligands and to immobilize proteins, and so forth. Especially in an industrial setting the compatibility of agarose with strong alkali for sanitation purposes is greatly appreciated.

The trend in bioseparation toward high-performance separation calls for smaller and smaller particles. Unfortunately this comes at a cost, namely the pressure drop over the separation column that according to Darcy's law (Coulson et al. 1991) is proportional to the squared inverse of the particle diameter ($\Delta_p = k/d_p^2$). Thus, reducing the particle diameter with a factor of 5 increases the pressure drop with a factor of 25, a fact that has made it difficult to use agarose as a base material in high performance systems due to its limited pressure resistance.

One way to improve the chromatographic performance of agarose gels without having problems with high backpressure is described here. Instead of reducing their size, the particles are provided with superpores that are large enough to allow part of the chromatographic flow to pass through each individual particle. This is a real bonus for the mass transfer, since the pore flow will transport the biomolecules rapidly to the interior of the particles, leaving only very short distances to be covered by diffusion, which is a much slower process than convective flow. Thus, superporous particles combine the beneficial properties of small, standard particles (rapid equilibration with the mobile phase) with the beneficial properties of large particles (low pressure drop, easier handling; Afeyan et al. 1990; Gustavsson and Larsson 1996; Heeter and Liapis 1996; McCoy et al. 1996; Rodrigues 1997; Nash and Chase 1998). Figure 7.1 gives an immediate understanding of the pore flow concept.

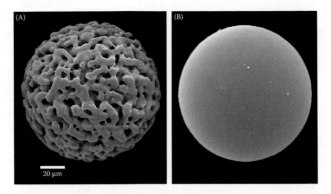

FIGURE 7.1
SEM picture of superporous (A) and homogeneous (B) agarose beads.

It is easy to envisage how a substantial part of the chromatographic flow will pass through beads provided with large superpores and at the same time transport the molecules of interest to the interior of the bead. The pore flow concept in beads has been described extensively for materials other than agarose, for example, polystyrene (Afeyan et al. 1990) and has also resulted in commercial products (POROS supports based on polystyrene [Perceptive Biosystems]). Several other available separation particles should also exhibit pore flow as judged from the ratio between their pore diameter and particle diameter. Examples are porous silica such as LiChrospher Si 4000 (Merck) and the polystyrene-based Source (GE Healthcare). Chapter 4 in this volume covers so-called cryogels that contain flow pores and also may be prepared in bead form from a variety of base materials, including agarose.

The superporous agarose described here is suitable for preparation of beads as well as monoliths. It can be used as chromatography support for separation and isolation of a wide range of molecules and particles (small molecules, proteins, plasmids, viruses, and cells). It is an excellent support for immobilization of enzymes and it can be a functional part of composite materials used for a variety of purposes.

7.2 Preparation of Superporous Gels

7.2.1 Superporous Particles

Superporous agarose particles are prepared by a double emulsification procedure (Gustavsson and Larsson 1996; Larsson 1998). Many variations (choice of solvents and surfactants, concentration of agarose, volume ratios, temperature, stirring rate, etc.) are possible. The variations will influence the properties of the superporous beads (particle size, porosity, superpore diameter). To achieve consistent results the manufacturing conditions must be carefully controlled. It should be pointed out that the procedures involve flammable solvents at elevated temperature, calling for safety measures. A typical recipe is as follows:

Step 1. Warm aqueous agarose solution (100 ml; 6% w/v) is mixed in a thermostated reactor (60°C) with 50 ml of a water-immiscible solvent (cyclohexane) containing an oil-in-water surfactant (3 ml of Tween 80). The mixture is stirred vigorously (1000 rpm) and a thick emulsion is formed (Emulsion 1). The precise recipe influences the properties of the final beads. For example, a lower stirring rate and a higher surfactant concentration increase the superpore diameter. The quality of Emulsion 1 may be checked under the microscope provided with a heated stage. Two main phases should be visible: a continuous agarose phase and a continuous organic phase.

Step 2. Emulsion 1 is now poured into 300 ml of a thermostated solution of cyclohexane (60°C) containing 12 ml of an water-in-oil surfactant (Span 85) while stirring (600 rpm). Droplets of Emulsion 1 are now formed (Emulsion 2). The stirring rate influences the size of droplets (higher stirring rate gives smaller droplets and ultimately smaller bead size).

Step 3. Emulsion 2 is now cooled while stirring. At a temperature below 40°C the agarose solidifies and the superporous beads are formed.

Step 4. The obtained superporous beads are washed with water, ethanol-water (50%) and water. The particles are finally sized in different fractions (e.g., 108–180 micrometer) by wet sieving using graded metal screens.

7.2.2 Superporous Monoliths

Superporous monoliths are easy to prepare and may be obtained in many different shapes as exemplified in Figure 7.2.

Step 1. Emulsion 1 is prepared as described for superporous particles (see above steps).

Step 2. Emulsion 1 is poured into a thermostated mold, (e.g., a stoppered chromatography column; Figure 7.3). In some instances it is preferable to make a final, slow stirring of Emulsion 1 after its transfer to the mold. The reason is that pouring the emulsion may lead to some stratification of the superpores, ultimately resulting in uneven flow through the chromatography column. In other instances the stratification of the superpores may be desired,

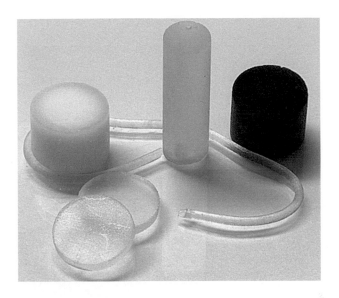

FIGURE 7.2
Examples of monolithic agarose: rods, membranes, and fibers. The black rod is a composite containing carbon black as active ingredient. The thick membranes were prepared by casting the superporous agarose between two glass plates and then punching out circular membranes. The spaghetti-like fiber was prepared by filling plastic tubing with Emulsion 1 (see text), cooling the tubing, and then pressing out the fiber with tap water. (From Gustavsson, P.-E., and Larsson, P.-O., *J. Chromatogr. A*, 832, 29–39, 1999. With permission.)

   making transport faster in one direction compared to that in the perpendicular direction.

Step 3. The mold is cooled and the agarose will solidify and form a superporous monolith.

Step 4. The ends of the monolith are trimmed, flow adaptors are attached, and the organic phase is washed away by pumping water, ethanol-water (50%) and water through the bed.

 Large monolithic gels (diameter > 3 cm) may suffer from uneven superpore diameter, namely larger pores in the center of the gel. This is normally undesirable, since it will lead to an uneven flow profile through the monolith—the fastest flow will occur in the center. The reason for the large pores in the center is as follows. In step 3 the mold is cooled and the agarose solidifies. However, the cooling requires transport of heat, a process that is not instantaneous. Especially in large diameter monoliths (above 3 cm), the center part will solidify considerably later than the outer part. This will influence the superpore diameter, since Emulsion 1 is inherently unstable and as soon as the stirring ceases, the two phases will start to separate, a process that continues until gelling has occurred. The longer the time before gelling, the larger the superpores will become. Thus, the center part of the gel will contain larger superpores than

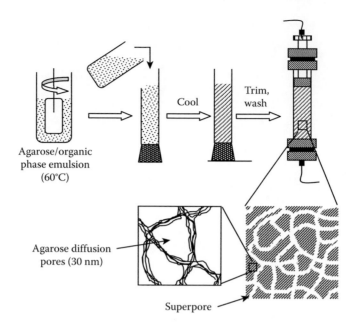

FIGURE 7.3
Preparation of superporous agarose monoliths. The details of the process are given in the text. (From Gustavsson and Larsson, *J. Chromatogr. Library.*, 67, 121–141, 2003. With permission.)

the peripheral part. One way of avoiding the consequences of this problem is to operate large monolithic gels in a radial mode (Gustavsson and Larsson 2001). An alternative remedy is to prepare thick, circular (0.5 cm) superporous membranes and stack them to a suitable height in a column (Gustavsson and Larsson 1999).

It is interesting to note also that the preparation of other monolithic materials may have similar problems with flow pore homogeneity. Thus, the manufacture of convention interaction media (CIM) discs (see Chapter 12) involves the polymerization of methacrylate monomers, a process that generates a lot of heat that must be effectively dissipated to avoid inhomogeneous pores. Also the production of cryogels (see Chapter 2) relies on the effective transport of heat to get an even growth of pore-forming ice crystals.

7.2.3 Superporous Composite Gels

It is easy to prepare composite materials based on superporous agarose. In case of superporous beads the extra component is mixed with Emulsion 1, a prerequisite being that the extra component should spontaneously localize to the agarose phase or possibly to the interface between the agarose and the organic phase. Thus, it is possible to prepare superporous beads loaded with smaller, functional particles (e.g., carbon black, ion-exchange particles, hydrophobic particles, or affinity particles). The ultimate composite will harbor the functional particles behind a thin layer of protecting agarose, preventing the functional particles to come into direct contact with the chromatographic stream. Substances to be adsorbed must therefore first diffuse through a thin agarose layer before binding can occur. This will prevent very large molecules and cells to contact the adsorbent—a restricted access material (RAM).

Similarly, monolithic agarose beds can be provided with additional elements giving them improved properties. For example, Emulsion 1 may be cast in the presence of a structure that improves the stability of the resulting composite or provides it with special properties (e.g., electric conductivity), which is of interest for sensor applications (Khayyami et al. 1997).

7.3 Characterization of Superporous Gels

As indicated above, the properties of superporous gels are very much dependent on their composition as well as the conditions used during manufacture, for example the stirring speed. Any change in the manufacturing protocol might change the final properties of the gel. Thus, it is prudent to have a toolbox of techniques for characterization of the superporous gels. Interesting parameters are bead diameter, superpore diameter, superpore shape, distribution of superpores, pore flow, bed stability, and, of course, chromatographic performance. Several techniques are used for such characterization, notably some imaging techniques.

7.3.1 Microscopy Characterization

An immediate view of the properties of superporous agarose beads may be conveniently obtained with a light microscope (Gustavsson and Larsson 1996). Since superporous beads often have a diameter around 100 micrometers, the superpores will usually have a diameter of 5–10 micrometers and the pores are fairly easy to observe. Nevertheless, it is beneficial to improve the contrast between the liquid-filled pores and the surrounding agarose phase. One way to achieve this is to allow the particles to dry slightly before examination. Thus, it is easy to get an immediate check of the fundamental properties of the gel beads: particle diameter/distribution as well as a rough estimate of the superpore diameter. Figure 7.4 gives two examples of light microscopy images of superporous gels, particles, and a slice of monolithic agarose. To obtain higher resolution images of the superpore architecture in beads and in monoliths, an electron microscopy is necessary, see Figure 7.1. However, imaging of low percentage agarose structures is notoriously difficult due to the high water content. Artifacts and distortion of the structure are easily introduced during sample preparation. One rather laborious way

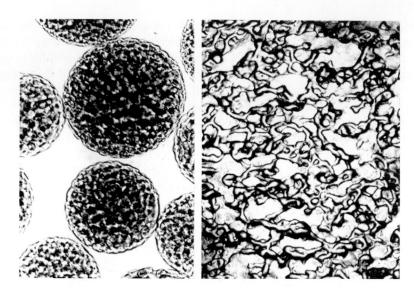

FIGURE 7.4
Light microscope pictures of superporous agarose beads (left panel) and the surface of a superporous membrane (right panel).

to circumvent these difficulties is illustrated in Figure 7.5 (Gustavsson and
Larsson 1996). Here the superpores of a monolithic agarose gel were filled
with an acrylate monomer mix that was hydrophobic enough not to enter the
agarose phase. The monomers were allowed to polymerize. Then the composite was immersed in hot water, causing the agarose to melt away, leaving
an acrylate replica suitable for electron microscopy. The figure seems to indicate that the pores in the sample were not completely random but stratified
in two directions.

7.3.2 Measuring Pore Flow

The performance of superporous particles depends on the pore flow. If the
flow is too low the particles will behave as standard particles. In one sense
the ideal flow velocity inside the particle should be the same as outside the
particle. However, this is unrealistic since this would require flow pores with
the same diameter as the interstitial pores (i.e., the pores formed between
the particles in the chromatographic bed). A more realistic superpore flow
velocity is 1–10% that of the interstitial pores. It can easily be shown that the
relative flow velocity (i.e., superpore flow velocity/interstitial pore velocity),
is proportional to the squared superpore diameter/interstitial pore diameter
ratio (Gustavsson, Axelsson, and Larsson 1998)

$$\text{Relative velocity (\%)} = (d_s/d_i)^2 \times 100$$

$(d_s = \text{superpore diameter}; d_i = \text{interstitial pore diameter}).$

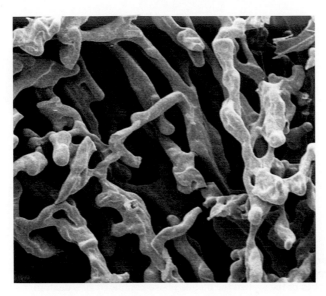

FIGURE 7.5
A SEM picture of a replica of the superpores in superporous agarose. The distance across the picture is approximately 200 μm.

Knowing the superpore size it is possible to calculate the superpore fluid velocity and vice versa (Gustavsson, Axelsson, and Larsson 1998). Also important for a good performance is an even distribution of the superpores. From the method of preparation and from images (Figures 7.1, 7.4, 7.5) one could assume an even distribution.

It is possible to carry out experiments to directly measure superpore flow velocities and the distribution of flow paths. Figure 7.6 illustrates such an experiment. A special column was packed with a mixture of superporous beads and standard beads. A microscope provided with a video camera was focused on particles four layers down into the bed and the flow inside and outside the superporous particles were measured by observing the movement of suspended microparticles (dyed yeast cells) through the column. The results were most convincing. Some microparticles entered the superporous particles and continued inside the superpores along a tortuous path until they emerged on the other side and were swept away. Figure 7.7 illustrates this further. In the left panel the location of one microparticle is shown at one second intervals. As expected the fluid velocity outside the superporous particle (A, E) is considerably higher than inside (C) or close to the particle surface (B, D). In the right panel the flow path of 12 different microparticles are shown. The distribution of flow channels is reasonably even. Clearly, only flow paths essentially parallel to the gross direction of flow will be discovered in this way.

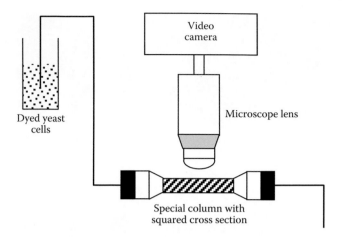

FIGURE 7.6

Direct measurement of pore flow in superporous agarose particles. The optical system was focused on beads four layers down in the bed to avoid column wall effects. To improve the optical properties of the system the mobile phase contained 20% dextran, making the super-porous beads more transparent (matching refractive index). Pore distribution, pore flow veloc-ity, and pore turtuosity was determined from video recordings showing the flow paths of dyed yeast cells. (From Gustavsson, P.-E., Axelsson, A., and Larsson, P.-O., *J. Chromatogr. A*, 795, 199–210, 1998. With permission.)

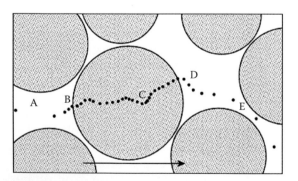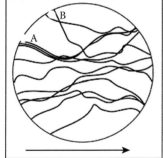

FIGURE 7.7

Direct measurement of pore flow in a superporous bead. The left panel shows the travel of one yeast cell during a 40-second period. The location of the cell is given at one second inter-vals. The right panel shows the travel paths of 12 yeast cells through one 300 μm superporous bead. Due to the limitation of the depth of focus only flow paths in the center of the bead are shown. Flow paths that are almost coinciding are probably situated in the same superpore. The arrow indicates the longitudinal column direction. (From Gustavsson, P.-E., Axelsson, A., and Larsson, P.-O., *J. Chromatogr. A*, 795, 199–210, 1998. With permission.)

7.3.3 Chromatographic Characterization Height Equivalent to a Theoretical Plate (HETP)

A convincing way of proving the presence of a pore flow in the superpores is to check the chromatographic performance of a few differently sized molecules at different flow rates. The performance is measured by the HETP value (height equivalent to one theoretical plate; plate height). Examples of suitable substances to chromatography are

- A low molecular weight substance, for example, sodium azide (high diffusion coefficient, full access to the superpores as well as the agarose phase).
- A high molecular weight substance, for example, bovine serum albumin (low diffusion coefficient, full access to superpores, good access to the agarose phase).
- A very high molecular weight substance, for example, 0.4 micrometer latex particles (very low diffusion coefficient, full access to the superpores, no access to the agarose phase).

Chromatography of such substances at a few flow rates should give characteristic results (Gustavsson and Larsson 1996) as shown in Figure 7.8, where superporous particles are compared with homogeneous particles of the same size. The left panel shows that superporous particles perform better than homogeneous particles, but not dramatically so with a low molecular weight substance such as NaN$_3$. Low molecular weight substances diffuse rapidly and the advantage gained by the superpores is therefore limited. In the center panel 0.4 micrometer latex particles are chromatographed. They are so large that they will not enter the agarose phase. Thus, no benefit is obtained from the superpores, since any

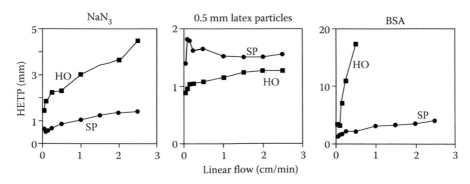

FIGURE 7.8
Size-exclusion chromatography of low molecular weight (left panel), very high molecular weight (middle panel), and high molecular weight (right panel) substances. HETP is given as a function of the linear flow velocity. SP = superporous agarose beads, HO = homogeneous agarose beads. (From Gustavsson, P.-E., and Larsson, P.-O., *J. Chromatogr. A*, 734, 231–40, 1996. With permission.)

diffusion in the agarose phase is not on the agenda. In fact, the superpores give some additional band spreading (higher HETP value), since the microparticles that enter the superpores migrate slower than microparticles traveling outside the particles. In the right panel, on the other hand, a dramatic benefit is seen with the superporous particles. Here a protein is chromatographed, having a low diffusion coefficient (additionally diminished in agarose). Especially at higher flow rates, the superporous beads are many times more efficient than the homogeneous beads. Clearly, here the shorter diffusion distance offered by the superporous beads makes a huge difference in chromatographic performance.

A telltale behavior like the one depicted in Figure 7.8, will be strong proof of the presence of superpores and pore flow.

7.4 Derivatization of Superporous Gels

There exists a plethora of methods for providing agarose with ligands of various kinds (Porath and Axén 1976; Carlsson, Janson, and Sparrman 1989). All are applicable to superporous agarose particles. In rare cases some precautions must be observed though, since the superpores induce a certain mechanical weakness. Thus, when introducing a high density of ion exchange ligands, the repulsive forces presented at low ionic strength may disrupt the superporous particles if no countermeasures are taken. This is in contrast to the corresponding homogeneous particles that usually will react only by swelling. A standard countermeasure is to apply a cross-linking agent (e.g., epichlorhydrin) prior to the introduction of the ion exchange ligand (Tiainen et al. 2007).

7.5 Applications of Superporous Agarose

Superporous agarose beads have been used for chromatography-separation of a variety of molecules and particles. In most cases the primary objective has been to use the beneficial effect of short residual diffusion distances in the agarose phase. In other cases the large superpores have been used for chromatography of particles such as cells or very large molecules such as plasmids, in which case the superpore walls themselves have been utilized as an extended adsorption surface. Other applications involve the use of superporous agarose as a support for immobilized enzymes or as a part of composite materials. Tables 7.1 and 7.2 give a number of examples. Table 7.1 refers to chromatographic applications, while Table 7.2 refers to other applications (e.g., biosensors).

TABLE 7.1

Chromatographic Applications of Superporous Agarose

Format	Mode	Ligand	Target Substance(s)	References
Beads	GPC	—	Particles, BSA, N_aN_3	Gustavsson and Larsson 1996
Beads	IEX	PEI	BSA, OVA	Gustavsson and Larsson 1996
Beads	IEX	PEI, QA	Plasmids	Tiainen et al. 2007
Beads	—	—	yeast cells	Gustavsson, Axelsson, and Larsson 1998
Beads	AC	anti-Factor VIII	Factor VIII	Pålsson et al 1999
Beads	AC	NAD-analog	LDH	Gustavsson et al. 1997
Beads	AC	WGA/Red Blood Cells	Cytochalasin B	Gottschalk et al. 2003
Beads	AC	Protein A	IgG	Gustavsson et al. 1997
Beads	HIC	Phenyl	RNase, Lyz, BSA	Gustavsson, Axelsson, and Larsson 1999
Monolith	AC	NAD-analog	LDH	Gustavsson and Larsson 1999
Monolith	AC	Cibacron Blue	LDH	Gustavsson and Larsson 2001
Monolith composite	HA	Hydroxyapatite particles	BSA, Lyz, CytC	Gustavsson and Larsson 1999

Abbreviations: AC = Affinity chromatography, AcChE = Acetylcholine esterase, BSA = Bovine serum albumin, GPC = Gel permeation chromatography, CytC = Cytochrome C, HIC = Hydrophobic Interaction Chromatography, IEX = Ion exchange chromatography, LDH = Lactate dehydrogenase, Lyz = Lysozyme, OVA = Ovalbumin, PEI = Polyethyleneimine, QA = Quaternary amine, RNase = Ribonuclease, RVC = Reticulated vitreous carbon.

TABLE 7.2

Nonchromatographic Applications of Superporous Agarose

Format	Application	Active Components	Reference
Monolith	Adsorbent in FIA system	Anti-β-galactosidase	Nandakumar et al. 2000
Monolith	Lactose conversion	Lactase	Gustavsson and Larsson 2001
Monolith	Electrophoresis	Internal cooling by solvents	Gustavsson and Larsson 1999
Monolithic composite	Biosensor	AcChE, RVC	Khayyami et al. 1997

Abbreviations: AcChE = Acetylcholine esterase, FIA = Flow injection analysis, RVC = Reticulated vitreous carbon.

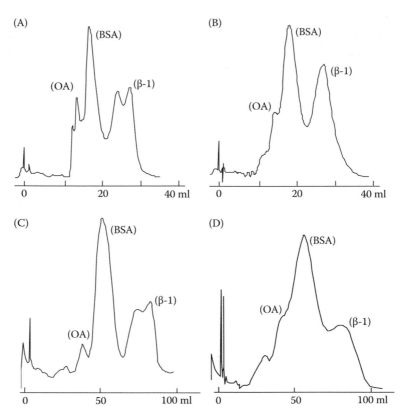

FIGURE 7.9

Ion exchange chromatographic comparison of homogeneous and superporous support at two elution rates. A = Superporous support at 76 cm/h. B = Homogeneous support at 76 cm/h. C = Superporous support at 305 cm/h. D = Homogeneous support at 305 cm/h. Bead size 75–106 μm. Sample, 1 ml protein mixture containing 0.25 mg borine serum albumin (BSA), 0.25 mg ovalbumin (OA) and 0.25 mg β-lactoglobulin (β-1) in starting buffer. (From Gustavsson, P.-E., and Larsson, P.-O., *J. Chromatogr. A.*, 734, 231–40, 1996. With permission.)

7.5.1 Superporous Agarose for Protein Separation

As discussed above, proteins are large molecules that diffuse rather slowly. Thus, their chromatographic performance will benefit considerably from adsorbents where the diffusion distances are short, such as in very small particles or in pore flow particles such as superporous agarose. The positive effect will be especially pronounced at high flow velocities, where the diffusion might become a severely limiting factor. Figure 7.9 gives a clear example of this effect. A model mixture of proteins was separated by ion exchange chromatography with two kinds of supports: superporous agarose (panels A and C) and homogeneous agarose (panels B and D), both having the same particle size. Low flow rate (upper panels) as well as high flow rate (lower panels) were used. The figure shows that the conditions were such that both kinds of adsorbents were performing better at low flow rates. However, the super-porous material gave a much better resolution than the homogeneous material regardless of flow rate, illustrating the benefits of short diffusion distances.

Figure 7.10 describes purification of a recombinant factor VIII protein (r-VIII SQ) intended for pharmaceutical use (Pålsson et al. 1999). The r-VIII SQ protein was obtained from a CHO (Chinese Hamster Ovary) cell culture and the figure describes the purification from a crude, filtered cultivation broth. The ligand used was a strong-binding monoclonal antibody ($K_{diss} = 3 \times 10^{-10}$ M), necessary for the efficient binding of r-VIII SQ from the very dilute solutions that cultivation broths constitute (about 10^{-9} M). In the run depicted in Figure 7.10, the column was loaded to near saturation (300 column volumes). After washing, the target was eluted as a very narrow peak in high concentration (up to 1,500 times more concentrated than the inlet). In scouting experiments preceding the separation shown in Figure 7.10, homogeneous reference beads and commercially available beads had been compared. The superporous beads had about the same binding capacity as the homogeneous reference beads but a much better binding kinetic (steeper breakthrough curves) as expected from a pore flow material. The commercial material (preactivated Fast Flow support; GE Healthcare) showed an unexpectedly low capacity but very good binding kinetics. Fluorescence microscopy studies indicated that the reason for this was a pronounced asymmetric ligand distribution in the commercial gel. The ligands (anti-rVIII) were located in a rather thin outer layer explaining the effective kinetics (short diffusion distances) and low capacity (crowding). Superporous gels, in contrast, showed an even ligand

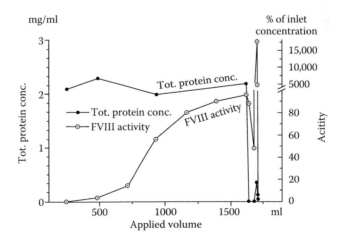

FIGURE 7.10

Chromatogram for a superporous immunoaffinity gel showing total protein concentration and factor VIII activity during sample application (100 cm/h), washing (100 cm/h), and elution (15 cm/h). Sample: filtered cell culture broth. Column dimensions: 25 × 10 mm (length × I.D.). (From Pålsson, E., Smeds, A.-L., Peterson, A., and Larsson, P.-O., *J. Chromatogr. A*, 840, 39–50, 1999. With permission.)

distribution thanks to the superpores that promoted a rapid transport of activating chemicals as well as ligands through the whole bead.

7.5.2 Superporous Agarose for Plasmid Isolation

Plasmids are interesting vectors for future gene therapy programs. New and more effective methods for their isolation are therefore being developed. A particularly challenging aspect of plasmid purification is their size. Commonly used 5 kbp plasmids (MW $= 3.3 \times 10^6$) have a hydrodynamic radius of about 0.15 µm (1500 Å), that is about 20 times larger than most proteins (Tiainen et al. 2007). Thus, plasmids have essentially no access to the diffusion pores present in most adsorbents designed for protein separation (pore size about 300 Å). The flow pores in superporous agarose, on the other hand, are large enough to allow the facile transport of plasmids. The walls of the flow pores will also provide a large area for binding the plasmids, whereas the underlying agarose phase with its diffusion pores will be largely inaccessible.

Figure 7.11 gives an example of plasmid isolation from a clarified alkaline lysate, using superporous agarose beads provided with quaternary amine ligands. To achieve a good plasmid capacity, particles with rather narrow superpores (four micrometer) were prepared in order to increase the binding area (Tiainen et al. 2007). A clarified alkaline lysate was diluted slightly to obtain the proper ion strength and applied to the column (A). Preruns had shown that an ion strength equivalent to 0.2–0.4 M NaCl both improved binding capacity and plasmid recovery. The presence of salts presumably prevented a substantial part of the RNA and the protein to bind and possibly prevented the plasmid to get trapped in an overly tight binding with the

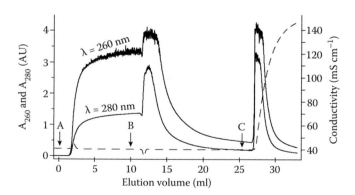

FIGURE 7.11
Purification of plasmids (pJV4) from a clarified alkaline lysate with anion exchange chromatography. Bead type: 45–75 µm agarose beads with 4 µm superpores and quaternary amine ligand. Column: 37 × 5 mm (L × I.D.). Arrow A: Injection of 10 ml lysate. Arrow B: Washing. Arrow C: Elution with 2 M NaCl. The broken line indicates conductivity. (From Tiainen, P., Ljunglöf, A., Gustavsson, P.-E. and Larsson, P.-O., *J. Chromatogr.* A, 1138, 84–94, 2007. With permission.)

adsorbent. Washing with 0.4 M NaCl (B) eluted some remaining RNA and protein. The plasmids were finally eluted with 2 M NaCl (C). Subsequent analysis of the eluted plasmid fraction showed a very good plasmid recovery and a good reduction of RNA and proteins.

In separate binding capacity experiments it was shown that the capacity of superporous beads was considerably higher than that of corresponding homogeneous beads. In fact, it was found to be a good correlation between the available surface area and the amount of plasmid bound, regardless if the adsorbent was based on homogeneous beads (only outer surface) or super-porous beads (outer surface plus superpore wall surface). These studies were supported by confocal microscopy of beads removed from a chromatography column. The bound plasmids were visualized by labeling *in situ* with the fluorescent dye TOTO-3 (Figure 7.12). The figure shows an optical section through the middle of the particle. Clearly the plasmids are located in the available surfaces—the outer surface as well as the superpore walls. Possibly a very limited penetration of the plasmid into agarose phase close to the surface occurs.

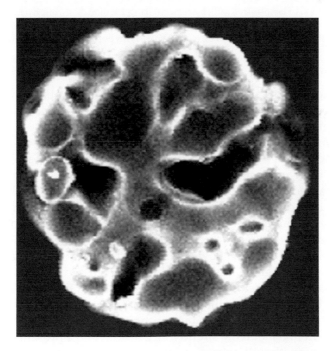

FIGURE 7.12
Confocal image of a superporous anion-exchange bead after adsorption of plasmid DNA under dynamic (chromatographic) conditions. The figure shows an optical section through the middle of an 120 μm particle with 20 μm superpores. Ligand: poly(ethyleneimine). The bright areas indicate fluorescence that is, the presence of plasmids labeled with TOTO-3. (From Tiainen, P., Ljunglöf, A., Gustavsson, P.-E. and Larsson, P.-O., *J. Chromatogr. A*, 1138, 84–94, 2007. With permission.)

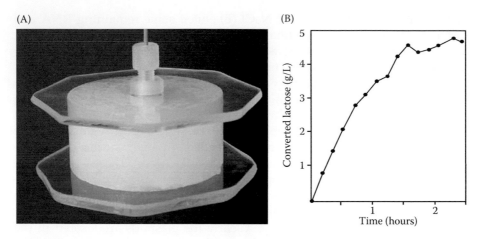

FIGURE 7.13
(A) A simple radial column construction based on an agarose monolith. The radial column was operated by immersing it into the liquid to be processed (reagents, washing buffer, enzyme, or enzyme substrate) and sucking the liquid through the bed by a pump attached to the outlet tubing. (B) Production of lactose-free milk by recirculating 1 L of milk through a radially operated bed (65 ml) of superporous agarose derivatized with β-galactosidase. (From Gustavsson, P.-E., and Larsson, P.-O., *J. Chromatogr. A*, 925, 69–78, 2001. With permission.)

7.5.3 Enzyme Reactors

Superporous agarose, in bead form or as monoliths, can be activated with a large number of agents and the activated resin used to immobilize enzymes. In one application β-galactosidase from *Kluyveromyces lactis* was immobilized on a monolithic superporous agarose column (Gustavsson and Larsson 2001). All activation (coupling and washing steps were conveniently carried out on a column and resulted in a matrix with 1.8 mg/ml of β-galactosidase), expressed an activity of about 1400 units. Several types of radial columns were used including the uncomplicated construction shown in Figure 7.13. Two oversized glass plates sealed the ends of the agarose monolith and the liquid was drawn through the agarose block by attaching a pump to the center of the column. The radial flow had an inward direction. During operation the device was immersed in the liquid to be processed. A 65 ml monolithic column was used for lactose hydrolysis of skim milk. Figure 7.13B shows the conversion of 1 L of skim milk recirculated through the radial column. As pointed out earlier, large agarose monoliths are preferably operated in radial mode, since in this mode the slightly larger flow pores present in the center part of the monolith will not affect the flow profile in contrast to what happens with axial flow.

7.5.4 Superporous Agarose Applications in Bioanalysis and Biosensors

Small superporous monolithic columns are easy to prepare and may be conveniently incorporated in bioanalytical devices (Nandakumar et al. 2000).

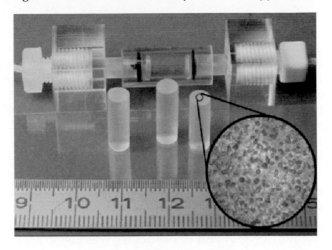

FIGURE 7.14
Superporous agarose miniplugs (15 × 5 mm) for analytical devices. The insert shows the distribution of flow pores (dark structures) in the gel. The flow pore diameter is about 50 μm. (From Nandakumar, M. P., Pålsson, E., Gustavsson, P.-E., Larsson, P.-O., and Mattiasson, B., *Bioseparation.*, 9, 193–202, 2000. With permission.)

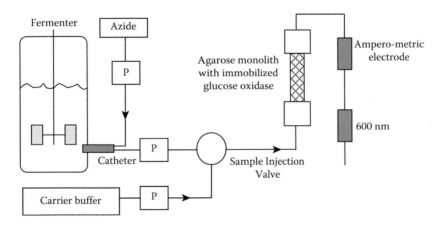

FIGURE 7.15
Set-up for monitoring glucose and cell mass in a fermenter. (From Nandakumar, M. P., Pålsson, E., Gustavsson, P.-E., Larsson, P.-O., and Mattiasson, B., *Bioseparation.*, 9, 193–202, 2000. With permission.)

The small monoliths may be carriers of adsorbents for the analyte or of enzymes participating in the detection process. Figure 7.14 shows an example of such gel plugs containing trapping antibodies and glucose oxidase. The plugs could be easily inserted into flow systems using the mini column also shown.

The mini plug with immobilized glucose oxidase was used for sensitive on-line monitoring of glucose in a fermenter. Figure 7.15 shows the set-up for simultaneous monitoring glucose as well as cell mass. Via a double-lumen catheter arrangement, a sample stream was continuously withdrawn from the fermenter without risk of contamination. Discrete samples were injected into a FIA (Flow Injection Analysis) stream, eventually reaching the super-porous agarose plug with immobilized glucose oxidase, where the glucose was converted to gluconolactone and H_2O_2. The hydrogen peroxide was measured by an amperometric electrode.

The superporous agarose plugs worked most reliably. One important aspect in this context was the superpore structure of the plugs used. The superpore diameter was about 50 micrometer, allowing unhindered passage of microbial cells (*Escherichia coli*) and of any debris. No contamination of the plugs was noticed, one contributing factor probably being the smooth surface of the flow pores.

A biosensor for the detection of pesticides was constructed based on a composite material consisting of superporous agarose, RVC (Reticulated Vitreous Carbon) and immobilized acetylcholine esterase (Khayyami et al. 1997). RVC is a hard, electrically conducting material suitable as an electrode in biosensors. It has a very high porosity (97%) and is available in a number of pore sizes ranging from 0.25 to 5 mm. Figure 7.16 shows the design of a flow-through biosensor based on this composite material. The working electrode of the sensor was formed by pouring Emulsion 1 (see steps above) into a small column supplied with a tight-fitting RVC

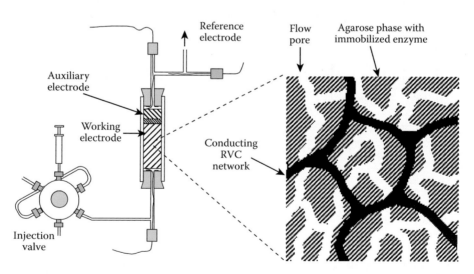

FIGURE 7.16

Amperometric biosensor for detection of acetylcholine esterase inhibitors. The working electrode consisted of an RVC—superporous agarose composite (see insert) with immobilized acetylcholine esterase. The composite was connected to a potentiostat via a platinum wire seen at the bottom of the figure. (From Khayyami, M., Perez Pita, M. T., Pena Garcia, N., Johansson, G., Danielsson, B., and Larsson, P.-O., *Talanta.*, 45, 557–63, 1997. With permission.)

cylinder. The emulsion easily filled the RVC pore structure. After cooling, gelling, and washing a composite was obtained that contained three continuous phases: the agarose phase, the superpore phase, and the electrically conducting RVC phase (see insert in Figure 7.16). After activation of the agarose phase with CNBr and coupling of acetylcholine esterase, an amperometric biosensor could be assembled as shown in Figure 7.16. The biosensor was operated by pumping the enzyme substrate acetylthiocholine continuously through the electrode together with a mediator (Meldola blue). The immobilized enzyme hydrolyzed the substrate to the electroactive product thiocholine, which reduced the mediator that in turn was reoxidized on the RVC electrode giving an amperometric signal. Whenever a sample containing pesticide was introduced in the flow via the injection valve, the enzyme was correspondingly inhibited resulting in a lower signal.

The biosensor worked effectively and was quite stable—the response was still 60% of the original response after one month. The positive properties can be ascribed to the close contact between reagents, catalysts, and electrodes thanks to the effective distribution and entanglement of the three continuous phases: RVC, agarose, and superpores.

7.5.5 Superporous Agarose Employed as Internally Cooled Electrophoresis Gels

Electrophoresis is a high resolution analytical method abundantly used for separation and analysis of complex protein mixtures and nucleic acid mixtures. One problem with the technique is the generation of heat in the separation gel, leading to temperature gradients. Such gradients influence the migration rate and may seriously compromise the resolution. An effective solution to this problem is the use of capillary electrophoresis, where the separation gels are so thin that temperature gradients, in practice, do not arise.

Preparative electrophoresis requires thick gels to achieve sufficient separation capacity. However, thick gels are, in practice, impossible to use due to the formation of severe temperature gradients. To circumvent this problem we investigated the use of superporous agarose as an electrophoresis gel. The principal idea is that the electrophoretic separation will occur in the agarose phase, while a water-immiscible coolant is pumped through the superpores and effectively removes the heat generated in the agarose phase (Figure 7.17).

Experiments with a preliminary design (Gustavsson and Larsson 1999) proved that the principal of internal cooling works. A substantial diminishing of the temperature gradient was observed (e.g., in one case a reduction of the temperature at the center of the separation gels with as much as 20°C

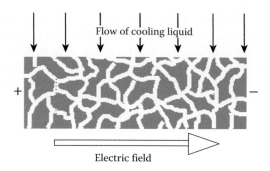

FIGURE 7.17
Principle of heat removal in thick electrophoresis gels by using an internal, circulating coolant.

was noted). Although several difficulties must be addressed, the prospect of turning a high-resolution method into a high-resolution and high-capacity method is certainly appealing.

7.6 Concluding Remarks

Superporous agarose may be prepared by a double emulsion technique that provides particles and monoliths of almost any desired shape. Superporous agarose exhibits a distinctly bimodal pore structure with (a) diffusion pores with a diameter of approximately 30 nm and (b) flow pores with a diameter at least 100 times as large, typically 5–50 micrometers. The presence of such flow pores provides the gel particles with unique properties that can be utilized advantageously in bioseparation, biocatalysis, and bioanalysis, the result being faster separation, higher catalytic activity, and so on.

The superporous agarose may be provided with various ligands and it may be combined with other materials to form composites with unique properties. The superporous agarose have been practically used for the effective separation of many biocompounds such as proteins and nucleic acids. The material is also well suited as a support for enzymes and antibodies and may be incorporated as a functioning unit in bioanalytical systems.

Acknowledgment

The financial support by CBioSep (Swedish Centre for BioSeparation) is gratefully acknowledged.

References

Afeyan, N. B., Gordon, N. F., Mazsaroff I., Varady, L., Fulton, S. P., Yang, Y. B., and Regnier, F. E. 1990. Flow-through particles for the high-performance liquid-chromatographic separation of biomolecules: Perfusion chromatography. *J. Chromatogr.* 519:1–29.

Carlsson, J., Janson, J.-C., and Sparrman, M. 1989. Affinity chromatography. In *Protein purification*, ed. J.-C. Janson and L. Rydén, 375–442. Wiley-VCH: New York.

Coulson, J. M., Richardson J. F., Backhurst J. R., and Harker, J. H. 1991. *Chemical engineering* vol. 2, 4th ed. Pergamon Press: Oxford.

Gottschalk, I., Gustavsson, P.-E., Ersson, B., and Lundahl, P. 2003. Improved lectin-mediated immobilization of human red blood cells in superporous agarose beads. *J. Chromatogr.* 784:203–8.

Gustavsson, P.-E., Axelsson, A., and Larsson, P.-O. 1998. Direct measurement of convective fluid velocities in superporous agarose beads. *J. Chromatogr. A* 795:199–210.

Gustavsson, P.-E., Axelsson, A., and Larsson, P.-O. 1999. Superporous agarose beads as a hydrophobic interaction chromatography support. *J. Chromatogr. A* 830:275–84.

Gustavsson, P.-E., and Larsson, P.-O. 1996. Superporous agarose, a new material for chromatography. *J. Chromatogr. A* 734:231–40.

Gustavsson, P.-E., and Larsson, P.-O. 1999. Continuous superporous agarose beds for chromatography and electrophoresis. *J. Chromatogr. A* 832:29–39.

Gustavsson, P.-E. and Larsson, P.-O. 2003. Monolithic polysaccharide materials *J. Chromatogr. Library.* 67:121–141.

Gustavsson, P.-E., Mosbach, K., Nilsson, K. and Larsson, P.-O. 1997. Superporous agarose as an affinity chromatography support. *J. Chromatogr. A* 776:197–203.

Khayyami, M., Perez Pita, M. T., Pena Garcia, N., Johansson, G., Danielsson, B., and Larsson, P.-O. 1997. Development of an amperometric biosensor based on acetylcholine esterase covalently bound to a new support material. *Talanta* 45:557–63.

Larsson, P.-O. 1998 March. Super porous polysaccharide gels. US Patent 5723601.

McCoy, M. A., Kalghatgi, K., Regnier, F. E., and Afeyan, N. B. 1996. Perfusion chromatography: Characterization of column packings for chromatography of proteins. *J. Chromatogr. A* 743:221–29.

Medin, A. S. 1995. *Studies on structure and properties of agarose.* Doctoral thesis, Department of Biochemistry, Uppsala University, Sweden.

Nandakumar, M. P., Pålsson, E., Gustavsson, P.-E., Larsson, P.-O., and Mattiasson, B. 2000. Superporous agarose monoliths as mini-reactors in flow injection systems. *Bioseparation* 9:193–202.

Nash, D. C., and Chase, H. A. 1998. Perfusion chromatography: Characterization of column packings for chromatography of proteins. *J Chromatogr. A* 807:185–207.

Pålsson, E., Smeds, A.-L., Peterson, A., and Larsson, P.-O. 1999. Faster isolation of recombinant factor VIII SQ, with a superporous agarose matrix. *J. Chromatogr. A* 840:39–50.

Porath, J., and Axén, R. 1976. Immobilization of enzymes to agar, agarose and sephadex support. In *Methods in Enzymology*, Vol. 44., ed. K. Mosbach. Academic Press: San Diego.

Rodrigues, A. E. 1997. Permeable packings and perfusion chromatography in protein separation. *J Chromatogr B* 699:47–61.

Sofer, G. K., and Hagel, L. 1997. *Handbook of process chromatography*. Academic Press: San Diego.

Tiainen, P., Ljunglöf, A., Gustavsson, P.-E. and Larsson, P.-O. 2007. Superporous agaros anion exchangers for plasmid isolation, *J. Chromatogr.* A 1138: 84–94.

8

Fast-Responsive Macroporous Hydrogels

Hossein Omidian and Kinam Park

CONTENTS

8.1 Introduction

In 1960, the history of synthetic hydrogels witnessed the first generation of its application in the biomedical field. Wichterle and Lim developed the first biomaterial hydrogel based on 2-hydroxyethyl methacrylate [1]. Since then, more and more hydrogels have been introduced with biomedical and pharmaceutical applications. Interestingly, all these developments have been more associated with the microstructure than chemical structure of these hydrogels. Hydrogel is generally a two-phase composite network of water and solid hydrogel. The hydrogel itself is a synthetic or natural-based hydrophilic polymer with the strong ability to interact intra- and intermolecularly with itself and with water. These two interactions are generally controlled by the composite makeup, in other words, the hydrogel structure and the amount of water inside the hydrogel. As a result, the composite is provided with properties that are known as swelling thermodynamics (capacity), swelling kinetics (rate), as well as mechanical properties.

Swelling is determined by the nature and the hydrophilic lipophilic balance (HLB) of the hydrogel that is the function of the amounts and nature of functional and side groups as well as ions. The higher the HLB value of the hydrogel, the more hydrophilic the structure is, and hence the superior will be the swelling properties. This property dictates how a hydrogel chain should approach and interact with water as the second component. As the HLB value of the hydrogel increases, affinity to the water molecules increases; this in turn increases the amount of bound water inside the hydrogel structure. Although the solid phase of the composite (the hydrogel) supplies the swelling forces and mechanical properties, the liquid phase of the composite (water) dictates the quality of these properties.

Water normally acts as a plasticizer for the solid hydrogel that increases the diffusion of a given component (e.g., drug, oxygen, etc.) into the solid matrix. Water controls the distance between the polymer chains by which the rate of diffusion into the hydrogel structure will be controlled. Water generally acts as a mass transfer medium to control the amount of transferring material and the rate of transfer process. On the other hand, although more water facilitates the transfer process, it adversely affects the hydrogel mechanical properties. Water resides between the hydrated polymer chains and provides weak regions with less resistance to the external forces.

Thermodynamically, a hydrogel as a system is required to respond to the surroundings, which is water or an aqueous medium. Although hydrogels are entitled to respond to water because of their structure, their response rate has been challenging for many applications. By far, the most interesting but hidden property of a hydrogel is its ability to change a one-dimensional transport process to a two-dimensional process. Transport process in a solid hydrogel may be exampled to driving a car on the road, which is one-dimensional and its degree of freedom is small. Adding water into the system increases the degree of freedom and hence facilitates the transport process. An analogy to this is if a car and road are replaced by a ship and sea respectively. For the applications in which a fast diffusion is very desirable, even these hydrogels fail to perform desirably. This requires an addition of another dimension to the transport process, which then requires an addition of another phase to the composite.

A three-dimensional transport process can be achieved by adding or incorporating air into the composite. Air provides a path for an airplane in which it can move at a much faster rate and with infinite degrees of freedom. Air does the same thing in a hydrogel composite comprising solid polymer, water and air. Air provides an additional transport path or diffusion path into the hydrogel structure, by which diffusion or the transport process will be expedited to a significant level. On the other hand, air replaces some portion of the hydrogel body, which is responsible for the hydrogel mechanical and swelling properties. It will be no surprise that the addition of air will be accompanied by an inferior swelling capacity and mechanical properties. What contributes the most in determining the swelling and mechanical properties is now partially replaced by air, which eventually weakens the intermolecular interactions.

For many pharmaceutical applications, fast swelling kinetics or a fast diffusion process is more desirable than improved mechanical properties. All smart hydrogels with the ability to respond to the changes in pH, temperature, solvent composition, or any other stimulants can benefit from this modification. In other words, air addition to a two-phase solid hydrogel/water composite is the simplest and the most practical approach to change a slow responsive system to a fast responsive system regardless of the type of the stimulation. For instance, in the pharmaceutical area, hydrogels are used to control the release of the drug. For this purpose, drugs are dispersed or dissolved in a hydrogel and their release is controlled by the molecular diffusion of the drug and relaxation of the polymer chains. However, the technological success of these applications has so far been limited by the low efficiency and slow rate of response to stimulants in the surrounding environment.

8.2 Super Water-Absorbent Polymers (SAPs)

About three decades ago, super-absorbent polymers (SAPs) were introduced into agriculture and later in diaper industries as a product with the ability of water retention. The SAPs are structurally cross-linked hydrophilic polymers, which have the ability to absorb substantial amounts of water or aqueous fluids (200–700 times their original weight or volume) in a relatively short period of time [2,3]. Depending on the manufacturing process and the materials used during their preparation, the swelling rate of SAPs ranges from a fraction of a minute to hours. The swelling kinetics, however, is mainly determined by the particle sizes of the SAP products. Although porosity can be incorporated into the SAP structure, this will weaken their position for the applications where good mechanical property in the swollen state is very much desirable. Different generations of SAPs were evolved with the intention of enhancing mechanical properties of the swollen SAP particles. The SAP industry is not worried about the swelling as the SAP formulators can control the swelling kinetics by manipulating the particle size. Although SAPs prepared by inverse suspension are intrinsically porous, there is no special and immediate need to make the SAP structure porous. The SAPs are responsive hydrogels due to the fact that their structure is generally ionic, which enables them to respond to the pH. For the same reason, they are also responsive to the solvent composition. In summary, the SAPs are typical fast-responsive hydrogels, which are stimuli sensitive and react fast to the stimulant because of their size. This superior swelling or water-retention property of SAP materials received attention in pharmaceutical areas for gastric retention application. In 1998, superporous hydrogels (SPHs) were introduced as a different category of SAPs with interconnected macroporous structure.

8.3 Responsive Hydrogels

Depending on the situation in which they are used, hydrogels are referenced by a variety of terms. Hydrogels may be classified as ionic, nonionic, and hydrophobically modified depending on their starting materials and structure. In terms of solubility in water, they are further classified as soluble and swellable. Swellable hydrogels are in general chemically cross-linked hydrogels. Gelling property is a term that is practically used for the hydrophobically modified hydrogels, which display sol gel transition. Table 8.1 shows how swellability, and gelling properties of hydrogels change with stimulants such as pH, temperature, and solvent composition. If the change in swellability is proportional to the change in stimulant, the response is called direct, otherwise it is the reverse.

Ionic hydrogels are generally made of ionic monomers such as acrylic acid, methacrylic acid, and their sodium, potassium, or ammonium salts. These polymers in their noncross-linked form dissolve in water at a higher pH. On the other hand, cationic polymers based on cationic monomers such as diallyldimethylammonium chloride dissolve better at low pH medium. Exact solubility behavior of these polymers is determined by the pH of the solution and the hydrogel pKa. Solubility of these hydrogels is also dependent on the ionic strength of the solution and the solvent composition. As ionic strength of the solution increases or as the portion of nonsolvent increases, so does the insolubility of the hydrogel.

Nonionic hydrogel polymers are on the other hand almost insensitive to the pH and temperature changes. Instead, they are very sensitive to the solvent composition. Since they are generally soluble in water, the addition of a nonsolvent such as methanol, ethanol, or acetone will eventually lead to the precipitation of the polymer. Hydrophobically modified hydrogels are hydrogels with hydrophobic side groups. These groups have the ability to aggregate or associate at a certain temperature. At this particular temperature, polymer-polymer interactions are favored at the expense of polymer-water interactions. This change in intermolecular interactions results in gelation or sometimes precipitation.

TABLE 8.1

Swelling Response of Hydrogels to Environmental Stimulants

Ionic polymers	Increases with increase in pH	Direct response, anionic polymers; poly(acrylic acid) and its salts, alginate, pectin
	Decreases with increase in pH	Reverse response, cationic polymers; polyethyleneimine, poly(diallyldimethylammonium chloride), chitosan
Hydrophobically-associated structures (sol-gel transition)	Increases with decrease in temperature	
		Reverse response, PEO-PPO-PEO triblock polymers, polyNIPAM, cellulose polymers (methyl cellulose, hydroxypropyl methylcellulose)
Nonionic polymers	Increases with solvent increase; addition of more nonsolvent (ethanol, methanol, acetone) leads to shrinkage, precipitation	Direct response, polyacrylamide, poly(hydroxyethyl acrylate), poly(hydroxyethyl methacrylate), poly(ethylene oxide)

It should be pointed out that factors responsible for solubility are directly responsible for the swellability. Polymer solvent interactions, electrostatic forces, and osmotic pressure are determining factors in the solubility of a polymer in water or in an aqueous medium. A polymer with good solubility behavior will potentially swell more in that solvent. The only factor that differentiates the solubility and the swellability is the elastic force that is provided by the presence of cross-links. The hydrogel swellability varies with its cross-link density. Table 8.1 shows how different responsive polymers react to their corresponding stimulants.

Regardless of their structure, hydrogels respond to a given stimulant by absorbing or desorbing the surrounding fluid. Therefore, the hydrogel response is a mass transport process that is determined by the diffusion of the solvent and relaxation of the polymer chains. In either case, as the diffusion path becomes shorter, the overall mass transport process becomes faster. As mentioned earlier, the most practical way to shorten the diffusion path is to introduce porosity into the hydrogel structure. A responsive hydrogel with the porous structure is called responsive porous hydrogel. Depending on the average size of the pores, the porous hydrogel is called macroporous (50 μm to a few millimeters) or microporous (1–50 μm).

8.4 Responsive Macroporous Hydrogels

Although pores are basically air pockets and should theoretically have no influence on the swelling properties, both swelling and mechanical properties of hydrogels are affected by the air component of the composite. On the weight basis, the porous and nonporous forms of a hydrogel should theoretically have the same swelling capacity. Pores within the hydrogel structure act like external reservoirs with the ability to hold more water. The water located within the pores is not bound to the structure and can easily be removed under pressure. This extra nonbound water also weakens the hydrogel structure as it reduces the solid content of the hydrogel. On the volume basis, both swelling capacity and swelling rate are influenced by the amount of air within the hydrogel. Another important factor is the pore size and pore size distribution. To achieve homogenous response by the whole hydrogel matrix, the pores need to be monodispersed or similar in size. Monodispersed pores provide very narrow pore size distribution that positively affects the hydrogel strength. However, if the gas blowing technique is used to make porous hydrogels, the monodispersed feature will be very difficult to achieve. Since foaming reaction will occur as gelling reaction proceeds, a poly dispersed porosity and an open-closed cell structure will be practical and more common to achieve. This feature will depend on particle size of the foaming agent, its solubility in the reaction medium, water content, hydrophilicity-hydrophobicity of the reacting mixture, temperature, dispersibility, pH, existence of a dispersing agent, and viscosity of the medium in which the foaming agent is dispersed. By far, the most important factor in achieving a close-to-monodisperse status is a synchronized foaming and gelling reaction.

Although techniques such as freeze-drying [4], porogenation [5–7], microemulsion formation [8], and phase separation [9] are used to prepare porous hydrogels, the most practical approach has so far been the gas blowing technique utilizing acid-induced decomposition of a bicarbonate compound [10]. In the synthesis of macroporous hydrogels, the monomers are simultaneously polymerized and cross-linked using a redox initiating system in solution. The polymerization reaction may happen in the presence of air or under a blanket atmosphere. At a desirable time when the gelling mass is sufficiently thin, an acidic foaming aid reacts with a bicarbonate compound and carbon dioxide gases are generated. As gelling reaction or polymerization and cross-linking reaction proceeds, the reacting mass becomes viscous and generated gases will be entrapped within the hydrogel matrix. The foam structure will simultaneously be stabilized using a foam stabilizer. The key to obtaining homogenous porosity or monodisperse porosity is to have the foaming reaction almost completed before the reacting mass becomes viscous. Moreover, a complete foaming reaction should immediately be followed by a very fast gelling reaction, which enables the matrix to quickly and adequately entrap the gases within the hydrogel matrix. At this point, when foaming is completed and the temperature of the reaction rises, the foaming gelling system tends to collapse because of high temperature and the surface tension of the water inside the hydrogel matrix. As soon as the foaming and gelling reactions reach a plateau as indicated by no change in foam height and temperature, the foam hydrogel is further stabilized by immersing into a nonsolvent, usually ethanol. Complete dehydration results in a porous product, which is white in color because of air dispersion in the hydrogel. Dehydration in alcohol will freeze the growing chains and any other ongoing reactions; this will conclude the hydrogel properties based on the events during the gelling reaction. In other words, the length of the polymer chains (polymer molecular weight) and the level of cross-linking reaction (cross-link density) just prior to dehydration process will determine the final hydrogel properties. This is why factors affecting the gelling reaction need to be carefully identified and evaluated. Typical materials needed to prepare a macroporous responsive hydrogel (superporous hydrogel) are shown in Table 8.2 with their corresponding effects on general synthesis and hydrogel properties.

8.5 Development of Superporous Hydrogels (SPHs)

The macroporous hydrogel prepared via simultaneous polymerization and cross-linking and synchronized foaming-gelling reactions is a high swelling polymer with almost no mechanical strength. This hydrogel can swell up to 100–1,000 times its own weight in distilled water depending on the monomer and the cross-linking process. The concentration of hydrogel in the swollen mass of hydrogel and water will be about 0.1–1.0%, which justifies why mechanical properties are so poor. The hydrogel will be very desirable for applications where a high and fast swelling property is needed. These features make the hydrogel a good candidate for conventional super water-absorbing applications. However, other properties such as good compressive or fatigue properties may be required for biomedical and pharmaceutical applications. Macroporous hydrogels intended for cell culture or gastric retention are required to be strong and tough in their swollen state. As a result, the desire for tailor-made swelling and mechanical properties in particular applications has evolved into three different generations of SPHs. Conventional or first generation SPHs are identified by fast swelling, high swelling capacity, and weak or no mechanical properties. The Superporous

TABLE 8.2

Ingredients Necessary to Make Macroporous (or Superporous) Hydrogels

Ingredient	Role
Monomer	Building block of the polymer, accelerates the gelling reaction
Diluent	Helps produce a smooth and controllable reaction; retards and suppresses the exothermic reaction
Cross-linker	Links the linear polymer chains; accelerates the gelling reaction
Foaming aid	Generally acidic; reacts with foaming agent to produce gases; retards the gelling reaction; increases the foam volume
Foam stabilizer	No sensible effect on gelling reaction, contributes to a well-formed cellular structure
Comonomer (ionic)	Retards the gelling reaction; helps better bicarbonate dispersion; modifies swelling and mechanical properties
Hybrid agent	Generally retards the gelling reaction; leads to poorer bicarbonate dispersion; significantly contributes in mechanical properties
Reductant and oxidant	Redox system; accelerates reaction; leads to less control on the reaction, which leads to more heterogeneity in hydrogel
Cation (III), very reactive	Ionotropic gelation; contributes in SPH rigidity and toughness; decreases swelling; provides brittle fracture mechanism
Cation (III), moderately reactive	Contributes to SPH flexibility; has almost no effect on swelling compared to more reactive cations; provides average mechanical property
Cation (II)	Contributes to flexible SPH; has almost no effect on swelling; provides weak mechanical property
Nonsolvent	Replaces water in the SPH; stabilizes the foam structure

hydrogel (SPH) composites or the second generation SPHs are characterized by fast swelling, medium swelling capacity, and improved mechanical properties. The third generation SPH or SPH hybrids possess elastic properties that can be highly useful in the development of gastrointestinal devices, as well as in other pharmaceutical and biomedical applications. A micrograph of a typical SPH structure is shown in Figure 8.1. Unique SPH swelling and mechanical properties are partially accounted for in terms of surface microwrinkles, surface micropores, and internal interconnected micropores.

8.6 The First Generation SPHs: Conventional SPHs

The first generation SPHs are similar to the currently marketed superabsorbent polymers except that their swelling capacity is independent of their

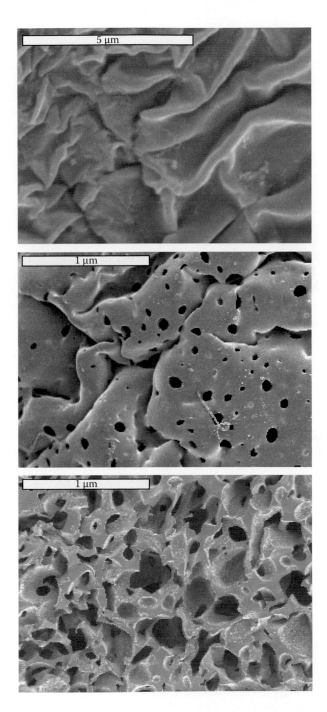

FIGURE 8.1
Fast responsive macroporous hydrogels with surface microwrinkles (top), surface micropores (middle), and internal interconnected micropores (bottom).

size in the dry state. Due to their structural superporosity, the small or large SPH products can reach to their ultimate or equilibrium swelling capacity almost at the same time. Since the objective has been to achieve ultimate swelling properties, the formulation design requires using ionic (acrylic acid, methacrylic acid and their sodium, potassium, or ammonium salts) or non-ionic but very hydrophilic monomers (acrylamide). Moreover, the amount of cross-linker in the formulation should be kept very low. One very important formulation aspect in preparation of macroporous hydrogels is the minimum cross-linker concentration (MCC). Below this value, although favoring the swelling capacity, the hydrogel cannot maintain its porosity during the foaming-gelling reaction, which severely affects its swelling kinetics. The MCC needs to be determined based on the hydrogel formulation and the reaction conditions.

Since ionic monomers are the monomers of choice for the first generation SPH, the SPH product either in its dried or swollen state will be rigid and brittle. In its dry state, the rigidity of the SPH will be high because of high glass transition temperature of the polymer. On the other hand, the ionic structure or very hydrophilic structure of the SPH has the ability to quickly absorb moisture from the surrounding environment. If there is no provision made to prevent the SPH from absorbing moisture, the SPH will be significantly plasticized to a level that may irrecoverably lead to losing the porous structure. One important step in stabilizing the porous structure of the SPHs is to minimize its water content by dehydrating it in alcohol. The lower surface tension of alcohol prevents the porous structure of the SPH from collapsing during drying. In the swollen state, the solid content of the swollen hydrogel mass will be very low, which leads to very poor mechanical properties. If the application requires the swollen hydrogel to possess certain mechanical properties, such as cell culture, gastric retention, or diet aid, a stronger swollen hydrogel would be needed. The desire for better mechanical properties triggered the further development of the later SPH generations.

8.7 The Second Generation SPHs: SPH Composites

The first attempt to make stronger SPH was to use active water-swellable fillers in the conventional SPH formulations. The filler is selected among conventional pharmaceutical superdisintegrants including cross-linked carboxymethylcellulose (CMC) or cross-linked poly (vinyl pyrrolidone). The filler in its dry state is added into the SPH formulation before foaming-gelling reactions starts. Once added, depending on the filler swelling capacity, certain amounts of the reacting mixture containing monomer, initiator, cross-linker, and foaming agent will be absorbed into the filler structure. Polymerization will occur in the free and entrapped reacting mixtures simultaneously. After drying, the entrapped chains of the SPH hydrogel will increasingly interact with the filler chains, which reinforce the whole SPH composite. In fact, swellable filler acts as reinforcing filler in the SPH composite, which increases the modulus of the hydrogel. Under mechanical stress, a higher modulus hydrogel will fail or break under brittle fracture mechanism, which prevents the hydrogel from repeated use. In other words, upon developing a craze or crack under certain mechanical force, the failed area will grow almost instantly to the entire bulk of the product, which results in complete disintegration. If the application requires multiple uses of hydrogel or multiple loading or unloading cycles, the second generation SPH will not meet the requirements. They still swell to a large size with mechanical properties improved to a minor extent. For many years, these macroporous hydrogels have been an attractive research tool for developing peroral and intestinal drug delivery systems [11–13].

More approaches to make second generation SPHs are acidification, impregnation (interchain complexation), addition of latex into SPH formulation, addition of gelatin followed by its cross-linking, ionotropic gelation of nonpolysaccharides added into the SPH formulation (polyvinyl acetate or polyvinyl alcohol treated with sodium tetraborate decahydrate), ionotropic gelation of complexable monomers (acrylic acid, acrylamide), surface cross-linking (using a cationic resin of polyamidoamine-epichlorohydrine adduct or glycerin), and thermogelation of the protein incorporated into the SPH formulation (egg white ovalbumin protein).

Gastric retention application requires the swollen macroporous hydrogel to resist multiple loading and unloading cycles of stresses during contractions and expansions of the stomach. This requires the hydrogel to be either tough or elastic. In both cases, the swollen hydrogel will eventually fail desirably under the ductile failure mechanism. This means if a craze or crack is developed at the surface or inside the structure, it will not grow quickly and the whole platform will have a much longer service life under applied stresses. This feature and desire led to development of the third generation SPH with superior mechanical properties.

8.8 The Third Generation SPHs: SPH Hybrids

To synthesize SPHs with very high mechanical or elastic properties, the conventional SPH formulation is hybridized with a polymer having complexation ability. As opposed to SPH composites wherein a cross-linked, water-swellable filler is used, the SPH hybrids are benefiting from a water-soluble polymer, a hybrid agent that undergoes cross-linking followed by SPH formation. Depending on the monomer and its compat-ibility with the hybrid agent, polysaccharides such as sodium alginate, chitosan, CMC can successfully be used to develop a third generation SPH. The advantage of hybrids over composites is that the former pro-duces an interpenetrated homogeneous network structure while the lat-ter produces interpenetrated heterogeneous network structure. Elastic or resilient properties of the acrylamide-alginate hydrogel hybrids have been shown by Omidian et al. [14]. A polyacrylamide hydrogel interpen-etrated with calcium-treated sodium alginate at high cation concentration displays rubbery properties in its swollen state. The swollen hydrogel can resist static compressive forces of about 25 N and can withstand loading and unloading cycles. Omidian et al. [15] have shown that a successful hydrogel hybrid with superior mechanical properties can be prepared if the hybrid agent can undergo a very effective ionotropic gelation. Alginate, chitosan, and CMC in the presence of calcium, sodium tripoly-phosphate, and iron, respectively, can potentially be used to hybridize the conventional SPHs.

With an alginate approach, the alginate viscosity (low viscous, medium viscous, and high viscous), alginate grade (M/G ratio), and the cation type (II valence or III valence) are the critical factors that need to be considered. Iron-treated alginate generally results in a hydrogel hybrid with a very porous skin, which enables the hydrogel to absorb water very quickly. The rate of ionotropic gelation reaction can be controlled by the reactivity of the cation (iron much greater than aluminum) and its compound (cal-cium chloride much greater than calcium gluconate). The same discussion is valid for the CMC as a hybrid agent. Since of its anionic nature, non-ionic monomers like acrylamide can effectively and safely be hybridized with sodium alginate and CMC. In the case of ionic monomers, these two hybrid agents can be cautiously used at a lower concentration. Similar to alginate and CMC, the chitosan can also undergo a very effective gelation in the presence of phosphates. Chitosan as a cationic hydrocolloid can effectively be used with nonionic monomers but its use with ionic mono-mers should cautiously be limited to a very low concentration.

The advantage of utilizing this approach is the SPH hybrids with a vast variety of mechanical properties can be prepared by manipulating the func-tional parameters. Among those, the type and concentration of the hybrid agent, and the type and concentration of the ion are the most effective ones. For instance, CMC can be cross-linked with calcium and aluminum. A soft pliable and rigid brittle hydrogel hybrid can be obtained if the hydrogel is treated with calcium and aluminum, respectively. Omidian et al. [16] used a cation combination to manipulate the hydrogel mechanical proper-ties. Different macroporous hydrogel preparations are shown in Table 8.3. Figure 8.2 clearly shows how different SPH generations differ in their swell-ing and mechanical properties.

TABLE 8.3

Example of Formulations to Prepare Different Generations of SPHs

	SAP	SPH 1st	SPH 2nd	SPH 3rd	SPH 3rd-Modified
Acrylamide (50 wt %), aq. µg	600	600	600	600	600
Bisacrylamide (1 wt %), aq. µg	100	100	100	100	100
Pluronic F127 (10 wt %), aq. µg	—	200	200	200	200
Distilled water, µg	500–1,500	500–1,500	500–1,500	500–1,500	500–1,500
Composite agent, cross-linked CMC powder, mg	—	—	10–50	—	—
Hybrid agent, Linear CMC, aq. 2 wt % Solution, µg	—	—	—	500	500
Acetic acid, µg	40	40	40	40	40
Tetramethylethylene-diamine (40 v/v %), aq. µg	50	50	50	50	50
Ammonium persulfate (20 wt %), aq. µg	50	50	50	50	50
Sodium bicarbonate, mg	30	30	30	30	30
Cation (II), 10 wt % aq.	—	—	—	+	+
Cation (III), 10 wt % aq.	—	—	—	—	+

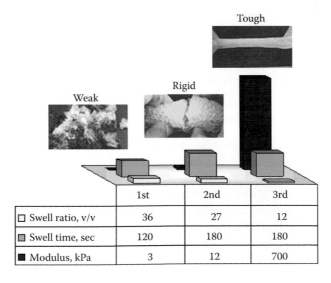

	1st	2nd	3rd
☐ Swell ratio, v/v	36	27	12
▨ Swell time, sec	120	180	180
■ Modulus, kPa	3	12	700

FIGURE 8.2

Comparison of the three typical SPH generations.

8.9 SPH Properties

8.9.1 Swelling Property

The SPHs in general are characterized by their swelling and mechanical properties in different media. Swelling is generally measured by weight, volume, or dimension. With nonporous hydrogels or less porous hydrogels, swelling measurement by weight is almost accurate and is good for comparative studies. As porosity within the structure increases and pore morphology changes from closed to interconnected, weight measurement will be more challenging. These features result in less swelling values because even the SPH weight can cause unbound water inside the gel to be removed via capillary desorption mechanism. In these circumstances, volume or dimensional swelling ratios are preferred. Depending on the application, the swelling properties should be measured in its corresponding liquid. For instance, gastric retention application requires swelling in a low pH of gastric medium. Intestinal absorption requires higher pH medium. Urine or blood absorptions require swelling medium containing salts and so forth. During swelling measurement, the homogeneity of the pores can also be evaluated. The gels are opaque in the dry state and become transparent in the swollen state. Transition from opaque to transparent is an indication of pore homogeneity and morphology. With monodisperse interconnected pores, this transition occurs in a very short period of time. As pores become polydispersed and closed, transition requires a longer time [17].

For some applications, the swelling property of the gel needs to be simultaneously measured under stress. The stress can be compression, tension, twisting, bending, or even fatigue depending on the application. Gastric retention for instance requires the gel to retain its integrity in the swollen state under constant static and dynamic forces. Swelling property of a typical fast-responsive macroporous hydrogel is shown in Figure 8.3.

8.9.2 Mechanical Property

The SPHs in their swollen state are generally weak and contain pores of different sizes ranging from 100 to 1,000 μm. The hydrogel regions surrounded by pores of different sizes behave differently in response to an applied mechanical stress. Regular mechanical testers and texture analyzers are generally used to evaluate the SPH mechanical properties. From a typical mechanogram, three important mechanical properties can be extracted, e.g., gel modulus, gel failure or breaking point, and finally the mode of gel failure (brittle or ductile). These types of measurements provide selective or local assessment because a certain area of a swollen gel is exposed to the probe of the mechanical tester. If other unexamined parts of

FIGURE 8.3
Examples of swelling with fast-responsive superporous hydrogel in distilled water.

the sample contain a defective structure, it will be overlooked during the measurement. Real mechanical response of the hydrogel will be obtained if the whole platform is exposed to the real stresses existing in the application environment [18,19].

8.10 SPH Stability

The final properties of the SPH product will be dependent on the materials, method of polymerization, reaction conditions, and more importantly its porous structure. Since the advantage of SPHs over SAPs is the kinetic of swelling, this property needs to be maintained during storage and on the shelf. The SPH as a final product can be used by itself or in encapsulated form. If used alone, the SPH can be stored in a very dry condition in the presence of moisture absorbing materials like silica gel. Although SPH itself can also absorb moisture from the environment, silica gel materials can absorb less moisture but at a much higher rate, which is very desirable for the SPH during storage. By keeping the surrounding environment dry, the pore morphology of the prepared SPH can most likely be maintained. This assures stability of the SPH product in terms of its swelling properties. Another stability aspect will be appearance and extractables. By appearance, the SPH material is required to maintain its color (original color, off-white to white) over the stability period. Any stain or discoloration requires careful observation.

Another very important stability issue is the amount of extractables in the final SPH product. Based on an in-house specification that needs to be set up according to the toxicity values of the ingredients, the final purified SPH should contain impurities less than or equivalent to the acceptable values.

High performance liquid chromatography and gas chromatography are the common methods to determine impurities in the final product. The level of impurities can be changed if the SPH product is exposed to high temperature and a high humid environment. Moisture and temperature can plasticize the SPH structure, reduce its glass transition temperature, and increase permeation of moisture and oxygen into the structure. Besides, the UV light might also adversely affect the SPH structure depending on the monomers used. These interactions may either increase the levels of impurities to the non-accepted values or change the accepted properties to an unacceptable level. Apparently, storing at a low temperature and dry condition away from light will significantly enhance the SPH stability for the duration of storage. The situation will be more complicated if SPH is encapsulated in an orally administrable capsule.

For the encapsulation process, the SPH structure needs to be adequately flexible. This is automatically achieved with low glass transition polymers (acrylic esters), but it requires moisture activation if a high glass transition polymer (acrylic acid, acrylic salt) is used. Under both circumstances, polymer chains will gain more mobility that increases their intermolecular interactions. This in turn will change the SPH swelling properties and the swelling kinetics in particular. Although change in pore morphology will negatively affect the swelling rate, it will enhance the SPH mechanical properties. Any change in pore morphology is irrecoverable and changes in SPH properties will hence be permanent. Therefore, strict care needs to be taken to preserve the desirable porosity of the SPH structure. Overall, it would be anticipated that an encapsulated SPH product might have a shorter half-life and requires stringent storing regulations.

Another very important stability issue is related to the moisture content of the SPH and the capsule. The gelatin-based and hydroxypropyl methylcellulose (HPMC)-based capsules contain 13–16% and 4–6% moisture, respectively. The amount of moisture within the hydrogel product needs to be adjusted based on the type of the capsule. At low moisture content, the SPH would be able to absorb moisture from the gelatin capsule, which results in capsule rigidity and brittleness. On the other hand, if the SPH contains high moisture, the moisture can migrate to the HPMC capsule through which the capsule will be softened and becomes less stable. If the moisture content of the capsule and the SPH is not equilibrated, the drug stability might be affected unless protected against moisture.

In performing a stability study as a necessary step to conduct a clinical trial, one needs the information listed in Table 8.4 to be incorporated into the stability protocol. Although more information can always be included, this will depend on the timeline and the budget assigned to the activity. Nevertheless, all data collected throughout the stability period will be used to justify the use of potentially stable and effective SPH for the clinical human trial.

TABLE 8.4

Information Necessary for Stability Study before Conducting Human Study

Supplies	SPH (item code, lot number, material, moisture content); Capsule (item code, size, material, manufacturer, moisture content); container (type, material, capacity, manufacturer); silica gel (type, manufacturer, item code)
Appearance	SPH (stain, discoloration, uneven surface, uneven diameter across the length); Encapsulated SPH (discoloration, stain, crack on capsule, SPH stickiness to the capsule, capsule rigidness)
Swelling property	SPH (pre- and postswelling dimensions at time zero and after full stability period); for the encapsulated SPH, pre- and postswelling numbers are generally lower; average of at least five samples needs to be examined
Extractable	SPH (pre- and poststability values; monomers, initiators, cross-linker, foaming aid (acetic or citric acid), dehydrant (alcohol); no need to examine capsule
Stability condition	Room temperature of $25 \pm 2°C$ and $60 \pm 5\%$ relative humidity

8.11 SPH Scale Up

A successful SPH preparation in a glass tube, a beaker, or in a tray does not necessarily mean the SPH can easily be scaled up. For any scale up, a number of parameters need to be optimized and some of the important factors are listed in Table 8.5.

8.12 SPH Safety and Nontoxicity

Safety and nontoxicity of the SPH products either alone or in encapsulated forms should be verified in animal models before conducting a human clinical trial. A SPH product will be regarded safe if it offers no treatment during administration for which the esophagus obstruction will be the most concerning issue. It has to be verified that with multiple administration of SPHs, the esophagus area will not be obstructed. On the other hand, the SPH product should contain less than permitted values of unwanted ingredients either from the synthesis such as residual monomer, initiator, and cross-linker or from the dehydration step such as alcohols. It also needs to be confirmed that the SPH retention inside the GI tract will not produce any other unwanted compounds such as hydrolysis product in the harsh acidic conditions. Apparently, the hydrolysis reaction will be favored at lower pHs and also at higher temperatures.

Another safety aspect of hydrogel formulations for human clinical trial is transmissible spongiform encephalopathies (TSEs) also known as prion

TABLE 8.5

Parameters to Consider for Scale-Up Production of SPHs

Fill-up Ratio	The optimum amounts of the reacting solution in the reactor
Aspect ratio	Ratio of the diameter to the height of the reacting solution that determines the surface/volume ratio of the reacting mixture; affects the heat transfer
	The aspect ratio needs to be optimized based on the optimum fill-up ratio
Heat	During scale up, volume significantly increases and hence the level of reaction exothermicity
Gas dispersion	Best sparged bubble column (SBC) dispersion technique needs to be practiced
	Nonuniform dispersion results in generating hot spots (with SBC lumps) nonuniform dispersion results in heterogeneous macroporous mass
Synchronized foaming and gelation	If foaming rate is higher than the gelling rate, gases will escape faster, which results in a lower porous hydrogel
	If foaming rate is lower than the gelling rate, less gas will be generated, which results in a less porous hydrogel
Purity	Change in the scale can affect the purity of the product because of the change in pore content; more porous hydrogel will be more pure

diseases. The disease that is originated from animals can affect the brain and nervous system and can be transferred to the other species. This issue requires that every material that is used in hydrogel preparation be provided with a TSE certificate. The certificate should clearly state that no raw material used in the manufacturing of a particular ingredient or product is derived from or has been exposed to any animal sources.

8.13 Hydrogel Characterization

A macroporous hydrogel can be characterized in terms of its structure, pore morphology (size, shape, closed cell, open cell, interconnected) as well as its swelling and mechanical properties. Thermal properties will be evaluated when the hydrogel is thermosensitive or thermoresponsive. Table 8.6 shows techniques and equipment which have so far been used in the characterization of porous hydrogels.

8.14 Applications

A major constraint in oral-controlled release drug delivery is that not all drug candidates are absorbed evenly throughout the gastrointestinal (GI) tract. Some are only absorbed in a particular portion or are absorbed to

TABLE 8.6

Techniques Used for Characterization of Hydrogels

Property	Technique	Reference
Gelation	Temperature measurement	31
Porosity	SEM	10,32,33,34, 35,36
	Mercury porositometry	37
	Cryogenic transmission electron microscopy	38
Pore structure, pore size, and surface morphology	SEM	2,39,40,41, 37,42,43,44,45,46,47,48
	Size exclusion chromatography	49
	Mercury intrusion porositometry (MIP), helium pycnometry, X-ray microtomography (XMT), SEM	50
	Cryo-transmission electron, laser scanning confocal microscopies	51,52
	Coulter porositometer	53
Swelling	SEM	54,24
	Video monitoring, conductivity measurements	55
	Volume and weight measurements	14
	Atomic force microscopy (AFM)	53
Hydrogel morphology	Optical microscopy, SEM	56
	X-ray photoelectron spectroscopy, atomic force microscopy, FTIR and surface plasmon resonance (SPR)	57,53
Hydrogel micropattern and microstructure	Optical and atomic force microscopy	58
	Laser scanning confocal microscopy, ultra small angle neutron scattering	38
Thermal stability	TGA, DSC	32,59,60,61, 62,46
Chemical structure and composition	FTIR	32,54,61,46, 50,47,53
	Solid state ^{13}C-NMR	24, 63
Mechanical properties	Rheometric studies	64
	Mechanical tester	31
Drug release	UV spectroscopy	56
Hydrogel-protein interaction	FTIR	28
Hydrogel-cell interaction	SEM, phase-contrast microscope	36
Hydrogel-filler interaction	FTIR	65
Hydrogel functional property	Immunoaffinity chromatography	66
State of water	DSC	48

different extents in various segments of the GI tract. Such drugs are said to have an absorption window. These drugs generally suffer from low bioavailability because they are simply provided with a short residence time at the site of absorption, normally in the small intestine area. The idea of extending the gastric residence time of drugs with low absorption window is a *holy grail* in oral drug delivery. If a drug having a low absorption window passes the absorption segment, it will be wasted away with no further absorption across the GI tract. Harsh stomach environment, diet status (fed or fasted), intra and intersubject variability, stomach motility, dynamic change of the stomach pH, food effect, and so on are all accounted for a gastric retention program to be a very challenging one.

Even though a few concepts including swelling and unfolding have been utilized in designing gastric retention platforms, no technology has so far proved to be commercially feasible and therapeutically effective in enhancing the absorption and bioavailability of the drugs with a low absorption window. Gastric retention will help to increase the drug absorption and its bioavailability by prolonging the residence time of the drug in the stomach. It can also help drugs with local action in the upper part of the small intestine, for example in the treatment of peptic ulcer disease. On the other hand, there are certain situations where gastric retention is undesirable. Aspirin and non-steroidal, anti-inflammatory drugs are known to cause gastric lesions, and hence slow release of such drugs in the stomach is unwanted. Drugs that may irritate the stomach lining or are unstable in acidic environments should not be formulated in gastroretentive systems. Drugs, such as isosorbide dinitrate that are absorbed equally well throughout the GI tract, will not benefit from a gastric retention system. From a patient prospective, gastric retention is not recommended for individuals with gastric hypomotility (gastroparesis).

To achieve gastric retention, the platform must withstand the forces caused by peristaltic waves in the stomach and by the constant grinding and churning processes. A gastric retention platform must resist premature gastric emptying, and once the purpose has been served, it should be removed from the stomach with ease. Various approaches have been pursued to increase the retention of an oral dosage form in the stomach. These include bioadhesive systems, high-density, low-density floatable systems, expandable, and swellable systems. Regardless of the type of technology to prepare gastric retention platforms, there are certain requirements that need to be met by the platform as shown in Table 8.7.

In summary, a successful gastric retention platform is expected to stay in the stomach, preferably in fasted state for a few hours (6–8 hrs), to release its imbedded drug in a controlled manner and finally to disintegrate for a safe removal from the body. Overall, a potential gastric retention platform should generally meet the requirements including pharmaceutical acceptability, mechanical properties, retention, ease of processing, and disintegration.

To be utilized as a gastroretentive platform, a typical macroporous hydrogel should have a specific shape such as a cylinder, should swell in the

TABLE 0.7

Requirements for Developing Gastric Retention Platforms

Mechanical properties	A gastric retention platform has to be able to resist strong stomach forces during the housekeeping wave sweeping period of stomach motility.
Physical properties	A gastric retention platform should be orally administrable.
Gastric pH	A gastric retention platform should be robust (intact and stable) to pH fluctuations. The gastric pH under fast conditions is 1 to 2, while under fed conditions is about 3 to 4 following the meal.
Food effects	A gastric retention platform should not interfere with the normal physiological gastric emptying of food.
Elimination	A gastric retention platform should be eliminated following its complete retention in the stomach. It can be biodegraded in the intestine or can be disintegrated anywhere in the GI tract.
Pharmaceutical acceptability	A gastric retention platform should contain materials that are generally recommended as safe (GRAS).
Toxicity and safety	A gastric retention platform should not threaten the life or health of a subject. The challenging issues will be the esophagus obstruction and residual toxic materials.

stomach fluid within minutes to a size of about 20 times its volume, and should resist pressures ranging from 0.5–2.0 N/cm^2 presumably in the fed state. It also has to be flexible for the encapsulation process, should be disintegrated in a controlled manner, should have the drug loading capability, and should be stable and pharmaceutically acceptable. With the hope that all these requirements are met, the strong and elastic SPH hybrids have been considered and studied for this application. The SPH is prepared as a reservoir that houses the drug or drug delivery system within. The drug itself can be formulated as a single unit (tablet or wax) or multiple units (microparticles or macroparticles). The platform is encapsulated in a regular capsule and is orally administered. Feasibility of these platforms for the solid dose (tablet) and semisolid (wax) drug delivery systems [20,21], peroral peptide delivery systems [12,13,22–29], and fast dissolving tablets [30] has so far been studied.

A very electrifying but far-reaching application for macroporous SPHs is a platform that induces satiety in humans. Because of their high and fast swelling properties, the SPHs can presumably occupy a significant portion of the stomach volume, leaving less space for food, and hence suppress the appetite [30]. This may potentially benefit obese people. To achieve satiety, a swollen SPH should occupy at least 400 ml of the stomach volume. A HPMC or gelatin capsule can accommodate at most 1 gram of the hydrogel. This can be translated to a swelling capacity of at least 400 ml/g for a given hydrogel having density of one. The authors cannot recall any hydrogel with this level of swelling capacity in the very harsh acidic stomach conditions. Moreover, the hydrogel is expected to reside in the gastric area for at least a few hours. The most challenging part will be if the high swelling hydrogel can resist the phase III of the stomach motility which occurs almost every two hours in the fasted state. This requires the hydrogel to be at least formulated with other excipients, which can potentially suppress the stomach activities.

8.15 Researches on SPH

The past 10 years has witnessed a great number of research activities on macroporous SPHs. The research has focused on different aspects of synthesis, characterization, development, processing, applications, safety, preclinical, and clinical studies. A summary of these activities is compiled in Table 8.8.

TABLE 8.8

Some of the Recent Research Activities on Superporous Hydrogels

Hydrogel Properties	
Cross-linking and polymerization of acrylamide and acrylic acid in the presence of NaHCO$_3$; a pH-sensitive hydrogel intended for general water-absorbent applications	40
Study surface morphology by SEM, measure porosity using mercury porositometry; interconnected pores of a few hundred micrometers; swelling feature is predominantly determined by internal pores and not surface pores	37
Modify hydrogel swelling and mechanical properties; interpenetrating polymer network via interchain complexation of poly(acrylamide-co-acrylic acid)/polyethylenimine	67
Development of Superporous Hydrogels	
Use of superdisintegrants in preparing SPH composites from conventional SPHs; high swelling and good mechanical properties, gastric retention tested in fasted and fed dogs; Gastric retention for 2–3 hrs in fasted and for 24 hrs in fed conditions	68
Structural characterization of SPH composites; solid state ^{13}C NMR to study polymer structure; SPH composites are less porous, have lower swelling ratio but better mechanical property than conventional SPHs	24
Effect of acidification on swelling and mechanical properties of poly(acrylamide-co-acrylic acid) SPHs	69
Use of Taguchi experimental design in the synthesis of mechanically strong macroporous SPH; examine the effects of the formulation parameters on the gelation status	31
Combined chemical cross-linking and physical ionotropic gelation to make very tough elastic SPHs intended for gastric retention application; SPH based on interpenetrating network of acrylamide and calcium alginate	14
Perform freeze-thawing and drying to achieve mechanically strong SPH of Glycol chitosan/PVOH; number of freezing/thawing cycles were found to be much more effective than freezing time on hydrogel property	70
Polyacrylamide SPH grafted with acrylic acid used to absorb low MW (Cu(II)) and high MW (lysozyme) ligands	71

(Continued)

TABLE 8.8 (Continued)

Polyacrylamide SPH grafted with different water-insoluble monomers like glycidyl methacrylate and t-butyl acrylamide; grafting of water-soluble monomers proceed better in aqueous media; grafting of water-insoluble monomers proceed better in water-DMSO solutions	72
Introduce ion equilibration as a novel approach to synthesize strong SPHs	16
Review chemistry, structure, preparation, and processing of superabsorbent polymers and SPH generations	73
Review swelling devices based on SPHs	74
Evaluate fatigue properties of SPHs	18
Introduce a novel mechanical tester that simulates real service conditions	19, 75
Evaluate swelling properties of SPHs in different swelling media	76
Safety and toxicity studies on SPHs using swine emesis model	77
Structural and calorimetric characterization of SPH hybrids	78
Morphology, hydration and mechanical properties of SPH hybrids	79
Acrylamide-based SPH composite containing hydroxyapatite; hydroxyapatite had no effect on pore structure, but increased the mechanical property to almost tenfold; it is cytocompatible toward fibroblasts; intended for bone tissue engineering	80
Poly(acrylic acid) SPH particles as tablet superdisintegrant; evaluate the effect of SPH on tablet swelling and mechanical properties	81

Drug Delivery

To enhance transport of drugs across porcine intestinal epithelium and to study attachment to the gut walls; intended for intestinal protein and peptide delivery	22
Study mechanism of opening in tight junctions; using SPH as permeation enhancer for peptide delivery	23, 26
Scintigraphy using Tc-99 to study the SPH movement throughout the GI tract; intended for targeted intestinal protein and peptide delivery	27
Release rate and mechanism, drug stability under different environmental conditions, drug-platform interaction; intended for peroral peptide delivery of buserelin, octreotide, and insulin	28
In vivo pig study of using platforms to improve intestinal absorption of insulin; intended for intestinal protein delivery	12
To improve kinetics of swelling and shrinking of conventional hydrogels; study swelling dependency on pH and ionic strength; evaluate single versus multiple unit drug delivery systems	82
Study compression effect on swelling property and pore structure; intended for gastric retention application	83
Review concepts, developments and approaches to extend gastric residence time of orally administrable dosage form	84
Review development of different SPH generations intended for gastric retention and intestinal absorption of proteins	17
Highly pH-sensitive hydrogels based on chitosan and glycol chitosan with mechanical stability over swelling and de-swelling cycles; intended for gastric retention applications	85
Utilize SPH composites to improve intestinal transport of the peptide drug desmopressin *in vitro*	11
Poly(acrylamide-co-acrylic acid) interpenetrated with chitosan or glycol chitosan; better swelling properties in SGF than in water with increase in chitosan concentration; glycol chitosan provides better swelling kinetic because of its higher affinity to water	86

(Continued)

TABLE 8.8   (Continued)

Carbopol-containing SPH composites with different responsive properties; Internal interconnected porous and external nonporous structure with the ability to adhere faster and better to the intestinal mucosal than regular SPHs; intended as potential candidate for transmucosal delivery	35
Carbopol-containing SPHs studied in various salt and pH solutions, their biocompatibility evaluated using tissue damage and cytotoxicity studies; claimed to be an effective carrier for peroral delivery of peptide and protein drugs	87
Review gastroretentive floating and SPHs intended to fight H. pylori	88
Poly(acrylic acid-co-acrylamide)/O-carboxymethyl chitosan full interpenetrating network with mucoadhesive properties intended for peroral absorption of insulin	89, 47
Anion exchanger agarose prepared by incorporating ethylene imines and quaternary amine functionality; with the ability to allow large molecules such as plasmids to transport into their interior; plasmid binding capacity was found much higher with smaller diameter superporous agarose beads	90
Tissue Engineering	
Poly(vinyl alcohol) and poly(vinyl pyrrolidone) prepared by double emulsion process; hydrogel emulsions were physically cross-linked by freeze-thaw cycles; intended as articular cartilage repair that should have good mechanical properties to function synergistically with healthy cartilage and porous enough to allow for tissue integration	91

Summary

Because of their porous structure, the responsive macroporous hydrogels can respond to their surrounding environment very quickly. The basic properties of these hydrogels include swelling capacity, swelling rate, and mechanical strength and are critically dependent on the pore size and morphology. This feature has opened a new era in the hydrogel history that can extend their application into the pharmaceutical and biomedical area. In this article, the basic concepts, formulation, development, scale up, characterization, properties, safety, stability, and applications of the most recent type of responsive macroporous hydrogels are examined.

References

1. Wichterle, O., and D. Lim. 1960. Hydrophilic gels for biological use. *Nature* 185:117–18.
2. Chen, J., H. Park, and K. Park. 1999. Synthesis of superporous hydrogels: Hydrogels with fast swelling and superabsorbent properties. *Journal of Biomedical Materials Research* 44 (1):53–62.

3. Askari, F., S. Nafisi, H. Omidian, and S. A. Hashemi. 1993. Synthesis and characterization of acrylic-based superabsorbents. *Journal of Applied Polymer Science* 50:1851–55.

4. Patel, V. R., and M. M. Amiji. 1996. Preparation and characterization of freeze-dried chitosan-poly(ethylene oxide) hydrogels for site-specific antibiotic delivery in the stomach. *Pharmaceutical Research* 13:588–93.

5. Oxley H. R., P. H. Corkhill, J. H. Fitton, and B. J. Tighe. 1993. Macroporous hydrogels for biomedical applications: Methods and morphology. *Biomaterials* 14:1064–72.

6. Kon M., A. C. De Visser. 1981. A poly(HEMA) sponge for restoration of articular cartilage defects. *Plastic and Reconstructive Surgery* 67:288–94.

7. Badiger M. V., M. E. McNeil, and N. B. Graham. 1993. Progens in the preparation of microporous hydrogels based on poly(ethylene oxide). *Biomaterials* 14:1059–63.

8. Bennett D. J., R. P. Burford, T. P. Davis, and H. J. Tilley. 1995. Synthesis of porous hydrogel structure by polymerizing the continuous phase of a microemulsion. *Polymer International* 36:219–26.

9. Chirila T. V., I. J. Constable, G. J. Crawford, S. Vijayasekaran, D. E. Thompson, Y.-C. Chen, W. A. Fletcher, and B. J. Griffin. 1993. Poly(2-hydroxyethyl methacrylate) sponges as implant materials: *in vivo* and *in vitro* evaluation of cellular invasion. *Biomaterials* 14:26–38.

10. Kabiri, K., H. Omidian, and M. J. Zohuriaan-Mehr. 2003. Novel approach to highly porous superabsorbent hydrogels: Synergistic effect of porogens on porosity and swelling rate. *Polymer International* 52 (7):1158–64.

11. Polnok, A., J. Coos Verhoef, G. Borchard, N. Sarisuta, and H. E. Junginger. 2004. *In vitro* evaluation of intestinal absorption of desmopressin using drug-delivery systems based on superporous hydrogels. *International Journal of Pharmaceutics* 269 (2):303–10.

12. Dorkoosh, F. A., J. Coos Verhoef, G. Borchard, M. Rafiee-Tehrani, J. H. M. Verheijden, and H. E. Junginger. 2002. Intestinal absorption of human insulin in pigs using delivery systems based on superporous hydrogel polymers. *International Journal of Pharmaceutics* 247 (1–2):47–55.

13. Dorkoosh, F. A., J. Coos Verhoef, J. H. M. Verheijden, M. Rafiee-Tehrani, G. Borchard, and H. E. Junginger. 2002. Peroral absorption of octreotide in pigs formulated in delivery systems on the basis of superporous hydrogel polymers. *Pharmaceutical Research* 19 (10):1532–36.

14. Omidian, H., J. G. Rocca, and K. Park. 2006. Elastic, superporous hydrogel hybrids of polyacrylamide and sodium alginate. *Macromolecular Bioscience* 6 (9):703–10.

15. Omidian, H., Y. Qiu, S. Yang, D. Kim, H. Park, and K. Park. 2005. *Hydrogels having enhanced elasticity and mechanical strength properties.* U.S. Patent 6,960,617.

16. Omidian, H., and J. G. Rocca. 2006. *Formation of strong superporous hydrogels.* U.S. Patent 7,056,957.

17. Omidian, H., K. Park, and J. G. Rocca. 2007. Recent developments in superporous hydrogels. *Journal of Pharmacy and Pharmacology* 59 (3):317–27.

18. Gavrilas, C., H. Omidian, and J. G. Rocca. 2005. Dynamic mechanical properties of superporous hydrogels. Presented at the 8th U.S.-Japan Symposium on Drug Delivery Systems, Hawaii.

19. Gavrilas, C., H. Omidian, and J. G. Rocca. 2005. A novel simulator to evaluate fatigue properties of superporous hydrogels. Presented at the 8th U.S.-Japan Symposium on Drug Delivery Systems, Hawaii.
20. Rocca, J. G., K. Shah, and H. Omidian. 2004. Superporous hydrogels containing solid and semi-solid carriers. *Gattefosse Technical Bulletin* 97:73–84.
21. Li, G., H. Omidian, and J. G. Rocca. 2005. Wax-loaded superporous hydrogel platforms. Presented at the 32nd Annual Meeting of the Controlled Release Society, Florida.
22. Dorkoosh, F. A., G. Borchard, M. Rafiee-Tehrani, J. Coos Verhoef, and H. E. Junginger. 2002. Evaluation of superporous hydrogel (SPH) and SPH composite in porcine intestine ex-vivo: Assessment of drug transport, morphology effect, and mechanical fixation to intestinal wall. *European Journal of Pharmaceutics and Biopharmaceutics* 53 (2):161–66.
23. Dorkoosh, F. A., C. A. N. Broekhuizen, G. Borchard, M. Rafiee-Tehrani, J. C. Verhoef, and H.E. Junginger. 2004. Transport of octreotide and evaluation of mechanism of opening the paracellular tight junctions using superporous hydrogel polymers in Caco-2 cell monolayers. *Journal of Pharmaceutical Sciences* 93 (3):743–52.
24. Dorkoosh, F. A., J. Brussee, J. Coos Verhoef, G. Borchard, M. Rafiee-Tehrani, and H. E. Junginger. 2000. Preparation and NMR characterization of superporous hydrogels (SPH) and SPH composites. *Polymer* 41 (23):8213–20.
25. Dorkoosh, F. A., J. Coos Verhoef, Matheus H. C. Ambagts, M. Rafiee-Tehrani, G. Borchard, and H. E. Junginger. 2002. Peroral delivery systems based on super-porous hydrogel polymers: release characteristics for the peptide drugs buserelin, octreotide and insulin. *European Journal of Pharmaceutical Sciences* 15 (5):433–39.
26. Dorkoosh, F. A., D. Setyaningsih, G. Borchard, M. Rafiee-Tehrani, J. Coos Verhoef, and H. E. Junginger. 2002. Effects of superporous hydrogels on para-cellular drug permeability and cytotoxicity studies in Caco-2 cell monolayers. *International Journal of Pharmaceutics* 241 (1):35–45.
27. Dorkoosh, F. A., M. P. M. Stokkel, D. Blok, G. Borchard, M. Rafiee-Tehrani, J. C. Verhoef, H. E. 2004. Junginger Feasibility study on the retention of superporous hydrogel composite polymer in the intestinal tract of man using scintigraphy. *Journal of Controlled Release*. 99 (2):199–206.
28. Dorkoosh, F. A., M. P. M. Stokkel, D. Blok, G. Borchard, M. Rafiee-Tehrani, J. Coos Verhoef, and H. E. Junginger. 2002. Peroral delivery systems based on super-porous hydrogel polymers: release characteristics for the peptide drugs buserelin, octreotide and insulin. *European Journal of Pharmaceutical Sciences* 15 (5):433–39.
29. Dorkoosh, F. A., J. Coos Verhoef, G. Borchard, M. Rafiee-Tehrani, and H. E. Junginger. 2001. Development and characterization of a novel peroral peptide drug delivery system. *Journal of Controlled Release* 71 (3):307–18.
30. Park, H. 2002. Superporous hydrogels for pharmaceutical & other applications. *Drug Delivery Technology* 2:38–44.
31. Omidian, H., and K. Park. 2002. Experimental design for the synthesis of poly-acrylamide superporous hydrogels. *Journal of Bioactive and Compatible Polymers* 17 (6):433–50.
32. Abd El-Mohdy, H. L., E. S. A. Hegazy, and H. A. Abd El-Rehim. 2006. Characterization of starch/acrylic acid super-absorbent hydrogels prepared by ionizing radiation. *Journal of Macromolecular Science Part a-Pure and Applied Chemistry* 43 (7):1051–63.

33. Ramos, R., V. Carvalho, and M. Gama. 2006. Novel hydrogel obtained by chitosan and dextrin-VA co-polymerization. *Biotechnology Letters* 28 (16):1279–84.
34. Park, Y. J., J. Liang, Z. Yang, and V. C. Yang. 2001. Controlled release of clotdissolving tissue-type plasminogen activator from a poly(L-glutamic acid) semi-interpenetrating polymer network hydrogel. *Journal of Controlled Release* 75 (1–2):37–44.
35. Tang, C., C. Yin, Y. Pei, M. Zhang, and L. Wu. 2005. New superporous hydrogels composites based on aqueous carbopol((R)) solution (SPHCcs): Synthesis, characterization and *in vitro* bioadhesive force studies. *European Polymer Journal* 41 (3):557–62.
36. Tian, W. M., S. P. Hou, J. Ma, C. L. Zhang, Q. Y. Xu, I. S. Lee, H. D. Li, M. Spector, and F. Z. Cui. 2005. Hyaluronic acid-poly-D-lysine-based three-dimensional hydrogel for traumatic brain injury. *Tissue Engineering* 11(3–4):513–25.
37. Gemeinhart, R. A., H. Park, and K. Park. 2000. Pore structure of superporous hydrogels. *Polymers for Advanced Technologies* 11(8–12):617–25.
38. Pakstis, L. M., B. Ozbas, K. D. Hales, A. P. Nowak, T. J. Deming, and D. Pochan. 2004. Effect of chemistry and morphology on the biofunctionality of selfassembling diblock copolypeptide hydrogels. *Biomacromolecules* 5 (2):312–18.
39. Abd El-Rehim, H. A., E. S. A. Hegazy, and D. A. Diaa. 2006. Characterization of super-absorbent material based on carboxymethylcellulose sodium salt prepared by electron beam irradiation. *Journal of Macromolecular Science-Pure and Applied Chemistry* A43 (1):101–13.
40. Bajpai, S. K., M. Bajpai, and L. Sharma. 2006. Investigation of water uptake behavior and mechanical properties of superporous hydrogels. *Journal of Macromolecular Science-Pure and Applied Chemistry* A43 (3):507–24.
41. El-Mohdy, H. L. A., and A. Safrany. 2008. Preparation of fast response superabsorbent hydrogels by radiation polymerization and crosslinking of N-isopropylacrylamide in solution. *Radiation Physics and Chemistry* 77:273–79.
42. Goraltchouk, A., T. Freier, and M. S. Shoichet. 2005. Synthesis of degradable poly(L-lactide-co-ethylene glycol) porous tubes by liquid-liquid centrifugal casting for use as nerve guidance channels. *Biomaterials* 26 (36):–755563.
43. Guilherme, M. R., et al., Thermo-responsive sandwiched-like membranes of IPN-PNIPAAm/PAAm hydrogels. *Journal of Membrane Science*, 2006. 275(1-2): p. 187-194.
44. Kim, J. H., S. J. Sim, D. H. Lee, D. Kim, Y. K. Lee, and J. Kim. 2004. Preparation and properties of biodegradable hydrogels based on glutaraldehyde-crosslinked poly(2-hydroxyethyl aspartamide). *Journal of Industrial and Engineering Chemistry* 10 (2):278–82.
45. Lakhiari, H., T. Okano, N. Nurdin, C. Luthi, P. Descouts, D. Muller, and J. Jozefonvicz. 1998. Temperature-responsive size-exclusion chromatography using poly(N-isopropylacrylamide) grafted silica. *Biochimica et Biophysica Acta-General Subjects* 1379 (3):303–13.
46. Namkung, S., and C. C. Chu. 2006. Effect of solvent mixture on the properties of temperature- and pH-sensitive polysaccharide-based hydrogels. *Journal of Biomaterials Science-Polymer Edition* 17 (5):519–46.
47. Yin, L. C., L. Fei, F. Cui, C. Tang, C. Yin. 2007. Superporous hydrogels containing poly(acrylic acid-co-acrylamide)/O-carboxymethyl chitosan interpenetrating polymer networks. *Biomaterials* 28 (6):1258–66.

48. Zhang, K. P., Y. L. Luo, and Z. Q. Li. 2007. Synthesis and characterization of a pH- and ionic strength-responsive hydrogel. *Soft Materials* 5:183–95.
49. Gustavsson, P. E., and P. O. Larsson. 1999. Continuous superporous agarose beds for chromatography and electrophoresis. *Journal of Chromatography A* 832 (1–2):29–39.
50. Partap, S., A. Muthutantri, I. U. Rehman, G. R. Davis, and J. A. Darr. 2007. Preparation and characterisation of controlled porosity alginate hydrogels made via a simultaneous micelle templating and internal gelation process. *Journal of Materials Science* 42 (10):3502–07.
51. Schneider, J.P., D. J. Pochan, B. Ozbas, K. Rajagopal, L. Pakstis, and J. Kretsinger. 2002. Responsive hydrogels from the intramolecular folding and self-assembly of a designed peptide. *Journal of the American Chemical Society* 124 (50):15030–37.
52. Sosnik, A., D. Cohn, J. S. Roman, G. A. Abraham. 2003. Crosslinkable PEO-PPO-PEO-based reverse thermo-responsive gels as potentially injectable materials. *Journal of Biomaterials Science-Polymer Edition* 14 (3):227–39.
53. Ying, L., E. T. Kang, K. G. Neoh, K. Kato, and H. Iwata. 2004. Drug permeation through temperature-sensitive membranes prepared from poly(vinylidene fluoride) with grafted poly(N-isopropylacrylamide) chains. *Journal of Membrane Science* 243 (1–2):253–62.
54. Bajpai, A. K. 2007. Blood protein adsorption onto macroporous semi-interpenetrating polymer networks (IPNs) of poly(ethylene glycol) (PEG) and poly(2-hydroxyethyl methacrylate) (PHEMA) and assessment of *in vitro* blood compatibility. *Polymer International* 56 (2):231–44.
55. Low, L. M., S. Seetharaman, K. He, and M. J. Madou. 2000. Microactuators toward microwaves for responsive controlled drug delivery. *Sensors and Actuators B-Chemical* 67 (1–2):149–60.
56. Arun, A., and B. S. R. Reddy. 2005. *In vitro* drug release studies of 2-hydroxyethyl acrylate or 2-hydroxypropyl methacrylate-4-{(1E,4E)-5-[4-(acryloyloxy) phenyl]3-oxopenta-1,4-dienyl]phenyl acrylate copolymer beads. *Journal of Biomedical Materials Research Part B-Applied Biomaterials* 73B (2):291–300.
57. Hirata, I., M. Okazaki, and H. Iwata. 2004. Simple method for preparation of ultra-thin poly (N-isopropylacrylamide) hydrogel layers and characterization of their thermo-responsive properties. *Polymer* 45 (16):5569–78.
58. Ito, Y. 1999. Pattern change of stimuli-responsive polymers. *Kobunshi Ronbunshu* 56 (10):617–25.
59. Badiger, M. V., M. E. McNeill, and N. B. Graham. 1993. Porogens in the preparation of microporous hydrogels based on poly(ethylene oxides). *Biomaterials* 14 (14):1059–63.
60. Bajpai, A. K., and S. Kankane. 2007. Preparation and characterization of macroporous poly(2-hydroxyethyl methacrylate)-based biomaterials: Water sorption property and *in vitro* blood compatibility. *Journal of Applied Polymer Science* 104 (3):1559–71.
61. Ling, Y. D., and M. G. Lu. 2008. Fabrication of poly (N-isopropylacrylamide-co-itaconic acid) hydrogels in DMSO/water mixtures and their characterization. *Iranian Polymer Journal* 17 (2):155–66.
62. Motta, A., C. Migliaresi, F. Faccioni, P. Torricelli, M. Fini, and R. Giardino. 2004. Fibroin hydrogels for biomedical applications: preparation, characterization and *in vitro* cell culture studies. *Journal of Biomaterials Science-Polymer Edition* 15 (7):851–64.

63. Sun, L. F., R. X. Zhuo, and Z. L. Liu. 2003. Studies on the synthesis and properties of temperature responsive and biodegradable hydrogels. *Macromolecular Bioscience* 3 (12):725–28.
64. Cohn, D., A. Sosnik, and S. Garty. 2005. Smart hydrogels for *in situ* generated implants. *Biomacromolecules* 6 (3):1168–75.
65. Kabiri, K., and M. J. Zohuriaan-Mehr. 2003. Superabsorbent hydrogel composites. *Polymers for Advanced Technologies* 14 (6):438–44.
66. Palsson, E., A.-L. Smeds, A. Petersson, and P. Larsson. 1999. Faster isolation of recombinant factor VIII SQ, with a superporous agarose matrix. *Journal of Chromatography A* 840 (1):39–50.
67. Kim, D., and K. Park. 2004. Swelling and mechanical properties of superporous hydrogels of poly (acrylamide-co-acrylic acid)/polyethylenimine interpenetrating polymer networks. *Polymer* 45 (1):189–96.
68. Chen, J., W. E. Blevins, H. Park, and K. Park. 2000. Gastric retention properties of superporous hydrogel composites. *Journal of Controlled Release* 64 (1–3):39–51.
69. Kim, D., K. Seo, and K. Park. 2004. Polymer composition and acidification effects on the swelling and mechanical properties of poly(acrylamide-co-acrylic acid) superporous hydrogels. *Journal of Biomaterials Science-Polymer Edition* 15 (2):189–99.
70. Park, H., and D. Kim. 2006. Swelling and mechanical properties of glycol chitosan/poly(vinyl alcohol) IPN-type superporous hydrogels. *Journal of Biomedical Materials Research Part A* 78A (4):662–67.
71. Savina, I. N., B. Mattiasson, and I. Y. Galaev. 2005. Graft polymerization of acrylic acid onto macroporous polyacrylamide gel (cryogel) initiated by potassium diperiodatocuprate. *Polymer* 46 (23):9596–9603.
72. Savina, I. N., B. Mattiasson, and I. Y. Galaev. 2006. Graft polymerization of vinyl monomers inside macroporous polyacrylamide gel, cryogel, in aqueous and aqueous-organic media initiated by diperiodatocuprate(III) complexes. *Journal of Polymer Science Part A-Polymer Chemistry* 44 (6):1952–63.
73. Omidian, H., J. G. Rocca, and K. Park. 2005. Advances in superporous hydrogels. *Journal of Controlled Release* 102 (1):3–12.
74. Omidian, H., and K. Park. 2008. Swelling agents and devices in oral drug delivery. *Journal of Drug Delivery Science and Technology* 18 (2):83–93.
75. Omidian, H., C. Gavrilas, and J. G. Rocca. 2005. Mechanical properties of gastroretentive platforms using a novel simulator. Presented at the 32nd Annual Meeting of the Controlled Release Society, Florida.
76. Li, G., H. Omidian, and J. G. Rocca. 2005. Solvent effects on the swelling properties of superporous hydrogels. Presented at the American Association of Pharmaceutical Scientists, Tennessee.
77. Townsend, R., J. G. Rocca, and H. Omidian. 2005. Safety and toxicity studies of a novel gastroretentive platform administered orally in a swine emesis model. Presented at the 32nd Annual Meeting of the Controlled Release Society, Florida.
78. Thangamathesvaran, P. M., H. Omidian, A. Rathinavelu, and J. G. Rocca. 2004. Metal-chelated superporous hydrogels; Part 2: Structural and calorimetric characterization. Presented at the AAPS Pharmaceutics and Drug Delivery Conference, Philadelphia, PA.
79. Thangamathesvaran, P. M., H. Omidian, A. Rathinavelu, and J. G. Rocca. 2004. Metal-chelated superporous hydrogels; Part 1: Morphology, hydration and mechanical properties., Presented at the AAPS Pharmaceutics and Drug Delivery Conference, Philadelphia, PA.

80. Tolga Demirtas, T., A. G. Karakecili, and M. Gumusderelioglu. 2008. Hydroxyapatite containing superporous hydrogel composites: Synthesis and *in vitro* characterization. *Journal of Materials Science: Materials in Medicine* 19 (2):729–35.
81. Yang, S. C., Y. Fu, S. H. Jeong, and K. Park. 2004. Application of poly(acrylic acid) superporous hydrogel microparticles as a super-disintegrant in fast-disintegrating tablets. *Journal of Pharmacy and Pharmacology* 56 (4):429–36.
82. Gemeinhart, R. A., J. Chen, H. Park, and K. Park. 2000. pH-sensitivity of fast responsive superporous hydrogels. *Journal of Biomaterials Science-Polymer Edition* 11 (12):1371–80.
83. Gemeinhart, R. A., H. Park, and K. Park. 2001. Effect of compression on fast swelling of poly(acrylamide-co-acrylic acid) superporous hydrogels. *Journal of Biomedical Materials Research* 55 (1):54–62.
84. Hwang, S. J., H. Park, and K. Park. 1998. Gastric retentive drug-delivery systems. *Critical Reviews in Therapeutic Drug Carrier Systems* 15 (3):243–84.
85. Park, H., K. Park, and D. Kim. 2006. Preparation and swelling behavior of chitosan-based superporous hydrogels for gastric retention application. *Journal of Biomedical Materials Research Part A* 76A (1):144–50.
86. Seo, K. W., D. J. Kim, and K. N. Park. 2004. Swelling properties of poly(AM-co-AA)/chitosan pH sensitive superporous hydrogels. *Journal of Industrial and Engineering Chemistry* 10 (5):794–800.
87. Tang, C., L. Yin, J. Yu, C. Yin, and Y. Pei. 2007. Swelling behavior and biocompatibility of carbopol-containing superporous hydrogel composites. *Journal of Applied Polymer Science* 104 (5):2785–91.
88. Bardonnet, P. L., V. Faivre, W. J. Pugh, J. C. Piffaretti, and F. Falson. 2006. Gastroretentive dosage forms: Overview and special case of *Helicobacter pylori*. *Journal of Controlled Release* 111 (1–2):1–18.
89. Yin, L. C., J. Ding, L. Fei, M. He, F. Cui, C. Tang, and C. Yin. 2008. Beneficial properties for insulin absorption using superporous hydrogel containing interpenetrating polymer network as oral delivery vehicles. *International Journal of Pharmaceutics* 350 (1–2):220–29.
90. Tiainen, P., P. E. Gustavsson, A. Ljunglöf, and P. O. Larsson. 2007. Superporous agarose anion exchangers for plasmid isolation. *Journal of Chromatography A* 1138 (1–2):84–94.
91. Spiller, K. L., S. J. Laurencin, D. Charlton, S. A. Maher, and A. M. Lowman. 2008. Superporous hydrogels for cartilage repair: Evaluation of the morphological and mechanical properties. *Acta Biomaterialia* 4:17–25.

Section II

Characterization of Macroporous Polymers

9

Characterization of Macroporous Gels

Irina N. Savina, Paul E. Tomlins,
Sergey V. Mikhalovsky, and Igor Yu. Galaev

CONTENTS

9.1 Introduction

Hydrogels represent a special class of hydrophilic polymer materials and are defined as two- or multicomponent systems consisting of a water or aqueous solution, filled three-dimensional network of polymer chains (Hoffman 2002; Rosiak et al. 2002). Hydrogels adsorb significant amounts of water, from 10 to 20% to several thousand times their dry weight (Hoffman 2002). The amount of water adsorbed depends on the type of polymer and cross-link as well as the cross-link density (Hoffman 2002). The ability of biocompatible hydrogels to adsorb and hold aqueous solutions has been exploited in many biotechnological, biomedical, and pharmaceutical applications over the past 40 years (Peppas et al. 2000; Lozinsky et al. 2003). Hydrogels are commonly used for soft contact lenses, wound dressings, drug delivery systems, and superabsorbents (Lee and Mooney 2001; Drury and Mooney 2003; Lozinsky et al. 2003; Chaterji, Kwon, and Park 2007; Hoare and Kohane 2008).

Hydrogels consist of a network of polymer chains that are linked by either a physical or chemical coupling. The voids or pores formed within the network can range in size from 10 nm to micrometers. This range of pore sizes can be classified into two groups: micropores spanning the range 0.001 to 0.1 μm and macropores that describe pores larger than 0.1 μm. The porosity and pore interconnectivity of a hydrogel determines many of its performance characteristics (swelling capacity and molecular permeability). High levels of porosity and interconnectivity are essential for manufacturing hydrogels intended for use in tissue engineering (Woerly 1997; Gomes et al. 2006; Oh et al. 2007), as separation matrices (Arvidsson et al. 2002; Kumar et al. 2003; Gomes et al. 2006; Hanora et al. 2006; Plieva, Galaev, and Mattiasson 2007), or for making bioreactors (Bansal et al. 2006).

A range of different synthetic and natural biocompatible polymer systems are used for hydrogel manufacturing. In many cases the final material selection will be influenced by the need for specific biological response, mechanical performance, and chemical stability—or instability as gels used in regenerative medicine are usually designed to degrade over a fairly short timescale. The overall porosity, pore size, pore size distribution, and interconnectivity of the pores can be tailored to meet the requirements of a particular application. However, such structural optimization can only be carried out if the hydrogel is suitably characterized. This is not an easy task as many of the commonly used characterization techniques can only be used with dry materials. Drying hydrogels can significantly alter their structure causing shrinkage, deformation, closure of pores, or even collapse of the network (Spiller et al. 2008). The primary cause of such damage is due to the effects of surface tension. As water has a high surface tension to air, then air (evaporative) drying may cause significant damage to hydrogel structure. A number of alternative approaches have been developed for drying these materials, none of which provide an ideal solution to this problem. The most common alternative to air drying involves removing the water by sublimation after rapid freezing followed by exposure to a partial vacuum at low temperature (−70°C or below), through freeze-drying. The rapid freezing approach minimizes the size of any ice crystals that form and thereby reduces their potential to damage or distort the polymer chain network. An alternative approach is to remove the water in a series of steps replacing it by a fluid that has a lower surface tension (e.g., ethanol; Trieu and Qutubuddin 1994) followed by a critical point drying. Drying at the critical point allows reduction of surface tension to be kept to a minimum (http://www.quorumtech.com/Manuals/Current_Technical_Briefs/CPD-Technical-Brief.pdf2002). The critical point of a liquid-gas system is the temperature and pressure at which the densities of liquid and gas are equal and the transition from liquid to gas takes place without an interface. If the sample is totally immersed in a liquid below its critical point and then exposed to conditions above the critical point the sample will be immersed in gas. Hence it can be dried without being exposed to the damaging surface tension forces. In order to carry out the procedure at a convenient temperature and pressure, critical point drying is carried out in carbon dioxide, which has a critical point at 31.1°C at 7.39 MPa. Water is not suitable as its critical point occurs at a high temperature (374°C) and pressure (22.06 MPa) that will damage or destroy the sample. Before critical point drying the liquid in the hydrogel (water) has to be replaced with carbon dioxide. As water is not miscible with carbon dioxide it is usually replaced with ethanol miscible with carbon dioxide, and then ethanol is replaced with carbon dioxide. Although the freezing and critical point techniques were developed to eliminate the potentially damaging effects of drying the material, it remains questionable as to whether the structure of a dried hydrogel is comparable with that of its hydrated equivalent.

Porous materials can contain three different types of pores: enclosed pores, open- or through-pores, and blind-end pores (Figure 9.1) that are defined according to their interconnectivity or lack of it:

- Enclosed pores are isolated within a matrix.
- Open- or through-pores connect to the outer surface of the material and permit flow of liquid (or gas) from one side of the structure to the other.
- Blind-end pores have contact with an exposed internal or external surface through a single orifice.

Terminology can be applied to hydrogels that typically consist of voids within a mesh-like structure. Depending on the manufacturing method for

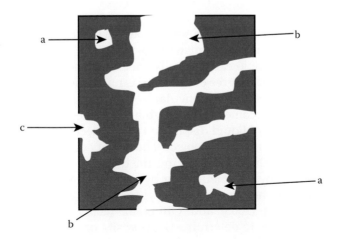

FIGURE 9.1
Schematic representation of the different pore types that can exist within porous materials: (a) enclosed pores, (b) open- or through-pores, and (c) blind-end pores.

producing porous hydrogels structures that have closed, open- or through-pores with different degrees of interconnectivity can be formed (Kang, Lee, and Lee 2005; Perez et al. 2008). The relative proportion of these different pore types found in a structure will have a significant effect on its suitability for a particular application. Recent studies have concentrated on the development of approaches to porous gel characterization that will be fast, accurate, and nondestructive, while providing a comprehensive overview of various morphological and architectural characteristics. Of particular interest is a technique that allows the study of hydrogels in their natural, hydrated state.

This chapter provides an overview of the techniques currently used for the characterization of porous hydrogels by describing the potential of each particular method and its limitations (Table 9.1). New approaches for obtaining structural information for macroporous hydrogels and methods for quantifying these data are also discussed.

9.2 Determination of Porosity by Gravimetric Methods

Gravimetric methods are based on assessing the porosity of hydrogels by measuring the ratio of the mass/volume of a polymer network and liquid (water) that fills the pores within the gel. The density of the material is defined as mass/volume therefore a measure of porosity can be obtained by measuring the weight and volume of a porous sample. The porosity (P) is then calculated according to

$$P = (1 - \rho_p/\rho_s) \times 100\%,$$

where ρ_p is the density of the porous material and ρ_s is the density of the solid material from which the porous material is made.

The mass of the sample can be easily measured to a high level of accuracy. However, the same degree of accuracy is much more difficult to achieve when measuring its volume, particularly for irregularly shaped samples. This issue can be overcome by employing Archimedes's Principle, which states that the volume of liquid displaced by an object completely immersed in it is equal to the volume of the object. The density of an unknown material can be determined either by measuring the decrease in weight of the body due to buoyancy (buoyancy method) or by measuring the weight of the fluid or gas displaced (e.g., helium pycnometry; Tomlins et al. 2005).

Ramay and Zhang (2003) and Oh et al. (2007) have both used the Archimedes's principle to determine the porosity of porous matrices. A dry specimen of weight (W) is immersed into a liquid (nonswelling) of a known volume (V_1). The total volume of liquid and specimen is recorded as V_2. The volume difference ($V_2 - V_1$) is the volume of the material. The specimen is

TABLE 9.1

Techniques Used to Investigate the Porous Structure of Hydrogels

Method	Sample Preparation	Measures	Main Drawbacks
Gravimetric methods	Dependant on the technique used	Pore volume (porosity)	Determination of hydrogel network density
Mercury porosimetry (MP)	Drying required	Porosity, pore size, interconnectivity, surface area, pore size distribution	Insensitive to closed pores. May cause sample distortion. Toxic and destructive assay
Gas pycnometry	Drying required	Pore volume (porosity)	Insensitive to closed pores
Scanning electron microscopy (SEM)	Drying required, samples need to be fixed and sputter coated		Image analysis can be ambiguous, destructive
		Pore size and morphology, wall thickness, pore size distribution	
Cryo-SEM	Drying not needed, hydrated samples are frozen in slush nitrogen (T ~ 210°C)		Image analysis can be ambiguous
X-ray micro-computer tomography	Drying required, contrast agent necessary	Porosity, pore size and morphology, interconnectivity, surface area, wall thickness, pore size distribution	Image analysis can be ambiguous, can be resolution limited, beam hardening
Confocal laser scanning microscopy	Drying not needed, staining with fluorescent marker	Porosity, pore size and morphology, surface area, wall thickness, pore size distribution	Limited laser penetration depth due to scattering. Image analysis can be ambiguous

then removed from the liquid and the residual liquid volume is measured as V_3. The total volume of the specimen, V, apparent density, ρ, and porosity P is then given by

$$V = V_2 - V_3;$$

$$\rho = W/(V_2 - V_3);$$

$$P = (V_1 - V_3)/(V_2 - V_3).$$

Knowing the gel wall swelling capacity ($m_{swollen\ gel\ walls}$) and the total volume of the specimen ($V_{swollen\ gel}$), the amount of liquid (water) in the macropores ($m_{water\ in\ macropores}$) or volume of macropores ($V_{macropores}$) can then be calculated as follows

$$m_{water\ in\ macropores} = m_{swollen\ gel} - m_{swollen\ gel\ walls},$$
$$V_{macropores} = m_{water\ in\ macropores}/\rho_w \times V_{swollen\ gel} \times 100\%,$$

where ρ_w is the density of liquid (water) in the pores.

The swelling capacity of the gel walls can be estimated by squeezing out the capillary water from the pores (Plieva, Andersson et al. 2004) or by re-swelling the dried gel in a high humidity atmosphere (Plieva, Savina, et al. 2004; Plieva et al. 2005a). Another approach is to grind the dried gel to form fine particles that are smaller than the pore dimensions. The swelling capacity of the polymer powder is then a measure of the swelling behavior of the material matrix within the gel.

9.2.1 Mercury Porosimetry (MP)

MP is an established method for studying the porosity of materials. The basis of the technique is to progressively flood a sample with nonwetting mercury by applying pressure. The pressure is related to the pore diameter via the Washburn equation

$$D = (-4\gamma \cos \theta)/p,$$

where D is the pore diameter at pressure p, θ is the surface tension of mercury and γ is the contact angle of mercury with the material. For polymer surfaces a contact angle of 140° and surface tension of 0.48 Nm^{-1} is used (Blacher et al. 2001; Egger et al. 2008).

MP, although a relatively simple and quick technique, has its limitations: the pore size distribution obtained from the Washburn equation is a representation of the actual structure (Tomlins et al. 2005). The calculated porosity is, of course, not sensitive to closed pores and this method is also unable to distinguish between through- and blind-end pores. There are also limitations on the size of pores that can be determined, which is due to the lack of mercury penetration (<7 nm) or flow into very large pores that occurs outside the measurable pressure range.

The applicability of MP for studying hydrogel structure is under discussion. Obviously samples need to be dried before the technique can be used, which is likely to distort the matrix. There are also concerns that additional distortion or collapse of the structure will occur at the high pressures used by mercury porosimeters. However, some authors who have used MP for studying the structure of hydrogels reported negligible compressibility (Gemeinhart, Park, and Park 2000; Maia et al. 2005; Spiller et al. 2008). Others have corrected porosimetry data for sample compression based on the compressibility of a nonporous sample (Dziubla and Lowman 2004).

9.2.2 Gas Pycnometry

This technique detects the change in pressure when the specimen displaces a gas. These data can be used to calculate the specimen volume, density, and porosity (Partap et al. 2007; Van Vlierberghe, 2007, 2008). A specimen of unknown volume is placed in a sample chamber of a known volume, which is then filled with gas and the pressure measured. Helium is commonly used because of its small molecular size (it makes an excellent penetrant and its inert nature). A reference chamber is also filled with gas but at a higher pressure. This chamber is separated from the sample chamber by a closed valve. Both chambers are allowed to equilibrate before the valve separating the two chambers is opened; the resultant pressure is then measured. The calculated sample volume is based on the pressure differential. This information can be used to calculate the density of the sample provided its mass is known. The porosity of the sample can be determined from the following

$$P = (V_b - V_s)/V_b.$$

The skeletal volume (V_s) of the porous material is also measured by this method and its bulk volume (V_b) has been determined by gravimetric methods.

Gas pycnometry is also insensitive to closed pores. Problems can be encountered in accurately measuring the dimensions (length, width, and height) of the sample that is usually prepared as a cube (Ho and Hutmacher 2006). Moreover, gas pycnometry does not provide any additional information with respect to other parameters (e.g., surface area to volume ratio, interconnectivity, and pore size).

9.2.3 Imaging

Imaging is the most popular approach in studying the morphology of macroporous gels and for obtaining quantifiable data on pore sizes and the overall porosity. A variety of methods can be used to image the macroporous gel, such as light microscopy (LM), scanning electron microscopy (SEM), Cryo-SEM, X-ray microcomputer tomography (micro-CT), and confocal laser scanning microscopy (CLSM). The 'validity' of the quality of information obtained using these techniques depends on the type of material and the structure of the gel. Factors such as structural distortion during sample preparation, instrument resolution, and available contrast all have an impact on the quality of the resultant images.

9.2.4 Scanning Electron Microscopy

SEM is widely used imaging technique for characterizing the morphology of porous gels (Kim and Chu 2000; Matzelle and Reichelt 2008; Perez et al. 2008).

Convectional SEM uses a primary beam of electrons that interact with the specimen of interest in a vacuum environment, resulting in the emission of different types of electrons and electromagnetic waves. The secondary electrons ejected from the specimen surface are collected and displayed to provide a high-resolution micrograph. A SEM sample preparation of a hydrogel involves fixing the sample, followed by drying, attachment to a metallic stub, and then coating with an electron dense metal film prior to data collection. The metallic coating is usually applied by sputteting, and is typically 20 to 30 nm thick. Common conductive metals that are used to form the coating gold, platinum, or a gold/palladium alloy.

The SEM provides high-resolution images of hydrogels (Figure 9.2) but the need to dry the specimen is likely to alter the morphology of the gel (Fergg, Keil, and Quader 2001). Alternative techniques such as Cryo-SEM and environmental SEM (ESEM) can be used, which are less demanding in terms of sample preparation and allow the hydrated sample to be studied during dehydration (Plieva et al. 2005a, 2005b; Pratoomsoota et al. 2008). Samples for Cryo-SEM are frozen as rapidly as possible at a very low temperature to minimize damage from ice crystal formation and inhibit ice sublimation (Apkarian and Wright 2005). The sample is cryo-fixed, generally by plunging it into semi-solid nitrogen (slush nitrogen) close to its freezing point of −210°C, then the sample is transferred *in vacuo* to the cold-stage of the SEM cryo-preparation chamber, where fracturing can be performed if necessary.

100 μm

FIGURE 9.2
Scanning electron microscopy image of a cross-section of 10% HEMA macroporous hydrogel prepared at −12°C.

Then the temperature is raised and water present in the sample begins to sublime revealing the gel structure at the surface of the sample. The sample is examined below ambient temperature (typically between –100°C and –175°C). Images can be captured during the entire sublimation process. Figure 9.3 presents Cryo-SEM images made after 5, 23, 35, and 53 min of water sublimation from macroporous gel prepared by cryopolymerization of 2-hydroxyethyl methacrylate (HEMA; for details of cryopolymerization see Chapter 2). The most accessible water, namely water at the surface of the specimen and inside the larger pores, starts to sublime first revealing the underlying morphology (Figure 9.3 a, b, and c). At longer times (>53 minutes) the water sublimes away from the polymer walls resulting in complete drying of the hydrogel and collapse of the pore structure (Figure 9.3 d). This series of images can be used to examine the structure of the hydrogel as it degrades with the data collected at intermediate stages of dehydration, that presumably is representative of the un-perturbed morphology. Another advantage of Cryo-SEM is that it can be used for observing difficult samples,

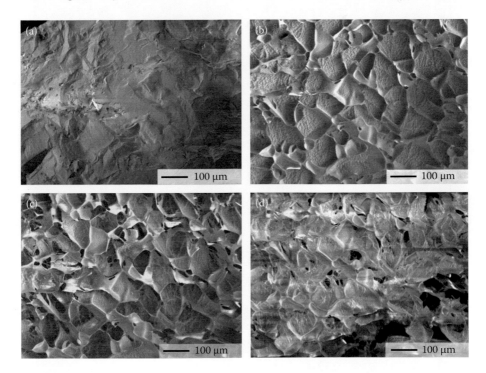

FIGURE 9.3
Cryo-SEM images of the HEMA macroporous hydrogel prepared at –12°C and initial monomer concentration 10%. Images were captured after 5, 23, 35, and 53 min of water sublimation at –95°C (Figure 3 a, b, c, and d, respectively). The initial sample is a frozen block (a), after 23 min of sublimation the pore structure started to appear. Most of the water in the pores had sublimed after 35 min and prolonged sublimation results in dehydration of the polymer walls leading to the collapse of the hydrogel.

such as those with greater beam sensitivity (see http://www.mwrn.com/vendors/ebsciences/cryo_sem.pdf). The technique may also be used to study dynamic processes through a series of time resolved images.

In ESEM, images are captured from samples exposed to a low-pressure gaseous environment (e.g., 1–50 Torr) and high relative humidity (up to 100%). Neither freezing, drying or coating with metal or carbon of the samples is required i.e., hydrogels can be studied in an unaltered, hydrated state albeit under a reduced pressure (Spiller et al. 2008). The internal pressure in the sample chamber can be controlled within fine limits if required to monitor changes in the structure of the gel as it dehydrates (Plieva et al. 2005a). No pores are visible in the fully hydrated acrylamide (AAm) macroporous hydrogel sample as the pores are filled with water (Figure 9.4a). As the water evaporates, the surface details become more visible (Figure 9.4b, c, and d). The ESEM micrographs at a high degree of dehydration show a macroporous structure with pores of hundreds of micrometers in size and dense thin walls (Figure 9.4d).

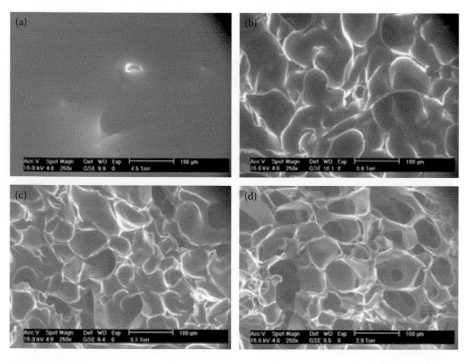

FIGURE 9.4
ESEM images of a polyacrylamide macroporous hydrogel. The hydrogel was prepared by radical polymerization of monomers at −12°C from 10% monomer solution. Images of a fully hydrated sample were captured at different degrees of dehydration at chamber pressures of (a) 4.5, (b) 3.8, (c) 3.1, and (d) 2.9 Torr. (Reprinted from Plieva, F. M., Karlsson, M., Aguilar, M.-R., Gomez, D., Mikhalovsky, S., and Galaev, I. Yu., *Soft Matter*, 1, 303, 2005. Reproduced by permission of The Royal Society of Chemistry.)

SEM, Cryo-SEM, and ESEM provide 2-dimensional images (2-D) showing the 3-dimensional (3-D) morphological structures of gels (Gemeinhart et al. 2000; Blacher et al. 2001; Ghasemi-Mobarakeh, Semnani, and Morshed 2007; Pratoomsoota et al. 2008). Although pore size, pore size distribution, and strut thickness can be assessed (Kim and Chu 2000), it is difficult to make quantitative estimations of pore interconnectivity (Ferreira et al. 2006; Spiller et al. 2008) using such data. However, such limitations can be overcome by using stereo pairs to generate 3D images.

9.2.5 X-Ray Microcomputer Tomography

X-ray microcomputer tomography (x-ray micro-CT) has been recently employed to characterize porous polymer biomaterials (Hedberg et al. 2005; Dubruel et al. 2007; Van Vlierberghe et al. 2007, 2008). X-ray micro-CT is a nondestructive method that, depending on the instrument resolution, can be used to generate 3-D images in a relatively short timeframe. It uses x-rays to create a series of cross-sectional, 2-D images of a sample that are then rendered to create a complete 3-D image (Figure 9.5). Using 3-D analysis software, the sample porosity, pore size, pore size distribution, surface area, and interconnectivity can be determined.

Samples for x-ray micro-CT need to be dry to get the best quality images. The contrast between hydrogels that are mainly composed of low electron density elements such as carbon, hydrogen, oxygen, and air is not high (Savina, Cnudde et al. 2007). This limits the quality of the images that can be obtained. This limitation can be overcome by pretreating the samples with

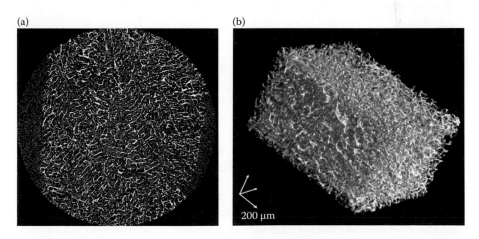

(a)   (b)

200 µm

FIGURE 9.5
X-ray micro-CT images of the cross-section (ca. 2.5 mm in diameter) of 6% HEMA macropous hydrogel prepared by (a) radical polymerization of monomers at –12°C and (b) its 3-D reconstruction. (Reprinted from Savina, I. N., Cnudde, V., D'Hollander, S., Van Hoorebeke, L., Mattiasson, B., Galaev, I. Yu., and Du Prez, F., *Soft Matter*, 3, 1176–84, 2007. Reproduced by permission of The Royal Society of Chemistry.)

an electron dense contrast enhancer that contains iodine or a heavy metal (Darling and Sun 2004).

Large blocks of porous gel of micrometer or even millimeter size can be used in x-ray micro-CT, depending on the resolution required. Typically, most instruments are able to resolve within the 5–10 µm range. This requires a sample of approximately 125 mm^3. Higher resolution instruments are available (<1 µm) but these require much smaller sample sizes (approximately 1 mm^3). Preparing such small samples from highly porous, fragile materials can be impractical particularly for samples that contain large >100 µm cavities. Therefore there is a trade-off between imaging 'large' pores and being able to resolve fine detail.

9.2.6 Magnetic Resonance Imaging (MRI)

Analogous to x-ray micro-CT, magnetic resonance imaging (MRI), can be used to obtain 3-D images of large samples and used to characterize the porosity and pore interconnectivity of hydrogels (Mather et al. 2008). The methods uses a powerful magnetic field to align the nuclear magnetization of hydrogen atoms in the water contained within the material. Radio frequency fields are used to systematically alter the alignment of this magnetization, causing the hydrogen nuclei to produce a rotating magnetic field detectable by the scanner. This signal can be manipulated by additional magnetic fields to build up enough information to construct an image of the material. The ability of MRI to study hydrogels in hydrated state is a major advantage. However, the MRI has relatively low resolution and thus suitable for imaging large pores and being less sensible for finer structures.

9.2.7 Light Microscopy and Confocal Laser Scanning Microscopy

Light microscopy and confocal laser scanning microscopy study the hydrogel in its hydrated state. The major disadvantage with conventional light microscopy (LM) lies in the illumination pattern. As the entire specimen is illuminated, both in-focus and out-of-focus information points contribute to the image resulting in blurring and poor contrast images. There is also a sharp decline in image quality with increasing sample thickness, resulting in the need for thinly sectioned samples (O'Brien et al. 2005; Madaghiele et al. 2008).

Confocal laser scanning microscopy (CLSM) provides almost blur-free optical sectioning of the specimen by eliminating the out-of-focus information through spatial filtering and by using a point source of light. The CLSM is convenient as no special sample preparation is needed apart from staining with fluorescent dye, which for systems with autofluorescence may not be required. Hydrogel samples can be imaged *in situ*, which is a distinct advantage over such techniques as micro-CT and SEM where the drying and freezing/etching process required to prepare sample are likely to damage the original gel structure. The 2-D images can be generated by scanning points across the focal plane of the specimen and subsequently rendered to give a detailed 3-D image (Figure 9.6; Savina, Tuncel et al. 2007). Using image analysis software, the overall characteristics of the pore structure of macroporous gels can be obtained. The truly outstanding feature of CLSM is that the porous structure of hydrogels can be studied in its native hydrated state (Hanthamrongwit et al. 1996; Fergg, Keil, and Quader 2001). Examples of CLSM showing the surface micromorphology of porous polymer matrices can be found in (Semler, Tjia, and Moghe 1997; Savina, Tuncel et al. 2007). CLSM can also be used to track the changes in polymer micromorphology that occur during degradation, for example (Tjia 1998).

The depth of penetration for CLSM is limited to approximately 50–500 µm depending on losses due to scattering even when used with long wavelength in the near infrared, which has more penetrating power than visible light. Thus the information that can be obtained from hydrogels using CLSM is limited to near the surface. This depth limitation can be overcome by sectioning the samples to build up a more complete picture but this approach will

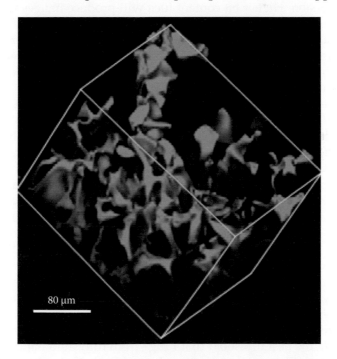

80 µm

FIGURE 9.6

Three-dimensional reconstruction of the two-dimensional CLSM images of acrylamide macroporous hydrogel. The hydrogel was prepared by radical polymerization of monomers (6 w/v %) at –12°C. Fluorescein amine was used to stain the hydrogel for CLSM and images were generated by optical sectioning (500 optical sections) in z-distance of 160 µm. (From Savina, I. N., Tuncel, M., Tuncel, A., Galaev, I. Yu., and Mattiasson, B., *eXPRESS Polymer Letters*, 1, 189–96, 2007.)

introduce artifacts. For hydrogels with pore sizes in the range of 50–300 μm, information on the total porosity, pore shape, size, distribution, and on the wall thickness can still be obtained from the analysis of orthogonal 2-D images (Dainiak et al. 2009; Savina et al. 2009). However it is challenging to make inferences on the existence of closed pores and determine the level of pore intercon-nectivity as analysis of larger cells blocks with dimensions of at least 300 μm, is required. As CLSM is collecting information based on the distribution of fluorescent markers, special care should be taken when staining the gels. The fluorescent marker should be evenly distributed in the polymer network to fully represent the original structure.

9.2.8 Image Analysis

As images will almost certainly be captured, stored, and analyzed in digital format, care should be taken to ensure that the process preserves as much information as possible (ASTM 2603-06 2007). Images captured generally involves digital conversion of an analogue signal that can be done at 8-bit or 16-bit conversion. Higher bit content files will produce images with a much broader range of tonalities. This additional information is highly beneficial if the contrast needs to be enhanced prior to thresholding. However the file size for 16-bit images is much larger than for 8-bit images and for most purposes, 8-bit conversion is suitable and preferable. Moreover, some image analysis programs have limited options for manipulating 16-bit files.

A 256 greyscale monochrome image will have tonalities that range from pure black (greyscale value 0) to pure white (greyscale value 255). In an ideal situation, dark level 0 regions will correspond to the pores within the scaffold and the light grey walls will approach level 255. However, in practice, images show a range of tonalities (Figure 9.7a) which need to be resolved into those are associated with the pores and walls respectively. After the threshold has been applied the data are converted into a binary file where the pores are pure black and the walls pure white. The challenge of segmentation or thresholding is to divide the greyscale image into regions associated with pores (pure black) and walls (pure white), respectively (Figure 9.7b). Porosity calculations are significantly affected by the greyscale threshold value chosen for the segmentation (Mather et al. 2008). Figure 9.8 is illustrating the effect of varying the threshold values in calculation of the porosity of a macroporous poly(HEMA) hydrogel.

There are a number of options available in image analysis software for determining the best threshold that range from fully automatic to a manual *by eye* approach (Rajagopalan, Yaszemski, and Robb 2004). The latter method, although subjective is actually quite consistent even between different individuals (Rajagopalan, Yaszemski, and Robb 2005). Selecting the threshold manually allows the user to explore different options with some software packages providing feedback as to which areas of the image are affected by the changes. The objective methodologies used in automated routines are

(a)   (b)

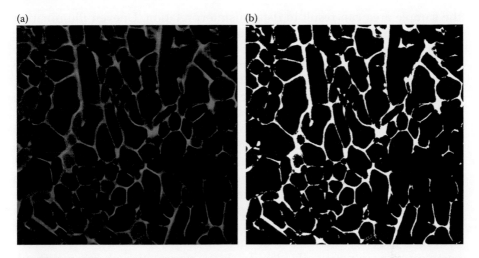

FIGURE 9.7
(a) A 256 greyscale monochrome image and (b) binary image after segmentation or applying threshold.

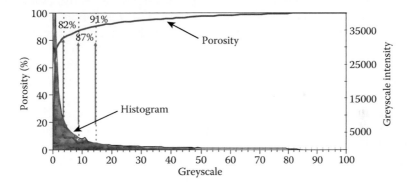

FIGURE 9.8
The dependence of calculated gel porosity on the threshold value: the porosity for threshold for greyscale values of 4, 9, and 17 was 82, 87, and 91%, respectively.

based on a statistical analysis of the histogram interpreting the histogram, which is interpreted as two overlapping normal distributions of the tonalities associated with the pores and walls, respectively.

Images can also be sharpened, smoothed, de-blurred, or restored to improve the contrast between the polymer walls and the pores to derive quantitative information (Blacher et al. 2001). Such enhancements help to improve image quality by eliminating noise or other artifacts, however, it may also introduce additional errors in the calculation of pore characteristics (Tomlins et al. 2003).

Image analysis is based on the interpretation of differences in greyscale intensity or hue and therefore any unevenness in illumination over the sample area can result in data misinterpretation. Uneven illumination occurs due to the source itself, due to a nonlinear response of the detector, or due to poor sample alignment. The effect of nonlinear illumination can be removed by subtracting a background captured in the absence of a sample or by applying a gradient filter.

Once the threshold of the greyscale value has been defined, the greyscale image is reduced to a binary black and white image (Figure 9.7b). Tonalities between pure black and the threshold, are reduced to black, level 0, and those between the threshold and pure white are converted to level 255 (pure white). The binary image that is now unambiguously divided into pores and walls, can be analyzed in terms of the ratio of pore area (or volume) to the sample area (or volume) to obtain the porosity. The dimensions and shape of the pores can also be analyzed to produce a set of quantifiable measures that represent the structure. Analysis of reconstructed 3-D representation of hydrogels provides information about pore shape, tortuosity, and interconnectivity and can potentially identify closed pores although this is difficult to do in highly porous interconnected structures.

Conventional light microscopy, fluorescent microscopy, and SEM produce 2-D images (i.e., 2-D representation of 3-D structures can be difficult to interpret). For instance, SEM produces images of high resolution and contrast, however extraction of quantitative data from SEM images is difficult as the images contain information not only from the exposed surface but also from the regions below it (Mather et al. 2008). The analysis of thinly sectioned samples improve the quality of the resulted calculation, however this method requires preparation and drying of thin slices of the hydrogel which may result in significant alteration of the structure. The x-ray micro-CT, MRI, and CLSM offer significant advantages over other imaging techniques as they allow 3-D virtual images of the structure to be created by assembling a series of 2-D images obtained at different locations within the sample. This allows not only pore shape, pore size, and strut-wall thickness but also the total pore volume, surface area, and levels of interconnectivity to be determined.

9.3 Determination of the Nanoporous Structure of Hydrogels Using ¹H NMR Spectroscopy and Thermally Stimulated Depolarization (TSD)

Investigating the behavior of water in hydrogels using ¹H NMR spectroscopy and thermally stimulated depolarization (TSD) current can provide useful information about their nanoporous structure. Both techniques are based on studying the properties of water in the hydrogel with layer-by-layer freezing out of bulk and interfacial water. Three types of water have been classified in hydrogels: nonfreezing or bound water, freezing interfacial or intermediate water, and free or bulk water (Fyfe and Blazek 1997; Baumgartner et al. 2003; Plieva et al. 2005a). Unlike bulk water which freezes at 0°C, there is a fraction of water which remains liquid at temperatures below the normal freezing point. This water is known as bound or structured water. It remains liquid at temperature below 0°C because of interaction with a solid surface or polar solute molecules and ions, which disturb the hydrogen bond network in interfacial water preventing ice formation. Characteristics of the adsorbed water or water confined to pores can be explored using temperature dependence of the ^1H NMR and TSD current signal intensity of the unfrozen interfacial water at temperatures below 0°C. ^1H NMR spectroscopy records the proton signal (particularly, water protons), which depends on both the local environment and how the water interacts with the polymer network. At the temperature below 0°C, the proton signals from ice and macromolecules did not contribute to the recorded ^1H NMR spectra because of the much shorter time ($\sim 10^{-6}$ s) of the transverse relaxation of protons in solids in comparison with that of mobile water (~ 2.4 s for bulk water and 0.01–0.5 s for bound water; Gun'ko et al. 2005). Thus at a temperature below 0°C, ^1H NMR signal corresponds to mobile (liquid water), which can be quantified. Comparing these data with thermodynamic properties of unfrozen water provide information on the nanoporous structure of hydrogels. Thermodynamically, the lowering of the freezing point of interfacial water below 0°C with the perturbed hydrogen bond network is defined by reduction of Gibbs free energy due to the interactions with surroundings. The narrower the pores, the greater the depression of the freezing point of water. Using the dependence of Gibbs free energies on pore size and the relationship between the amount of non-frozen water and its freezing point, the size and volume of the hydrogel pores filled with the non-frozen water could be calculated (Mikhalovska et al. 2006).

Measurement of the TSD current is another technique that has been successfully used to study the behavior of water at the surface. The TSD measures dielectric properties of water related to relaxation of bound (dipoles of water molecules) and mobile (protons) charges. It can be effectively used to analyze the associativity and mobility of the interfacial water because the measured temperature dependence of depolarization characterizes the environment of water molecules. It can be used to study the interaction of these molecules with their neighbors and to measure the number of hydrogen bonds per molecule: their strength, mobility of bound charges (water dipoles), free charges (protons), and the size of water clusters. The pore size distribution measurement is based on the relaxation energy of dipoles (as bound charges), which depends on temperature assuming that relaxation of weaker bound dipoles occurs at a lower temperature, and the smaller number of the hydrogen bonds per water molecule, the smaller the water cluster is and its location in the smaller pore. Thus the state and amount of unfrozen water as a function of temperature below the normal freezing point could be studied by TSD current methods and the nanoporous structure of the hydrogel could be assessed in the way analogous to NMR (Mikhalovska et al. 2006).

9.3.1 Gel Permeability

Hydrogel permeability is a very important parameter as it characterizes the diffusion and transport of the solutes through the matrix. It depends on a number of parameters including the nature of the solutes, temperature, pH, ionic strength, and solute concentration, but mostly on the interconnectivity of pores, pore orientation, pore size, and distribution within the hydrogel. Hydrogels with high levels of porosity are not necessarily highly permeable, as permeability is highly dependent on the degree of interconnectivity. Thus measuring probe solute permeation through the matrix is a useful metric for assessing permeability.

This can be measured via a direct permeation experiment that measures the rate of water flow through the gel under a certain (usually 1 meter of water column) hydrostatic pressure (Arvidsson et al. 2002; Plieva, Savina et al. 2004; Plieva et al. 2006).

A permeability coefficient, Pc, can be determined for hydrogel matrices that is derived from Darcy's law (Agrawal et al. 2000; Yao et al. 2006) where;

$$Pc = \mu FL/hA,$$

where F is the flow rate, L the thickness of the specimen in the flow direction, h the pressure drop across specimen, A the cross-sectional area of specimen, and μ the fluid viscosity.

Probes of different molecular weights or particle sizes are used to investigate pore size in gels. Dextran or proteins labeled with fluores cein, are used as molecular probes. Experiments can be conducted in two ways: tracking the flow of probe molecules through the matrix or by studying the diffusion of the probe molecules through the gel along a concentration gradient. Solutes of different sizes will diffuse in the gel at different rates depending on their size, flexibility, and interactions with the hydrogel network. There will be always a portion of the imbibed water in a gel that is not accessible to probes due to pore dead ends, small pores with the diameter less than that of the probe molecule, and probe-polymer matrix interactions.

If the diffusion through the open pores is straightforward with molecules moving to reduce the concentration gradient, then the process is described by Fick's first law

$$J = -D\Delta c$$

where the flux, J depends on the concentration gradient, Δc over a specific distance through a proportional constant, the diffusion coefficient D.

The applicability of this simple relationship to porous material depends on the effective path length of an open pore that is often substantially longer than the sample dimensions, due to its tortuosity. In this case, the diffusion coefficient measured for straight pore diffusion needs to be reduced by an appropriate amount

$$D_{porous} = (\varepsilon/\tau) \times D_{free},$$

where ε/τ is the pore tortuosity factor. The tortuosity factor in porous material will differ for different molecular species due to both size and charge effects.

There is a relationship between the diffusion coefficient and the molecular mass of solute; the larger the molecules the slower their motion (Mikhalovsky et al. 2006). Studying the penetration of the proteins through a collagen scaffold demonstrated that proteins with different molecular weights could penetrate through the collagen hydrogel without significant retardation. However, the diffusion of a portion of protein molecules through the hydrogel is slower than in the aqueous solution because of the protein adsorption within the matrix as well as retardation of the diffusion for the next portion of molecules through pores because adsorbed protein molecules block the pore cross-section (Mikhalovsky et al. 2006).

Beside permeation measurements, permeability, k, can also be derived from known porosity, tortuosity, and the average specific surface area. In these calculations, the pore network is represented by a series of parallel channels. Tortuosity is defined as the ratio for the length of the actual flow path connecting two points to the straight-line distance between the points. The average specific surface area is defined as the ratio of the pore area to the scaffold volume (Ho and Hutmacher 2006).

$$k = P^3/T^2S^2,$$

where P is the porosity, T the tortuosity and S the average specific area.

9.4 Concluding Remarks

Characterization of the porous hydrogels remains a challenging task. Special attention should be paid to the sample preparation method (drying methods, fixation, metal coating, slicing) as this may have a sigificant effect on the hydrogel morphology resulting in significant structural distortion. The analysis of the dried hydrogels remains controversial, as the structure of dried hydrogel could be altered during the drying process. Thus the techniques of studying hydrogels in a hydrated state such as light and fluorescence microscopy, ESEM, MRI, and CLSM are currently attracting attention and these techniques have been often used together with commonly used ones (SEM and MP) to prove the results obtained.

Imaging techniques remain a powerful method for the hydrogel morphology analysis. The pore size, shape, and distribution could be assessed or calculated using image analysis techniques. The resolution depends on the technique and instrument used and may require the special sample preparation (fixation, drying, sputter coating, or treatment with contrast enhancer). Despite image analysis being sensitive to the thresholding methodology and affected by unevenness in illumination, it is a useful technique to get quantitative data on the porosity. 2-D images representing 3-D structures of the hydrogels, because of interfering with the intermediate plane above or below the plane of interest, are particularly difficult for analysis and require special precautions with the data derived. Analysis of thinly sectioned samples improves the quality of data obtained, however this is a destructive method, which may result in distortion of hydrogel porosity. Micro-CT, MRI, and CLSM techniques capable of nondestructive sectioning and collecting information from one plane of interest possess the key advantage over other techniques. They allow recreation of a 3-D macroporous hydrogel model and make the porosity calculation of the model. Thus, the analysis of the 3-D model assesses quantitatively interconnectivity of the pores.

Analysis of only total porosity is not particularly useful information. The modern analysis should provide information on the pore size, shape, tortuosity, distribution, as well as interconnectivity. For many applications, especially when hydrogels are used as a scaffold or matrix for separation, the interconnectivity becomes a key feature. The interconnectivity can be assessed by the permeability test, however the 3-D analysis image analysis provide proper characterization as it finds and estimates the number of closed-enclosed pores as well.

At the moment there are not many techniques that can provide the whole spectra of hydrogel porosity characteristics and a combination of different techniques should be used to get a valid and reliable hydrogel porosity analysis.

References

Agrawal, C. M., McKinney, J. S., Lanctot, D., and Athanasiou, K. A. 2000. Effects of fluid flow on the *in vitro* degradation kinetics of biodegradable scaffolds for tissue engineering. *Biomaterials* 21:2443–52.

Apkarian, R. P., and Wright, E. R. 2005. Cryo and cryo-etch methods or quality preservation of hydrogels imaged at high magnification by low temperature SEM. *Microscopy Microanalysis* 11:1088–89.

Arvidsson, P., Plieva, F. M., Savina, I. N., Lozinsky, V. I., Fexby, S., Bulow, L., Galaev, I. Y. and Mattiasson, B. 2002. Chromatography of microbial cells using continuous supermacroporous affinity and ion-exchange columns. *Journal of Chromatography A* 977:27–38.

ASTM F 2603-06 Standard Guide for Interpreting Images of Polymeric Tissue Scaffolds. *no. (West Conshohocken, PA: ASTM Interpretation)*

Bansal, V., Roychoudhury, P. K., Mattiasson, B., and Kumar, A. 2006. Recovery of urokinase from integrated mammalian cell culture cryogel bioreactor and purification of the enzyme using *p*-aminobenzamidine affinity chromatography. *Journal of Molecular Recognition* 19:332–39.

Baumgartner, S., Lahajnar, G., Sepe, A., and Kristl, J. 2003. Investigation of the state and dynamics of water in hydrogels of cellulose ethers by ^{1}H NMR spectroscopy. *AAPS PharmSciTech*, online at http://www.aapspharmscitech.org/default/issueView.asp?vol=03&issue=04 3:36.

Blacher, S., Maquet, V., Pirard, R., Pirard, P. J., and Jérôme, R. 2001. Image analysis, impedance spectroscopy and mercury porosimetry characterisation of freeze-drying porous materials. *Colloids and Surfaces A: Physicochemical and Engineering Aspects* 187–188:375–83.

Chaterji, S., Kwon, I. K., and Park, K. 2007. Smart polymeric gels: Redefining the limits of biomedical devices. *Progress in Polymer Science* 32:1083–1122.

Dainiak, M., Allan, I., Savina, I. N., Cornelio, L., James, L., James, S., Mikhalovsky, S., Jungvid, H. and Galaev, I. Y. 2009. Gelatin-fibrinogen cryogel dermal matrices for wound repair I: Preparation, optimisation and *in vitro* study. Submitted.

Darling, A. L., and Sun, W. 2004. 3D microtomographic characterization of precision extruded poly-ε-caprolactone scaffolds. *Journal of Biomedical Materials Research Part B: Applied Biomaterials* 70B:311–17.

Drury, J. L., and Mooney, D. J. 2003. Hydrogels for tissue engineering: Scaffold design variables and applications. *Biomaterials* 24:4337–51.

Dubruel, P., Unger, R., Van Vlierberghe, S., Cnudde, V., Jacobs, P. J. S., Schacht, E., and Kirkpatrick, C. J. 2007. Porous gelatin hydrogels: 2. *In vitro* cell interaction study. *Biomacromolecules* 8:338–44.

Dziubla, T. D., and Lowman, A. M. 2004. Vascularization of PEG-grafted macroporous hydrogel sponges: A three-dimensional *in vitro* angiogenesis model using human microvascular endothelial cells. *Journal of Biomedical Materials Research Part A* 68A:603–14.

Egger, C. C., du Fresne, C., Raman, V. I., Schädler, V., Frechen, T., Roth, S. V., and Müller-Buschbaum, P. 2008. Characterization of highly porous polymeric materials with pore diameters larger than 100 nm by mercury porosimetry and X-ray scattering methods. *Langmuir* 24:5877–87.

Fergg, F., Keil, F. J., and Quader, H. 2001. Investigations of the microscopic structure of poly(vinyl alcohol) hydrogels by confocal laser scanning microscopy. *Colloid Polymer Science* 279:61–67.

Ferreira, L., Figueiredo, M. M., Gil, M. H., and Ramos, M. A. 2006. Structural analysis of dextran-based hydrogels obtained chemoenzymatically. *Journal of Biomedical Materials Research Part B: Applied Biomaterials* 77B:55–64.

Fyfe, C. A., and Blazek, A. I. 1997. Investigation of hydrogel formation from hydroxypropylmethylcellulose (HPMC) by NMR spectroscopy and NMR imaging techniques. *Macromolecules* 30:6230–37.

Gemeinhart, R. A., Chen, J., Park, H., and Park, K. 2000. pH-sensitivity of fast responsive superporous hydrogels. *Journal of Biomaterials Science Polymer Edition* 11:1371–80.

Gemeinhart, R. A., Park, H., and Park, K. 2000. Pore structure of superporous hydrogels. *Polymers for Advanced Technologies* 11:617–25.

Ghasemi-Mobarakeh, L., Semnani, D., and Morshed, M. 2007. A novel method for porosity measurement of various surface layers of nanofibers mat using image analysis for tissue engineering applications. *Journal of Applied Polymer Science* 106:2536–42.

Gomes, M. E., Holtorf, H. L., Reis, R. L., and Mikos, A. G. 2006. Influence of the porosity of starch-based fiber mesh scaffolds on the proliferation and osteogenic differentiation of bone marrow stromal cells cultured in a flow perfusion bioreactor. *Tissue Engineering* 12:801–9.

Gun'ko, V. M., Turov, V. V., Bogatyrev, V. M., Zarko, V. I., Leboda, R., Goncharuk, E. V., Novza, A. A., Turov, A. V., and Chuiko, A. A. 2005. Unusual properties of water at hydrophilic/hydrophobic interfaces. *Advances in Colloid and Interface Science* 118:125–72.

Hanora, A., Savina, I., Plieva, F. M., Izumrudov, V. A., Mattiasson, B., and Galaev, I. Y. 2006. Direct capture of plasmid DNA from non-clarified bacterial lysate using polycation-grafted monoliths. *Journal of Biotechnology* 123:343–55.

Hanthamrongwit, M., Wilkinson, R., Osborne, C., Reid, W. H., and Grant, M. H. 1996. Confocal laser-scanning microscopy for determining the structure of and keratinocyte infiltration through collagen sponges. *Journal of Biomedical Materials Research, Part A* 30:331–39.

Hedberg, E. L., Shih, C. K., Lemoine, J. J., Timmer, M. D., Liebschner, M. A., Jansen, J. A., and Mikos, A. G. 2005. *In vitro* degradation of porous poly(propylene fumarate)/poly(D,L-lactic-*co*-glycolic acid) composite scaffolds. *Biomaterials* 26:3215–25.

Ho, S. T., and Hutmacher, D. W. 2006. A comparison of micro CT with other techniques used in the characterization of scaffolds. *Biomaterials* 27:1362–76.

Hoare, T. R., and Kohane, D. S. 2008. Hydrogels in drug delivery: Progress and challenges. *Polymer* 49:1993–2007.

Hoffman, A. S. 2002. Hydrogels for biomedical applications. *Advanced Drug Delivery Reviews* 43:3–12.

Kang, H. G., Lee, S. B., and Lee, Y. M. 2005. Novel preparative method for porous hydrogels using overrun process. *Polymer International* 54:537–43.

Kim, S.-H., and Chu, C.-C. 2000. Pore structure analysis of swollen dextran-methacrylate hydrogels by SEM and mercury intrusion porosimetry. *Journal of Biomedical Materials Research Part B: Applied Biomaterials* 53:258–66.

Kumar, A., Plieva, F. M., Galaev, I. Y., and Mattiasson, B. 2003. Affinity fractionation of lymphocytes using a monolithic cryogel. *Journal of Immunological Methods* 283:185–94.

Lee, K. Y., and Mooney, D. J. 2001. Hydrogels for tissue engineering. *Chemical Reviews* 101:1869–79.

Lozinsky, V. I., Galaev, I. Y., Plieva, F. M., Savina, I. N., Jungvid, H., and Mattiasson, B. 2003. Polymeric cryogels as promising materials of biotechnological interest *Trends in Biotechnology* 21:445-451.

Madaghiele, M., Sannino, A., Yannas, I. V., and Spector, M. 2008. Collagen-based matrices with axially oriented pores. *Journal of Biomedical Materials Research Part A* 85:757-767.

Maia, J., Ferreira, L., Carvalho, R., Ramos, M. A., and Gil, M. H. 2005. Synthesis and characterization of new injectable and degradable dextran-based hydrogels. *Polymer* 46:9604–14.

Mather, M. L., Morgan, S. P., White, L. J., Tai, H., Kockenberger, W., Howdle, S. M., Shakesheff, K. M., and Crowe, J. A. 2008. Image-based characterization of foamed polymeric tissue scaffolds. *Biomedical Materials* 3:1–11.

Matzelle, T., and Reichelt, R. 2008. Review: Hydro-, micro- and nanogels studied by complementary measurements based on SEM and SFM. *Acta Microscopica* 17:45–61.

Mikhalovska, L. I., Gun'ko, V. M., Turov, V. V., Zarko, V. I., James, S. L., Vadgama, P., Tomlins, P. E., and Mikhalovsky, S. V. 2006. Characterisation of the nanoporous structure of collagen-glycosaminoglycan hydrogels by freezing-out of bulk and bound water. *Biomaterials* 27:3599–3607.

Mikhalovsky, S. V., Mikhalovska, L. I., James, S. L., Tomlins, P. E., Grant, P. V., Vadgama, P., and Gun'ko, V. M. 2006. Characterization of hard and soft porous materials and tissue scaffolds. In *Combined and hybrid adsorbents, fundamentals and applications, NATO security through science series C: Environmental security*, eds. J. M. Loureiro and M. T. Kartel. Ukraine: Springer 309–20.

O'Brien, F. J., Harley, B. A., Yannas, I. V., and Gibson, L. J. 2005. The effect of pore size on cell adhesion in collagen-GAG scaffolds. *Biomaterials* 26:433–41.

Oh, S. H., Park, I. K., Kim, J. M., and Lee, J. H. 2007. *In vitro* and *in vivo* characteristics of PCL scaffolds with pore size gradient fabricated by a centrifugation method. *Biomaterials* 28:1664–71.

Partap, S., Muthutantri, A., Rehman, I. U., Davis, G. R., and Darr, J. A. 2007. Preparation and characterisation of controlled porosity alginate hydrogels made via a simultaneous micelle templating and internal gelation process. *Journal of Materials Science* 42:3502–7.

Peppas, N. A., Bures, P., Leobandung, W., and Ichikawa, H. 2000. Hydrogels in pharmaceutical formulations. *European Journal of Pharmaceutics and Biopharmaceutics* 50:27–46.

Perez, P., Plieva, F. M., Gallardo, A., San Roman, J., Aguilar, M. R., Morfin, I., Ehrburger-Dolle, F., et al. 2008. Bioresorbable and nonresorbable macroporous thermosensitive hydrogels prepared by cryopolymerization. Role of the cross-linking agent. *Biomacromolecules* 9:66–74.

Plieva, F. M., Andersson, J., Galaev, I. Y., and Mattiasson, B. 2004. Characterization of polyacrylamide based monolithic columns. *Journal of Separation Science* 27:828–36.

Plieva, F. M., Galaev, I. Y., and Mattiasson, B. 2007. Monolithic gels prepared at subzero temperatures as novel materials for chromatography of particulate-containing fluids and cell culture applications. *Journal of Separation Science* 30:1657–71.

Plieva, F. M., Huiting, X., Galaev, I. Y., Bergenstahl, B., and Mattiasson, B. 2006. Macroporous elastic polyacrylamide gels prepared at subzero temperatures: control of poros structure. *Journal of Materials Chemistry* 16:4065–73.

Plieva, F. M., Karlsson, M., Aguilar, M.-R., Gomez, D., Mikhalovsky, S., and Galaev, I. Y. 2005a. Pore structure in supermacroporous polyacrylamide based cryogels. *Soft Matter* 1:303–9.

Plieva, F. M., Karlsson, M., Aguilar, M.-R., Gomez, D., Mikhalovsky, S., Galaev, I. Y., and Mattiasson, B. 2005b. Pore structure of macroporous monolithic cryogel prepared from poly(vinyl alcohol). *Journal of Applied Polymer Science* 100:1057–66.

Plieva, F. M., Savina, I. N., Deraz, S., Andersson, J., Galaev, I. Y., and Mattiasson, B. 2004. Characterization of supermacroporous monolithic polyacrylamide based matrices designed for chromatography of bioparticles. *Journal of Chromatography B* 807:129–37.

Pratoomsoot, C., Tanioka, H., Hori, K., Kawasaki, S., Kinoshita, S., Tighe, P. J., Dua, H., Shakesheff, K. M., and Rose, F. R. A. J. 2008. A thermoreversible hydrogel as a biosynthetic bandage for corneal wound repair. *Biomaterials* 29:272–81.

Rajagopalan, S., Lu, L., Yaszemski, M. J., and Robb, R. A. 2005. Optimal segmentation of microcomputed tomographic images of porous tissue-engineering scaffolds. *Journal of Biomedical Materials Research Part A* 75A:877–87.

Rajagopalan, S., Yaszemski, M. J., and Robb, R. (2004). Evaluation of thresholding techniques for segmenting scaffold images in tissue engineering. Proceedings paper of the society of photooptical instrumentation engineers (SPIE).

Ramay, H. R., and Zhang, M. 2003. Preparation of porous hydroxyapatite scaffolds by combination of the gel-casting and polymer sponge methods. *Biomaterials* 24:3293–3302.

Rosiak, J. M., Janik, I., Kadlubowski, S., Kozicki, M., Kujawa, P., Stasica, P., and Ulanskia, P. (2002). *Radiation formation of hydrogels for biomedical applications.* Oxford: Elsevier.

Savina, I. N., Cnudde, V., D'Hollander, S., Van Hoorebeke, L., Mattiasson, B., Galaev, I. Y., and Du Prez, F. 2007. Cryogels from poly(2-hydroxyethyl methacrylate): macroporous, interconnected materials with potential as cell scaffolds. *Soft Matter* 3:1176–84.

Savina, I. N., Dainiak, M., Phillips, G., Galaev, I. Y., and Mikhalovsky, S. 2009. Nondestructive, quantitative method for the 3D characterisation of porous hydrogel scaffolds. Submitted.

Savina, I. N., Tuncel, M., Tuncel, A., Galaev, I. Y., and Mattiasson, B. 2007. Analysis of polymer grafted inside the porous hydrogel using confocal laser scanning microscopy. *eXPRESS Polymer Letters* 1:189–96.

Semler, E. J., Tjia, J. S., and Moghe, V. P. 1997. Analysis of surface microtopography of biodegradable polymer matrices using confocal reflection microscopy. *Biotechnology Progress* 13:630–34.

Spiller, K., Laurencin, S., Charlton, D., Maher, S., and Lowman, A. 2008. Superporous hydrogels for cartilage repair: Evaluation of the morphological and mechanical properties. *Acta Biomaterialia* 4:17–25.

Tjia, J. S. 1998. Analysis of 3-D microstructure of porous poly(lactide-glycolide) matrices using confocal microscopy. *Journal of Biomedical Materials Research* 43:291–99.

Tomlins, P., Grant, P., Mikhalovsky, S., Mikhalovska, L., James, S., and Vadgama, P. 2005. Characterization of polymeric tissue scaffolds. *Measurement Good Practice Guide* No. 89.

Tomlins, P., Grant, P. V., Mikhalovsky, S., James, S., and Mikhalovska, L. 2003. Measurement of pore size and porosity of tissue scaffolds. *ASTM Special Technical Publications* 1452:3–11.

Trieu, H. H., and Qutubuddin, S. 1994. Polyvinyl alcohol hydrogels I. Microscopic structure by freeze-etching and critical point drying techniques. *Colloid and Polymer Science* 272:301–9.

Van Vlierberghe, S., Cnudde, V., Dubruel, P., Masschaele, B., Cosijns, A., Paepe, I. D., Jacobs, P. J. S., Van Hoorebeke, L., Remon, J. P., and Schacht, E. 2007. Porous gelatin hydrogels: 1. Cryogenic formation and structure analysis *Biomacromolecules* 8:331–37.

Van Vlierberghe, S., Dubruel, P., Lippens, E., Masschaele, B., Van Hoorebeke, L., Cornelissen, M., Unger, R., Kirkpatrick, C. J., and Schacht, E. 2008. Toward modulating the architecture of hydrogel scaffolds: Curtains versus channels. *Journal of Materials Science: Materials in Medicine* 19:1459–66.

Woerly, S. 1997. Porous hydrogels for neural tissue engineering. *Material Science Forum*, online at http://www.scientific.net 250:53–68.

Yao, K., Yun, J., Shen, S., Wang, L., He, X., and Yu, X. 2006. Characterization of a novel continuous supermacroporous monolithic cryogel embedded with nanoparticles for protein chromatography. *Journal of Chromatography A* 1109:103–10.

10

Macroporous Polymeric Materials: Synthetic Strategies and Morphological Characterizations

Anil K. Bajpai and Sandeep K. Shukla

CONTENTS

10.1 Introduction

A pore, defined in general sense as a limited space or spatial confinement, is known to form the basis of modern materials science and contributes significantly to a wide spectrum of novel applications ranging from tissue engineering, immunodiagnostics, and chromatographic support materials to combinatorial chemistry (Zhao 2006). The understanding, design, and precise control over pore dimensions have significantly advanced science and technology, and are playing prime roles in the exploration and application of new technologies. As per the recent IUPAC conventions, pores are classified into three categories: micropores (less than 2 nm pore size), mesopores (between 2 and 50 nm pore size), and macropores (larger than 50 nm pore size). No matter that porous materials are nonpolymeric, polymeric, or hybrid in origin, they are all of great scientific and technological significance because of the ability of the pore wall to interact with atoms, ions, molecules, and supermolecules. Moreover, the porous materials may be made tailorable to specific functional groups for the desired application, which also enables these materials to be highly attractive in frontier research (Hollister 2005).

Among porous materials of various chemical and compositional types, the macroporous materials owe a prime position in the area of materials science due to advancements in the design and processing of these materials driven by the rapid growth of emerging applications like energy conversion and storage, environment friendly catalysis, sensors, tissue engineering, DNA sequencing, drug delivery, cell markers, and photonics (Omidian, Rocca, and Park 2005). All of these technological applications require a high level of control over the dimension, structure, and properties of the pores and porous materials, which can easily be achieved by strategically designing a sitemap of the synthesis part of the material.

10.2 Applications of Macroporous Materials

In many biotechnological and biomedical applications a biocompatible and hydrophilic structure with desired porosity is required, which is best provided by a macroporous matrix of polymeric nature. Thus, some specific prominent applications of macroporous materials are discussed below.

10.2.1 Enzyme Immobilization

Immobilization of enzymes has attained significant interest for separation of the enzymes from reaction mixtures. It imparts greater stability to the enzyme, allowing greater use in the development of continuous processing, enabling better control of the catalytic processes, and an economical utilization of an otherwise cost-prohibitive enzyme application. Many porous polymers have been studied extensively as support for immobilization as the porosity of these polymers have a direct effect on diffusion of substrates and products through them, which affects the rate of enzymatic conversion. It is well recognized that the presence of meso or macropores is a key requirement to minimizing diffusional limitations (Hogde 1988). Reactive functional groups of the macroporous matrix assist the formation of permanent covalent bonds between the support and enzyme without affecting its tertiary structure. A number of such macroporous polymer supports are cited in literature with different types of enzymes (Bhushan et al. 2007).

10.2.2 Tissue Engineering

The incidence of organ and tissue loss or failure is increasing steadily, whereas immune rejection and the number of donors limit the traditional surgical treatment of implantation of a healthy organ from a donor. As an application of tissue engineering, the use of cell transplantation is now being investigated as an alternative therapeutic strategy for tissue repair and organ replacement (Liu and Ma 2004). In culturing cells, shape of the scaffold, a temporary substrate to allow growth and specialization of the cell culture plays an important role (Yang et al. 2001). Polymeric scaffolds must be porous enough to allow a high density of cells to be seeded, yet also possess sufficient mechanical stability and a well-defined network of interconnected pores to permit growth into the implanted structure. Thus, macroporous polymer materials meet all the desired qualities and, therefore, have been extensively utilized in tissue engineering (Sun et al. 2005).

10.2.3 Chromatographic Support Materials

In a search for an enhanced and simple way to prepare chromatographic separation media, several groups found that macroporous polymers in the shape of a continuous rod could be a useful alternative to columns packed with particles (Fields 1996). Such a porous polymer rod, often called a monolith, is characterized by a system of interconnected pores with a bimodal distribution: the small pores provide the desired surface area required for the specific interactions, while the larger channels allow achieving a high flow rate at moderate pressures. Therefore, major advantages of using such monoliths in chromatography include an increased speed, capacity, and resolution. Moreover, they eliminate most of the problems related to the column packing procedure.

10.2.4 Controlled Drug Delivery

Both the conventional macroporous polymers and the recently designed high internal phase emulsion polymers (HIPE), have been used in achieving zero order release dynamics for a variety of drugs (Meese et al. 2002). Whereas

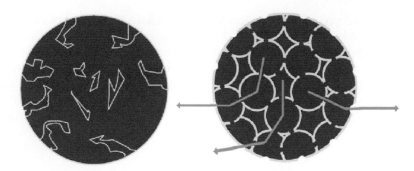

FIGURE 10.1
Conventional macroporous polymer containing irregular pores of Å dimensions that terminate within a solid matrix, and high internal phase emulsion polymer with large cavities of μm dimensions that are interconnected. (From Landgraf, W., Li, N.-H., and Benson, J., *Drug Del. Tech.*, 5 (2), 48–55, 2005. With permission.)

the former architecture contains irregular pores of Ångström dimensions that terminate within the solid matrix, the later one contains large cavities of micrometer dimensions. Another significant difference is that in the conventional macroporous matrix total porosity is typically 50% while in the later matrix it is more than 70%. The structural difference between the two matrices is shown in Figure 10.1.

Apart from these applications, the macroporous polymers find a multitude of applications in the biomedical and pharmaceutical field.

10.3 Desired Characteristics of Macroporous Materials

The intrinsic material properties such as mechanical (Gong et al. 2006) or thermal (Mathieu et al. 2006) and structural features like pore size, porosity, pore morphology, and so forth, play a vital role in shaping suitability of the macroporous materials for a specific application. For instance, from a tissue engineering viewpoint, the optimal pore size should be 20 μm for the ingrowths of fibroblasts and hepatocytes, between 20 and 150 μm for the skin regeneration, and in the range of 100–150 μm for bone regeneration (Ma 2004). Similarly when the pore size of polymer particles is in the range of 100–150 Å, a longer time will be required to separate the macromolecular components because of their slow diffusion rate (Sherrington 1998). For chromatographic separation, it is recommended that the size of the pores must be almost 10 to 20 times the solute molecular size.

For designing macroporous polymers as scaffolds in tissue engineering (Hollister, Maddox, and Taboas 2002) the macro- and micro-structural properties that have to be taken into consideration may be summarized as shown in Table 10.1.

The macroporous polymer should possess an open-pore geometry with a highly porous surface and microstructure that allows cell ingrowth and reorganization *in vitro* and provides the necessary space for neovascularization from surrounding tissues *in vivo*.

As mentioned earlier, the ultimate application of porous material decides the size of the pores that the materials should have to replace or restore tissue or organ function(s). Since the pore size and internal surface area are related to each other, a compromise between the two properties is to be determined by looking to the application of the materials. Good mechanical strength of macroporous materials is an essential prerequisite for many applications where the matrix has to withstand a constant stress of surrounding tissues or biological fluids. It is important to mention here that the degradation of matrix should be tuned such that it retains sufficient structural integrity

TABLE 10.1

Structural Properties of Macroporous Polymers required in Tissue Engineering Applications

Property	Remarks	Application	Reference
Porosity	Highly porous microstructure with interconnected pores network	Uniform cell distribution, cell survival and proliferation, and migration	LeGeros and LeGeros 1995
	More than 90% porosity and pore interconnectivity	Removal of metabolic waste and byproducts from cells	Mikos et al. 1993
	Large polymer, high porosity, and pore interconnectivity	Bioreactors	Hutmacher 2000
Pore size	Pore size in the range 200 to 400 and 500 μm	Bone tissue regeneration	Boyan et al. 1996
	Pore size between 20 and 125 μm	Regeneration of adult mammalian skin	Yannas et al. 1989
	Pore size between 45 to 150 μm	Regeneration of liver tissues	Yannas et al. 1989
	Pore sizes are too small	Cellular penetration and matrix elaboration	Rout et al., 1988
Surface properties	High internal surface area to volume ratios	Chromatography and tissue engineering	
	Morphology and physiochemistry of the materials	cell attachment, migration, and intracellular signaling	McClary, Ugarova, and Grainger 2000
Mechanical properties	Structural integrity	Regenerating load-bearing tissues such as cartilage, bone, etc.	Liu and Ma 2004

until the newly grown tissue has replaced the matrix supporting function (Braddock et al. 2001).

10.4 Synthesis of Macroporous Polymers

Designing macroporous materials by various physical and chemical routes is not like conventional synthesis, as in the former case where an intimate relationship does exist between the method adopted for synthesis and structural and morphological features of the material to be synthesized. Precisely speaking, a high level of accurate control is desired over macro- and microstructural properties such as spatial form, mechanical strength, density, porosity, pore size, pore size distribution, and pore interconnectivity, respectively. Some of the prime methods of macroporous polymers synthesis are briefly discussed below.

10.4.1 Macroporous Polymer Foams Produced by Hydrocarbon Templating

Polymer foams are utilized in a range of applications such as mechanical dampeners, thermal, acoustic, electrical insulators, and so on (Gibson and Ashby 1997). These foams have been produced by a dispersion of a gaseous phase in a fluid polymer phase, leaching of a water-soluble inorganic fugitive phase, or phase separation methods (Di Maio et al. 2005). However, these processes do not generally offer optimal control over pore structure and bulk characteristics. Recently an attempt has been made to achieve enhanced control over both porosity and bulk properties by combining two distinct foaming processes: leaching of a fugitive phase with polymer precipitation. This was achieved by using a nonwater-soluble particulate hydrocarbon fugitive phase derived from waxes, which allowed for the formation of pores with concomitant precipitation of the polymer phase. The macroporosity of the polymer foam was determined by the hydrocarbon fugitive phase (porogen), which also functioned as a template for the rapid precipitation of the polymer. Bulk properties of the foam could be manipulated independently of the macroporosity and pore size by the incorporation of inorganic and organic fillers into a highly viscous polymer phase.

According to the proposed method, in the first step a multicomponent system consists of a viscous polymer solution and a particulate hydrocarbon porogen is compacted in a Teflon mold. In the subsequent step the polymer-solvent porogen is subjected to extraction in a hydrocarbon solvent (such as pentane or hexane) that is a nonsolvent for the polymer but miscible with the polymer solvent. As a result the porogen is extracted

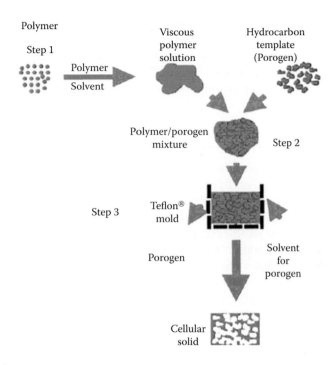

FIGURE 10.2
A schematic presentation of the steps involved in the preparation of macroporous polymer by leaching of fugitive phase. (From Shastri, V. P., Martin, I., and Langer, R., *Proc. Natl. Acad. Sci., USA*, 97 (5), 1970–75, 2000. With permission.)

with a simultaneous rapid precipitation of the polymer phase and the formation of the network of the same. The whole process may schematically be shown in Figure 10.2.

10.4.2 Macroporous Polymers Produced by Radiation

Synthesis of macroporous polymer materials by free radical polymerization is a rather simple technique and provides opportunities to regulate morphology of the produced polymer material. The variables that control pore size are the percentage of the cross-linker, the type and volume of the porogen, the concentration of the free radical initiator in the reaction mixture, and the reaction temperature. Photopolymerization initiation can also be used in which case the synthesis is much faster and can be achieved even at a low temperature (Rohr et al. 2003). The subject has been nicely reviewed in the recent past (Svec 2004).

Another more efficient and advantageous method of polymerization may be the use of ionizing radiations (Mesquita et al. 2002). This method is beneficial from the viewpoint that the radicals can be generated directly thus avoiding the use of an initiator. Moreover, simultaneous cross-linking of growing macroradical chains takes place thus avoiding the use of toxic cross-linking agents. The polymerization may be carried out at any temperature and within very short time periods. Owing to the larger penetration depths of ionizing radiations compared to photoinitiated polymerization, the shape and size of porous polymers can be readily optimized for specific applications.

10.4.3 Surfactant Reverse Micelles Swelling Method

A variety of methods have been developed for producing pores with large sizes in a porous polymer. In one of the most common techniques a soluble polymer is used as porogen (Horak, Labsky, and Pilar 1993). The method, however, suffers from the problem that it is not always easy to wash the polymer porogen out of the polymeric particles. In another method, called nanoparticles agglomeration method (Whitney et al. 1998), it is very difficult to control the pore sizes and reproducibility was not good. Several workers employed inorganic particles as porogen to produce macroporous particles. For instance, poly(glycidyl methacrylate-co-ethylene dimethacrylate) microspheres were prepared using calcium carbonate granules and organic diluents as mixed porogen (Wu, Bai, and Sun 2003). The authors obtained porous materials with pores of different sizes depending on the nature of the porogen.

Ma and coworkers (Zhou et al., 2007) developed a simple method of surfactant reverse micelles swelling method to prepare poly(styrene-divinylbenzene) microspheres with a pore size of 500 nm. The advantage of this method is that the preparation process is very easy. The oil phase contained monomer styrene, cross-linking agent divinylbenzene, surfactant, diluent, and initiator (benzoyl peroxide). Due to the high surfactant concentration, a lot of reverse micelles were formed in the oil phase. After the oil phase is dispersed in the aqueous phase, the reverse micelles in the oil droplets could absorb water from the aqueous phase and formed bicontinuous structure. The water phase in the oil droplets formed pores after polymerization. The authors observed that whereas the surfactant was important for the formation of larger pores, the smaller pores were related to the nature and amount of diluents.

10.4.4 Solid Freeform Fabrication (SFF) Technique

Although the methods discussed previously have their own advantages and disadvantages, there are certain limitations, which cannot be overlooked and deserve attention. For example, almost all synthetic procedures are labor intensive, result in highly inconsistent macro- and micro-structural and material properties, involve toxic organic solvents, tedious and time consuming, and above all they do show shape limitations such as limited pore sizes and insufficiently interconnected pores. Thus, an approach must be engineered that could promise not only to resolve the limitation problems effectively but also fabricate macroporous polymers of desired architecture and properties.

The introduction of solid freeform fabrication (SFF) technologies signals the start of a new revolutionary era for product design and manufacturing industries (Chua et al. 2003). Also called rapid prototyping (RP), the SFF fabrication technologies represent a group of techniques that model a macroporous scaffold directly from a computer-aided design (CAD) data set. These methods are also regarded as watershed events (Kochan 1992) as they deliver tremendous time and cost savings. The SFF techniques are computerized fabrication techniques that can rapidly produce highly complex 3-D physical objects using data generated by the CAD system, computer-based medical imaging modalities, digitizers, and other data makers (Chua and Leong 1997). They build up a specific body shape by the selective layer-by-layer addition of material guided by a computer program (Chua, Chou, and Wong 1998). The step-by-step construction of macroporous polymers facilitates improved reproducibility. In contrast to conventional fabrication methods, parameters such as porosity, pore size, and geometric stability can be controlled more precisely (Chua et al. 2005). Unlike conventional computerized machining processes that involve the removal of materials from a stock, SFF techniques use the underlying concept of layered manufacturing (Kruth 1991) whereby 3-D objects are fabricated with layer-by-layer building via the processing of solid sheet, liquid, or powder material stocks. The typical process chain for all SFF techniques is presented in Figure 10.3.

The SFF techniques are quite outstanding and offer enormous capabilities of manufacturing, which enable them to be employed for biomedical applications ranging from the production of scale replicas of human bones (D'Urso et al. 2000) and body organs (Sanghera et al. 2001) to advanced customized drug delivery systems (Leong et al. 2001) and other areas of medical sciences including anthropology (Recheis et al. 1999), paleontology (Zollikofer and De Leon 1995) and medical forensics (Vanezi et al. 2000). It is worth mentioning here that although the applications of SFF for biomaterials formation is not yet widespread, its immense potential for producing porous biomedical polymers with highly complex macro- and microstructures is widely recognized and is receiving vast interest and attention from many researchers.

10.4.4.1 Advantages

Some significant advantages of SFF techniques are outlined below.

> **Specific designs:** Making use of CAD models it is possible to fabricate biomaterials with a complex design and the specific requirements of the patients.
>
> **Controlled fabrication:** Since the technique makes use of automated computerized fabrication, it is easy to produce high throughput production with minimal manpower requirements. High

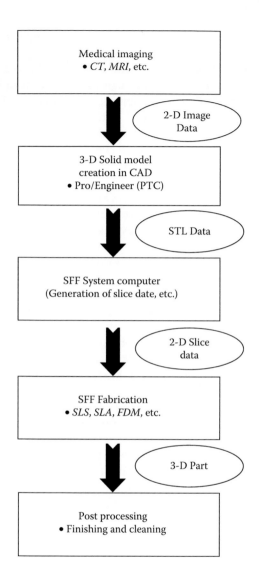

FIGURE 10.3

A schematic representation of solid freeform fabrication process. (CT- computer tomography, MRI – magnetic resonance imaging, CAD - computer aided design, SLS - selective laser sintering, FDM - fused deposition modeling) (From Leong, K. F., Cheah, C. M., and Chua, C. K., *Biomaterials*, 24, 2363–78, 2003. With permission.)

structure resolution of the SFF technique with the ability to control individual process parameters makes them an attractive tool to fabricate complex biomedical architectures with consistent pore morphologies.

Amicable processing conditions: The novelty of the SFF technique rests upon the fact that a wide variety of process conditions are employed and none of the process involves either organic solvents or other toxic chemicals. Moreover, fabrication is carried out at room temperature so that the bioactive agents may be safely impregnated into the biomaterial devices (Low et al. 2001).

Anisotropic microstructure: The SFF and CAD techniques fabricate a porous material, which could have a different microstructure at different regions so that its multifunctional performance could be achieved and not be realized by other conventional techniques. Thus having an anisotropic microstructural design is advantageous from the viewpoint of many specific applications of porous materials.

10.4.4.2 Types of SFF Techniques

The major SFF techniques used in the fabrication of porous polymers, especially as scaffolds for biomaterial applications, is briefly discussed below:

10.4.4.2.1 Three-Dimensional Printing (3D-P)

In 3-D printing technology, the shaped bodies are created by a layered printing process with adhesive bonding using powder as a base material (Sachs et al. 1989). In the beginning, a layer of powder is spread over a building platform. Then a binder solution is deposited precisely by an ink-jet print head on the powder layer to join the single powder particles, resulting in a 2-D layer profile. A fresh layer of powder is laid down and the process is repeated. After drying the binder, the nonjoined powder is removed by an air jet flow (Pfister et al. 2004) and the finished scaffold retrieved. The resolution of 3D-P is influenced by the particle size of the powder, the nozzle size, and the degree of control over the position controller that defines print-head movement. An achievable channel diameter of 500 μm has been reported (Leukers et al. 2005).

10.4.4.2.2 Fused Deposition Modeling (FDM)

In this technique the compressed polymer melts are the starting materials for the preparation of porous matrices using a 3-D fiber deposition technique (Ang et al. 2006; Malda et al. 2005). It involves the concept of melt extrusion to deposit a parallel series of material rods that forms a material layer. In essence, a polymer melt is formed into a fiber with a temperature-controlled extruder. Its nozzle deposits the fiber or the filaments on a motor driven x-y-z table. The machine first builds a layer of fibers with a well-defined distance. On the top, fiber layers are deposited which form a 3-D scaffold with an exact porous morphology and 100% interconnectivity. The positional control of the table is often computer aided and allows for the construction of 3-D structures. The FDM technique was employed for producing poly(ε-caprolactone) (PCL) scaffolds with different geometrically consistent honeycomb-like patterns and fully interconnected porous channels (Zein et al. 2002).

10.4.4.2.3 Selective Laser Sintering

This technique uses deflected laser beams (infrared laser, carbon dioxide) to sinter thin layers (Tan et al. 2005a) of powdered materials, such as wax, synthetic polymers (polycarbonate (Curodeau, Sachs, and Caldarise 2000), polyamides (Giordano et al. 1996), or mixtures-coating with polymers, such as ultra HMPE, poly(etheretherketone)/hydroxyapatite (Tan et al. 2005b) or polyvinyl alcohol/hydroxyapatite (Rimell and Marquis, 2000)) to prepare solid 3-D scaffolds. The laser beam selectively scans over the powder surface. The interaction of the laser beam with the powder elevates the powder temperature just beyond the glass transition temperature but below the melting temperature of the used material. The powder particles that are in contact are deformed and fused together. New layers of powder are deposited by a roller, building a new sintered layer on top of the previous one.

10.4.4.3 Comparative Limitations of SFF Techniques

The SFF techniques discussed above have shown promise in fabricating macroporous scaffolds such as carnio- and maxillo-facial reconstructions, however, the flexibility, processing capabilities of SFF techniques, and microstructures of SFF fabrication methods have some inherent limitations. Limitations of the 3D-P technique include the fact that the pore sizes of fabricated scaffolds are dependent on the particle size of the stock powder. The mechanical properties and accuracy of the fabricated 3D-P scaffolds are another consideration that needs attention. The limitations of FDM include the fact that the pore openings for the scaffolds are not consistent. Moreover, FDM does not permit much variation in pore morphology within the same scaffold volume. In the case of the selective laser sintering (SLS) technique, the limitations include the fact that the pores created in SLS fabricated materials are dependent on the particle size of the powder stock used and the compaction pressure exerted onto the powder bed while depositing the powder layers. Another serious limitation is that since the SLS technique involves a high processing temperature; the technique is limited to the processing of only thermally stable polymers.

10.4.5 Thermally Induced Phase Separation Method

Thermally induced phase separation was first applied to poly(D,L-lactic acid) scaffolds (Schugens et al. 1996a) although several authors have also applied this technique to composite scaffolds (Roether et al. 2002). The technique consists of inducing a solid–liquid or liquid–liquid phase separation, which is done by dissolving the polymer in a solvent and quenching the solution at a certain temperature. Quenching induces a phase separation into a polymer-rich and a polymer-poor phase. The solvent must then be removed from the phase separated solutions either by freeze-drying or by solvent extraction. The main advantage of the phase separation method is that pore morphology and orientation can be tailored by altering the thermodynamic and kinetic parameters of the processing. Its disadvantages include the use of potentially toxic solvents and a high degree of anisotropy of the porosity. The latter may actually be beneficial for certain biomedical and industrial applications such as nerve regeneration, filtration membranes, mechanically damping materials, packaging, and so on (Boccaccini, et al. 2003).

10.4.6 Freeze–Thaw Method

Macroporous hydrogels prepared by a repeated freeze-thaw process have special properties that make them more effective in biomedical applications (Saini 2007). The hydrogels that were prepared from the repeated freeze-thawing process were spongy, rubbery, elastic, and displayed good mechanical strength (Hassan and Peppas 2000a). For the entrapment and encapsulation of labile bioactive substance and living cells, physically cross-linked gels are of great interest especially when the gel formation occurs under mild conditions in the absence of organic solvents.

Poly(vinyl alcohol) (PVA) gels that are prepared by freezing-thawing techniques have shown many improved properties over hydrogels prepared by traditional chemical techniques. Repeated cycles of freezing at –20°C and thawing at 25°C result in the formation of crystalline regions that remain intact upon being placed in contact with water or biological fluids at 37°C. These PVA hydrogels show increased mechanical strength over most hydrogels because the crystalline regions are capable of better distributing a given mechanical load or stress. Additionally, the gels show high elasticity and are capable of being extended to five or six times their initial length. Because of these characteristics, they are obviously used for a variety of biomedical and pharmaceutical applications (Hassan and Peppas 2000b). Porous cryogels with excellent blood compatibility have been prepared by freeze-thawing methods (Bajpai and Saini 2005; Bajpai and Saini 2006) from PVA, casein, and egg albumin.

Another example of macroporous gels are polyacrylamide cryogels (pAAm cryogels; Plieva et al. 2004). The pAAm cryogels (from the Greek *krios* [*kryos*] meaning frost or ice) are produced by radical polymerization of acrylamide and cross-linker N,N-methylene-bis-acrylamide when the polymerization system is partially frozen after the onset of polymerization reaction. The ice crystals formed after partial freezing perform as porogen, while the dissolved monomers and initiator are concentrated in a small fraction of a nonfrozen fluid in which polymerization proceeds. Despite looking like a solid ice block, the polymerization proceeds in a very concentrated (nonfrozen) monomer solution and still unfrozen as a dense polymer gel is formed. After melting, a continuous system of pores is formed in place of ice crystals; the pores are separated by walls of a dense polymer gel. Macroscopically the gel has a sponge-like morphology. Cryogels are considered in Chapters 2,4 and 6 in more detail.

10.5 Morphology of Macroporous Polymers

The morphology and internal structure of macroporous materials are intimately related to and regulated by types of synthetic strategies adopted for matrix fabrication. This part of the chapter focuses on the influence or mode of fabrication for polymers on the morphology and structural features of the macroporous polymers, which are the key parameters to predicting the suitability of the material for a specific application.

The process of hydrocarbon templating is applicable to a wide range of polymer systems including water-soluble polymers, as long as the following conditions are satisfied: (1) the hydrocarbon porogen is extracted below the melting temperature of the polymer to ensure isotropy in the properties of the resulting foam, (2) the polymer has good solubility in a solvent that is a poor solvent for the porogen to obtain a viscous polymer solution wherein the porogen can be distributed uniformly, and (3) the polymer has a molecular weight of at least 40,000 D to ensure structural stability of the resulting foam.

The hydrocarbon templating method was employed by Shastri and coworkers (2000) who prepared a series of foams that were synthesized of poly(L-lactic acid) (PLLA) and poly(lactide-*co*-glycolic acid) (PLGA) of both spherical and polyhedral morphologies and studied them by the scaning electron microscopy (SEM) technique as shown in Figure 10.4. It is clear from the SEM images that the foams revealed the presence of a bicontinuous

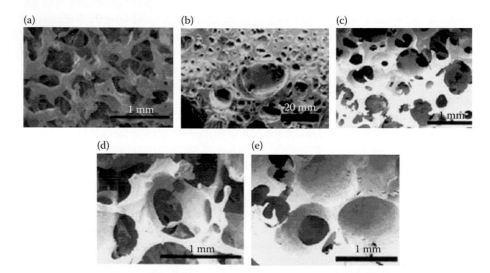

FIGURE 10.4
Scanning electron microscopic characterization of poly(lactide-*co*-glycolic acid) (PLGA) and poly(L-lactic acid) (PLLA) produced using particulate hydrocarbon porogen. (From Shastri, V. P., Martin, I., and Langer, R., *Proc. Natl. Acad. Sci., USA*, 97 (5), 1970–75, 2000. With permission.)

network of the polymer and void with two distinct pore architectures. The geometry and size of larger pores in the foam were nearly identical to those of the particulate hydrocarbon phase. Spherical hydrocarbon particles resulted in pores with spherical morphology (C to E) whereas polyhedral hydrocarbon particles resulted in pores with irregular morphology (A).

Using the hydrocarbon templating approach, one can alter the degradation of polymer foams by blending a water-soluble immiscible phase into the parent polymer phase. Such blending is really difficult using a process like melt foaming due to the change in viscosity and interfacial dynamics. Current methods for the production of porous polymeric structures for tissue engineering are generally limited to the production of membrane <4 mm in thickness. This limits their utility in the engineering of structurally complex tissues such as musculoskeletal reconstruction. Because the present process allows for the production of large monolithic foams, porous polymeric structures of various geometries can be easily fabricated.

Safrany and coworkers (2005) polymerized diethyleneglycol dimethacrylate by gamma radiation of varying doses using various aliphatic alcohols as porogen. The authors studied morphology of the prepared porous materials and found that the size of the pores decreased with an increasing concentration of monomer. It was also noticed that the nature of organic porogen had a pronounced effect on the pore size as evident from the scanning electron micrographs shown in Figure 10.5.

Dose rate has a pronounced effect on the morphology of the macroporous polymers. An increase in the dose rate, while keeping all other parameters constant, leads to an increased rate of free radical formation, which in turn results in an increase in the rate of polymerization and cross-linking, therefore, to earlier phase separation. Earlier phase separation leads to bigger nuclei and bigger pores between them. Thus, the larger the dose of irradiation, the greater the pore size of the polymer matrix.

In the method of surfactant reverse micelles swelling, the morphology of microspheres depends on many factors such as nature of surfactant, diluent, and cross-linking agent. It has been observed that when the concentration of surfactant is raised above CMC, different types of micelles would be formed in the oil phase such as spherical, clubbed, hexagonal, and lamellar. It has been observed that the samples prepared with Span 85 were microporous (ca 10 nm) although some locations on the particles showed macropores (>200 nm). Probably, Span 85 was too hydrophobic in nature while with Span 83, the particles were spherical and their average pore size was 93 nm.

The potential of the FDM technique in making porous materials is demonstrated in Figure 10.6, which illustrates the micrographs of different scaffold microstructures (Hutmacher et al. 2001). The specimens shown were fabricated on a commercial FDM system, using custom-made, high-density polyethylene filament stocks. The pore interconnectivity of a FDM-fabricated structure is observed from the morphology of a sliced structure taken at a 45° angle to the z-axis as shown in Figure 10.7.

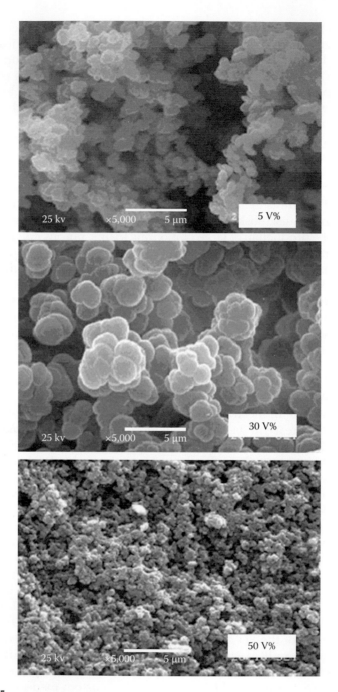

FIGURE 10.5
SEM images of macroporous polymers from methanol solutions with different monomer content. The incubation was done at 25°C, with a dose rate of 16 kGy/h and with a total dose of 30 kG. (From Safrany, A., Beiler, B., Laszlo, K., and Svec, F., *Polymer*, 46, 2862–71, 2005. With permission.)

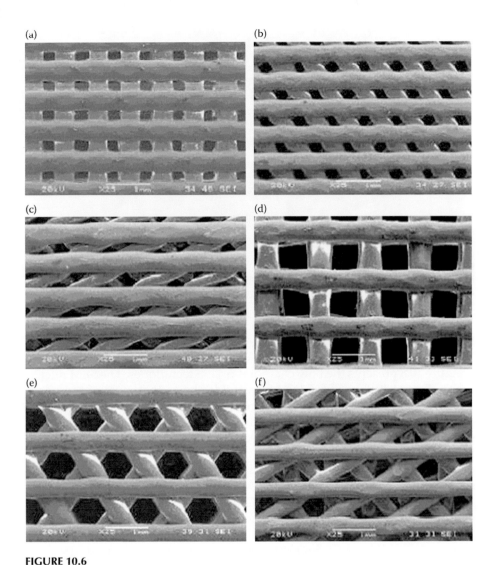

FIGURE 10.6
Variation in FDM fabricated high-density polyethylene scaffold microstructures produced using different raster gap settings and lay down pattern. (From Leong, K. F., Cheah, C. M., and Chua, C. K., *Biomaterials*, 24, 2363–78, 2003. With permission.)

The TIPS technique provides a scaffold with a uniform pore size and high degree of interconnection, good mechanical properties, and various morphologies. The morphology can be controlled by several experimental parameters such as the quenching temperature, quenching rate, quenching period or aging time, polymer concentration, solvent to nonsolvent ratio, molecular structure, and added surfactant or porogen (Hua et al. 2002). The regular and highly interconnected porous PLGA/nanohydroxyapatite (NHA) scaffolds

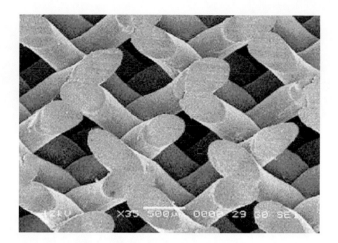

FIGURE 10.7
Micrograph showing 3-D pore interconnectivity within structure fabricated by fused deposition modeling (FDM). (From Leong, K. F., Cheah, C. M., and Chua, C. K., *Biomaterials*, 24, 2363–78, 2003. With permission.)

have been fabricated with a pore size suitable for bone tissue engineering (Huang et al. 2008).

Figure 10.8 presents the flow chart of the TIPS techniques. The SEM micrograph (Figure 10.9) of PLGA/NHA scaffolds shows that most of the pores are connected with their neighbors and have many small pores on the big pore's wall, which entitles the foam's novel connectivity and makes it more suitable to serve as scaffolds for cell seeding and nutrients transporting in tissue engineering.

A regular and highly interconnected macroporous PLLA scaffold was fabricated (Kim et al. 2004) from a PLLA-dioxane-water ternary system with added polyethylene glycol (PEG)–PLLA diblock copolymer. The morphology of the scaffold was investigated in detail by controlling the following TIPS parameters: quenching temperature, aging time, polymer concentration, molecular structure, and diblocks concentration. The phase diagram was assessed visually on the basis of the turbidity. The cloud-point curve shifted to higher temperatures with increasing PEG content in the diblock copolymer, due to a stronger interaction between PEG and water in solution. The addition of diblock copolymers (0.5 wt % in solution) stabilized interconnections of pores at a later stage without segregation or sedimentation. The pore size of the scaffold could be easily controlled in the range of 50–300 μm. The MC3T3-E1 cells (an osteoblast-like cell) proliferated successfully in the produced scaffolds in four weeks.

The regular and highly interconnected macroporous scaffolds with pores ranging in size from 50 to 150 μm were fabricated from PLGA-dioxane-water ternary systems without any surfactant or other additive (Hua, Park, and Lee 2003). The effect of scaffold morphology on processing parameters including

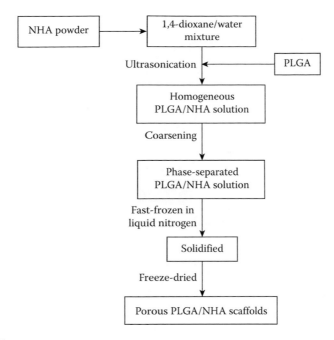

FIGURE 10.8
The preparation of poly(lactide-*co*-glycolic acid; PLGA)/nanohydroxyapatite (NHA) scaffolds by TIPS. (From Huang, Y. X., Ren, J., Chen, C., Ren, T. B., and Zhou, X. Y., *J. Biomater. Appl.*, 22, 409–32, 2008. With permission.)

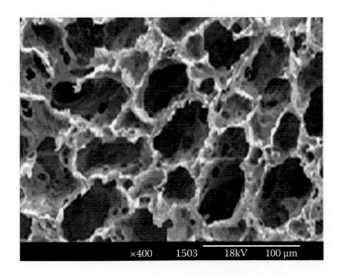

FIGURE 10.9
SEM micrographs of poly(lactide-co-glycolic acid; PLGA)/nanohydroxyapatite (NHA) scaffolds prepared from PLGA-dioxane-water. (From Huang, Y. X., Ren, J., Chen, C., Ren, T. B., and Zhou, X. Y., *J.Biomater. Appl.*, 22, 409–32, 2008. With permission.)

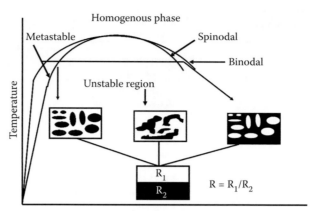

FIGURE 10.10
Schematic presentation of a typical polymer-solvent-nonsolvent ternary phase diagram (R1: polymer lean phase, R2: polymer rich phase, Fv: volume ratio. (From Hua, F. J., Park, T. G., and Lee, D. S., *Polymer*, 44, 1911–20, 2003. With permission.)

quenching temperature, polymer concentration, solvent composition, and molecular weight was investigated as a function time.

In principle, the final porous morphology is thought to rely on the thermodynamic state of the solution to be quenched, as schematized in the temperature-composition phase diagram in Figure 10.10. Nucleation and growth of the phase separated polymer results in a poorly connected, stringy, or beady structure. In contrast, spinodal phase separation is expected to give rise to a highly interconnected structure. The initial pore size is determined by the concentration fluctuations induced by the quenching, the polymer concentration in the solution, and fluctuations associated with thermal flow. Therefore, the final porous morphology can be controlled to some extent either by varying processing parameters such as the quenching rate, quench temperature, and quenching period or by varying formulation parameters such as the polymer concentration or solvent composition.

The PLLA was blended with PLGA to increase the viscosity of polymer solution; a block copolymer of PEG with PLGA was added as a surfactant to decrease the interfacial tension between the polymer-rich and polymer-lean phases. The effect of TIPS parameters including the concentration of diblocks copolymer and PLGA/PLLA ratio was also studied (Shin et al. 2005). By these fabrication methods, regular and well-connected macroporous matrices of pore sizes ranging from 50 to 200 μm were prepared and their morphology could be controlled by adjusting process parameters such as quenching temperature, aging time, polymer concentration, composition, and length of PEG and PLGA blocks in the added copolymer. The combination of these two methods yield a new method to prepare open, regular, and well-interconnected macroporous scaffolds of PLGA (Figure 10.11).

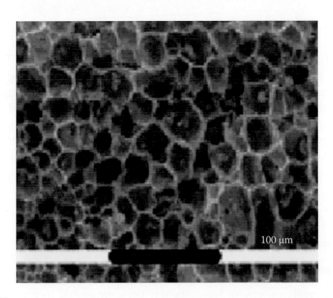

FIGURE 10.11
SEM images of the scaffold prepared from poly(lactide-*co*-glycolic acid) (PLGA) and poly(L-lactic acid) (PLLA) solution in dioxane-water solution with the addition of block copolymer of PLGA with poly(ethylene glycol) (PEG). (From Shin, K. C., Kim, B. S., Kim, J. H., Park, T. G., Nam, J. D., and Lee, D. S., *Polymer*, 46, 3801–8, 2005. With permission.)

Freeze-drying the polylactide solutions in 1,4-dioxane has been studied as a way to produce microcellular foams (Schugens et al. 1996b). The thermally induced phase separation has been studied in relation to several processing and formulation parameters. The effects such as polymer concentration, chain stereoregularity, polymer molecular weight, and cooling rate have been investigated in connection with the porous morphology and the physicomechanical characteristics of the final foams. As a rule, bundles of channels are formed with a diameter of about 100 µm. The channels have a preferential orientation that fits the cooling direction. A porous substructure (10 µm) is observed in the internal walls of the tubular macropores.

The TIPS process is also used for the fabrication of porous foams based on biodegradable polymers like PLLA and its copolymers with D-lactic acid or glycolic acid (Nam and Park 1999). Diverse foam morphologies were obtained by systematically changing several parameters involved in the TIPS process, such as polymer type and concentration, coarsening conditions, solvent-nonsolvent composition, and the presence of an additive. The produced foams had microcellular structures with average pore diameters ranging from 1 to 30 µm depending on the process parameters, which were characterized by a SEM and a mercury intrusion porosimetry. Additionally, Pluronic F127 was used as an additional porogen to control the pore geometry and size. In particular, the addition of polymeric surfactant in the TIPS formulation enhanced the size of pores and improved their interconnectivity. The prepared foams could find applications in controlled drug delivery.

Macroporous PLLA scaffolds with regular and highly interconnected (Hua, Nam, and Lee 2001) pores with sizes ranging from 50 to 300 µm were fabricated from a PLLA/1,4-dioxane-water ternary system via TIPS in the presence of small amounts of NaCl. The addition of salt has raised the cloud-point temperature because of a salting-out effect. Spinodal decomposition was induced at a higher temperature with a large quench depth. The crystal-lization of PLLA prevented gelation to a large extent, which consequently led to the creation of highly interconnected macropores that are essential for effective cell penetration. An optimal quenching route was found for the fabrication of well-designed macroporous structures.

10.6 Current Challenges and Future Prospects

The macroporous polymers have shown their potential in biomedical and allied fields and emerged as one of the most promising materials with high performance applications. However, at present the regenerative biology has shown promise for one of the biomedical revolutions of the current century. It includes implantation of bioartificial tissues, cell transplantation, and stimulation of regeneration from residual tissues *in vivo*. To achieve these approaches, macroporous polymeric materials with tailor-made functional groups could be of great use and may enable researchers to explore the inter-faces of living tissues and biomaterials. Thus, designing new materials with multifunctional properties possess challenges to polymer chemists and engi-neers who have to be aware of the intimate relation between the synthetic approaches of macroporous matrices and their structural architectures, which are the ultimate factors to controlling and deciding the performance of the material. Thus, the field of macroporous materials deserves careful considerations of the following points:

1. Development of highly efficient and cost effective methods to fabri-cate the porous materials.
2. Precise control over the internal molecular setup and morphology of the porous materials.
3. To impart the highest order of mechanical strength, biocompatibil-ity, immunogenicity, and biodegradability to the synthesized porous matrices.
4. A close resemblance of the material to the natural tissues in respect to their overall physicochemical and biological properties if the material is being designed for biomedical purposes.

References

Ang, K. C., Leong, K. F., Chua, C. K., and Chandrasekaran, M. 2006. Investigation of the mechanical properties and porosity relationship in fused deposition modelling—fabricated porous structure. *Rapid Prototyping J.* 12 (2):100–105.

Bajpai, A. K., and Saini, R. 2005. Preparation and characterization of biocompatible spongy cryogels of poly(vinyl alcohol)-gelatin and study of water sorption behavior. *Polym. Int.* 154 (5):796–806.

Bajpai, A. K., and Saini, R. 2006. Preparation and characterization of novel biocompatible cryogels of polyvinyl alcohol and egg albumin and their water sorption study. *J. Mater. Sci. Mater. Med.* 17 (1):49–61.

Bhushan, I., Parshad, R., Qazi, G. N., Ingavle, G. Jamalpure, T. M. Rajan, C. R., Ponrathnam, S., and Gupta, V. K. 2007. Macroporous beads for lipase immobilization: Kinetic resolution of a racemic drug intermediate. *J. Bioact. Compat. Polym.* 22:174–94.

Boccaccini, A. R., Notingher, I., Maquet, V., and Jerome, R. 2003. Bioresorbable and bioactive composite materials based on polylactide foams filled with and coated by bioglass particles for tissue engineering application. *J. Mater. Sci. Mater. Med.* 14:443–50.

Boyan, B. D., Hummert, T. W., Dean, D. D., and Schwartz, Z. 1996. Role of material surfaces in regulating bone and cartilage cell response. *Biomaterials* 17 (2):137–46.

Braddock, M., Houston, P., Campbell, C., and Ashcroft, P. 2001. Tissue engineering for bone repair. *News Physiol. Sci.* 16:208–13.

Chua, C. K., Chou, S. M., and Wong, T. S. 1998. A study of the state-of-the-art rapid prototyping technologies. *Int. J. Adv. Manufact. Tech.* 14 (2):146–52.

Chua, C. K., Feng, C., Lee, C. W., and Ang, G. O. 2005. Rapid investment casting: Direct and indirect approaches via model maker II. *Int. J. Adv. Manufact. Tech.* 25 (1–2):26–32.

Chua, C. K., and Leong, K. F. 1997. Rapid prototyping in Singapore: 1988 to 1997. *Rapid Prototyping J.* 3:116–19.

Chua, C. K., Leong, K. F., Cheah, C. M., and Chua, S. W. 2003. Development of a tissue engineering scaffold structure library for rapid prototyping. Part I: Investigation and classification. *Int. J. Adv. Manufact. Tech.* 21 (4):291–301.

Curodeau, A., Sachs, E., and Caldarise, S. 2000. Design and fabrication of cost orthopedic implants with free from surface textures from 3D printed ceramic shell. *J. Biomed. Mater. Res.* 53 (5):525–35.

Di Maio, E., Mensitieri, G., Iannace, S., Nicolais, L., Li, W., and Flumerfelt, R. W. 2005. Structure optimization of polycaprolactone foams by using mixture of CO_2 and N_2 as blowing agents. *Polym. Eng. Sci.* 45 (3):432–44.

D'Urso, P. S., Earwaker, W. J., Barker, T. M., Redmond, M. J., Thompson, R. G., Effeney, D. J., and Tomlinson, F. H. 2000. Custom carnioplasty using steredithography and acrylic. *Br. J. Plast. Surg.* 53 (3):200–204.

Fields, S. M. 1996. Silica xerogels as a continuous column support for high performance liquid chromatography. *Anal. Chem.* 68:2709–12.

Gibson, L., and Ashby, M. 1997. *Cellular solids: Structure and properties.* Oxford: Pergamon.

Giordano, R. A., Wu, B. M., Borland, S. W., Cima, L. G., Sachs, F. M., and Cima, M. J. 1996. Mechanical properties of dense of polylactic acid structures fabricated three dimensional properties. *J. Biomater. Sci. Polym. Ed.* 8 (1):63–75.

Gong, S., Wang, H., Sun, Q., Xue, S.-T., and Wang, J.-Y. 2006. Mechanical properties and *in vitro* biocompatibility of porous zeiss scaffold. *Biomaterials* 27 (20):3793–99.

Hassan, C. M., and Peppas, N. A. 2000a. Structure and morphology of freeze/thawed PVA hydrogels. *Macromolecules* 33:2472–79.

Hassan, C. M., and Peppas, N. A. 2000b. Cellular freeze/thawed PVA hydrogels. *J. Appl. Polym. Sci.* 76:2075–79.

Hogde, P. 1988. *Synthesis and separation using functional polymers*, ed. P. Hogde, D. C. Sherrington, 43–122. Chichester: Wiley.

Hollister, S. J. 2005. Porous scaffolds design for tissue engineering. *Nat. Mater.* 4:518–24.

Hollister, S. J., Maddox, R. D., and Taboas, J. M. 2002. Optimal design and fabrication of scaffolds to mimic tissue properties and satisfy biological constraints. *Biomaterials* 23 (20):4095–4103.

Horak, D., Labsky, J., and Pilar, J. 1993. The effect of polymeric porogen on the properties of macroporous poly(glycidyl methacrylate-*co*-ethylene dimethacrylate). *Polymer* 34 (16):3481–89.

Hua, F. J., Kim, G. E., Lee, J. D., Son, Y. K., and Lee, D. S. 2002. Macroporous poly(L-lactide) scaffold: Preparation of a macroporous scaffold by liquid-liquid phase separation of a PLLA-dioxane-water system. *J. Biomed. Mater. Res.* 63:161–67.

Hua, F. J., Nam, J. D., and Lee, D. S. 2001. Preparation of a macroporous poly(L-lactide) scaffold by liquid-liquid phase separation of a PLLA/1,4-dioxane/water ternary system in the presence of NaCl. *Macromol. Rapid Commun.* 22 (13):1053–57.

Hua, F. J., Park, T. G., and Lee, D. S. 2003. A facile preparation of highly interconnected macroporous poly(D, L-lactic acid-*co*-glycolic acid) (PLGA) scaffolds by liquid–liquid phase separation of a PLGA–dioxane–water ternary system. *Polymer* 44:1911–20.

Huang, Y. X., Ren, J., Chen, C., Ren, T. B., and Zhou, X. Y. 2008. Preparation and properties of poly(lactide-*co*-glycolide) (PLGA)/nano-hydroxyapatite (NHA) scaffolds by thermally induced phase separation and rabbit MSCs culture on scaffolds. *J. Biomater. Appl.* 22:409–32.

Hutmacher, D. W. 2000. Scaffold in tissue engineering bone and cartilage. *Biomaterials* 21 (24):2529–43.

Hutmacher, D. W., Schantz, T., Zein, I., Ng, K. W., Tech, S. H., and Tan, K. C. 2001. Mechanical properties and cell cultural response of polycaprolactone scaffolds designed and fabricated via fused deposition modeling. *J. Biomed. Mater. Res.* 55 (2):203–16.

Kim, H. D., Bae, E. H., Kwon, I. C., Pal, R. R., Nam, J. D., and Lee, D. S. 2004. Effect of PEG–PLLA diblock copolymer on macroporous PLLA scaffolds by thermally induced phase separation. *Biomaterials* 25 (12):2319–29.

Kochan, D. 1992. Solid freedom manufacturing possibilities and restrictions. *Comput. Ind.* 20 (2):133–40.

Kruth, J. P. 1991. Material incress manufacturing by rapid prototyping techniques: *CIRP Ann.* 40 (2):603–14.

Landgraf, W., Li, N.-H., and Benson, J. 2005. New polymer enables near zero-order release of drugs. *Drug Del. Tech.* 5 (2 February):48–55.

LeGeros, R. Z., and LeGeros, J. P. 1995. Calcium phosphate biomaterials: Preparation, properties and biodegradation. In *Encyclopedia handbook of biomaterials and bioengineering part A: Materials*, eds. D. L. Wise, D. J. Trantolo, D. E. Altobelli, M. J. Yaszemslci, J. D. Gresser, and E. R. Schaartz. Vol. 2, 1429–63, New York: Marcel Dekker.

Leong, K. F., Cheah, C. M., and Chua, C. K. 2003. Solid freeform fabrication of three-dimensional scaffolds for engineering replacement tissues and organs. *Biomaterials* 24:2363–78.

Leong, K. F., Phua, K. K., Chua, C. K., Du, Z. H., and Teo, K. O. 2001. Fabrication of porous polymeric matrix drug delivery devices using the selective laser sintering technique. *Pro. Inst. Mech. Eng. H.* 215:191–201.

Leukers, B., Gülkan, H., Irsen, S. H., Milz, S., Tille, C., Schieker, M., and Seitz, H. 2005. Hydroxyapatite scaffold for bone tissue engineering made by 3D printing. *J. Mater. Sci. Mater. Med.* 16 (12):1121–24.

Liu, X., and Ma, P. X. 2004. Polymeric scaffolds for bone tissue engineering: *Ann. Biomed. Eng.* 32:477–86.

Low, K. H., Leong, K. F., Chua, C. K., Du, Z. H., and Cheah, C. M. 2001. Characterization of SLS parts for drug delivery devices. *Rapid Prototyping J.* 7 (5):262–67.

Ma, P. X. 2004. Scaffolds for tissue fabrication. *Mater. Today* 2 (5):30–40.

Malda, J., Woodfield, T. B. F., Van der Vloodt, F., Wilson, C., Martens, D. E., Tramper, J., van Blitterswijk, C. A., and Riesle, J. 2005. The effect of PEGT/PBT scaffolds architecture on the composition of tissue engineering cartilage. *Biomaterials* 26 (1):63–72.

Mathieu, L. M., Mueller, T. L., Bourban, P.-E., Pioletti, D. P., Mueller, R., and Manson, J.-AE. 2006. Architecture and properties of an isotropic polymer composite scaffold for bone tissue engineering. *Biomaterials* 27 (6):905–16.

McClary, K. B., Ugarova, T., and Grainger, D. W. 2000. Modulating fibroblast adhesion spreading and proliferation using self-assembled monolayer films of alkylthiolates on gold. *J. Biomed. Mater. Res.* 50 (3):429–39.

Meese, T. M., Hu, Y., Nowak, R. W., and Marra, K. G. 2002. Surface studies of coated polymer microspheres and protein release from tissue-engineered scaffolds. *J. Biomater. Sci. Polym. Ed.* 13 (2):141–45.

Mesquita, A. C., Mori, M. N., Veira, G. M., and Andrade e Silva, L. G. 2002. Vinyl acetate polymerization by ionizing radiation. *Radiat. Physics Chem.* 63 (3–6):465–68.

Mikos, A. G., Sarakinos, G., Lyman, M. D., Ingber, D. E., Vacanti, J. P., and Langer, R. 1993. Prevascularization of porous biodegradable polymers. *Biotechnol. Bioeng.* 42:716–23.

Nam, Y. S., and Park, T. G. 1999. Biodegradable polymeric microcellular foams by modified thermally induced phase separation method. *Biomaterials* 20:1783–90.

Omidian, H., Rocca, J. G., and Park, K. 2005. Advances in super porous hydrogels. *J. Controll. Rel.* 102 (2):3–12.

Pfister, A., Landers, R., Laib, A., Huebner, U., Schmelzeisen, R., and Muelhaupr, R. 2004. Biofunctional rapid prototyping for tissue engineering application: 3D bioplotting versus 3D printing. *J. Polym. Sci. Part A: Polym. Chem.* 42 (3):624–38.

Plieva, F. M., Savina, I. N., Deraz, S., Andersson, J., Galaev, I.Yu., and Mattiasson, B. 2004. Characterization of supermacroporous monolithic polyacrylamide based matrices designed for chromatography of bioparticles. *J. Chromatog. B* 807 (1):129–37.

Recheis, W., Weber, G. W., Schafer, K., Knapp, R., and Seidler, H. 1999. Virtual reality and anthropology. *Eur. J. Radiol.* 31 (2):88–96.

Rimell, J. T., and Marquis, P. M. 2000. Selective laser sentering of ultra high molecular weight polyethylene for clinical applications. *J. Biomed. Mater. Res.* 53 (4):414–20.

Roether, J. A., Boccaccini, A. R., Henech, L. L., Maquet, V., Gautier, S., and Jérôme, R., 2002. Development and *in vitro* characterisation of novel bioresorbable and bio-active composite materials based on polylactide foams and bioglass for tissue engineering applications. *Biomaterials* 23:3871–78.

Rohr, T., Hilder, E. F., Donovan, J. J., Svec, F., and Frechet, J. M. J. 2003. Photografting and the control of surface chemistry in three-dimensional porous polymer monoliths. *Macromolecules* 36:1677–84.

Rout, P. G. J., Tarrant, S. F., Frame, J. W., and Davies, J. E. 1988. *Interaction between primary bone cell cultures and biomaterials. Part 3, a comparison of dense and macro porous hydroxyapatite: Bioceramics and clinical applications*, eds. A. Pizzoferratto, P. G. Ravaglioli, and A. J. C. Lee, 591–96. Amsterdam: Elsevier.

Sachs, E. M., Haggerty, J. S., Cima, M. J., and Williams, P. A. 1989. Three dimensional printing techniques. U.S. Patent #5204055.

Safrany, A., Beiler, B., Laszlo, K., and Svec, F. 2005. Control of pore formation in macroporous polymers synthesized by single-step γ-radiation-initiated polymerization and cross-linking. *Polymer* 46:2862–71.

Saini, R. 2007. Polyvinyl alcohol based biocompatible macromolecular network and their water sorption study. Ph.D. diss., Rani Drugawati University, Jabalpur, India.

Sanghera, B., Naique, S., Papaharilaou, Y., and Amis, A. 2001. Preliminary study of rapid prototype medical models. *Rapid Prototyping J.* 7 (5):275–84.

Schugens, C., Maquet, V., Grandfils, C., Jerome, R., and Teyssie, P. 1996a. Polylactide macroporous biodegradable implants for cell transplantation. II. Preparation of polylactide foams by liquid-liquid phase separation. *J. Biomed. Mater. Res.* 30 (4):449–61.

Schugens, C., Maquet, V., Grandfils, C., Jerome, R., and Teyssie, P. 1996b. Biodegradable and macroporous polylactide implants for cell transplantation: I. Preparation of macroporous polylactide supports by solid-liquid phase separation. *Polymer* 37 (6):1027–38.

Shastri, V. P., Martin, I., and Langer, R. 2000. Macroporous polymer foams by hydro-carbon templating. *Proc. Natl. Acad. Sci. USA* 97 (5):1970–75.

Sherrington, D. C. 1998. Preparation structure and morphology of supports. *Chem. Commun.* 21:2275–86.

Shin, K. C., Kim, B. S., Kim, J. H., Park, T. G., Nam, J. D., and Lee, D. S. 2005. A facile preparation of highly interconnected macroporous PLGA scaffolds by liquid–liquid phase separation II. *Polymer* 46:3801–8.

Sun, J., Wu, J., Li, H., and Chang, J. 2005. Macroporous poly[3-hydroxybutyrate–co-3-hydroxyvalerate] matrices for cartilage tissue engineering. *Eur. Polym. J.* 41:2443–49.

Svec, F. 2004. Preparation and HPLC applications of rigid macroporous organic polymer monoliths. *J. Sep. Sci.* 27:747–66.

Tan, K. H., Chua, C. K., Leong, K. F., Cheah, C. M., Gui, W. S., Tan, W. S., and Wiria, F. E. 2005a. Selective laser sintering of biocompatible polymers for applications in tissue engineering. *Bio-Med. Mater. Eng.* 15 (1–2):113–24.

Tan, K. H., Chua, C. K., Leong, K. F., Naing, M. W., and Cheah, C. M. 2005b. Fabrication and characterization of three dimensional of poly(ether-ether–ketone)/hydroxyapatite biocompatible scaffolds using laser sentering. *Proc. Inst. Mech. Eng. Part H.J. Eng. In Medicine* 219 (H3):183–94.

Vanezi, P., Vanezis, M., McCombe, G., and Niblett, T. 2000. Facial reconstruction using 3D computer graphics. *Forensic. Sci. Int.* 108 (2):81–95.

Whitney, D., McCoy, M., Gordon, N., and Afegan, N. B. 1998. Characterization of large pore polymeric supports for use in perfusion biochromatography. *J. Chromatogr. A.* 807 (2):165–84.

Wu, L., Bai, S., and Sun, Y. 2003. Development of rigid bidisperse porous microspheres for high speed protein chromatography. *Biotechnol. Prog.* 19 (4):1300–1306.

Yang, S., Leong, K. F., Du, Z., and Chua, C. K. 2001. The design of scaffolds for use in tissues engineering. Part I. Traditional factors. *Tissue Eng.* 7 (6):679–86.

Yannas, I. V., Lee, E., Orgill, D. P., Skrabut, E. M., and Murphy, G. F. 1989. Synthesis and characterization of a model extra cellular matrix that induces partial regeneration of adult mammalian skin. *Proc. Natl. Acad. Sci. USA* 86:933–37.

Zein, I., Hutmachen, D. W., Tan, K. C., and Teo, H. S. 2002. Fused deposition modeling of novel scaffolds architectures for tissue engineering application. *Biomaterials* 23 (4):1169–85.

Zhao, X. S. 2006. Novel porous materials for emerging applications. *J. Mater. Chem.* 16:623–25.

Zhou, W. R., Gu, T. Y., Su, Z. G., and Ma, G. H. 2007. Synthesis of macroporous poly(styrene-divinyl benzene) microspheres by surfactant reverse micelles swelling method. *Polymer* 48 (7):1981–88.

Zollikofer, C. P. E., and De Leon, M. S. P. 1995. Tools for rapid prototyping in biosciences. *IEEE Comput. Graph.* 15 (6):48–55.

Section III

Application of
Macroporous Polymers

11

Macroporous Gels for Isolation of Small Molecules from the Solutions Containing Suspended Material

Bo Mattiasson, Fatima M. Plieva, and Igor Yu. Galaev

CONTENTS

11.1 Introduction

Biotechnological processes often result in relatively dilute product streams from where the product molecules have to be isolated. The amount of water is often so large that it is unrealistic to remove the water by evaporation thereby concentrating the target molecule. Adsorption offers an interesting alternative.

When using adsorbents one can operate in a batch adsorption mode or in chromatographic mode. The latter is more efficient, especially if separation of molecules will take place. However, conventional chromatographic separation methods are based on handling clear solutions free from particulate matter. If particles appear in the feed to the chromatographic column, one needs to remove the particles by, for example, centrifugation or filtration prior to feeding the adsorption column. Otherwise, clogging problems may appear and the column will eventually be completely blocked. Several strategies have been applied in order to avoid such clogging problems. One method that was successful was the use of expanded bed chromatography where the chromatographic particles are kept suspended in an expanded bed chromatographic mode [1]. By such an arrangement it was possible to maintain a chromatographic bed, but with reasonable distances between the chromatographic particles so that unwanted material could pass freely between them without causing clogging. However, when passing particulate matter through ion exchange columns, aggregation of the adsorbent particles was observed. Coating of the adsorbent was one way to eliminate such complications [2,3]. Other strategies to treat particulate containing feeds have involved the use of macrobeads so that the interstitial spaces in the columns were bigger and then more easily let foreign particulate matter pass through the column [4,5]. The multicolumn concept of chromatography (known as simulated moving bed, SMB chromatography) is an economically viable process technology that allows for less buffer consumption compared to convectional chromatography [6]. Isolation of proteins from crude feed streams of proteins from potato juice was evaluated using expanded bed adsorption in the SMB concept of chromatography [7], thus represented some steps toward an economically viable process technology with crude feeds.

The most intensely studied strategy to cope with chromatography of particulate-containing material has been to use macroporous adsorbents. One interesting recent development concerning adsorbents is the introduction of supermacroporous hydrogels. Such gels are characterized by having large pores allowing, for example, microbial cells to pass through the gels. The gels are produced as cryogels (i.e., the gel formation has taken place when the liquid has been apparently frozen). The ice crystals form an interconnected network, and when thawed a network of pores is formed. The pore dimensions are defined by the sizes of the ice crystals formed during the cryogel formation process. There have been several different adsorbent configurations presented and evaluated [8–13]. Each adsorbent has its pros and cons. The larger the pores, the lower the capacity, but also the higher the probability that the adsorbent will be suitable for processing fluids containing large suspended particles. This chapter is summarizing efforts to use supermacroporous adsorbents to treat particulate-containing liquids for the isolation of small molecules (proteins are treated separately in this volume).

11.2 Supermacroporous Cryogel Media

The cryogelation platform renders it possible to prepare macroporous materials from practically any gel forming system and with a wide range of porosities [9,10]. The porosity depends often on the polymer precursor used. For example, using poly(vinyl alcohol) (PVA, one of the most studied synthetic polymers) of different grades (i.e., with a different degree of hydrolysis), it is possible to form PVA cryogels with pore size up to 1–2 µm [14,15] or with pores up to 100 µm [16]. The first type of PVA cryogels is prepared mainly in the shape of beads or membranes [12,13,15,17] and the second type is prepared as high flow-path monolithic columns [16,18].

The PVA cryogel beads (with pore size up to 1–2 µm) can be used to capture small molecules from crude feeds when operating in well stirred bioreactors. For example, the PVA cryogels with immobilized bacteria and yeast cells were used for biosorption of metals [19,20]. However, these cryogels, prepared as monolithic columns, had high flow resistance and back pressure [16] due to the too small size of macropores. The presence of so-called flow through pores with dimensions of at least 1.5–2 µm [21] is needed to keep low flow resistance through a stationary monolithic medium.

In contrast, the cryogels with much larger pore size, up to 200 µm, are prepared as high flow-path monoliths [10,22–24]. Due to the continuous large pores in the cryogel monoliths, the mass transfer is mainly due to convection [23], resulting in a very low back pressure through the monoliths [23,25,26] allowing for scaling–up composed monoliths [22].

The presence of large continuous pores in the macroporous cryogel monoliths allows particulate matter to pass through (e.g., microbial cells can pass through a bed of cryogels) without being retained [27–30]. The open porous structure has many advantages, but there is the drawback of a lower capacity. There are different ways to compensate for this drawback. One of the possibilities to increase the capacity is to use the macroporous gels with small pore sizes [18,31,32]. Another approach is to graft the polymer chains on the surface of the pores thus allowing for the increased number of binding sites with targets [33–37]. Another interesting approach is the double-freezing approach when the new cryogel network is synthesized inside the interconnected pores of the preformed cryogel [18,38]. Formation of the composite systems is an excellent approach for the development of macroporous cryogel systems of increased capacity [39–42]. All these approaches will be described in the coming sections.

11.2.1 Cryogels with Decreased Pore Sizes

There are methods to shrink the pore size in cryogel monoliths by, for example, modifying the freezing protocol. By a quicker freezing, more and smaller ice crystals are formed and this will lead to formation of smaller pores. The influence of the freezing conditions on the final properties of cryogels was intensively studied for the polyacrylamide (pAAm) [31,42–48], poly(N-isopropylacrylamide) (pNIPA) [49–52] and poly(2-hydroxyethyl methacrylate) pHEMA cryogels [18,53,54].

The pAAm cryogels are by far the most studied cryogel system. The pAAm cryogels are produced via free radical polymerization at subzero temperature in the partly frozen aqueous medium. The pAAm cryogel is a hetero-phase system where the supermacropores of 10–200 µm in size (macrostructure) is filled with solvent and surrounded by the dense pore walls (microstructure). The pore walls in pAAm cryogels are very dense because they are composed of the concentrated polymer phase [24,31,43]. The presence of large continuous pores in cryogel monoliths with convective flow-through and very low back pressure allows for fast processing of feed streams at applied high flow rates [22,23,26,55]. However, the presence of large pores in cryogel monoliths results in a low available surface area for binding a target molecule and reduced capacity. One of the approaches to increase surface area and, as a result, increase binding capacity with target molecules is to decrease the size of pores in cryogels. The pore sizes in the pAAm cryogel decrease with the increase in the concentration of monomers in the initial reaction mixture. That results in changing the physical-chemical and porous properties of pAAm-cryogel monoliths [31]. Thus, the low-swelling degree for the pAAm cryogels prepared from 6 to 22% monomer concentration indicates that the pore walls were formed from the very high polymer concentration (Table 11.1, reproduced from [31]). The differential scanning colorimetry (DSC) measurements showed clearly that water in pAAm cryogels is composed of water presented in supermacropores (or free water) and water bound to a polymer network

TABLE 11.1

Properties of pAAm-cryogels Produced at –12°C from Feeds with Different Monomer Concentration

Monomer Concentration[a] (%)	Notation	Swelling Degree/g H_2O per g Dried Polymer (for Gels)	Swelling Degree/g H_2O per g Dried Polymer (for Cryogels)	Polymer Bound Water (% of Total Water in Cryogel)[b]	Water Flow Resistance of Cryogel/ cm h^{-1}[c]
6	6-pAAm	14.7	5.5	4.0	870
10	10-pAAm	9.2	4.5	7.2	370
15	15-pAAm	5.9	3.9	11.3	81
18	18-pAAm	4.4	3.5	14.0	61
22	22-pAAm	3.9	3.2	17.3	20

Source: Reproduced from F. M. Plieva, M. Karlsson, M.-R. Aguilar, D. Gomez, S. Mikhalovsky, and I.Yu. Galaev, *Soft Matter*, 2005. With permission.

[a] Vinyl (AAm + AGE)/divinyl (cross-linker) ratio was 15:1 (mol/mol).

[b] The amount of polymer bound water was determined from water vapor adsorption by dry cryogels and presented as percentage of total water in cryogel.

[c] Water flow resistance of pAAm-cryogels was measured as a flow rate through pAAm-cryogel columns at constant hydrostatic pressure equal to 1 m water column.

[18,31,43]. The pore size in the pAAm cryogel decreased and the thickness of pore walls increased with increasing concentration of monomers in the initial solution [31]. Decreasing pore size in pAAm cryogels influenced the hydrodynamic properties of these monoliths. The back pressure increased for the pAAm cryogels prepared from the feed with 15% monomer concentration compared to pAAm cryogels prepared from the feed with 6% monomer concentration (unpublished data).

The functional abilities of the pAAm-cryogel monoliths prepared from the feed with different monomer concentrations were changed as well. In order to estimate the accessibility of the covalently coupled ligands in the pAAm cryogels prepared from feeds with different monomer concentration, the functional epoxy groups were converted into metal chelating iminodiacetate groups via reaction with iminodiacetic acid. The advantage of this ligand is its ability to bind easily assayed Cu(II) ions (low molecular weight target) and when loaded with Cu(II) ions to bind a model protein (lysozyme, molecular weight of 14.4 kDa, high molecular weight target). The amount of Cu(II) ions bound by the pAAm cryogels increased linearly with the increase of monomer concentration in the feed (Figure 11.1) [31]. The result was pretty straight forward as the total amount of AGE copolymerized in the cryogel increased with increased monomer concentration in the feed, hence increasing the total amount of iminodiacetic acid (IDA) groups present in the cryogel. IDA groups, even those buried inside the pore walls, were easily accessible to small Cu(II) ions. On the contrary, a high molecular weight target molecule, lysozyme, probably could not penetrate inside the dense pore walls and hence bound only to the Cu(II)-IDA present at the surface of the pore walls. The total increase in IDA groups with increasing monomer concentration in the feed was counterbalanced

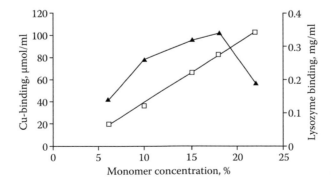

FIGURE 11.1
Binding of Cu(II) ions (open squares) and lysozyme (closed triangles) by pAAm cryogel with covalently coupled IDA prepared at –12°C from feeds with different monomer concentration. (Reproduced from F. M. Plieva, M. Karlsson, M.-R. Aguilar, D. Gomez, S. Mikhalovsky, and I.Yu. Galaev, *Soft Matter*, 1, 303–9, 2005. With permission.)

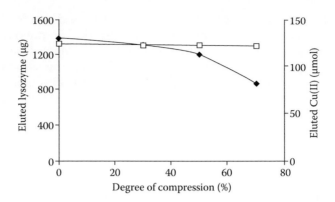

FIGURE 11.2
Elution of lysozyme (closed diamonds) and Cu(II) ions (open squares) from the noncompressed and compressed Cu(II)-IDA-pAAm cryogel monoliths at different degrees of compression of the cryogel monoliths. (Reproduced from F. M. Plieva, E. De Seta, I.Yu. Galaev, and B. Mattiasson, *Sep. Purif. Techno.*, 65(1), 110–16, 2009. With permission.)

by the decreased pore size and volume and hence the area of pores accessible for lysozyme. As a result, the amount of lysozyme bound by different cryogels had a clear maximum at about 18% monomer concentration in the feed (Figure 11.1) [31].

One of the unique features of pAAm cryogels is their elasticity, a feature that allows for a simple scale-up of cryogel monoliths, for example, by placing the cryogel monoliths-discs on top of each other or for processing compressed cryogel monoliths [22]. Due to the extensive uniaxial compression of IDA-pAAm cryogel monoliths, it was possible to elute the bound small targets like Cu(II) ions with high recovery in a small elution volume (Figure 11.2) [22]. It is notable that the degree of compression of IDA-pAAm cryogel monoliths did not influence the recovery of small molecules as Cu(II) ions, however, the recovery of a high molecular weight target such as lysozyme was decreased at cryogel compression more than 60% (Figure 11.2) [22]. Thus, the compression of cryogel monoliths allowed for decreasing the bed volume and elution of the targets as more concentrated solutions [22].

The IDA containing pAAm cryogels prepared from feeds with different concentrations of monomers and arranged inside the open-ended plastic core (so-called protective housing) showed an efficient capture of small molecules like Cu(II) ions from the aqueous solution [11]. Capacity for binding Cu(II) ions was higher for the IDA-pAAm cryogels prepared from higher monomer concentration that is, when the porosity of prepared IDA-cryogels was decreased [11]. This novel arrangement of cryogels inside protective shells (defined as macroporous gel particles, MGPs) prevented the MGPs from attrition at stirring, thus allowing for the use of MGPs in the well-stirred bioreactors [11,56]. The MGPs will be described in more detail later in this chapter.

11.2.2 Grafting Polymer Chains to the Pore Surface of Cryogel Monoliths

One way to increase the surface area of an adsorbent is to graft polymer chains to the pore walls. The grafting of functional polymers to the pore surface of pAAm cryogel monoliths was found to be an efficient and convenient method for the preparation of cryogels with both controlled extent of functional groups incorporated and tailored surface chemistries. Anion-exchange polymer chains of poly(2-[dimethylamino]ethyl methacrylate) (pDMAEMA) and poly([2-{methacryloyloxy}ethyl]-trimethylammonium chloride; pMETA) and cation-exchange polymer chains of polyacrylate have been grafted onto pAAm-cryogel monoliths using potassium diperiodato-cuprate as an initiator in aqueous medium [34]. A variety of nonionic, poly-electrolyte, temperature sensitive, hydroxyl- and epoxy-carrying polymer chains were grafted onto pAAm cryogels in aqueous and aqueous-organic media [34,35,36]. The macroporous structure of pAAm cryogels promoted grafting by providing an ample surface for the reaction. The grafting is performed either by direct grafting in a solution of monomer and initiator or by using a two-step technique, namely, by saturating the matrix with initiator solution followed by replacing it with monomer solution. In the latter approach (two-step graft polymerization) a higher efficiency of graft polymerization was achieved [35].

The graft polymerization did not alter the macroporous structure of the pAAm cryogels, however the flow rate of solutes through the cryogel monolith decreased with increasing density of the grafted polymer. The sorption of low molecular weight targets (metal ions, dye) to the surface of pAAm cryogels increased linearly with increasing the grafting density. The binding of low molecular weight targets was independent of using the grafting approach (direct or two-step graft polymerization). It was possible to achieve the ion-exchange capacity up to 0.2–0.5 mmole/ml of wet cryogel (Figure 11.3) [35]. However, binding of high molecular weight targets like proteins on the grafted pAAm-cryogel columns showed the tentacle-type of protein binding to grafted polymer Figure 11.4), which depended on the architecture of the grafted polymer layer and took place after a certain degree of grafting has been reached. The binding of proteins by tentacle-like polymer chains allowed for increasing the binding capacity for proteins on the grafted pAAm cryogels up to 6–12 mg/ml (Figure 11.4) [33,57]. The observed difference in binding proteins on grafted pAAm cryogels could be due to the different architecture of grafted polymer layers that affects the accessibility of ion-exchange groups for protein binding

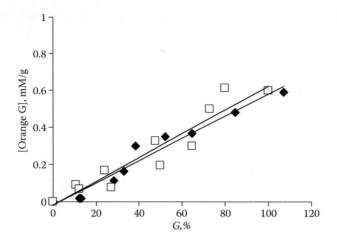

FIGURE 11.3

Binding of low molecular weight target (Orange G) to polyDMAEMA-grafted pAAm cryogel prepared by one-step procedure (closed diamonds) and two-step grafting (opened square), respectively. Pieces of about 2 mm size of polyDMAEMA-grafted pAAm cryogel were incubated with 5 mM Orange G solution in water, pH 5.0 for 40 hr at room temperature and constant shaking. The amount of Orange G adsorbed was estimated spectrophotometrically as decreasing of the concentration of Orange G in the solution before and after incubation with the cryogel. (Reproduced from I. N. Savina, I.Yu. Galaev, and B. Mattiasson, *J. Chromatogr. A.*, 1092, 199–205, 2005. With permission.)

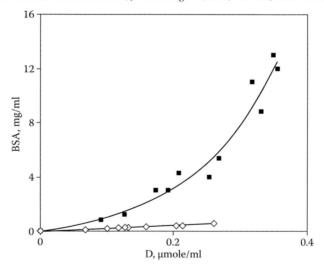

FIGURE 11.4

Binding of high molecular weight target, bovine serum albumin (BSA) by DMAEMA-grafted pAAm cryogels prepared by one-step technique (open rhombs) and two-step technique (closed squares), respectively. The monolith of pDMAEMA-grafted pAAm cryogel was saturated with BSA solution (1 mg/ml in running buffer, 20 mM Tris-HCl buffer, pH 7.0), unbound BSA was washed off with the running buffer and bound BSA was eluted with 1.5 M NaCl in the running buffer. (Reproduced from I. N. Savina, I.Yu. Galaev, and B. Mattiasson, *J. Mol. Recogn.*, 19, 313–21, 2006. With permission.)

(a)   (b)

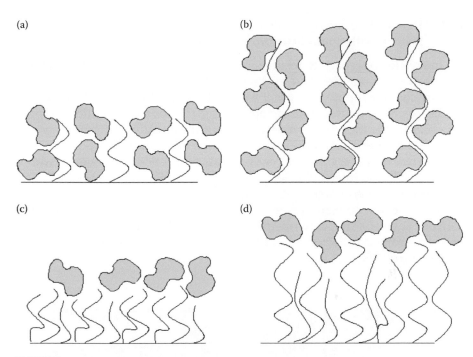

(c)   (d)

FIGURE 11.5
Schematic presentation of BSA sorption on the surface of pAAm cryogel grafted by using different techniques: (a) two-step techniques, low grafting degree, (b) two-step technique, high grafting degree, (c) one-step technique, low grafting degree, (d) one-step technique, high grafting degree. (Reproduced from I. N. Savina, I.Yu. Galaev, and B. Mattiasson, *J. Chromatogr. A.*, 1092, 199–205, 2005. With permission.)

(Figure 11.5). The same degree of grafting could be achieved either by grafting a large number of short polymer chains or by grafting a small number of long polymer chains. The binding of proteins by long polymer chains, separated from each other, will be higher as the ion-exchange groups will be much more accessible for protein binding compared to when short- and densely grafted polymer layers are used (Figure 11.5). The two-step procedure of grafting resulted in the growth of long grafted chains at the initial radical sites on the pAAm backbone (Figure 11.5). The long grafted polymer chains perform as a tentacle capable of multipoint interactions with negatively charged BSA molecules. The longer the chains (i.e., the higher the grafting density) the higher the protein-binding capacity. A smaller number of longer chains of grafted polymer could provide better BSA binding capacity as compared to a higher number of shorter grafted chains, as the long chains could perform as tentacles when binding protein molecules (Figure 11.5). In contrast, the high density of the grafted polymer chains has a very small effect on protein-binding capacity, which remains essentially the same as in the case of ion-exchange groups presented directly at the interface of cryogel surface (Figure 11.5).

11.2.3 Double-Freezing Approach to Form the Macroporous Systems with Controlled Porosity and Increased Capacity

An interesting approach to form macroporous monolith systems with high flow-through properties and an increased amount of binding sites is to prepare the macroporous cryogel monoliths with double continuous networks (so-called double-freezing approach). The formation of new cryogel networks inside the interconnected pores of the already preformed cryogel allows for a wide design of cryogel monoliths with controlled porosity and pore surface chemistry [18,38]. The synthesis of secondary cryogel networks occurs inside the interconnected pores of the preformed cryogels (or primary cryogels with pores of 10–100 μm in size and porosities higher than 90%). Elastic and sponge-like gels with a double continuous macroporous network have continuously interlacing macroporous networks of primary and secondary cryogels as visualized by confocal laser scanning microscopy (Figure 11.6).

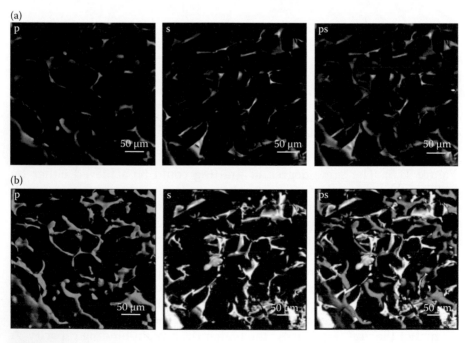

FIGURE 11.6

Confocal laser scanning microscopy images of double-continuous macroporous cryogels. (a) Two dimensional (2-D) image of primary PEG-cryogel stained with Rhodamine B (image p), secondary PEG-cryogel stained with FITC (image s) and DM PEG/PEG cryogel (image ps). (b) Three dimensional (3-D) reconstituted image of primary PEG-cryogel stained with Rhodamine B (image p), secondary PEG-cryogel stained with FITC (image s) and DM PEG/PEG cryogel (image ps). (Reproduced from F. M. Plieva, P. Ekström, I.Yu. Galaev, and B. Mattiasson, *Soft Matter*, 4, 2418–28, 2008. With permission.)

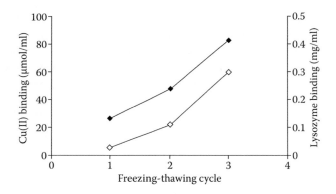

FIGURE 11.7
Binding of low molecular weight target, Cu(II) (closed squares) and high molecular weight target, lysozyme (open squares) by primary cryogel, and cryogels after two and three freezing-thawing cycles, respectively. (Reproduced from F. M. Plieva, P. Ekström, I.Yu. Galaev, and B. Mattiasson, *Soft Matter*, 4, 2418–28, 2008. With permission.)

Incorporation of functional reactive groups (or active ligands) into the cryogel network at each freezing-thawing cycle allows for the preparation of macroporous multilayered cryogels with increased amount of functional groups (bioactive ligands) for binding with the specific targets. Thus, pAAm-based macroporous cryogels bearing functional epoxy groups (which were then converted to IDA residues) were prepared via three sequential freezing-thawing cycles [18]. The behavior of the multilayered IDA-pAAm cryogels regarding the binding of small molecules was evaluated by binding Cu(II) ions and lysozyme [18]. The binding of both targets increased proportionally for the IDA-pAAm cryogels prepared after double and triple freezing-thawing cycles as compared to the binding by primary IDA-pAAm cryogels prepared via a single freezing-thawing cycle (Figure 11.7) [18]. The main reason for the same tendency of binding for low and high molecular weight targets (Cu(II) ions and lysozyme, respectively) was the formation of additional macroporous polymeric network at each freezing-thawing cycle presumably with the similar thickness of pore walls. That all resulted in the increased available surface area for binding with the model targets. Thus the sequential freezing-thawing is a versatile technique for the preparation of macroporous monolithic media with high flow-through properties and an increased amount of binding sites for the capture of small and large targets. It is noted that the sequential freezing-thawing concept also allowed for the preparation of macroporous high flow-through monoliths with mixed functionalities [38]. Thus, the pAAm-based monoliths with weak and strong cationic functionalities that is, bearing functional carboxy groups (weak cationic) in the primary cryogel and sulphopropyl groups (strong cationic) in the secondary cryogels effectively captured model target lysozyme from aqueous solution

[38]. Double freezing is also a versatile technique for the preparation of composite cryogel systems as was shown for secondary cryogel composed of pAAm with incorporated titania as a filler [42]. The incorporation of the filler into the cryogel pore walls resulted in the preparation of the macroporous system with high flow-through properties, increased mechanical strength, and efficient binding of the model protein, BSA [42].

11.2.4 Macroporous Composite Systems

One of the most efficient and straightforward ways to increase capacity of the macroporous cryogel adsorbents and keep the large size of macropores at the same time is the preparation of composite cryogel systems [11,22,32,39,41]. The composite cryogel systems (both prepared in beaded shape and arranged as monolithic columns) are especially promising for the capture of low molecular weight compounds. For example, composite Amberlite IRA/PVA cryogels were successfully used for the capture of a low molecular weight target from nonclarified fluids in the expanded bed mode. For these purposes, composite beads, composed of a PVA cryogel with entrapped particles of the strong anion-exchange resin Amberlite IRA-401 were prepared [32]. This composite material was used in expanded-bed, ion-exchange chromatography for the capture of negatively charged solutes like benzoate or lactate from a suspension of negatively charged cells. Due to hydrophilic properties of PVA the Amberlite IRA-PVA cryogel beads did not adsorb yeast cells. Finally, the composite beads were used for the direct capture of lactic acid directly from nonclarified fermentation medium of *Lactobacillus delbrueckii* [32] (Figure 11.8). Although the binding of lactic acid dropped due to the presence of other low molecular weight components in the fermentation medium, the bound lactic acid was recovered with 96% yield [32].

For the preparation of the composite cryogel monoliths, the main concern was to keep the large size of interconnected pores in the monoliths. Thus, the composite cryogel monoliths of high selectivity and low flow resistance were prepared via incorporation of the molecularly imprinted particles (MIP) into the solution of PVA (PVA with a degree of hydrolysis less than 90%) followed by freezing the reaction mixture at optimized regime [39]. When different fillers were incorporated into the polymer to be frozen, the distributed filler particles performed as ice nucleators (solid porous rough surfaces), thus ensuring the formation of uniform porous structure. Different micro- and nanosized fillers (e.g., MIP particles [39]; silica, titania, and hydroxyapatite particles [42]; or magnetic nanoparticles [41,42]) were used for the preparation of composite cryogel monoliths. The content of the filler in the system was typically in the range to keep (i) a high load of the functional filler in the reaction mixture and (ii) to keep high flow-rate properties of the composite cryogels prepared as monolithic columns. Essential

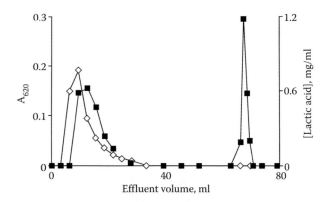

FIGURE 11.8
Capture of lactic acid by composite beaded PVA cryogel beads (with incorporated particles of Amberlite IRA-401) in expanded bed mode from nonclarified *Lactobacillus delbrueckii* fermentation broth. Experimental conditions: 2 ml of nonclarified broth was applied to the column with composite beaded PVA cryogel beads (4 ml settled volume) in an expanded bed mode at a linear velocity of 160 cm/h. The column was washed with deionized water in an expanded bed mode. The flow was interrupted allowing the adsorbent to settle and the elution was performed with 0.1 M HCl in a packed bed mode. The content of cells was analyzed by measuring absorbance at 620 nm in the effluent during adsorption, washing, and elution stages. Lactic acid was analyzed using HPLC. Open diamonds represent the cell content and closed squares represent the lactic acid concentration. (Reproduced from I. N. Savina, A. Hanora, F. M. Plieva, I.Yu. Galaev, B. Mattiasson, and V. I. Lozinsky, *J. Appl. Polym. Sci.*, 95, 529–38, 2005. With permission.)

is the accessibility of functional filler in a composite cryogel system for binding with a target molecule. For example, for the pAAm cryogels with embedded nanoparticles, the nonuniform distribution of nanoparticles was observed using transmission electron microscopy (TEM; Figure 11.9) [41]. In fact, only nanoparticles close to the pore surface were accessible for the adsorption of biomolecules [41].

The highly selective composite system was prepared on the basis of PVA cryogels with incorporated MIP particles. A combination of high selectivity in MIP and high flow path properties of cryogels allowed for the construction of an effective robust composite platform for the removal of contaminants at trace concentrations. Different polymer systems were used for the preparation of the composite MIP/cryogel systems (Table 11.2) [39]. It is worth noticing that the presence of the MIP particles in the reaction mixture resulted in the formation of the cryogel monolith systems with improved swelling after being dried (Table 11.2). For example, swelling of dried samples of the composite chitosan/MIP cryogel was much better as compared to the swelling of plain chitosan cryogels. The SEM analysis showed the microporosity of MIP and the porous structure of the composite MIP/PVA cryogels (Figure 11.10). Due to the presence of large pores in

(a) (b)

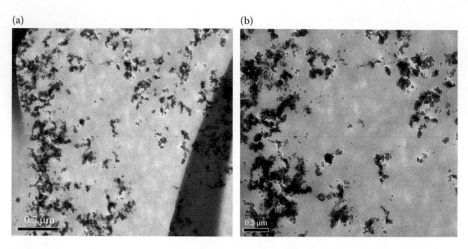

FIGURE 11.9
Transmission electron microscopy (TEM) photographs of the embedded Fe_3O_4 nanoparticles in the cryogel. (Reproduced from K. Yao, J. Yun, S. Shen, L. Wang, X. He, and X. Yu, *J. Chromatogr. A.*, 1109, 103–10, 2006. With permission.)

TABLE 11.2

Porous Properties of the Cryogels and Composite Cryogel/PVA Monoliths

Gel Precursors used for MG Preparation	Type of Monolith	Notation	Pore Volume[a] (%)	Water Flow Path through the Monolith Column[b] (cm/h)	Re-swelling after Drying (% of Initial Weight), Time
PVA (5.4%)	Control composite	PVA	88.6	620[c]	(100), within 1–2 min
		PVA/MIP	83.6	580	(100), within seconds
pAAm (7%)	Control composite	pAAm	90.2	820	(100), within seconds
		pAAm/ MIP	86.4	770	(100), within seconds
Chitosan (0.8%)	Control composite	Chs	Not checked	1100[c]	(16–22), hours
		Chs/MIP		835	(70–75), within 1–2 min

Source: Reproduced from M. L. Noir, F. M. Plieva, T. Hey, B. Quiesse, and B. Mattiasson, *J. Chromatogr. A.*, 1154, 158–64, 2007. With permission.

[a] Pore volume was estimated as described elsewhere [28].
[b] The water flow path through the monolith column was estimated at a hydrostatic pressure equivalent to 1 m water column.
[c] The PVA and chitosan monoliths were compressed up to 25% and 70% of their initial height, respectively.

FIGURE 11.10
(a) SEM images of molecularly imprinted particles (MIP) and (b and c) PVA/MIP composite at different magnifications. (Reproduced from M. L. Noir, F. M. Plieva, T. Hey, B. Quiesse, and B. Mattiasson, *J. Chromatogr. A.*, 1154, 158–64, 2007. With permission.)

the composite cryogel system, the back pressure through the composite cryogel cartridge was much less compared to that of a cartridge filled only with MIP particles (unpublished data). The prepared composite MIP/PVA columns (MIP prepared using the β-estradiol as a template) removed efficiently β-estradiol from aqueous solutions and wastewater effluents at trace concentration (2 µg/L) and flow rates up to 50 ml/min [39]. High efficiency of binding the β-estradiol was shown even in the presence of interfering compounds (Figure 11.11).

11.3 Macroporous Cryogels in Protective Shells

The formation of macroporous cryogels inside a plastic core (so-called protective housing) allows for expanding the potential applications of the

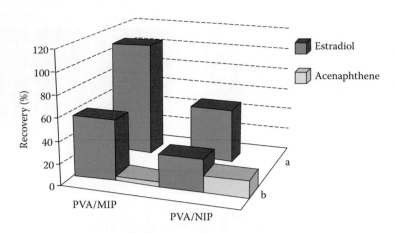

FIGURE 11.11
Recovery of 17β-estradiol and acenaphthene from PVA/MIP (black bars) and PVA/NIP (nonimprinted polymers; white bars) composites. Experimental conditions: (a) 2L of an aqueous solution containing 2 μg/L of 17β-estradiol or (b) 2L of an aqueous solution of 17β-estradiol, acenaphthene and humic acid at concentration of 2 μg/L each were applied to the composite monoliths at a flow rate of 50 mL/min. The 17β-estradiol was extracted from the composite monoliths with methanol: acetic acid (4:1 v/v) solution. (Reproduced from M. L. Noir, F. M. Plieva, T. Hey, B. Quiesse, and B. Mattiasson, *J. Chromatogr. A.*, 1154, 158–64, 2007. With permission.)

macroporous cryogels systems. By this design, the macroporous cryogels with 10–100 μm-sized pores can be used in well stirred bioreactors due to the protection against the impact caused by stirring [11]. Cryogels without the protection are easily destroyed under the same conditions. The cryogels inside the protective shells (known as macroporous gel particles, MGPs) were prepared from both low and high molecular weight precursors and with different functionalities [11]. Thus, the composite MIP/PVA MGPs were used for the capture of endocrine disrupting compounds (EDC) as β-estradiol, nonylphenol, and atrazine from water using MGPs in a moving bed reactor [40].

Differently designed protective shells can be used for the preparation of MGPs. The open-ended plastic carriers (known as Kaldnes carriers and used in environmental applications) were used for the formation of pAAm cryogels bearing metal chelate functionalities, iminodiacetate residues (IDA). The macroporous IDA-pAAm cryogels fully occupied the inner space of the Kadnes carriers (Figure 11.12), thus presenting the macroporous cryogel surrounded by a protective shell. The MGPs were shown to have high resistance to attrition at stirring [11], which depended on the porosity of the core cryogel formed inside the protective shell. The MGPs effectively removed heavy metals from water effluents at low concentration

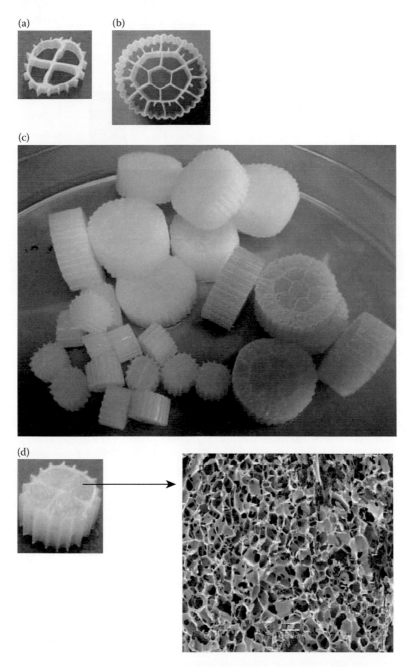

FIGURE 11.12
Picture of the Kaldnes carriers of (a) K1 and (b) K3 types where the (c) macroporous cryo-gels bearing different functionalities were formed; (d) SEM of pAAm cryogel formed inside the K1-carrier. (Reproduced from F. M. Plieva and B. Mattiasson, *Ind. Eng. Chem. Res.*, 47 (12), 4131–41, 2008. With permission.)

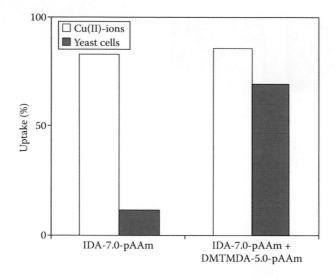

FIGURE 11.13
Uptake of Cu(II)-ions (white bars) and yeast cells (grey bars) by MGPs bearing a single imi-nodiacetate functionality (IDA-pAAm) or by MGPs with mixed functionalities (IDA-pAAm + DMTMDA-pAAm). (Reproduced from F. M. Plieva, and B. Mattiasson, *Ind. Eng. Chem. Res.*, 47 (12), 4131–41, 2008. With permission.)

[11,56]. The removal of small targets as Cu(II) ions was efficient even from the model mixtures with yeast cells (Figure 11.13). The use of a mixture of MGPs with different functionalities (IDA and anion exchange; IDA-pAAm and DMTMDA-pAAm, respectively, Figure 11.13) allowed for simultaneous removal of both, Cu(II) ions and yeast cells from the model mixtures (Figure 11.13). Interestingly, the IDA-MGPs became blue in color due to the captured Cu(II)-ions and were easily distinguished from the ion exchange MGPs that retained its native white color [11]. When the macroporous cryogels bearing different functionalities were prepared in housings of different sizes, the sorbents with captured targets differed both in color and in size allowing for simple sieving of the MGPs of different size and then the specific elution of individual targets (Figure 11.14). The possibility during one sorption run to adsorb and then elute separate different targets is one of the promising approaches in environmental separations when different low molecular weight contaminants should be removed. The use of MGP adsorbents is especially favorable for the diluted effluents, when the target molecules are presented in water at low concentration. Besides processing under stirring, the MGP adsorbents can be used in different column configurations, for example, as columns filled with MGPs [56] or in fluidized bed columns [40].

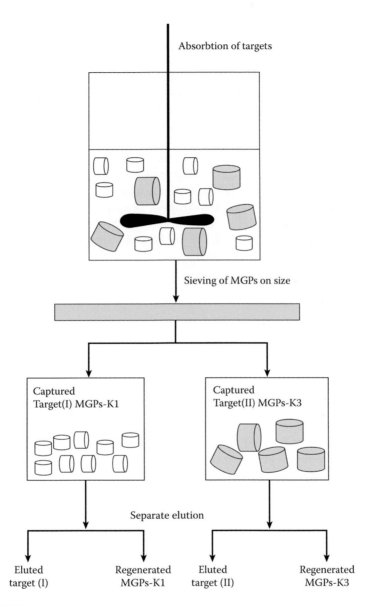

FIGURE 11.14
General approach for using the MGPs of different size and mixed functionality in sorption processes in well-stirred tanks. The MGPs prepared in different size housing (as K1 and K3) and bearing functionalities to specific targets (Target [I] and target [II]) will bind specifically the targets from complex mixture. Then the MGPs with captured targets can be separated through sieving (according to the size of MGPs) followed by elution of the bound targets using specific eluents. Thus it is possible during one run of sorption to adsorb and elute separately the targets (I) and (II) and regenerate the MGPs afterward. (Reproduced from F. M. Plieva, and B. Mattiasson, *Ind. Eng. Chem. Res.*, 47 (12), 4131–41, 2008. With permission.)

11.4 Concluding Remarks

Macroporous cryogels are very promising materials in industrial and environmental biotechnology as efficient means of processing particulate-containing streams. A unique combination of properties of macroporous cryogels like controlled macroporous structure and elasticity along with chemical and mechanical robustness of cryogels opens wide perspectives in designing the macroporous materials with superior performance for processing of complex fluids containing the suspended particles. The different format design of cryogels (beads, discs, monoliths, scaled discs, MGPs) allow for using all possible process designs. The development of cryogels in the protective housing (MGP) represents an advanced step in the preparation of robust macroporous media for use in well-stirred tanks with complex fluids containing suspended particles.

References

1. H. A. Chase. 1994. Purification of proteins by adsorption chromatography in expanded beds. *Trends Biotechnol.* 12: 296–303.
2. M. B. Dainiak, I.Yu. Galaev, and B. Mattiasson. 2002. Polyelectrolyte-coated ion exchangers for cell-resistant expanded bed adsorption. *Biotechnol. Progr.* 18:815–20.
3. M. E. Viloria-Cols, R. Hatti-Kaul, and B. Mattiasson. 2004. Agarose-coated anion exchanger prevents cell-adsorbent interactions. *J. Chromatogr. A* 1043:195–200.
4. A. E. Rodrigues, Z. P. Lu, and J. M. Loureiro. 1991. Residence time distribution of inert and linearly adsorbed species in a fixed bed containing "large-pore" supports: Applications in separation engineering. *Chem. Eng. Sci.* 46:2765–73.
5. A. Pai, S. Gondkar, and A. Lali. 2000. Enhanced performance of expanded bed chromatography on rigid superporous adsorbent matrix. *J. Chromatogr. A* 867:113–30.
6. J. Andersson, and B. Mattiasson. 2006. Simulated moving bed with a simplified approach: Purification of lactoperoxidase and lactoferrin from whey protein concentrate. *J. Chromatogr. A* 1107:88–95.
7. J. Andersson, D. Sahoo, and B. Mattiasson. 2008. Isolation of potato proteins using simulated moving bed technology. *Biotechnol. Bioeng.* DOI 10: 1002/ bit.22012.
8. V. I. Lozinsky, I.Yu. Galaev, F. M. Plieva, I. N. Savina, H. Jungvid, and B. Mattiasson. 2003. Polymeric cryogels as promising materials of biotechnological interest. *Trends Biotechnol.* 21:445–51.
9. V. I. Lozinsky, F. M. Plieva, I.Yu Galaev, and B. Mattiasson, 2002. The potential of polymeric cryogels in bioseparation. *Bioseparation* 10:163–88.

10. F. M. Plieva, I.Yu. Galaev, and B. Mattiasson. 2007. Macroporous gels prepared at subzero temperatures as novel materials for chromatography of particulate containing fluids and cell culture applications. *J. Sep. Sci.* 30:1657–71.
11. F. M. Plieva, and B. Mattiasson. 2008. Macroporous gel particles as novel sorbent materials: Rational design. *Ind. Eng. Chem. Res.* 47 (12):4131–41.
12. R. F. Martins, F. M. Plieva, A. Santos, and R. Hatti-Kaul. 2003. Integrated immobilized cell reactor-adsorption system for β-cyclodextrin production: A model study using PVA-cryogel entrapped *Bacillus agaradhaerens* cells. *Biotechnol. Lett.* 25:1537–43.
13. S. Das-Bradoo, I. Svensson, J. Santos, F. M. Plieva, B. Mattiasson, and R. Hatti-Kaul. 2004. Synthesis of alkylgalactosides using whole cells of *Bacillus pseudofirmus* species as catalysts. *J. Biotechnol.* 110:273–85.
14. V. I. Lozinsky. 1998. Cryotropic gelation of poly(vinyl alcohol) solutions. *Russ. Chem. Rev.* 67:573–86.
15. V. I. Lozinsky, and F. M. Plieva. 1998. Poly(vinyl alcohol) cryogels employed as matrices for cell immobilization. 3. Overview of recent research and developments. Enzyme Microb. Technol. 23:227–42.
16. F. M. Plieva, M. Karlsson, M.-R. Aguilar, D. Gomez, S. Mikhalovsky, I.Yu. Galaev, and B. Mattiasson. 2006. Pore structure of macroporous monolithic cryogels prepared from poly(vinyl alcohol). *J. Appl. Polym. Sci.* 100:1057–66.
17. E. N. Efremenko, O. V. Spiricheva, D. V. Veremeenko, A. V. Baibak, and V. I. Lozinsky. 2006. L(+)-Lactic acid production using poly(vinyl alcohol)-cryogel-entrapped *Rhizopus oryzae* fungal cells. *J. Chem. Technol. Biotechnol.* 81:519–22.
18. F. M. Plieva, P. Ekström, I.Yu. Galaev, and B. Mattiasson. 2008. Monolithic cryogels with open porous structure and unique double-continuous macroporous networks. *Soft Matter* 4: 2418–28.
19. J.-P. Ting, and G. Sun. 2000. Use of polyvinyl alcohol as a cell immobilization matrix for copper biosorption by yeast cells. *J. Chem. Technol. Biotechnol.* 75:541–46.
20. C. F. Degiorgi, R. A. Pizarro, E. E. Smolko, S. Lora, and M. Carenza. 2002. Hydrogels for immobilization of bacteria used in the treatment of metal-contaminated wastes. *Rad. Phys. Chem.* 63:109–13.
21. F. Svec. 2004. Preparation and HPLC applications of rigid macroporous organic polymer monoliths. *J. Sep. Sci.* 27:747–66.
22. F. M. Plieva, E. De Seta, I.Yu. Galaev, and B. Mattiasson. 2009. Macroporous elastic polyacrylamide monolith columns: Processing under compression and scale-up. *Sep. Pur. Tech.* 65 (1):110–16.
23. P. Persson, O. Baybak, F. Plieva, I.Yu. Galaev, B. Mattiasson, N. Nilsson, and A. Axelsson. 2004. Characterization of a continuous supermacroporous monolithic matrix for chromatographic separation of large bioparticles. *Biotechnol. Bioeng.* 88:224–36.
24. P. Arvidsson, F. M. Plieva, V. I. Lozinsky, I.Yu. Galaev, and B. Mattiasson. 2003. Direct chromatographic capture of enzyme from crude homogenate using immobilized metal affinity chromatography on a continuous supermacroporous adsorbent. *J. Chromatogr. A* 986:275–90.
25. F. M. Plieva, J. Andersson, I.Yu. Galaev, and B. Mattiasson. 2004. Characterization of polyacrylamide based monolithic columns. *J. Sep. Sci.* 27:828–36.

26. A. Hanora, F. M. Plieva, M. Hedström, I.Yu. Galaev, and B. Mattiasson. 2005. Capture of bacterial endotoxins using a supermacroporous monolithic matrix with immobilized polyethyleneimine, lysozyme or polymyxin B. *J. Biotechnol.* 118:421–33.

27. P. Arvidsson, F. M. Plieva, I. N. Savina, V. I. Lozinsky, S. Fexby, L. Bülow, I.Y. Galaev, and B. Mattiasson. 2002. Chromatography of microbial cells using continuous supermacroporous affinity and ion-exchange columns. *J. Chromatogr. A* 977:27–38.

28. F. M. Plieva, I. N. Savina, S. Deraz, J. Andersson, I.Yu. Galaev, and B. Mattiasson. 2004. Characterization of supermacroporous monolithic polyacrylamide based matrices designed for chromatography of bioparticles. *J. Chromatogr. B* 807:129–37.

29. W. Noppe, F. M. Plieva, K. Vanhoorelbeke, H. Deckmyn, M. Tuncel, A. Tuncel, I.Yu. Galaev, and B. Mattiasson. 2007. Macroporous monolithic gels, cryogels, with immobilized phages from phage-display library as a new platform for fast development of affinity adsorbent capable of target capture from crude feeds. *J. Biotechnol.* 131:293–99.

30. M. B. Dainiak, F. M. Plieva, I.Yu. Galaev, R. Hatti-Kaul, and B. Mattiasson. 2005. Cell chromatography. Separation of different microbial cells using IMAC super-macroporous monolithic columns. *Biotechol. Progr.* 21:644–49.

31. F. M. Plieva, M. Karlsson, M.-R. Aguilar, D. Gomez, S. Mikhalovsky, and I.Yu. Galaev. 2005. Pore structure in supermacroporous polyacrylamide based cryo-gels. *Soft Matter* 1:303–09.

32. I. N. Savina, A. Hanora, F. M. Plieva, I.Yu. Galaev, B. Mattiasson, and V. I. Lozinsky. 2005. Cryostructuration of polymer systems: XXIV. Poly(vinyl alcohol) cryogels filled with particles of a strong anion exchanger: Properties of the composite materials and potential applications. *J. Appl. Polym. Sci.* 95:529–38.

33. I. N. Savina, I.Yu. Galaev, and B. Mattiasson. 2005. Graft polymerization of acrylic acid onto macroporous polyacrylamide gel (cryogel) initiated by potas-sium diperiodatecuprate. *Polymer* 46:9596–9603.

34. I. N. Savina, I.Yu. Galaev, and B. Mattiasson. 2005. Anion-exchange super-macroporous monolithic matrices with grafted polymer brushes of N,N-dimethylaminoethyl-methacrylate. *J. Chromatogr. A* 1092:199–205.

35. I. N. Savina, I.Yu. Galaev, and B. Mattiasson. 2006. Ion-exchange macropo-rous hydrophilic gel monolith with grafted polymer brushes. *J. Mol. Recogn.* 19:313–21.

36. K. Yao, J. Yun, S. Shen, and F. Chen. 2007. In-situ graft-polymerization prepara-tion of cation-exchange supermacroporous cryogel with sulfo groups in glass columns. *J. Chromatogr. A* 1157:246–51.

37. J. Yun, S. Shen, F. Chen, and K. Yao. 2007. One-step isolation of adenosine triphosphate from crude fermentation culture of *Sacharomyces cerevisiae* using anion-exchange cryogel chromatography. *J. Chromatogr. B* 860:57–62.

38. F. M. Plieva, M. Hedström, I.Yu. Galaev, and B. Mattiasson. 2008. Cryogel monolith columns with double-continuous macroporous networks and mixed-functionalities. Unpublished manuscript.

39. M. L. Noir, F. M. Plieva, T. Hey, B. Quiesse, and B. Mattiasson. 2007. Macroporous molecularly imprinted polymer/cryogel composite systems for the removal of endocrine disrupting trace contaminants. *J. Chromatogr. A* 1154:158–64.

40. M. L. Noir, F. M. Plieva, and B. Mattiasson. 2009. Removal of endocrine-disrupting compounds from water using macroporous, molecularly imprinted cryogels in a moving-bed reactor. *J. Sep. Sci.* 32 (9):1471–79.

41. K. Yao, J. Yun, S. Shen, L. Wang, X. He, and X. Yu. 2006. Characterization of a novel continuous supermacroporous monolithic cryogel embedded with nanoparticles for protein chromatography. *J. Chromatogr. A* 1109:103–10.

42. F. M. Plieva, D. Pignetti, I.Yu. Galaev, and B. Mattiasson. 2008. Porous structure of supermacroporous composite cryogels. Unpublished manuscript.

43. F. Plieva, X. Huiting, I.Yu. Galaev, B. Bergenståhl, and B. Mattiasson. 2006. Macroporous elastic polyacrylamide gels prepared at subzero temperatures: Control of porous structure. *J. Mater. Chem.* 16:4065–73.

44. M. M. Ozmen, and O. Okay. 2005. Superfast responsive ionic hydrogels with controllable pore size. *Polymer* 46:8119–27.

45. D. Ceylan, M. M. Ozmen, and O. Okay. 2005. Swelling-deswelling kinetics of ionic poly(acrylamide) hydrogels and cryogels. *J. Appl. Polymer Sci.* 99:319–25.

46. M. M. Ozmen, M. V. Dinu, and O. Okay. 2008. Preparation of macroporous poly(acrylamide) hydrogels in DMSO/water mixture at subzero/temperatures. *Polymer Bull.* 60:169–80.

47. M. M. Ozmen, M. V. Dinu, E. S. Dragan, and O. Okay. 2007. Preparation of macroporous acrylamide-based hydrogels: cryogelation under isothermal conditions. *J. Macromol. Sci. Part A* 44:1195–1202.

48. K. Yao, S. Shen, J. Yun, L. Wang, X. He, and X. Yu. 2006. Preparation of poly-acrylamide-based supermacroporous monolithic cryogel bed under freezing-temperature variation conditions. *Chem. Eng. Sci.* 61:6701–08.

49. P. Perez, F. Plieva, A. Gallardo, J. S. Roman, M. Aguilar, I. Morfin, F. Ehrburger-Dolle, F. Bley, S. Mikhalovsky, I.Yu Galaev, and B. Mattiasson 2008. Bioresorbable and nonresorbable macroporous thermosensitive hydrogels prepared by cryopolymerization. Role of the cross-linking agent. *Biomacromolecules* 9:66–74.

50. I.Yu. Galaev, M. B. Dainiak, F. M. Plieva, and B. Mattiasson. 2007. Effect of matrix elasticity on affinity binding and release of bioparticles. Elution of bound cells by temperature-induced shrinkage of the smart macroporous hydrogel. *Langmuir* 23:35–40.

51. X.-Z. Zhang, and C.-C. Chu. 2003. Synthesis of temperature sensitive PNIPAAm cryogels in organic solvent with improved properties. *J. Material. Chem.* 13:2457–64.

52. A. Srivastava, E. Jain, and A. Kumar. 2007. The physical characterization of supermacroporous poly(N-isopropylacrylamide) cryogel: Mechanical strength and swelling/de-swelling kinetics. *Mater. Sci. Engin. A* 464:93–100.

53. I. N. Savina, V. Cnudde, S. D'Hollander, L. V. Hoorebeke, B. Mattiasson, I.Yu. Galaev, and F. D. Prez. 2007. Cryogels from poly(2-hydroxyethyl methacrylate): Macroporous, interconnected materials with potential as cell scaffolds. *Soft Matter* 3:1176–84.

54. M. Andac, F. M. Plieva, A. Denizli, I.Yu. Galaev, and B. Mattiasson. 2008. Poly(hydroxyethyl methacrylate)-based macroporous hydrogels with disulfide cross-linker. *Macromol. Chem. Phys.* 209:577–84.

55. A. Hanora, I. Savina, F. M. Plieva, V. A. Izumrudov, B. Mattiasson, and I.Yu. Galaev. 2006. Direct capture of plasmid DNA from non-clarified bacterial lysate using polycation-grafted monoliths. *J. Biotechnol.* 123:343–55.

56. L. Önnby, C. Giorgi, F. M. Plieva, and B. Mattiasson. 2008. Removal of heavy metals from water effluent using supermacroporous metal-chelate cryogels Unpublished manuscript.
57. I. N. Savina, B. Mattiasson, and I. Yu. Galaev. 2006. Graft polymerization of vinyl monomers inside macroporous polyacrylamide gel, cryogel, in aqueous and aqueous-organic media initiated by diperiodatocuprate (III) complexes. *J. Polym. Sci: Part A: Polym. Chem.* 44:1952–63.

12

Monolithic Macroporous Polymers as Chromatographic Matrices

Nika Lendero Krajnc, Franc Smrekar, Vida Frankovič, Aleš Štrancar, and Aleš Podgornik

CONTENTS

12.1 Introduction

The first reports about the monolithic supports were published in the late 1960s and early 1970s, but the real breakthrough occurred in the 1990s and still continues (Svec and Tennikova 2003). The first scientific paper of the second period was published by Hjerten and coworkers who introduced polyacrylamide gel (Hjerten, Liao, and Zhang 1989). This work was soon followed by the introduction of methacrylate (Tennikova, Belenkii, and Svec 1990) and silica monoliths (Nakanishi and Soga 1991; Minakuchi et al. 1996). Since then many other monoliths have been described, for example, monoliths prepared via methathesis polymerization (Mayr et al. 2001), monoliths cast from cellulose (Noel, Sanderson, and Spark 1993), superporous agarose monoliths (Gustavsson and Larsson 1999), cryogels from polyacrylamide (Arvidsson et al. 2002), and others. An excellent overview about the chromatographic monoliths can be found in the recently published book by Svec, Tennikova, and Deyl (2003).

Broad spectra of various monoliths can be attributed to several inherited properties (Podgornik and Štrancar 2005):

- transport based on convection
- extremely high porosity
- high capacity for extremely large molecules
- cheap production and easy column filling

All above mentioned properties are tightly related to the monolithic structure. In contrast to particulate supports, monoliths consist of a single piece of porous material. Pores are open and highly interconnected, forming a network of channels. The mobile phase is forced to flow through them transporting the molecules to be separated onto the active site by convection. Since there are normally no dead-end pores in the monoliths (the exception is porous silica monoliths mainly intended for separation of small molecules; Cabrera et al. 2000), there are no stagnant regions and the mass transfer between the stationary and the mobile phase is extremely fast (Zöchling et al. 2004) resulting in practically unaffected flow resolution and dynamic binding capacity (DBC; Mihelič et al. 2000). This is especially beneficial for purification of very large molecules having small mobility like proteins, polynucleotides, or viruses.

Another important consequence of the monolithic structure is high porosity, which strongly influences pressure drop at a given flow rate. The external porosity (a porosity consisting of voids between the particles) in packed beds reach a value up to 40%; the bed becomes unstable and collapses above 40%. High intraparticle porosity does not contribute to the decrease of a pressure drop since the liquid inside those pores is stagnant. The monoliths, on the other hand, exhibit external porosity up to 90% or more (Bandari et al. 2007; Bandari, Knolle, and Buchmeiser 2008). This is possible because the whole bed consists of a single skeleton and no free particles are present that might collapse.

Monoliths exhibit a very high-binding capacity for extremely large molecules. This is because the entire accessible surface is actually the surface of interconnected channels through which the sample is carried by the mobile phase. Assuming even a single layer adsorption, the surface covered by a single molecule increases with a square of its diameter while the molecule mass increases with a power of three (Endres et al. 2003). Therefore, the bigger the molecule, the higher the total mass can be adsorbed per surface unit. This theoretical prediction was recently experimentally confirmed (Yamamoto and Kita 2006).

The last two features are mainly related to the production of the monolithic columns. Particle shaped resins are normally prepared via suspension polymerization, while the monoliths are commonly produced via bulk polymerization. Once the beads are polymerized, they have to be sieved to obtain uniform particle size distribution. In the case of monoliths, an equivalent feature is a pore size distribution, which is defined already by polymerization conditions. In addition, bulk polymerization can be performed inside the column housing and therefore no packing procedure is required afterward. Because of this they are ideally suited for capillary columns or microfluid devices, for example.

Methacrylate monoliths were developed almost two decades ago (Tennikova, Belenkii, and Svec 1990) and presently represent one of the most important types of monoliths. One of the main reasons is their mechanically and chemically stable skeleton, which can withstand treatment of many chemical reagents and very high pressures. In addition, being conventionally prepared with glycidyl methacrylate as a monomer, they bear reactive epoxy moieties after polymerization that can be easily modified into virtually any functionality or used directly for immobilization of various biological ligands like proteins. Because of that they have been implemented almost in almost any field of separation science, being prepared in various formats, from microfluid devices (Logan et al. 2007; Mair et al. 2007; Levkin et al. 2008) and capillary columns (Svec 2004) up to several liters of preparative monolithic columns (Podgornik et al. 2000, 2004).

Due to a convective transport, the analysis time for large molecules can be extremely short and the entire separation can be completed within a few seconds (Štrancar et al. 1996). Taking into account that this is the range of response time of sensors routinely used in biotechnology such as pH and pO_2 probe, one can consider such a chromatographic system with the monolithic column as a chromatographic sensor (Barut et al. 2008). This feature perfectly matches recent Process Analytical Technology (PAT) guidelines. Besides extremely fast separations one can also benefit from unaffected dynamic binding capacity (DBC) flow, which enables significant increase of productivity. This becomes very important when large molecules are to be purified and therefore monolith structure has to be properly designed to enable free access for most of the molecules (Benčina et al. 2007).

Purification of macromolecules requires large volume monoliths. To prepare large methacrylate monoliths with a uniform pore structure, monolith polymerization in terms of reaction rate and heat dissipation should be understood and carefully controlled (Peters, Svec, and Fréchet 1997; Podgornik et al. 2000; Mihelič et al. 2001; Mihelič, Koloini, and Podgornik 2003). Currently, the monoliths of several liters were prepared (Figure 12.1; Podgornik et al. 2004) representing volumes sufficiently large for purification of plasmid DNA (pDNA), immunoglobulins M (IgM), or virus particles on a production scale. To be implemented under a strict good manufacturing practice (GMP) environment, monoliths have to be extensively characterized to provide necessary reproducibility and traceability. This article gives an overview about recent applications of the methacrylate monoliths trying to elucidate some interaction mechanism being involved as well as to summarize methods that have been implemented for the characterization of monolith.

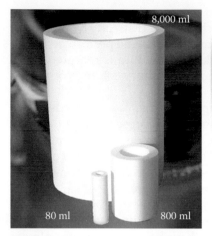

FIGURE 12.1
Large volume methacrylate monoliths (left) and appropriate housings (right).

12.2 Large Proteins

From the very beginning, proteins were the most studied topic of separation on the short bed monolithic columns. In fact, a development of such columns was driven by the investigation of a protein separation mechanism (Tennikova, Belenkii, and Svec 1990; Tennikova and Svec 1993; Tennikova and Reusch 2005). Needless to say, there is a plethora of publications dealing with studies of the monolith properties and their applications on various groups of proteins like plasma proteins (Josić et al. 1992; Josić, Bal, and Schwinn 1993; Josić et al. 1994; Štrancar et al. 1996; Štrancar et al. 1997; Amatschek et al. 2000; Branović et al. 2000, 2002; Pflegerl et al. 2002; Branović, Buchacher et al. 2003; Čerk Petrič et al. 2007), ligninases (Podgornik, Podgornik, and Perdih 1999; Podgornik et al. 2001; Milavec Žmak et al. 2003; Podgornik and Podgornik 2004), and others. More details can be found in recently reviewed articles (Barut et al. 2005; Podgornik and Štrancar 2005; Barut et al. 2008; Jungbauer and Hahn 2008). In this chapter we focus only on two large groups of proteins, namely, PEGylated proteins and IgM, which are the most recent topics of intensive study.

12.2.1 PEGylated Proteins

PEGylated proteins are not a group of similar proteins but rather different proteins on which polyethyleneglycol (PEG) molecule(s) are artificially attached. It is a widely used technology for modification of therapeutics, including Food and Drug Administration (FDA) approved proteins, but it is also implemented on other types of molecules like peptides and oligonucleotides. The main purpose of PEG attachment is to enhance the physical (such as solubility, hydrolytic stability, and aggregation) and biomedical properties (such as reduced antigenicity, proteolytic stability, prolonged serum circulation time, and ease of delivery) of the molecule. The PEG is a neutral polymer but it may affect the charge properties of proteins in three ways. First, the presence of the PEG conjugate may shield the surface charges of a protein, thereby weakening the binding to ion exchange resins. Second, conjugation to amino acid residues that alters their charge nature (e.g., conversion of amine groups to amides) hence affecting the pI. Third, protein surface localized PEGs may form hydrogen bonds with acidic or other groups and raise their pKa (Delgado, Malmsten, and Van Alstine 1997).

To obtain a pure target PEGylated protein without unwanted PEGylated forms and nonreacted native protein, a purification step must be implemented after PEGylation. Various PEGylated forms and a native protein can be very similar, consequently the separation might be a difficult and time-consuming task (Seely and Richey 2001; Fee and Van Alstine 2006). One the most commonly used techniques is ion-exchange chromatography due to its capability to separate similar proteins under very mild adsorption and elution conditions (Yamamoto, Nakanishi, and Matsuno 1988; Karlsson, Ryden, and Brewer 1998). Already a small change of protein surface charge distributions can affect both, the number of binding sites and the binding strength significantly. Therefore, it is important to understand how PEGylation affects the retention behavior of differently PEGylated protein forms (Yamamoto, Fujii et al. 2007).

The study of PEGylated lysozyme separation on strong cation exchange (SO_3^-) short monolithic columns (CEX) was performed by Akbarzaderaleh and co-workers (2007). They separated a mixture of mono, multiple, and non-PEGylated lysozyme molecules as a function of mobile phase pH value. Figure 12.2, demonstrates that at least nine peaks were obtained on the CEX monolithic column while five were separated on the Size Exclusion Chromatography (SEC) column (Figure 12.2b). All PEGylated lysozyme forms were eluted prior to a native lysozyme (Figure 12.2a). Analysis of SEC fractions on CEX monolithic column (confirmed also by SDS-PAGE analysis, Figure 12.2c) revealed that the number of attached PEG molecules of peaks one and two is larger than those in the peaks three to six (SEC fractions I and II, Figure 12.2d). Furthermore, CEX analysis of SEC fraction III demonstrated that peaks seven and eight were mono-PEGylated lysozyme (Figure 12.2e). Those results indicate that PEGylated proteins can be separated not only on the basis of the PEG-to-protein mass ratio but also by a difference in the PEGylation site. However, as it can be seen from Figure 12.2 very shallow gradients ($V_g = 100$ ml corresponds to 294 CVs) were needed for fine separations. This can be achieved in a reasonably short time with only monoliths applying high flow rates. As a consequence, faster separation in comparison to what is achieved with the SP-Sepharose HP media was obtained.

To further investigate the separation mechanism, a number of binding sites through which the molecule interacts with the matrix was determined (Yamamoto, Nakanishi, and Matsuno 1988). The results revealed that PEGylated and native lysozyme exhibited a comparable number of binding

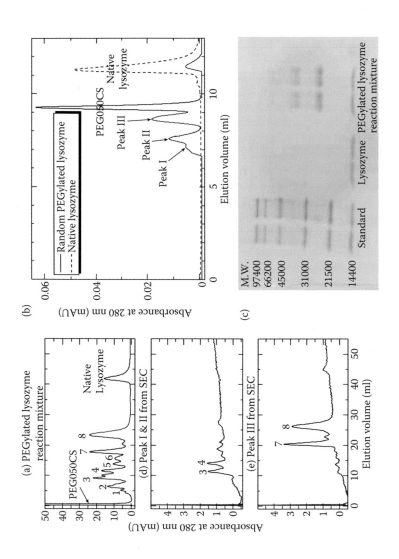

FIGURE 12.2

Separation of PEGylated lysozyme isoforms. (a) Separation of PEGylated reaction mixture on CIM SO3 disk monolithic column; (b) SEC chromatogram of PEGylated reaction mixture; (c) SDS-PAGE results of random PEGylated reaction mixture; (d) Separation of peak I&II from SEC on CIM SO3 disk monolithic column; (e) Separation of peak III from SEC on CIM SO3 disk monolithic column. (From Akbarzaderaleh, P., Abe, M., Yoshimoto, N., and Yamamoto, S., *Asia-Pacific Biochemical Engineering Conference*, Taiwan, November 4–7, 2007. With permission.)

sites. This demonstrates that the charge shielding effect is not due to the reduction of the binding site but probably due to steric hindrances that resulted in a weaker binding. Authors concluded for this particular case that PEGylation changed some physical and biochemical properties of the protein such as pH stability, surface charge density, and size, which enabled separation of the protein having a single PEG chain at a target site (Akbarzaderaleh et al. 2007).

A different study was performed by Hall, Wood, and Smith (2004). They compared efficiency of monolithic (CIM QA and SO3) and particulate based (Q-Sepharose HP and SP-Sepharose HP HiTrap) resins for separation of mono-PEGylated myelopoietin (MPO) from a native myelopoietin. Results demonstrated that the monolithic columns enabled five times faster separation combined with sharper, more symmetrical peaks, and better recovery on a preparative scale and a much shorter analysis time with comparable resolution.

Based on presented examples, one can conclude that monoliths exhibit a similar resolution as the particulate supports but at significantly shorter analysis times. Due to a possibility of implementing high flow rates, very shallow elution gradients can be implemented in a short time. This enables a separation of protein molecules with an equal number of PEG molecules, however, being attached on different positions.

12.2.2 IgMs

Currently, another very important group of proteins is IgMs. Due to a rapid progress in treatment for infectious diseases, cancer, and autoimmune diseases, the interest for a number of IgM monoclonal antibodies rose very fast (Gagnon, Richieri et al. 2007). The IgM proteins are by far the physically largest antibodies in the human circulatory system. Naturally they occur in a variety of forms with molecular weights around 1 MDa. The IgMs characteristics that can limit the various applications of standard purification processes are a large size, which corresponds to a low diffusion coefficient that leads to lower capacity and resolution of chromatographic columns, lower solubility, and conductivity than that of IgGs, susceptibly to denaturation at extreme pH, and poor tolerance of high salt concentrations. On the other hand IgMs are typically more charged than IgGs, which means that they bind more strongly than IgGs and most contaminants to ion-exchangers (Gagnon, Richieri et al. 2007). Various applications demand different purification levels of the immunoglobulins. The purification method of choice is dictated by the level of purity required and the intended use. While the therapeutic use of immunoglobulins requires highly purified molecules, some diagnostic applications require only the complete separation of IgG and IgM. Current standard methods for IgM purification are generally multistep, time-consuming processes, or they are not effective for removing all of the major impurities present in IgM samples (Gagnon, Richieri et al. 2007).

The ability of monoliths to separate IgM from other abundant proteins in human plasma like IgG and albumin was demonstrated by Brne and co-workers (2007). Different anion exchange short monolithic columns were tested and ethylene diamine (EDA) chemistry was found to give the highest resolution (Figure 12.3). The entire separation step was completed in less than three minutes including equilibration, loading, and elution, while separation itself lasted only about 90 seconds (Figure 12.3). When only the removal of IgM from IgG in human plasma is needed, quarternary amine (QA) chemistry can be selected since it gives the highest binding capacity of approximately 20 mg/ml. Robustness of this approach was tested on IgM removal from human plasma of different volunteers and very reproducible results were achieved. To avoid displacement that occurred with a real sample, the loading capacity was substantially lower, only about 2.5 mg/ml, but it was found the flow was unaffected. Such monolith can therefore be a good basis for the development of a diagnostic kit.

Another application of monolithic columns, this time for monitoring and purification of IgM monoclonal antibodies from mammalian cell culture was reported (Gagnon, Hensel et al. 2007; Gagnon, Hensel, and Richieri 2008;

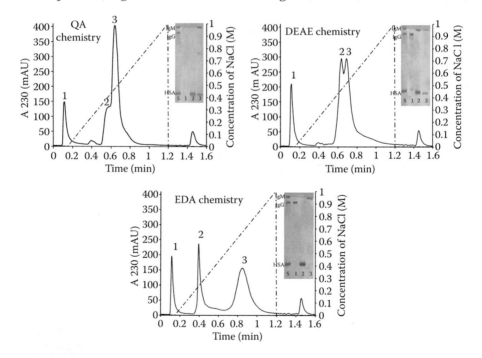

FIGURE 12.3
Separation of IgG, HSA, and IgM on CIM QA, CIM DEAE, and CIM EDA disk monolithic columns. Experimental conditions: injection volume: 20 μl; sample: 0.3 mg of IgG (peak 1), 1 mg of HSA (peak 2), and 1 mg of IgM (peak 3) dissolved in 1 ml of binding buffer; flow rate: 4 ml/ min; binding buffer: 20 mM phosphate buffer pH 7.2; elution buffer: 20 mM phosphate buffer pH 7.2 + 1 M NaCl. SDS-PAGE analysis: S: mixture of IgG, HSA, and IgM; numbers marking the lines correspond to marked peaks on chromatogram. (From Brne, P., Podgornik, A., Benčina, K., Gabor, B., Štrancar, A., and Peterka, M. *J. Chromatogr. A.*, 1144, 120–25, 2007. With permission.)

Gagnon, Richieri, and Hensel 2008). By proper adjustment of the chromatographic method they were able to monitor concentration of IgM in the media during the cultivation (Figure 12.4a). Using monolithic chromatographic supports these data were obtained more rapidly as with the SDS PAGE or immunological assays (Gagnon et al. 2007a) and fit perfectly with PAT directives of FDA. Alternatively, the same method can be implemented for in-process control of preparative purification process, monitoring IgM concentration during loading (Figure 12.4b), pre-elution (Figure 12.4c), elution concentration (Figure 12.4d), and during post-elution cleaning (Figure 12.4e).

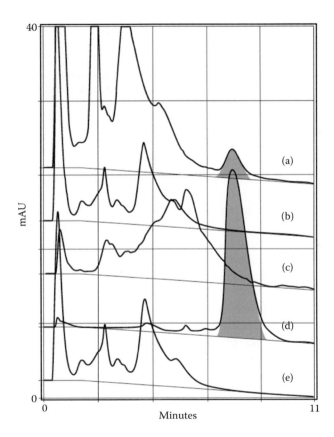

FIGURE 12.4
Application of CIM SO3 disk monolithic column for monitoring of IgM concentration in media during bioprocess and during preparative purification. Conditions: Equilibration with 10 mm sodium phosphate, pH 7.0. Elution; 10 minute (20 mL) linear gradient from 10 to 250 mM sodium phosphate, pH 7.0; Injection volume: 100 µL; Samples: (a) cell culture supernatant; (b) flow-through of hydroxyapatite column for cell culture supernatant loading; (c) Pre-elution wash of hydroxyapatite column; (d) IgM elution peak from hydroxyapatite column; (e) post-elution cleaning of hydroxyapatite column. IgM fraction is highlighted. (From Gagnon, P., Hensel, F., Andrews, P., and Richieri, R., *Poster presented at the Third Wilbio Conference on Purification of Biological Products,* Walthman, Massachusetts, September 24–26. 2007. With permission.)

TABLE 12.1

IgM Benchmark Purification Process Summary

	Cation Exchange	Anion Exchange	Hydroxyapatite
	8 mL Monolith	8 mL Monolith	10 mL Column
Sample volume, mL	250	25	2
Diluted volume, mL	1,250	250	250
Flow rate mL/min	20	20	3.34
Application time, min	62.5	12.5	75
Total buffer volume	1,950	950	650
Total step time, min	98	48	195
Recovery %	78(86*)	84	88
Purity %	~90	~95	~99

Source: From Gagnon, P., Hensel, F., and Richieri, R. *BioPharm International*, 1–7, 2008. http://biopharminternational.findpharma.com/biopharm/article/articleDetail.jsp?id=499148&sk=&date=&pageID=7. With permission.
*Includes the IgM that eluted prematurely in the wash.

The use of monolith was shown to be beneficial for preparative purification of clinical grade IgMs (Gagnon, Hensel, and Richieri 2008; Gagnon, Richieri, and Hensel 2008). A three-step, bind-elute chromatographic process, based on hydroxyapatite as a capturing step, followed by a two-step polishing performed on anion- and cation-exchange short monolithic columns was developed. Since IgM displays considerable structural and biochemical heterogeneity, the sequence of chromatographic steps can be exchanged and it is dependent on the individual IgM to be purified. The process parameters of the second purification step are summarized in Table 12.1. Due to a high binding capacity of the short monolithic columns for IgM (exceeding 30 mg/ml) where the flow is unaffected, the purification productivity can be very high. Based on experimental data, the productivity of 8 L monolithic column corresponds to the production of 3 kg of IgM per day.

Similar to the separation of PEGylated isoforms (also in the case of IgM monoliths) enable very fast separation and purification, in this case also combined with high dynamic binding capacity (DBC), resulting in a high productivity. Efficient purification can be performed either from human plasma or mammalian cell cultures.

12.3 Oligo and Polynucleotides

Another group of pharmaceutically important molecules are oligo and polynucleotides. Oligonucleotides gained importance with the rapid development of molecular biology in the past few decades. In experimental medicine they are used as antisense and antigen drugs, whereas in biochemistry as diagnostics (Agrawal 1996; Gold 1996; Sykora, Svec, and Fréchet 1999). They are also applied as probes for gene therapy or as linkers for gene splicing, DNA ligation and as primers in the polymerase chain reaction (PCR) for DNA amplification. Solid-phase synthesis on a large scale is already fully automated and multigram quantities can be produced in a relatively short period of time. In order to obtain a pure oligonucleotide after synthesis is completed, it has to be isolated from closely related impurities and other undesired oligonucleotides. This is commonly performed using chromatography (Sykora, Svec, and Fréchet 1999).

Being a rather small molecule in comparison to large proteins, DNA, or even viruses, it is interesting to investigate whether such molecules can be efficiently purified using short-bed monolithic supports. To achieve efficient separation on such columns, separation must be based on selective elution (Yamamoto, Nakanishi, and Matsuno 1988). This is the case when the molecules are strongly bound to the support, commonly via several interaction sites and is the reason why interaction mechanisms between oligonucleotides and the monolithic matrix was extensively studied.

Initials studies were performed by Podgornik and coworkers (1999a), who investigated the effect of oligonucleotide length on the number of binding sites under isocratic elution for 8–14mer oligonucleotide. They found very good linear correlation indicating that a strength of interaction between the oligonucleotide and the matrix varied substantially with the oligonucleotide length. This investigation was further extended (Yamamoto, Nakamura et al. 2007). Instead of isocratic they used the gradient elution mode, which enabled them to study oligonucleotides of up to 50mer length including double stranded ones. Interestingly, from Figure 12.5a it can be seen that there was no significant difference in the number of binding sites between single-strand and double-strand oligonucleotides meaning that the charge contribution to the interaction with the matrix is similar regardless of the internal structure of oligonucleotide. Similar to the previous findings, they reconfirmed linear correlation between oligonucleotide length and number of binding sites for shorter lengths, while for larger oligonucleotides charge contribution to the number of binding sites seems to decrease resulting in a nonlinear relationship (see Figure 12.5b). This seems to indicate that smaller oligonucleotides are completely extended over the surface resulting in an interaction of all phosphate groups while larger oligonucleotides preserve their 3-D conformation and the relative number of interaction sites decreases. The number of estimated binding sites is up to 50, which significantly exceeds typical values for proteins (see Figure 12.5). Another interesting observation was that elution salt concentration already reaches a plateau at around 0.8 M NaCl in oligonucleotides for which the number of binding sites still increases with the size (Figure 12.5a). Reasons for that are still unclear but it seems due to high charge density, the "activity" of each binding site decreases (one might

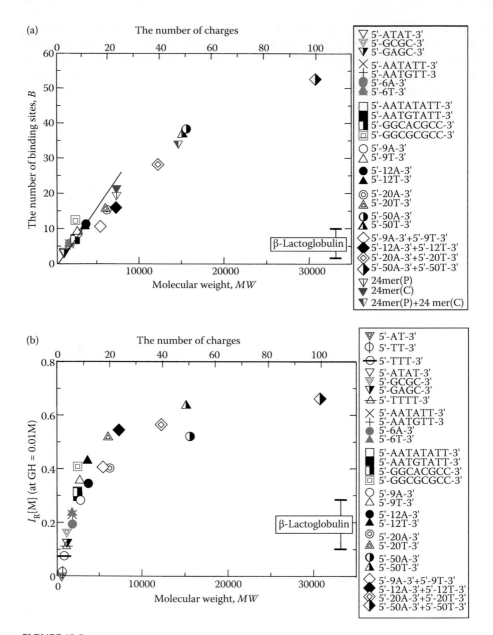

FIGURE 12.5
Relationships between the number of binding sites (a), of elution salt concentration (b), and the molecular weight or the number of charges of oligonucleotides. Stationary phase: CIM QA disk monolithic column; Values for β-lactoglobulin are shown for comparison. (From Yamamoto, S., Nakamura, M., Tarmann, C., and Jungbauer, A., *J. Chromatogr. A.*, 1144, 155–60, 2007. With permission.)

think in analogy with dissociation of acids—pH value can only approach zero regardless of acid concentration due to a decrease of activity coefficient). However, this is well above elution concentration of proteins indicating that the separation of oligonucleotides on the short monolithic columns should be possible. Another argument is high efficiency of the monolithic columns exhibiting HETP of 18 μm (Podgornik et al. 2002).

The demonstration of oligonucleotide separation was performed at the initial work (Podgornik et al. 1999a), which was further extended to the extremely thin monoliths of only 0.3 mm (Podgornik et al. 1999b). Yet, isocratic separation of four oligonucleotides was achieved. These data clearly indicate the potential of short monolithic columns for oligonucleotide separation. Almost simultaneously Sykora, Svec, and Fréchet (1999) studied the effect of group density on 12–24mer oligonucleotides separation using the same weak anion exchange chemistry. In contrast to previous works, they used longer methacrylate monolithic column, where the effect of ligand density is even more pronounced. They demonstrated that better separation of oligonucleotides is achieved at a higher group density being independent of the flow rate. Efficient separation on long monolithic columns was also demonstrated by Holdsvedova and coworkers who separated oligonucleotides up to 20mer on hydroxymethyl methacrylate-based monolithic column in a hydrophilic separation mode (2007).

While most of these works were intended for analytics, short monolithic columns (CIM) were also used for preparative purification of oligonucleotides together with a profilation of impurity residuals occurring during oligonucleotide synthesis (Figure 12.6; Štrancar et al. 2002). It is worth the

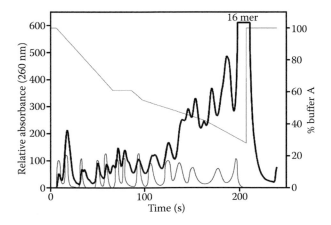

FIGURE 12.6
Semi-preparative purification of a 16-mer oligonucleotide on a CIM DEAE disk monolithic column. Sample: 16mer oligonucleotide from the reaction mixture—bold line; standards of 1, 2, 3, 4, 5, 6, 7, 9, 10, 11, 12, 14, 15, 16mer—thin line. Gradient of eluting buffer is shown in the Figure as thin dotted line. (From Štrancar, A., Podgornik, A., Barut, M., and Necina, R., *Advances in Biochemical Engineering/Biotechnology*, 50–85, 2002. With permission.)

time to emphasize that the entire purification step was completed in four minutes.

It is important to notice that all oligonucleotide separations were completed within a few minutes at room temperature. Such purification, therefore, represents an attractive alternative to conventional reverse phase (RP) separation, especially when volatile buffers like ammonium acetate would be used.

12.4 pDNA

In the field of polynucleotides, pDNA is the leading candidate for gene therapy, therefore most of the studies are directed into this area. Use of pDNA is increasing in different areas: from laboratory work, to vaccination, and gene therapy. For human applications, pDNA must fulfill all regulatory criteria in terms of purity, potency, identity, efficiency, and safety. Based on the success of ongoing clinical trials, predicted demand for clinical grade pDNA is in the range of several hundred grams or even kilograms (Urthaler et al. 2006). Because of that, several processes for pDNA purification were developed, all currently including chromatographic steps. Understanding of the interaction mechanism is therefore of the utmost importance to establishing an efficient purification process. Being that this is an extremely large, flexible molecule, its behavior might significantly differ from that of proteins and mechanical stability and could be an issue (Levy et al. 2000). To understand factors affecting DNA separation on monoliths, several works were published using pure DNA (Giovaninni, Freitag, and Tennikova 1998; Benčina, Podgornik, and Štrancar 2004; Zöchling et al. 2004; Urthaler et al. 2005; Danquah and Forde 2007). Screening of the mobile phases, ligands, and ligand densities were the main focus. Benčina and coworkers (2004) performed an extensive study about the effect of ionic strength, ethanol content, and pH value of mobile phase for recovery on weak and strong ion exchangers. It was found that the effect of the pH value is more pronounced on weak anion exchangers due to the different dissociation of the groups (Benčina, Podgornik, and Štrancar 2004). As already seen during the study of oligonucleotides, such molecules do interact with the matrix through the multiple binding sites. It might well be that the interaction is so strong that no complete elution can be performed. In fact, experiments demonstrated that the recovery of DNA seems to be higher at weaker interactions with the matrix either due to higher pH, higher ionic strength, or ethanol addition, reaching recovery values of up to 90%. Similar effect can be achieved by varying ligand density— the weaker the interaction the higher the recovery (Benčina, Podgornik, and Štrancar 2004). Such high recovery values achieved, even for genomic DNA (gDNA) (Benčina, Podgornik, and Štrancar 2004), demonstrated suitability of monolithic columns for DNA purification.

Another key topic of study was the estimation of DBC. Initial study was performed by Benčina and coworkers (2004) on weak anion exchange column. Effects of linear velocity, loading ionic strength, ligand density, and pore size was investigated. They demonstrated that flow velocity up to 700 cm/h has no effect on DBC. Further extension of this range up to 1000 cm/h performed by Urthaler and coworkers however, indicated a small decrease (2005). Similar results were reported by Zöchling and coworkers (2004). For separation of RNA and DNA, it was demonstrated that for flow velocity up to 1600 cm/h there is no change in resolution (Urthaler et al. 2005). Consequently capacity decrease can not be attributed to diffusion limitation. The authors offer two possible explanations for DBC decrease. One is the stretching of pDNA caused by flow-induced shear stress and another is DNA compaction on the surface that might act as a compacting agent like spermine and spermidine. Further work in this direction is needed to elucidate a precise mechanism (Zöchling et al. 2004).

The effect of the pore structure expectedly demonstrated that the larger the pore, the lower the capacity and vice versa. On the other hand, at the smallest tested pore sizes of 620 nm, break-through curves were significantly shallower indicating decrease of efficiency (Benčina et al. 2004). One possible explanation is partially restricted access to smaller pores that was recently demonstrated by Benčina and coworkers (2007). In this later work it was shown that too small pores might cause restricted access besides partial degradation of gDNA, but no data of DNA consistency were reported by Benčina and coworkers (2004). The effect of the ligand density was studied by Urthaler and coworkers (2005). They found good linear increase in DBC with increase in ligand density. However, one should mention that too high of ligand density might cause lower recovery and a reasonable compromise should therefore be made.

Since there is a pronounced effect of ligand density on capacity one would expect a similar effect of the mobile phase ionic strength of the DNA solution. Similar to the situation for proteins, a decrease of capacity with ionic strength increase is expected. However, Benčina, Podgornik, and Štrancar (2004) demonstrated that DBC of 0 M and 0.3 M NaCl concentration in gDNA loading solution was practically the same. Authors speculated that this might be due to a perpendicular orientation of DNA on the matrix. Recent studies of Frankovič and coworkers (2008) demonstrated similar behavior for pDNA in the range between 0 and 0.2 M NaCl, indicating a small optimum around 0.1 M NaCl. Interestingly, an identical trend was found for conventional and grafted monolith. Reported values of DBC for methacrylate monoliths vary with applied conditions and type of studied DNA but all are in the range between 7 to 14 mg/ml (Štrancar et al. 2002; Benčina, Podgornik, and Štrancar 2004; Zöchling et al. 2004; Urthaler et al. 2005; Benčina et al. 2007). An even higher value was recently reported for grafted monolith reaching a capacity of around 18 mg/ml (Frankovič et al. 2008). These values are very high, exceeding most of the particulate resins (Urthaler et al. 2005).

Zöchling and coworkers studied pDNA adsorption isotherm on ion exchange monolith (2004). Rectangular adsorption isotherm that is characteristic for a particle-based column was also obtained with methacrylate monolithic columns (Figure 12.7). The lowest tested concentration of 6 µg/ml pDNA already exhibited maximum DBC. This is very important from the process point of view, as in the case of low DNA concentration in solution, no additional step for preconcentration is needed. While presently, with increased pDNA titres, this does not seem to be a real advantage, it is extremely important for DNA removal in different processes of antibody and virus purification.

The application of pDNA for gene therapy requires removal of impurities like RNA, gDNA, proteins, plasmids, and also open circular and linear pDNA forms. Already in 1998 (Giovaninni et al. 1998), it was demonstrated that the short monolithic columns enable separation of pDNA isoforms. Combining this feature with high DBC it seems reasonable that short monolithic columns represent an attractive support for the purification of DNA, especially pDNA. This potential was first explored by Štrancar and coworkers (2002). They demonstrated that short methacrylate monolithic columns can be efficiently used for intermediate purification steps and exhibits the highest capacity of all tested resins (Štrancar et al. 2002). This initial work was followed by an extensive study by Urthaler and coworkers (2005). Among different anion exchangers, Diethylaminoethyl (DEAE) chemistry was selected due to the highest resolution, purity, and recovery approaching 100% (Table 12.2). The optimized process was scaled from 8 ml to 800 ml monolithic column (Figure 12.8; Urthaler et al. 2005) and recently to 8 L monolithic column (Paril 2008). It was

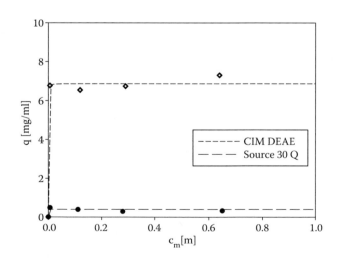

FIGURE 12.7
Adsorption isotherm of pDNA fitted with the Langmuir equation for a CIM DEAE disk monolithic column and SOURCE 30 Q. (From Zöchling, A., Hahn, R., Ahrer, K., Urthaler, J., and Jungabauer, A., *J. Separ. Sci.*, 27, 819–27, 2004. With permission.)

TABLE 12.2

DEAE, EDA, Two Weak Anion-Exchange Ligands and QA, a Strong One, Compared Regarding their Resolution, the Purity Parameters (Homogeneity, Estimated Purity), an Economic Parameter (Pool Yield), and a Purity Parameter (Recovery)

Ligand	Resolution, R_s	Pool Homogeneity (% ccc)	Estimated Purity (%)	Pool Yield (%)	Recovery (%)
DEAE (diethylaminoethyl)	1.31	78	92	100	100
EDA (ethylendiamin)	None	88	33	60[a]	92
QA (quarternary ammonium)	0.56	85	79	75	97

Source: From Urthaler, J., Schlegl, R., Podgornik, A., Štrancar, A., Jungbauer, A., and Necina, R., *J. Chromatogr. A.*, 1065, 93–106, 2005. With permission.

[a] Thirty-one percent of the loaded pDNA amount was found in the flow through fraction, due to overloading and breakthrough.

calculated that the usage of monolithic columns as compared to conventional packed bed columns increased productivity 15-fold.

While the described process implemented monolith only for the intermediate chromatographic step, Peterka and coworkers (2005) developed a purification process using two chromatographic steps with monolithic columns, namely monolith containing hydroxyl (OH) groups for a capture step and monolith containing DEAE groups as a second step. By proper adjustment of ammonium sulphate concentration in a loaded sample OH column enabled elimination of residual RNA without usage of RNAse. The pDNA purified with OH and DEAE monolithic columns was successfully used for transfection into cells, which confirms its high degree of purity. It is interesting to notice that the eluate from OH monolith can be loaded directly on the DEAE column without any adjustment. Both monoliths can be packed into a single housing and used in conjoint liquid chromatography (CLC) mode resulting in a simpler system and higher resolution (Figure 12.9).

These initial works were followed by several other applications. When required purity of pDNA is not very high, a single purification step is sufficient. Already Benčina and coworkers (2004) demonstrated successful purification using only CIM DEAE disk monolithic column of pDNA from *E. coli* and gDNA from *A. niger*. Forčič and coworkers used DEAE monolithic column to isolate gDNA from *E. coli* and mammalian cell cultures (Forčič et al. 2005a) and to concentrate residual gDNA from mammalian cell cultures lysate to enable PCR detection (Forčič et al. 2005b). They reported 50-fold concentration. A similar study was done by Jerman and coworkers (2005) in order to determine genetically modified DNA in food.

The RNA concentration was performed by Branović, Forčič and coworkers (2003). Their goal was to improve viral detection, in this particulate case for

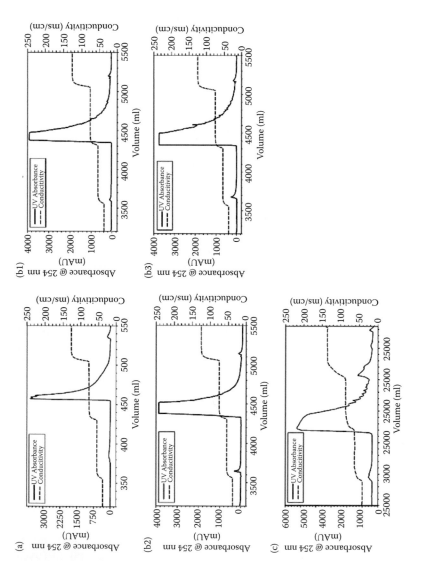

FIGURE 12.8

Comparable elution profiles of pDNA at different scales showing scalability and robustness of pDNA purification using CIM DEAE monolithic supports. (a) 8 ml tube monolithic column, (b1)–(b3) Three subsequent runs on the same 80 ml tube monolithic column, (c) 800 ml monolithic column. (From Urthaler, J., Schlegl, R., Podgornik, A., Štrancar, A., Jungbauer, A., and Necina, R., *J. Chromatogr. A.*, 1065, 93–106, 2005. With permission.)

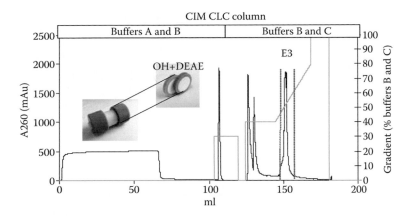

FIGURE 12.9
RNAse free purification of plasmid DNA using CIM columns in conjoint liquid chromatography (CLC) mode. Conditions: buffer A: 3 M Ammonium Sulphate, 50 mM Tris, 10 mM EDTA pH 7.2; buffer B: 50 mM Tris, 10 mM EDTA pH 7.2; buffer C: 50 mM Tris, 10 mM EDTA, 1.5 M NaCl pH 7.2.

measles and mumps. They demonstrated a fourfold increase in sensitivity using real time PCR. Another interesting application of RNA analysis using monolithic columns was described by Krajčič and coworkers (2007). They separated various forms of dsRNA to detect viral infection in a tomato plant (Figure 12.10). Separation was completed in 15 min instead of 2 hrs that standard purification protocol using CF-11 cellulose column requires. Moreover, using a monolithic column, dsRNA was completely purified of any residual DNA without usage of DNAse, while for standard procedure the usage of DNAse is obligatory.

A different application was developed in the work of Brgles and coworkers (2007) who took advantage of binding efficient DNA molecules on a monolithic column for quantitative analysis of DNA entrapment in liposomes. Liposomes are used as a drug delivery system for DNA by genetic vaccination and gene therapy, where DNA entrapped in liposomes has a higher transfection efficiency than the naked DNA. With an indirect method for quantitative determination of DNA encapsulated into liposomes, it is possible to avoid a complicated procedure of DNA labelling; DNA entrapped into liposomes could be used for further *in vivo* experiments.

Very recently some studies on large plasmids (20–60 kb) were presented (Lendero et al. 2006), which can potentially be used in the future for treatment of multigene diseases. Moreover there are demands for purification of different larger forms of DNA like bacterial artificial chromosome (BAC) or cosmids that are used in molecular biology laboratories. It was shown that monolithic columns enable processing of pDNA up to 62 kb without degradation (Figure 12.11) in spite of their tendency to degrade (Levy et al. 2000).

Similarly as for large proteins and oligonucleotides, flow-unaffected resolution and high DBC of monoliths for DNA can be transformed into high

Macroporous Polymers

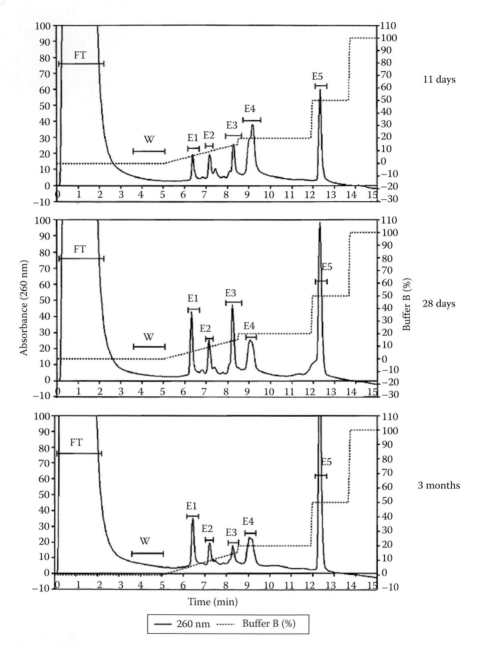

FIGURE 12.10
Application of chromatography on CIM DEAE disk in monitoring the post inoculation time-related dynamics of the viral (E2, E4) and satellite (E1, E3) RNAs. The peak designated as E5 derives form plant DNA. The post-inoculation time is expressed either in days or months. (From Krajčić, M., Ivancic-Jelecki, J., Forčić, D., Vrdoljak, A., and Škorić, D., *J. Chromatogr. A.,* 1144, 111–19, 2007. With permission.)

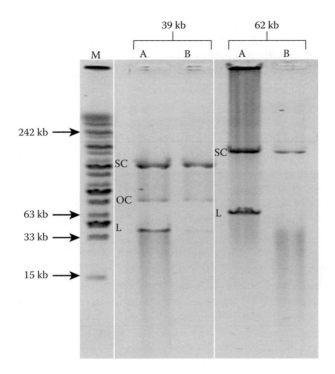

FIGURE 12.11
Pulsed field electrophoresis of two plasmids sizes 39 kb and 62 kb. M: Marker MidRange I PFG Marker (New England Biolabs). A: Plasmid DNA purified with isopropanol precipitation. B: Fraction of purified plasmid DNA with DEAE column. (From Lendero, N., Smrekar, F., Vidič, J., Peterka, M., Podgornik, A., and Štrancar, A., *Poster presented at Monolith Summer School*, Slovenia, Portoroz, May 28–31, 2006. With permission.)

productivity of the purification process. In contrast to the proteins however, DNA capacity also is preserved in the presence of NaCl up to 0.2 M. This is an attractive feature since several proteins do not bind under such conditions and consequently do not compete for the binding sites.

12.5 Viruses

For centuries, mankind led a battle against viruses, but now it is possible to use viruses for our benefit, both at a prophylactic level in the production of vaccines and at a therapeutic level in gene therapy (Rodrigues et al. 2007). Purification of viruses is mandatory to prevent toxicity, inflammation, and immune response. Different applications of the viruses require different levels of quality and biological potency of the therapeutic preparations (Rodrigues et al. 2007).

Furthermore virus production usually results in a low concentration of infective viral particles per ml of cell culture supernatant, which in most of the cases is insufficient for the final viral stocks required for a treatment. Concentration of viral particles is also recommended at the initial step of a downstream process, because it results in smaller process volumes, which means smaller equipment and shorter processing time (Segura, Kamen, and Garinier 2006). The demand for high-quality viral particles that meet standard requisites from the regulatory agencies (FDA and EMEA) is therefore increasing, as the technology has moved into clinical trials. Cost-efficient and scalable purification processes are essential for production of injectable-grade preparations to achieve final implementation of these viral particles as therapeutics (Rodrigues et al. 2007). A wide variety of drug products on a worldwide basis have been shown to be contaminated by viruses. This contamination has led to high industrial and societal costs, regulatory scrutiny, and ongoing public concerns over production safety (Brough et al. 2002). So in the manufacture of biopharmaceuticals, validation of virus clearance is critical (Czermak et al. 2008).

Conventional virus purification and concentration procedures are based on multiple cycles of differential centrifugation, sometimes after clarification with organic solvents and generally culminating in a density gradient centrifugation step. The methods are time-consuming, often inefficient, and the productivity is very low. In recent years, liquid chromatography has became the method of choice for efficient virus purification. Furthermore, the use of chromatography for virus removal from monoclonal antibodies (Mabs), for example, is increasing substantially. A large size of viruses requires specific supports to be used for their purification or efficient removal. Methacrylate monoliths therefore seem to be a good option. Due to the extreme size of viruses it is interesting to see whether chromatographic properties such as resolution and DBC remain unaffected by the flow. To investigate these phenomena Kalashnikova, Ivanova, and Tennikova (2007) used virus-mimicking synthetic particles with a size around 80 nm bearing different proteins at their outer surface. They confirmed that similar to the situation for proteins, the adsorption capacity for large particles and the resolution of separated zones are unaffected by the flow rate in both ion-exchange and affinity modes of chromatography at least up to the linear velocity of 530 cm/h.

Another important issue is virus capacity of the monolithic columns. Trilisky and Lenhoff (2007) has shown on conventional resins that capacity is dependent of NaCl presence in the loading sample and exhibits an optimum that is similar to results gained for plasmid purification (Frankovič et al. 2008), and therefore results might vary depending on applied conditions and virus type. Experiments with purified tomato mosaic virus (ToMV) loaded in low ionic strength buffer resulted in capacity of around 12 mg/ml, which equals to 2×10^{14} virus particles (Kramberger, Peterka et al. 2004). This value significantly exceeds results obtained from conventional particulate supports making monolith supports the choice for virus purification and might be even higher for loading solution with moderate ionic strength.

There are several applications of methacrylate monoliths for virus concentration and purification. Kramberger, Peterka, and coworkers (2004) purified and concentrated plant ToMV (Kramberger, Peterka et al. 2004, Kramberger, petrorič et al. 2004). Analysis of ToMV is important since its contaminant in irrigation water, however being present in concentrations, limit conventional detection methods. This problem is not limited only to ToMV, but to other more pathogenic viruses that represent even more serious problems. Traditional methods (centrifugation, precipitation, etc.) for virus concentration require long procedures with several steps and large volumes of samples. Applying a strong anion exchange methacrylate monolithic column, the Kramberger group achieved a concentration of ToMV enabling a lower detection limit compared to a nonconcentrated sample. Another important finding of this work was that concentrated virus preserved its infectivity. The next logical step was to investigate purification efficiency (Kramberger et al. 2007). By the optimization of conditions using a strong anion exchanger (QA), it was possible to achieve virus recovery of 98%. In addition more than 99% of DNA removal was obtained even without the use of deoxyribonuclease (e.g., DNAse or Benzonase; Figure 12.12). Protein removal checked with SDS-PAGE was similar to the removal obtained with conventionally purified virus particles. The entire procedure was approximately 60-fold faster compared to the conventional method, requiring only about two hours.

The recent application on virus concentration from environmental waters was reported by Gutierrez-Aguirre and coworkers (2008). In spite of being present in extremely low concentration, they could lead to pandemics or being misused as a bioterrorism weapon. With a concentration of rotavirus, they showed a significant decrease in limit of detection using PCR (Gutierrez-Aguirre et al. 2008). In fact, using an 8 ml CIM QA tube monolithic column they could detect down to 100 rotavirus particles in ml of water. Similar work was done by Kovač and coworkers (2008), who developed a method with monolithic columns for concentration of *Feline calicivirus* and *Hepatitis A* virus in water samples that enables detection with RT-PCR.

Monolithic columns were also implemented for purification of bacteriophages light (phages) (Smrekar et al. 2008). This application is important since phages are presently intensely studied as, for example, a substitute for antibiotics in phage therapy, as a delivery vehicle for gene therapy, or vaccination (Clark and March 2006). Fast purification was performed in a few minutes giving the reproducible recovery (measured by virus infectivity) in the range of 70% (Figure 12.13).

In addition to results from Kalashnikova and coworkers (2007) who presented usage of affinity mode chromatography for virus-mimicking synthetic particles, preliminary results from Cheeks and coworkers confirm the applicability of affinity chromatography on methacrylate monoliths for a real sample. They were using immobilize mental offinity chromotography (IMAC) monolith for purification of 6x histidine tagged lentivirus vector, which is used for gene therapy (Cheek et al. 2008).

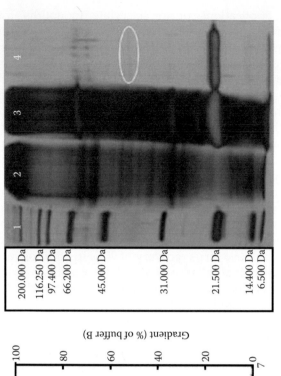

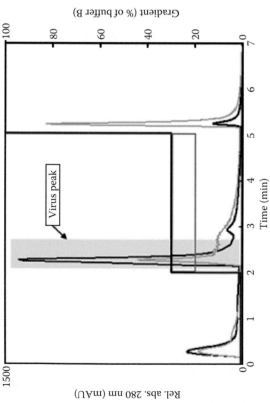

FIGURE 12.12

Purification of Tomato mosaic virus (ToMV) from 1 ml of homogenized and clarified plant material on CIM QA disk monolithic column. (Left) typical chromatogram; (Right) SDS–PAGE analysis of ToMV purification by conventional method: line 1 (marker proteins BioRad), line 2 (*N. clevelandii*), line 3 (*N. clevelandii* infected with ToMV), line 4 (purified ToMV; 40 kDa protein band encircled). Gradients of eluting buffers are shown as smooth lines whereas curved lines present elution profiles (A280). Black elution profile is result of buffers marked black and grey elution profile is result of buffers marked grey. (From Kramberger, P., Peterka, M., Boben, J., Ravnikar, M., and Štrancar, A., *J. Chromatogr. A.*, 1144, 143–49, 2007. With permission.)

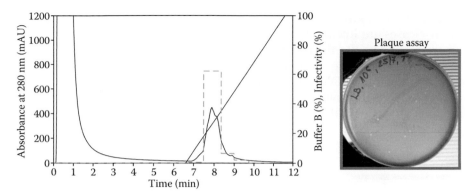

FIGURE 12.13
Gradient elution of phage T4 on the monolithic column (left) and plaque infectivity assay of purified phage (right). Conditions: mobile phase: buffer A 125mM NaH2PO4; buffer B 125mM NaH2PO4 + 1.5M NaCl; stationary phase: CIM QA disk monolithic column. Flow rate, 2 ml/min; gradient, 0–100% B (5 min). Sample, phage suspension; injection volume, 1 ml; detection,UV at 280 nm. Absorbance is shown as black curved line, gradient of eluting buffer as black smooth line and infectivity as dotted grey line. (From Smrekar, F., Ciringer, M., Peterka, M., Podgornik, A., and Štrancar, A.,*J. Chromatogr.* B. 861, 177–80, 2008. With permission.)

Maurer and coworkers recently reported purification of H1N1 and H5N1 influenza A viruses using strong ion-exchanger (IEX) monolithic chromatographic support (2008). A production platform for influenza virus with an efficient downstream part is extremely important, since high quantities of stable vaccine would be required in a very short time in case of epidemic outbreak. In addition, new downstream facilities should be developed since there is currently a lack of vaccine supply (Opitz et al. 2007). By using a strong anion exchange monolithic column for the intermediate step, 95% removal of DNA and protein was obtained combined with virus concentration. Because of assay variability of the infectivity test, recovery of virus is estimated as 50–100% in this step. Regardless of the precise value, this is still significantly higher than the other existing purification methods. An efficient pseudo-affinity method was developed for fast separation and sensitive analytical detection of influenza virus using methacrylate monoliths that could be used as a biochip (Kalashnikova, Ivanova, and Tennikova 2008).

A method was presented by Ball (2008) based on monolith for development and exemplification of an adenovirus purification platform, which has better sensitivity and is faster than compared to a particle-based column. A developed method enabled monitoring of the entire downstream process, which is in accordance to PAT regulative. Moreover Eglon and coworkers (2008) also showed purification of adenovirus using strong anion exchange monolith. They showed a high degree of protein removal and DBC around 10^{12} pfu/ml.

Reports about purification of various types of viruses either enveloped or nonenveloped ones, indicate versatility of monolith applications for virus concentration and purification. High infectivity after purification proves that viruses are not damaged during processing, an extremely important feature when viruses are intended to be used for vaccination or as delivery vectors, for example. Despite their size of several tens of nanometers there are no reports about blocking of monolithic columns when used for virus purification as reported for gDNA (Benčina et al. 2007). All monolith features such as flow independent resolution and high DBC were also confirmed in this case. Because of that, monoliths seem to be a matrix of choice for such applications.

12.6 Monolith Characterization

As described in previous sections, many attractive monolith features are a direct consequence of their particular structure. It has been shown that structure determines monolith efficiency (Tennikova et al. 1998; Hahn and Jungbauer 2001; Benčina et al. 2007) and should therefore be strictly controlled. However, this is not always a trivial task since monoliths consist of a single piece of material. This is different from particulate supports where a small representative sample can be analyzed. In the case of monoliths, in theory at least, every single piece of monolith has to be characterized and therefore an appropriate methodology has to be introduced. This is particularly true when implemented in industrial processes of biopharmaceutical products due to the regulatory agencies, which demand chromatographic columns in downstream processing not only to exhibit outstanding properties in terms of efficiency and productivity combined with high mechanical and chemical stability but also to be extensively characterized before, as well as throughout, the usage to assure constant quality of the purified product.

According to the four dimensions of chromatography (1) the chemistry of the stationary and mobile phase, (2) physical properties of the stationary phase, (3) operational mode, and (4) column and system hardware (Jungbauer and Hahn 2008), the two dimensions that are important for chromatographic media performance and should be controlled during or after the production process of the resin are chemical composition of the stationary phase surface and physical structure of the monolith (Tallarek, Leinweber, and Seidel-Morgenstern 2002; Podgornik et al. 2005). From the production point of view to assure the product quality it is important to choose the parameters that enable us to follow the repeatability of the product and the production process and are measurable using reliable methods that can be validated.

The monolith physical structure consists of nonporous (organic polymer monoliths) or (meso) porous skeleton (silica monoliths) with highly interconnected channels of macropores (>50 nm) that can all be reached by convective flow (Tallarek, Leinweber, and Seidel-Morgenstern 2002), and is usually described using terms like porosity, pore size, pore size distribution, specific surface area, and connectivity of the pores. In order to analyze the porous structure of monoliths, two types of techniques are generally used: direct and indirect techniques. Direct techniques (e.g., microscopy, electron microscopy, and X-ray analysis) provide actual images of the surface but no more than semiquantitative characterization of the surface area and pore volumes could be obtained. Indirect techniques (like gas adsorption and mercury penetration) are more appropriate for quantitative measurements, since they measure the macroscopic effects of phenomena occurring in the pore volume and on the pore surface. In addition a chromatographic method, such as inverse size-exclusion chromatography (ISEC), could also be a powerful method to quantitatively characterize the pore morphologies of the monolithic samples, especially for monoliths consisting of mesopores (Lubda et al. 2005). Furthermore, the NMR was suggested as a promising technique for evaluation of fluid transport properties (Hahn and Jungbauer 2001). The technique has already been used for determining the pore size distribution, connectivity, and tortuosity of the pore space as well as flow dispersion through the porous media on porous catalyst (Ren et al. 2003; Koptyug et al. 2007), zeolites (Bonardet et al. 1999), and particle bed resins (Stepišnik, Mohorič, and Duh 2001), but, according to our knowledge, no such experiments on monoliths have been published.

Often, especially at the beginning when the surface morphology of the monoliths was still unknown, direct imaging methods were used to visualize the surface of the media to help understand some hydrodynamic properties of the monoliths and to investigate and control the morphology-governing parameters during the synthesis of the bed (Svec, Tennikova and Deyl 2003). Due to its simplicity, scanning electron microscopy (SEM) is the most widespread imaging method used for studying the physical aspects of new monoliths. With the SEM, features down to 500 nm are readily imaged. However, detailed structures smaller than about 50 nm cannot be resolved (Cabral, Bandilla, and Skinner 2006). SEM images give a very good impression about the skeleton and surface structure, especially about the porosity, average pore diameter, the pore size distribution, the connectivity of the pores, the specific surface area (Krajnc et al. 2005; Mihelič et al. 2005; Cabral, Bandilla, and Skinner 2006; Junkar et al. 2007; Laschober, Sulyok, and Rosenberg 2007), and can be used as a semiquantitative method for determining those parameters and use them for further calculations (Lubda et al. 2005; Mihelič et al. 2005; Junkar et al. 2007).

A similar technique to SEM is transmission electron microscopy (TEM), which provides an image of very small structures down to the Ångstrom scale. Lubda and coworkers (2005) used this method to obtain the porosity data and compare them to the results obtained by other direct and indirect methods, especially for ISEC calculations.

Another imaging technique is high resolution optical microscopy (HROM) that was used by Kornyšova and coworkers to evaluate the morphology of a continuous bed without sample preparation and a possibility for *in situ* inspection of the bed homogeneity across the capillary length in dry and wet conditions (2002). However, since the HROM image again does not reach the nanometer scale range, they also used complementary atomic force microscopy (AFM) to investigate the surface of a continuous bed more closely, but because of poor resolution they were not able to evaluate mesopores or micropores, while macropores were consistently estimated by both techniques. The AFM seemed to be an appropriate method to characterize the organic monolith pore size in a wetted state as described by Cabral, Bandilla, and Skinner (2006). They optically confirmed that the volume of particles that make up the monolith changed with solvent type, but they were not able to observe mesopores in a wetted polymer since the pores most probably contracted to the size of an AFM tip.

From the above description, it is obvious that imaging methods do not give enough information about the monolith structure and the obtained images represent only partially inspected information about the entire monolith. Moreover, such methods are very hard to validate since setting up the convenient limits for images is very subjective. Nevertheless, they are excellent tools for investigating the monolithic morphology and useful to confirm and to help interpreting the results obtained with other, nonimaging methods.

Most frequently used nonimaging methods for determining the pore size and specific surface area of the monolith are mercury intrusion porosimetry (MIP; Ritter and Drake 1945), Brunauer-Emmet-Teller (BET; Brunauer, Emmett, and Teller 1938), and Barret-Joyner-Halenda (BJH; Barrett, Joyner, and Halenda 1951) gas adsorption-desorption isotherms, and ISEC (Aggebrandt and Samuelson 1964).

The MIP method is very convenient to determine the pore size distribution of macropores. The method is based on the penetration of mercury into the pores of certain sizes at a defined applied pressure. The pore diameter into which the mercury can penetrate is inversely proportional to the applied pressure: the higher the applied pressure, the smaller the measured pores. This proved to be a very suitable method for determining pores larger than several nm in diameter, so for the monoliths where the majority of the pore volume is presented by macropores (e.g., methacrylate monoliths; Podgornik et al. 2005). However, blindly trusting the results obtained by MIP measurements can be risky when the monolith structure and morphology are not taken into account. In such cases the use of another complementary method (e.g., imaging method or SEM) is crucial (Krajnc et al. 2005). From the pore size distribution results also show the specific surface area can be calculated, which can be less accurate, because a certain material structure has to be assumed for calculations, but this is still very informative when materials posses a similar structure (Podgornik et al. 2005).

For a more precise evaluation of the specific surface area and to measure the mesopore size distribution, gas sorption methods are used (Leinweber et al.

2002; Skudas et al. 2007). In a gas sorption experiment, the material is heated and degassed by vacuum force or inert gas purging to remove adsorbed foreign molecules. After that controlled doses of an inert gas, usually nitrogen, are introduced and the gas is adsorbed, or alternatively, withdrawn and desorbed. The amounts of gas molecules adsorbed or desorbed are determined by the pressure variations due to the adsorption or desorption of the gas molecules by the material (the adsorbent). Knowing the area occupied by one adsorbate molecule and using an adsorption model, the total surface area (using BET method) or the mesopore size distribution (using BJH method) of the material can be determined. However, the technique is not suitable to measure the size of the flow through pores. Therefore, to determine the size of the pores for monoliths with the macro- and mesopore structure, both techniques, gas sorption and MIP, should be used (Leinweber et al. 2002; Laschober, Sulyok, and Rosenberg 2007; Skudas et al. 2007).

As already mentioned, a very important chromatographic method for determining the pore size distribution is ISEC. The method is based on differences between individual probe molecules in diffusivity into the pores with appropriate pore diameter (Laurent and Killander 1964). It can be implemented on monoliths possessing mesopores that are filled with a "stagnant" mobile phase. Being a nondestructive method allowing *in situ* measurements under realistic chromatographic conditions, this is the method of choice for the extensive exploration of the accessible pore volume as a function of probe size (Yao and Lenhoff 2006). The ISEC is widely used to characterize a skeleton mesopore structure of silica monoliths (Leinweber and Tallarek 2003; Grimes et al. 2007) and together with MIP was also used to characterize the organic polymer-based monolithic capillary columns. The techniques were used complementary; the entire range of pore sizes in dry state was measured by MIP, while results from ISEC gave the size of the accessible mesopores in the swollen monoliths (Urban et al. 2008). On the other hand, because the size-exclusion effect can only be observed using media where mass transport is based on diffusion, the method cannot be used to characterize the polymeric monoliths consisting mainly of flow-through macropores, like methacrylate monoliths. Having a low percentage of micropores with diameters under 0.5 nm, but no mesopores, the only size exclusion effect can be obtained by using very small molecules, such as acetone, since the molecules like peptides or small proteins (e.g., lysozyme) are already too big to diffuse into available micropores (Hahn et al. 2007).

As one can see, each method mentioned above has advantages as well as disadvantages. Therefore, for a truthful description of the morphology and physical structure of the continuous bed, the use of more than one method is inevitable.

Before the column is used for the chromatographic purpose, the integrity of the monolithic structure must be inspected since the structure integrity influences the flow profile through the column substantially and so affects separation efficiency in a similar manner such as the quality of packing in the case of particle bed chromatographic columns (Hahn and Jungbauer 2001; Podgornik and Štrancar 2005). For monoliths, one proposed method was to inspect the resin's integrity by perfusion of the monolith with ferritin (Josić et al. 1992) and visually inspect the monolith surface. The internal flow pattern could be observed after cutting the monolith and seeing the inner surface. It is obvious that the method is destructive and that after the experiment the monolith is useless for further work. Hence another approach had to be utilized. Commonly, for both types of columns, monolithic as well as particle-based, pulse response and frontal analysis experiments are done to determine the flow profile through the column (Podgornik and Štrancar 2005). In the first case, acetone or NaCl solution injections are routinely performed (Rathore et al. 2003), but also a pulse response of small protein molecules, such as myoglobin, at nonbinding conditions, or recently proposed and more practical, a pulse of high ionic strength buffer injected into a low ionic strength buffer can be used (Podgornik et al. 2005). The main advantage of the later approach is that it can be performed with the buffers used for further application, so it can be done by the end user before each chromatographic run if needed. To obtain frontal analysis data, breakthrough profiles of the chosen molecules under binding conditions can be performed (Hahn and Jungbauer 2001). Irregularities in the monolithic structure or bed packing reflect in peak doubling or bimodal breakthrough curve, while uniformity of the monolith exhibit single peak or single step breakthrough curve (Hahn and Jungbauer 2001; Podgornik et al. 2005).

The analysis of the pulse response or frontal analysis experiments can also be used to evaluate the hydrodynamic dispersion of the uniform monolith, or in other words, the band broadening (Kaltenbrunner, Jungbauer, and Yamamoto 1997; Podgornik and Štrancar 2005; Podgornik et al. 2005; Hahn et al. 2007). The recorded peaks of a small sample pulse that is injected onto the column under nonbinding or binding conditions can be fitted by exponentially modified Gaussian (EMG) function (Grushka 1972), while in the case of frontal analysis data, the first derivative of the transition curve is used (Hahn et al. 2007). From the fitted peaks the first peak moment, equivalent to the retention time, and the second peak moment, representing the peak variance, can be calculated and further used for the calculation of HETP (height equivalent to theoretical plate) and dispersion of the resin and the system components. The last one cannot be measured directly, since each part of the chromatographic system, such as column housing, tubing connection, and detector contributes to the total band broadening. Therefore, to estimate the dispersion of the monolith itself, the extra column effects must be taken into account. Kaltenbruner and coworkers measured the pulse response on the monolithic columns with various column lengths, but the same extra column dispersion (1997) for methacrylate monoliths were obtained by placing a different number of CIM disks into the same housing. They calculated that for one CIM disk less than 10% of the total dispersion was caused by the monolith, which indicates that when using small column volumes, extra column broadening can present a problem, especially for separations performed at high linear velocities where the separation efficiency may be limited by the equipment design and not the monolithic column (Hahn and Jungbauer 2000). These results were confirmed by Mihelič, using numerical deconvolution of the experimental data, obtained by measuring the dispersion of the system with and without the monolith media (Mihelič 2002). Moreover, the hydrodynamic dispersion in larger monolithic columns (8 and 80 mL) operated in radial flow mode were similar to those estimated for small disks (operated in axial direction), as in all cases the mass transfer limitations of the methacrylate monoliths are negligible and the peak width does not change with an increase of flow rate (Hahn et al. 2007). However, extra column dispersion was found to be much lower using larger chromatography units. On the other hand, the hydrodynamic dispersion in silica monoliths is flow dependent, since diffusion-limited mass transfer of stagnant fluid entrained in the monoliths skeleton dominates the band spreading at higher flow rates (Leinweber et al. 2002).

Another very important term for the hydrodynamic characterization of monolithic materials is a pressure drop on the monolith and its relation to the particle bed columns. Because the structure of the monolith is quite different from the particle bed structure, the use of hydrodynamic equations developed for particle beds is questionable. Hence some new approaches were suggested; a pore-network model was used (Liapis, Meyers, and Crosser 1999; Meyers and Liapis 1999), for silica monoliths, Tallarek et al. introduced the equivalent particle diameter, which can be obtained by hydrodynamic permeability and hydrodynamic dispersion (Tallarek, Leinweber, and Seidel-Morgenstern 2002; Leinweber and Tallarek 2003), and later Vervoort with coworkers elaborated the model based on computational fluid dynamics simulations using Navier–Stokes equations (2003, 2004). Other present equations considering the shape and the porous structure of the packing material were developed by Miyabe and Guiochon (2002). The approach was recently tested for the evaluation of chromatographic performance of various types of separation media, including silica monolith (Miyabe 2008). It was quantitatively demonstrated that monolith and pellicular particle resin structures are the resins of choice for fast, high performance liquid chromatography providing sufficient efficiency for very fast separations.

While physical structure of silica monoliths is more symmetrical and can be described using the geometry of tetrahedral structure, the structure of methacrylate monoliths resembles the particle beds and is harder to be properly described by mathematic equations. Therefore, attempts have been made to characterize them by using the well-known Kozeny–Carman equation and

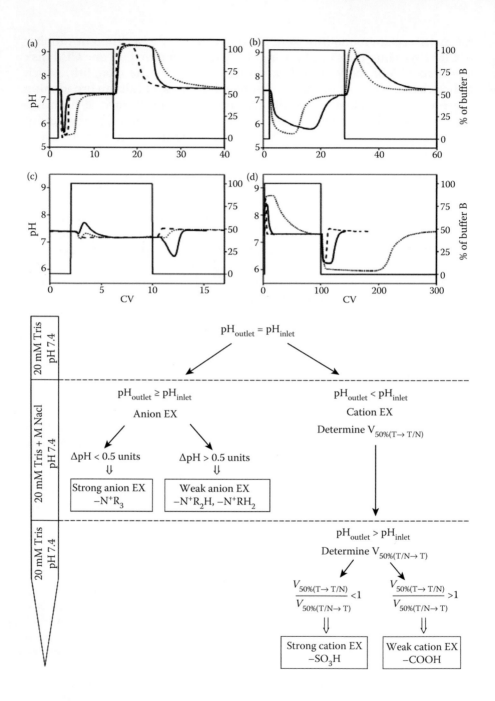

FIGURE 12.14   (Opposite)
pH transients on stepwise changes of ionic strength for (a) strong, (b) weak cation exchanger, and (c) strong and (d) weak anion exchangers. Effluent pH profiles of step changes between buffer A: 20mM Tris–HCl, pH 7.4 and buffer B: 20mM Tris–HCl + 1M NaCl, pH 7.4. Flow rate: 2 ml/min. Detection: pH. (a) Strong cation exchange columns: (———) PL SCX (1 ml) from Polymer Laboratories, (- - -) Resource S (1 ml) from GE Healthcare and (· · · ·· · ·) CIM SO3 disk monolithic column (0.34 ml) from BIA Separations. Method: 100% buffer A for 2 column volumes (CV), 100% buffer B for 12 CV, 100% buffer A for 30 CV. (b) Weak cation exchange columns: (———) Sartorius C15 (0.3 ml) from Sartorius and (· · · ·· · ·) CIM CM disk monolithic column (0.34 ml) from BIA Separations. Method: 100% buffer A for 2 CV, 100% buffer B for 27 CV, 100% buffer A for 40 CV. (c) Strong anion exchange columns: (———) Econo-Pac High Q cartridge (1 ml) form Bio-Rad Laboratories, (- - -) HiTrap Q HP (1 ml) from Amersham Biosciences and (· · · ·· · ·) CIM QA disk monolithic column (0.34 ml) from BIA Separations. Method: 100% buffer A for 2 CV, 100% buffer B for 8 CV, 100% buffer A for 10 CV. (d) Weak anion exchange columns: (———) Sartobind D75 (2.1 ml) from Sartorius, (- - -) Fractogel DEAE (1 ml) from Merck and (· · · ·· · ·) CIM QA disk monolithic column (0.34 ml) from BIA Separations. Method: 100% buffer A for 2 CV, 100% buffer B for 100 CV, 100% buffer A for 200 CV. (top Figure). On bottom Figure is appropriate scheme to distinguish among different active groups on most common ion exchangers . (From Lendero, N., Vidič, J., Brne, P., Frankovič, V., Štrancar, A., and Podgornik, A., J. Chromatogr. A., 1185, 59–70, 2008. With permission.)

calculations of the equivalent particle diameter from the pressure drop data have been made (Hahn and Jungbauer 2000; Zöhling et al. 2004). Recently, the pressure drop analysis of commercial CIM disk monolithic columns was presented implementing the Happel equation (Mihelič et al. 2005), and experimental results of pressure drop were compared to hydrodynamic models usually utilized for prediction of pressure drop in packed beds. The results of the comparison of the pressure drop between monolith and packed bed having the same porosity and specific surface area were surprising, as the pressure drop in a CIM disk monolithic column was 50% lower than in an adequate packed bed. Later another approach was developed for poly(high internal phase emulsion) (PolyHIPE) methacrylate monoliths employing the representative unit cell (RUC) model, which was originally derived for the prediction of pressure drop on catalytic foams (Junkar et al. 2007). The results indicated that the structure of those monoliths was hydrodynamically similar to foam structures.

Besides the structure, the surface chemistry is one of the most important parameters of the chromatographic media and also needs to be characterized. The characterization measurements of the surface chemistry and the capacity of the monolithic columns are usually performed on the end product. While in the case of particle bed resins, a small representative sample of particles can be taken from the entire batch and characterized by any kind of chemical or analytical test. This is not possible in the case of monoliths, since the entire bed consists of a single piece of material, which should not be contaminated or destroyed. Therefore, the characterization methods used for monoliths have to be nondestructive and noncontaminating.

The characterization of the active group type on the surface is sometimes demanded by end users as a part of income control to verify that the obtained resin is the expected one. For ion exchange resins the identification of the surface group type can be determined by measuring the pH profile during the step change between two buffer solutions varying in ionic strength but maintain the same pH. Using Tris buffers, with and without salt, the method was first developed for commercially available CIM ion exchange monolithic columns (Podgornik et al. 2005) and then generalized for ion exchange columns of any kind (Lendero et al. 2008; Figure 12.14).

The driving force in the optimization of the productivity in an industrial separation process, especially in the capture step, is the dynamic capacity of the chromatographic column (Jungbauer and Hahn 2008). To obtain that data the capacity is measured by frontal analysis experiments or sometimes, to save the sample, by continuously injecting a certain amount of the solution of clean target molecule (Benčina, Podgornik, and Štrancar 2004; Sugrue, Nesterenko, and Paull 2005; Brne et al. 2007; Kalashnikova et al. 2007). Since methacrylate monolithic columns are designed for purification and separation of large macromolecules, such as proteins, DNA, or viruses, this approach can be used only for gathering the data about the capacities for various molecules and sometimes to define optimal measuring conditions. However, it is very inconvenient and usually very expensive, especially for large volume columns to be used as an outlet control of the produced columns. Therefore, the DBC is usually given for test proteins (e.g., bovine serum albumin or lysozyme), which is determined by pumping protein solution through the chromatographic column under defined conditions (e.g., buffer type, pH, flow rate) until the breakthrough. Using proteins for the characterization of chromatographic media that are to be used for the purification of biopharmaceutical products can be problematic, as the proteins contaminate the matrix and without intensive validated cleaning, make it unusable for the separation or purification of target compounds. For that reason, a new method for measuring ion exchange capacity of the monolithic columns was developed (Lendero et al. 2005). The principle of the method is similar to the method used for identification tests and it is based on the transient pH formed at a sudden change in ionic strength of the used buffers. It was shown that the duration of this transient pH is in linear correlation with the total ionic capacity and also a good correlation with the protein dynamic capacity was obtained. Biologically compatible buffers are used to fulfill the regulatory requirements. Method accuracy can be adjusted by controlling the key parameters such as pH, concentration and temperature of the used buffer, and it can be validated for implementation in quality control of produced ion exchange chromatographic columns (Lendero et al. 2008). Furthermore, it enables end users to monitor the chromatographic properties of the column during its storage and its

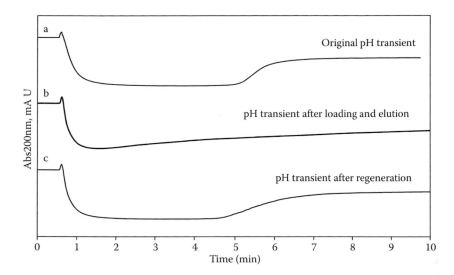

FIGURE 12.15
Monitoring of monolithic column regeneration via pH transient. Original column (a) was loaded with 40 kbp plasmid DNA, which was eluted afterward. Measurement via pH transient revealed that column was not entirely washed (b). After overnight regeneration with 1 M NaOH, original column ionic capacity was restored (c).

use, especially to monitor the ligand degradation or nonspecific binding. For example, if certain molecules remain bound to the monolith after elution, they cover a certain number of active groups that changes pH transient length (Figure 12.15b). After regeneration the initial column performance is established (Figure 12.15c).

12.7 Conclusions

Two decades later the monolith, developed from scientific curiosity to a key stationary phase in several fields of analytical chemistry, and also into a resin of choice for purification of large macromolecules. One can therefore anticipate that monoliths will become a preferred matrix to be applied when new purification processes of very large molecules will be established. Since several noninvasive methods for monolith on-line characterization were recently developed, they can be implemented on a lab scale but also under strict GMP conditions.

References

Aggebrandt, L. G., and Samuelson, O. 1964. Penetration of water-soluble polymers into cellulose fibers. *J. Appl. Polymer. Sci.* 8:2801–12.

Agrawal, S. 1996. Antisense oligonucleotides: Towards clinical trials. *Tibtech.* 14:376–87.

Akbarzaderaleh, P., Abe, M., Yoshimoto, N., and Yamamoto, S. 2007. Separation behavior of PEGylated proteins in relation to their physical and biochemical properties. Presented at the Asia-Pacific Biochemical Engineering Conference, Taiwan, November 4–7.

Amatschek, K., Necina, R., Hahn, R., Schallaun, E., Schwinn, H., Josić, Dj., and Jungbauer, A. 2000. Affinity chromatography of human blood coagulation factor viii on monoliths with peptides from a combinatorial library. *J. High. Resolut. Chrom.* 23:47–58.

Arvidsson, P., Plieva, F. M., Savina, I. N., Lozinsky, V. I., Fexby, S., Bulow, L., Galaev I. Yu., and Mattiasson, B. 2002. Chromatography of microbial cells using continuous supermacroporous affinity and ion-exchange columns. *J. Chromatogr. A* 977:27–38.

Ball, P. 2008. Rapid analysis of adenovirus with prototype CIM QA analytical HPLC columns to support the development of high titre manufacturing platform. Lecture presented at Monolith Summer School, Slovenia, Portorož, May 30–June 4.

Bandari, R., Knolle, W., and Buchmeiser, M. R. 2008. Comparative study on the separation behavior of monolithic columns prepared via ring-opening metathesis polymerization and via electron beam irradiation triggered free radical polymerization for proteins. *J. Chromatogr. A.* 1191:268–73.

Bandari, R., Knolle, W., Prager-Duschke, A., Gläsel, H. J., and Buchmeiser, M. R. 2007. Monolithic media prepared via electron beam curing for proteins separation and flow-through catalysis. *Macromol. Chem. Phys.* 208:1428–36.

Barrett, E. P., Joyner, L. G., and Halenda, P. P. 1951. The determination of pore volume and area distributions in porous substances. I. Computations from nitrogen isotherms. *J. Am. Chem. Soc.* 73:373–80.

Barut, M., Podgornik, A., Brne, P., and Štrancar, A. 2005. Convective interaction media short monolithic columns: Enabling chromatographic supports for the separation and purification of large biomolecules. *J. Separ. Sci.* 28:1876–92.

Barut, M., Podgornik, A., Urbas, L., Gabor, B., Brne, P., Vidič, J., Plevčak, S., and Štrancar, A. 2008. Methacrylate-based short monolithic columns: Enabling tools for rapid and efficient analyses of biomolecules and nanoparticles. *J. Separ. Sci.* 31:1867–80.

Benčina, K., Benčina, M., Podgornik, A., and Štrancar, A. 2007. Influence of the methacrylate monolith structure on genomic DNA mechanical degradation, enzymes activity and clogging. *J. Chromatogr. A.* 1160:176–83.

Benčina, M., Podgornik, A., and Štrancar, A. 2004. Characterization of methacrylate monoliths for purification of DNA molecules. *J. Separ. Sci.* 27:801–10.

Bonardet, J. L., Fraissard, J., Gedeon, A., and Springuel-Huet, M. A. 1999. Nuclear magnetic resonance of physisorbed Xe-129 used as a probe to investigate porous solids. *Catal. Rev. Sci. Eng.* 41:115–225.

Branović, K., Buchacher, A., Barut, M., Štrancar, A., and Josić, Dj. 2000. Application of monoliths for downstream processing of clotting factor IX. *J. Chromatogr. A.* 903:21–32.

Branović, K., Buchacher, A., Barut, M., Štrancar, A., and Josić, Dj. 2003. Application of semi-industrial monolithic columns for downstream processing of clotting factor IX. *J. Chromatogr. B. Analyt. Technol. Biomed. Life Sci.* 790:175–82.

Branović, K., Forčić, K., Ivancic, J., Štrancar, A., Barut, M., Kosutić-Gulija, T., Zgorelec, R., and Mazuran, R. 2003. Application of short monolithic columns for improved detection of viruses. *J. Virol. Meth.* 110:163–71.

Branović, K., Lattner, G., Barut, M., Štrancar, A., Josić, Dj., and Buchacher, A. 2002. Very fast analysis of impurities in immunoglobulin concentrates using conjoint liquid chromatography on short monolithic disks. *J. Immunol. Meth.* 271:47–58.

Brgles, M., Halassy, B., Tomašić, J., Santak, M., Forčić, D., Barut, M., and Štrancar, A. 2007. Determination of DNA entrapment into liposomes using short monolithic columns. *J. Chromatogr. A.* 1144:150–54.

Brne, P., Podgornik, A., Benčina, K., Gabor, B., Štrancar, A., and Peterka, M. 2007. Fast and efficient separation of immunoglobulin M from immunoglobulin G using short monolithic columns. *J. Chromatogr. A.* 1144:120–25.

Brough, H., Antoniou, C., Carter, J., Jakubik, J., Xu, Y., and Lutz, H. 2002. Performance of a Novel Viresolve NFR Virus Filter. *Biotechnol. Progr.* 18:782–95.

Brunauer, S., Emmett, P. H., and Teller, E. 1938. Adsorption of gases in multimolecular layers. *J. Am. Chem. Soc.* 60:309–19.

Cabral, J. L., Bandilla, D., and Skinner, C. D. 2006. Pore size characterization of monolith for electrochromatography via atomic force microscopy studies in air and liquid phase. *J. Chromatogr. A.* 1108:83–89.

Cabrera, K., Lubda, D., Eggenweiler, H.-M., Minakuchi, H., and Nakanishi, K. 2000. A new monolithic-type HPLC column for fast separaton. *J. High. Resol. Chromatogr.* 23:93–99.

Čerk Petrič, T., Brne, P., Gabor, B., Govednik, L., Barut, M., Štrancar, A., and Kralj, L. Z. 2007. Anion-exchange chromatography using short monolithic columns as a complementary technique for human serum albumin depletion prior to human plasma proteome analysis. *J. Pharmaceut. Biomed. Anal.* 43:243–49.

Cheek, M., Darling, D., Sorrell, A., Kamal, N., Farzeneh, F., and Slater N. K. H. 2008. IMAC processing of 6x histidine tagged lentiviral vectors using monolithic adsorbents. Lecture originally presented at Monolith Summer School, Slovenia, Portorož, May 30–June 4.

Clark, J. R., and March, J. B. 2006. Bacteriophages and biotechnology: Vaccines, gene therapy and antibacterials. *Trends Biotechnol.* 24:212–18.

Czermak, P., Grzenia, D. L., Wolf, A., Carlson, J. O., Specht, R., Han, B., and Wickramasinghe, S. R. 2008. Purification of the densonucleosis virus by tangential flow ultrafiltration and by ion exchange membranes. *Desalination.* 224:23–27.

Danquah, M. K., and Forde, G. M. 2007. The suitability of DEAE-Cl active groups on customized poly(GMA-co-EDMA) continuous stationary phase for fast enzyme-free isolation of plasmid DNA. *J. Chromatogr. B.* 853:38–46.

Delgado, C., Malmsten, M., and Van Alstine, J. M. 1997. Analytical partitioning of poly(ethylene glycol)-modified proteins. *J. Chromatogr. B.* 692:263–72.

Eglon, M., Banjac, M., Strappe, P. O'Brien, T., Lah, B., Strancar, A., and Peterka M. 2008. Development of a fast and reliable chromatography method for adenoviral vector purification using methacrylate monoliths. Poster originally presented at Monolith Summer School, Slovenia, Portoroz, May 30–June 4.

Endres, H. N., Johnson, J. A. C., Ross, C. A., Welp, J. K., and Etzel, M. R. 2003. Evaluation of an ion-exchange membrane for purification of plasmid DNA. *Biotechnol. Appl. Biochem.* 37:259–66.

Fee, C. J., and Van Alstine, J. M. 2006. PEG-proteins: Reaction engineering and separation issues. *Chem. Eng. Sci.* 61:934–39.

Forčič, D., Branović-Cakanic, K., Ivancic, J., Jug, R., Barut, M., and Štrancar, A. 2005a. Purification of genomic DNA by short monolithic columns. *J. Chromatogr. A.* 1065:115–20.

Forčič, D., Branović-Cakanic, K., Ivancic, J., Jug, R., Barut, M., Štrancar, A., and Mazuran, R. 2005b. Chromatographic detection of residual cellular DNA on short monolithic columns. *Anal. Biochem.* 336:273–78.

Frankovič, V., Podgornik, A., Lendero, N., Smrekar, F., Kranjc, P., and Štrancar, A., 2008. Characterization of grafted weak anion exchange methacrylate monoliths. Lecture presented at Monolith Summer School, Slovenia, Portoroz, May 30–June 4.

Gagnon, P., Hensel, F., Andrews, P., and Richieri, R. 2007. Recent advances in the purification of IgM monoclonal antibodies. Presented at the 3rd Wilbio Conference on Purification of Biological Products, Waltham, Massachusetts, September 24–26.

Gagnon, P., Hensel, F., and Richieri, R. 2008. Purification of IgM monoclonal antibodies, manufacturing challenges surround the use of IgM monoclonal. *BioPharmInternational.*1–7.http://biopharminternational.findpharma.com/biopharm/article/articleDetail.jsp?id=499148&sk=&date=&pageID=7

Gagnon, P., Richieri, R., and Hensel, F. 2008. IgM purification. The next generation. Presented at the 13th Annual Waterside Conference, Miami, February 4–6.

Gagnon, P., Richieri, R., Zaidi, S., Sevilla, R., and Brinkman, A. 2007. High Speed monolithic assays for IgM quantitation in cell culture production and purification process monitoring. Poster presented at the Third Wilbio Conference on Purification of Biological Products, Walthman, Massachusetts, September 24–26. http://www.validated.com/revalbio/pdffiles/PUR07.pdf

Giovannini, R., Freitag, R., and Tennikova, T. B. 1998. High-performance membrane chromatography of supercoiled plasmid DNA. *Anal. Chem.* 70:3348–54.

Gold, L. 1996. Oligonucleotides as research, diagnostic and therapeutic agents. *J. Biol. Chem.* 270:13581–84.

Grimes, B. A., Skudas, R., Unger, K. K., and Lubda, D. 2007. Pore structural characterization of monolithic silica columns by inverse size-exclusion chromatography. *J. Chromatogr. A.* 1144:14–29.

Grushka, E. 1972. Characterization of exponentially modified gaussian peaks in chromatography. *Anal. Chem.* 44:1733–38.

Gustavsson, P.-E., and Larsson, P.-O. 1999. Continuous superporous agarose beds for chromatography and electrophoresis. *J. Chromatogr. A* 832:29–39.

Gutierrez-Aguirre, I., Steyer, A., Poljšak-Prijatelj, M., Banjac, M., Peterka, M., Strancar, A., Boben, J., Gruden, K., and Ravnikar, M. 2008. Combination of qPCR and CIM technology to greatly improve the detection of different viruses. Lecture presented at Monolith Summer School. Slovenia, Portoroz, May 30–June 4.

Hahn, R., and Jungbauer, A. 2000. Peak broadening in protein chromatography with monoliths at very fast separations. *Anal. Chem.* 72:4853–58.

Hahn, R., and Jungbauer, A. 2001. Control method for integrity of continuous beds. *J. Chromatogr. A.* 908:179–84.

Hahn, R., Tscheliessnig, A., Bauerhansl, P., and Jungbauer, A. 2007. Dispersion effects in preparative polymethacrylate monoliths operated in radial-flow columns. *J. Biochem. Biophys. Meth.* 70:87–94.

Hall, T., Wood, D. C., and Smith, C. E. 2004. Preparative and analytical chromatography of pegylated myelopoietin using monolithic media. *J. Chromatogr. A.* 1041:87–93.

Hjerten, S., Liao, J.-L., and Zhang, R. 1989. High-performance liquid chromatography on continuous polymer beds. *J. Chromatogr.* 473:273–75.

Holdsvedova, P., Suchankova, J., Buncek, M., Backovska, V., and Coufal, P. 2007. Hydroxymethyl methacrylate-based monolithic columns designed for separation of oligonucleotides in hydrophilic-interaction capillary liquid chromatography. *J. Biochem. Biophys. Meth.* 70: 23–29.

Jerman, S., Podgornik, A., Cankar, K., Cadet., N., Skrt, M., Zel, J., and Raspor, P. 2005. Detection of processed genetically modified food using CIM monolithic columns for DNA isolation. *J. Chromatogr. A.* 1065:107–13.

Josić, D., Bal, F., and Schwinn, H. 1993. Isolation of plasma proteins from the clotting cascade by heparin affinity chromatography. *J. Chromatogr. A.* 632:1–10.

Josić, D., Lim, Y. P., Štrancar, A., and Reutter, W. 1994. Application of high-performance membrane chromatography for separation of annexins from the plasma membranes of liver and isolation of monospecific polyclonal antibodies. *J. Chromatogr. B.* 662:217–26.

Josić, D., Reusch, J., Löster, K., Baum, O., and Reutter, W. 1992. High-performance membrane chromatography of serum and plasma membrane proteins. *J. Chromatogr. A.* 590:59–76.

Jungbauer, A., and Hahn, R. 2008. Polymethacrylate monoliths for preparative and industrial separation of biomolecular assemblies. *J. Chromatogr. A.* 1184:62–79.

Junkar, I., Koloini, T., Krajnc, P., Nemec, D., Podgornik, A., and Štrancar, A. 2007. Pressure drop characteristics of poly(high internal phase emulsion) monoliths. *J. Chromatogr. A.* 1144:48–54.

Kalashnikova, I. V., Ivanova, N. D., Evseeva, T. G., Menshikova, A. Y., Valkh, E. G., and Tennikova, T. B. 2007. Study of dynamic adsorption behaviour of large-size protein-bearing particles. *J. Chromatogr. A.* 1144:40–47.

Kalashnikova, I. V., Ivanova, N. D., and Tennikova, T. B. 2008. Development of efficient pseudo-affinity methods for fast separation and sensitive analytical detection of influenza virus using monolithic media. Lecture presented at Monolith Summer School. Slovenia, Portoroz, May 30–June 4.

Kaltenbrunner, O., Jungbauer, A., and Yamamoto, S. 1997. Prediction of the preparative chromatography performance with a very small column. *J. Chromatogr. A.* 760:41–53.

Karlsson, E., Ryden, L., and Brewer, J. 1998. Ion-exchange chromatography. In *Protein purification: Principles, high resolution methods and applications*, ed. J. C. Janson and L. Ryden, 2nd ed., 145–205. New York: Wiley-VCH.

Koptyug, I. V., Kovtunov, K. V., Gerkema, E., Kiwi-Minsker, L., and Sagdeev, R. Z. 2007. NMR microimaging of fluid flow in model string-type reactors. *Chem. Eng. Sci.* 62:4459–68.

Kornyšova, O., Šurna, R., Snitka, V., Pyell, U., and Maruška, A. 2002. Polyrotaxane approach for synthesis of continuous beds for capillary electrochromatography. *J. Chromatogr. A.* 971:225–35.

Kovač, K., Guatierrez-Aguirre, I., and Ravnikar, M., Zimšek-Mijovski, J., Poljšak-Prijatelj, M., Banjac, M., Peterka, M., and Raspor, P. 2008. CIM QA monoliths as a tool for concentration of enteric viruses. Lecture presented at Monolith Summer School, Slovenia, Portoroz, May 30–June 4.

Krajčić, M., Ivancic-Jelecki, J., Forčić, D., Vrdoljak, A., and Škorić, D. 2007. Purification of plant viral and satellite double-stranded RNAs on DEAE monoliths. *J. Chromatogr. A.* 1144:111–19.

Krajnc, P., Leber, N., Štefanec, D., Kontrec, S., and Podgornik, A. 2005. Preparation and characterization of poly(high internal phase emulsion) methacrylate monoliths and their application as separation media. *J. Chromatogr. A.* 1065:69–73.

Kramberger, P., Peterka, M., Boben, J., Ravnikar, M., and Štrancar, A. 2007. Short monolithic columns: A breakthrough in purification and fast quantification of tomato mosaic virus. *J. Chromatogr. A.* 1144:143–49.

Kramberger, P., Peterka, M., Štrancar, A., Boben, J., Petrovič, N., Ravnikar, M., and Branovič, K. 2004. Virus concentration and separation using CIM supports. Lecture presented at Monolith Summer School, Slovenia, Portoroz, June 6–9.

Kramberger, P., Petrovic˘, N., Štrancar, A., and Ravnikar, M. 2004. Concentration of plant viruses using monolithic chromatographic supports. *J. Virol. Meth.* 120:51–57.

Laschober, S., Sulyok, M., and Rosenberg, E. 2007. Tailoring the macroporous structure of monolithic silica-based capillary columns with potential for liquid chromatography. *J. Chromatogr. A.* 1144:55–62.

Laurent, T. C., and Killander, J. 1964. A theory of gel filtration and its experimental verification. *J. Chromatogr.* 14:303–17.

Leinweber, F. C., Lubda, D., Cabrera, K., and Tallarek, U. 2002. Characterization of silica-based monoliths with bimodal pore size distribution. *Anal. Chem.* 74:2470–77.

Leinweber, F. C., and Tallarek, U. 2003. Chromatographic performance of monolithic and particulate stationary phases: Hydrodynamics and adsorption capacity. *J. Chromatogr. A.* 1006:207–28.

Lendero, N., Smrekar, F., Vidič, J., Peterka, M., Podgornik, A., and Štrancar, A. 2006. Purification of large plasmids from *E. coli*. Poster presented at Monolith Summer School, Slovenia, Portoroz, May 28–31.

Lendero, N., Vidič, J., Brne, P., Frankovič, V., Štrancar, A., and Podgornik, A. 2008. Characterization of ion exchange stationary phases via pH transition profiles. *J. Chromatogr. A.* 1185:59–70.

Lendero, N., Vidič, J., Brne, P., Podgornik, A., and Štrancar, A. 2005. A fast, simple and non-destructive method for determining the amount of ion exchange groups on resins. *J. Chromatogr. A.* 1065:29–38.

Levkin P. A., Eeltink, S., Stratton, T. R., Brennen, R., Robotti, K., Yin, H., Killeen, K., Svec, F., and Fréchet, J. M. J. 2008. Monolithic porous polymer stationary phases in polyimide chips for the fast high-performance liquid chromatography separation of proteins and peptides. *J. Chromatogr. A.* 1200:55–61.

Levy, M. S., O'Kennedy, R. D., Ayazi-Shamlou, P., and Dunnill, P. 2000. Biochemical engineering approaches to the challenges of producing pure plasmid DNA. *Tibtech.* 18:296–305.

Liapis, A. I., Meyers, J. J., and Crosser, O. K. 1999. Modeling and simulation of the dynamic behavior of monoliths. Effects of pore structure from pore network model analysis and comparison with columns packed with porous spherical particles. *J. Chromatogr. A.* 865:13–25.

Logan, T. C., Clark, D. S., Stachowiak, T. B., Svec, F., and Frıchet, J. M. J. 2007. Photo-patterning enzymes on polymer monoliths in microfluidic channels for steady-state kinetic analysis and spatially-separated multi-enzyme reactions. *Anal. Chem.* 79:6592–98.

Lubda, D., Lindner, W., Quaglia, M., von Hohenesche, C. F., and Unger, K. K. 2005. Comprehensive pore structure characterization of silica monoliths with controlled mesopore size and macropore size by nitrogen sorption, mercury porosimetry, transmission electron microscopy and inverse size exclusion chromatography. *J. Chromatogr. A.* 1083:14–22.

Mair, D. A., Rolandi, M., Snauko, M., Noroski, R., Frıchet, J. M. J., and Svec, F. 2007. Room temperature bonding for plastic high pressure microfluidic chips. *Anal. Chem.* 79:5097–5102.

Maurer, E., Peterka, M., Gassner, M., Seper, H., Gelgart, F., Banjac, M., Jarc, M., Lah, B., Kramberger, P., Štrancar, A., and Muster, T. 2008. Influenza vaccine purification platform. Lecture presented at Monolith Summer School, Slovenia, Portoroz, May 30–June 4.

Mayr, B., Tessadri, R., Post, E., and Buchmeiser, M. R. 2001. Metathesis-based monolith: Influence of polymerization conditions on the separation of biomolecules. *Anal. Chem.* 73:4071–78.

Meyers, J. J., and Liapis, A. I. 1999. Network modelling of the convective flow and diffusion of molecules adsorbing in monoliths and in porous particles packed in a chromatographic column. *J. Chromatogr. A.* 852:3–23.

Mihelič, I. 2002. Modeliranje polimerizacije, adsorpcije in hidrodinamike poli-(glicidilmetakrilatnih-etilen-dimetakrilatnih) monolitnih nosilcev. PhD Thesis, University of Ljubljana, Ljubljana, p. 162.

Mihelič, I., Koloini, T., and Podgornik, A. 2003. Temperature distribution effects during polymerization of methacrylate-based monoliths. *J. Appl. Polym. Sci.* 87:2326–34.

Mihelič, I., Koloini, T., Podgornik, A., and Štrancar, A. 2000. Dynamic capacity studies of CIM (Convective Interaction Media®) monolithic columns. *J. High Resol. Chromatogr.* 23:39–43.

Mihelič, I., Krajnc, M., Koloini, T., and Podgornik, A. 2001. Kinetic model of a methacrylate-based monolith polymerization. *Ind. Eng .Chem. Res.* 40:3495–3501.

Mihelič, I., Nemec, D., Podgornik, A., and Koloini, T. 2005. Pressure drop in CIM disk monolithic columns. *J. Chromatogr. A.* 1065:59–67.

Milavec Žmak, P., Podgornik, H., Jančar, J., Podgornik, A., and Štrancar, A. 2003. Transfer of gradient chromatographic methods on CIM monolithic columns. *J. Chromatogr. A.* 1006:195–205.

Minakuchi, H., Nakanishi, K., Soga, N., Ishizuka, N., and Tanaka, N. 1996 Octadecylsilylated porous silica rods as separation media for reversed-phase liquid chromatography. *Anal. Chem.* 68:3498–3501.

Miyabe, K. 2008. Evaluation of chromatographic performance of various packing materials having different structural characteristics as stationary phase for fast high performance liquid chromatography by new moment equations. *J. Chromatogr. A.* 1183:49–64.

Miyabe, K., and Guiochon, G. 2002. The moment equations of chromatography for monolithic stationary phases. *J. Phys. Chem. B*. 106:8898–8909.

Nakanishi, K., and Soga, N. 1991. Phase separation in gelling silica-organic polymer solution: Systems containing poly(sodium styrenesulfonate). *J. Am. Ceram. Soc.* 74:2518–30.

Noel, R., Sanderson, A., and Spark, L. 1993. In *Cellulosics: Materials for selective separations and other technologies*, eds. J. F. Kennedy, G. O. Philips, and P. A. Williams, 17–24. New York: E. Horwood.

Opitz, L., Salaklang, J., Buttner, H., Reichl, U., and Wolf, M. W. 2007. Lecitin-affinity chromatography for downstream processing of MDCK cell culture derived human influenza A viruses. *Vaccine*. 25:939–47.

Paril, H. 2008. Upscale from 800 mL to 8 L CIM. The next generation of pDNA purification. Lecture presented at Monolith Summer School, Slovenia, Portoroz, May 30–June 4.

Peterka, M., Glover, D., Kramberger, P., Banjac, M., Podgornik, A., Barut, M., and Štancar, A. 2005. Short monolithic columns: An enabling technology for the purification of proteins, DNA and viruses. *BioProc. J.* March–April:1–6.

Peters, E. C., Svec, F., and Fréchet, J. M. J. 1997. The preparation of large diameter "molded" porous polymer monoliths and the control of pore structure homogeneity. *Chem. Mater.* 9:1898–1902.

Pflegerl, K., Podgornik, A., Berger, E., and Jungbauer, A. 2002. Direct synthesis of peptides on convective interaction media monolithic columns for affinity chromatography. *J. Comb. Chem.* 4:33–37.

Podgornik, A., Barut, M., Jaksa, S., Jančar, J., and Štancar, A. 2002. Application of very short monolithic columns for separation of low and high molecular mass substance. *J. Liq. Chrom. Relat. Tech.* 25:3099–3116.

Podgornik, A., Barut, M., Jančar, J., and Štrancar, A. 1999a. Isocratic separations on thin glycidyl methacrylate-ethylenedimethacrylate monoliths. *J. Chromatogr. A.* 848:51–60.

Podgornik, A., Barut, M., Jančar, J., Štrancar, A., and Tennikova, T. 1999b. High-performance membrane chromatography of small molecules. *Anal. Chem.* 71:2986–91.

Podgornik, A., Barut, M., Štrancar, A., Josić, D., and Koloini, T. 2000. Construction of large-volume monolithic columns. *Anal. Chem.* 72:5693–99.

Podgornik, A., Jančar, J., Merhar, M., Kozamernik, S., Glover, D., Cucek, K., Barut, M., and Štrancar, A. 2004. Large-scale methacrylate monolithic columns: Design and properties. *J. Biochem. Biophys. Methods.* 60:179–89.

Podgornik, A., and Štrancar, A. 2005. Convective interaction media (CIM): Short layer monolithic chromatographic stationary phases. In *Biotechnology Annual Review,* ed. M. R. El-Gewely, 281–333. Amsterdam: Elsevier.

Podgornik, A., Vidič, J., Jančar, J., Lendero, N., Frankovič, V., and Štrancar, A. 2005. Noninvasive methods for characterization of large-volume monolithic chromatographic columns. *Chem. Eng. Technol.* 28:1435–41.

Podgornik, H., and Podgornik, A. 2004. Separation of manganese peroxidase isoenzymes on strong anion-exchange monolithic column using pH–salt gradient. *J. Chromatogr. B.* 799:343–47.

Podgornik, H., Podgornik, A., Milavec, P., and Perdih, A. 2001. The effect of agitation and nitrogen concentration on lignin peroxidase(LiP) isoform composition during fermentation of *phanerochaete chrysosporium*. *J. Biotechnol.* 88:173–76.

Podgornik, H., Podgornik, A., and Perdih, A. 1999. A method of fast separation of lignin peroxidases using convective interaction media disks. *Anal. Biochem.* 272:43–47.

Rathore, A. S., Kennedy, R. M., O'Donnell, J. K., Bemberis, I., and Kaltenbrunner, O. 2003. Qualification of a chromatographic column. Why and how to do it. *BioPharm Internat.* March:30–40.

Ren, X. H., Stapf, S., Kühn, H., Demco, D. E., and Blümich, B. 2003. Molecular mobility in fixed-bed reactors investigated by multiscale NMR techniques. *Magn. Reson. Imaging* 21:261–68.

Ritter, H. L., and Drake, L. C. 1945. Pore-size distribution in porous materials: Pressure porosimeter and determination of complete macropore-size distributions. *Ind. Eng. Chem. Anal. Ed.* 17:782–86.

Rodrigues, T., Carrondo, M. J. T., Avles, P. M., and Cruz, E. P. 2007. Purification of retroviral vectors for clinical application: Biological implications and technological challenges. *J. Biotechnol.* 127:520–41.

Seely, J. E., and Richey, C. W. 2001. Use of ion-exchange chromatography and hydrophobic interaction chromatography in the preparation and recovery of polyethylene glycol-linked proteins. *J. Chromatogr. A.* 908:235–41.

Segura, M. M., Kamen, A., and Garinier, A. 2006. Downstream processing of oncoretroviral and lentiviral gene therapy vectors. *Biotechnol. Adv.* 24:321–37.

Skudas, R., Grimes, B. A., Machtejevas, E., Kudirkaite, V., Hennessy, T. P., Lubda, D., and Unger, K. K. 2007. Impact of pore structural parameters on column performance and resolution of reversed-phase monolithic silica columns for peptides and proteins. *J. Chromatogr. A.* 1144:72–84.

Smrekar, F., Ciringer, M., Peterka, M., Podgornik, A., and Štrancar, A. 2008. Purification and concentration of bacteriophage T4 using monolithic chromatographic supports. *J. Chromatogr. B.* 861:177–80.

Stepišnik, J., Mohorič, A., and Duh, A. 2001. Diffusion and flow in a porous structure by the gradient-spin-echo spectral analysis. *Physica B.* 307:158–68.

Štrancar, A., Barut, M., Podgornik, A., Koselj, P., Schwinn, H., Raspor, P., and Josic, D. 1997. Application of compact porous tubes for preparative isolation of clotting factor VIII from human plasma. *J. Chromatogr. A.* 760:117–23.

Štrancar, A., Koselj, P., Schwinn, H., and Josić, D. 1996. Application of compact porous disks for fast separations of biopolymers and in-process control in biotechnology. *Anal. Chem.* 68:3483–88.

Štrancar, A., Podgornik, A., Barut, M., and Necina, R. 2002. Short monolithic columns as stationary phases for biochromatography. In *Advances in biochemical engineering/biotechnology*, ed. T. Scheper, 50–85. Berlin-Heidelberg: Springer-Verlag.

Sugrue, E., Nesterenko, P. N., and Paull, B. 2005. Fast ion chromatography of inorganic anions and cations on a lysinebonded porous silica monolith. *J. Chromatogr. A.* 1075:167–75.

Svec, F. 2004. Organic polymer monoliths as stationary phases for capillary HPLC. *J. Separ. Sci.* 27:1419–30.

Svec, F., and Tennikova, T. B. 2003. Historical review. In *Monolithic Materials: Preparation, properties, and applications*, eds. F. Svec, T. B. Tennikova, and Z. Deyl 1–15. Amsterdam: Elsevier.

Svec F., Tennikova T. B., and Deyl, Z. eds. 2003. *Monolithic materials: Preparation, properties and applications*. Amsterdam: Elsevier.

Sykora, D., Svec, F., and Fréchet, J. M. J. 1999. Separation of oligonucleotides on novel monolithic columns with ion-exchange functional surface. *J. Chromatogr. A.* 852:297–304.

Tallarek, U., Leinweber, F. C., and Seidel-Morgenstern, A. 2002. Fluid dynamics in monolithic adsorbents: Phenomenological approach to equivalent particle dimensions. *Chem. Eng. Technol.* 25:1177–81.

Tennikov, M. B., Gazdina, N. V., Tennikova, T. B., and Švec, F. 1998. Effect of porous structure of macroporous polymer supports on resolution in high-performance membrane chromatography of proteins. *J. Chromatogr.* 798: 55–64.

Tennikova, T. B., Belenkii, B. G., and Svec, F. 1990. High-performance membrane chromatography. A novel method of protein separation. *J. Liq. Chromatogr.* 13:63–70.

Tennikova, T. B., and Reusch, J. 2005. Short monolithic beds: History and introduction to the field. *J. Chromatogr. A.* 1065:13–17.

Tennikova, T. B., and Svec, F. 1993. High-performance membrane chromatography: Highly efficient separation method for proteins in ion-exchange, hydrophobic interaction and reversed-phase modes. *J. Chromatogr.* 646:279–88.

Trilisky, E. I., and Lenhoff, A. M. 2007. Sorption processes in ion-exchange chromatography of viruses. *J. Chromatogr. A.* 1142:2–12.

Urban, J., Eeltink, S., Jandera, P., and Schoenmakers, P. J. 2008. Characterization of polymer-based monolithic capillary columns by inverse size-exclusion chromatography and mercury-intrusion porosimetry. *J. Chromatogr. A.* 1182:161–68.

Urthaler, J., Ascher, C., Wohrer, H., and Necina, R. 2006. Automated alkaline lysis for industrial scale cGMP production of pharmaceutical grade plasmid-DNA. *J. Biotechnol.* 128:132–49.

Urthaler, J., Schlegl, R., Podgornik, A., Štrancar, A., Jungbauer, A., and Necina, R. 2005. Application of monoliths for plasmid DNA purification: Development and transfer to production. *J. Chromatogr. A.* 1065:93–106.

Vervoort, N., Gzil, P., Baron, G., and Desmet, G. 2003. A novel correlation for the pressure drop in monolithic silica columns. *Anal. Chem.* 75:843–50.

Vervoort, N., Gzil, P., Baron, G., and Desmet, G. 2004. Model column structure for the analysis of the flow and band-broadening characteristics of silica monoliths. *J. Chromatogr. A.* 1030:177–86.

Yamamoto, S., Fujii, S., Yoshimoto, N., and Akbarzadehlaleh, P. 2007. Effects of protein conformational changes on separation performance in electrostatic interaction chromatography: Unfolded proteins and PEGylated proteins. *J. Biotechnol.* 132:196–201.

Yamamoto, S., and Kita, A. 2006. Rational design calculation method for stepwise elution chromatography of proteins. *Food and Bioproducts Processing* 84:72–77.

Yamamoto, S., Nakamura, M., Tarmann, C., and Jungbauer, A. 2007. Retention studies of pDNA on anion-exchange monolith chromatography. Binding site and elution behaviour. *J. Chromatogr. A.* 1144:155–60.

Yamamoto, S., Nakanishi, K., and Matsuno, R. 1988. Ion-exchange chromatography of proteins. In *Chromatographic science series*, Marcel Dekker. New York: Basel 330–40.

Yao, Y., and Lenhoff, A. M. 2006. Pore size distributions of ion exchangers and relation to protein binding capacity. *J. Chromatogr. A.* 1126:107–19.

Zöchling, A., Hahn, R., Ahrer, K., Urthaler, J., and Jungabauer, A. 2004. Mass transfer characteristics of plasmids in monoliths. *J. Sep. Sci.* 27:819–27.

13

Chromatographic Separation of Plasmid DNA Using Macroporous Beads

Duarte M. de França Prazeres

CONTENTS

13.1 Introduction

13.1.1 Plasmid Biopharmaceuticals

Plasmids are closed, double stranded, DNA molecules that carry genes. These molecules are one of the cornerstones of recombinant DNA, a technology that has enabled the production of large amounts of proteins with clinical (e.g., insulin) and industrial (e.g., laundry enzymes) use. Additionally, biopharmaceuticals made up of plasmids have emerged in the scientific literature and biotechnology industry during the last 15 years, in the wake of the first scientific reports that anticipated the use of plasmids as vectors for the delivery of a therapeutic gene to specific target cells in the context of gene therapy (Wolff et al. 1990) and vaccination (Ulmer et al. 1993) applications. Although gene therapy was originally proposed for the treatment of single-gene disorders, the majority of current clinical trials involve the treatment of cancer and vascular disease (Edelstein, Abedi, and Wixon 2007). In the case of DNA vaccines, plasmids are used to express specific antigens on cell membranes, stimulating and enhancing the immune system response and memory, as a newer generation of vaccines—DNA vaccines (Gurunathan, Klinman, and Seder 2000). Several clinical trials are currently under way to assess the ability of DNA vaccines to raise protection against diseases such as AIDS and malaria. The first proof of the immunogenicity of a flu DNA vaccine in humans (Drape et al. 2006) also supports claims that DNA vaccines may be the best way to respond to an influenza pandemic (Forde 2005). In spite of the short history of this new class of medicinal agents, two veterinary DNA vaccines—one to protect horses against West Nile virus (Powell 2004) and another to protect farm-raised salmon against Infectious Haematopoietic Necrosis virus (Novartis 2005)—became the first plasmid biopharmaceuticals to gain FDA approval in 2005.

In another very recent landmark, the U.S. Department of Agriculture (USDA) has conditionally approved a therapeutic DNA vaccine designed to treat melanoma in dogs (Merial 2007). This extraordinary speed in bringing a product from a totally new class to the marketplace, together with the large number (more than 300) of plasmid biopharmaceuticals undergoing preclinical development and clinical trials makes it clear that other products will soon hit the market. Testing DNA vaccines and other plasmid biopharmaceuticals in clinical trials require large amounts of clinical-grade plasmid DNA. This demand is expected to increase as more products receive marketing approval. In the specific case of DNA vaccines, a surge in demand is also foreseen in the event of a pandemic influenza outbreak (Forde, 2005). Though dosage is still an open question, with 4 µg (Drape et al., 2006) to 4,000 µg single doses reported (www.clinicaltrials.gov; clinical trial identifier: NCT00408109), an annual demand for 900 million units (the world capacity for flu vaccine, Hoare et al. 2005) of a DNA vaccine would require anywhere between 3.6 to 3,600 kg of plasmid DNA. This is well over the multigram capacity claimed by contract manufacturing companies. Furthermore, plasmids are found almost exclusively in bacteria and present unique characteristics that translate into very specific problems that hamper product and process development and keep challenging scientists and engineers.

13.2 Downstream Processing of Plasmids

13.2.1 The Starting Material

The active component of a plasmid biopharmaceutical is produced in large quantities by growing a recombinant *Escherichia coli* host in bioreactors. The cultivation of *E. coli* cells has been pushed up to cell densities of 60 g dry weight/L and plasmid yields of 1 g/L (Carnes 2005). Such yields put a huge pressure on the downstream processing, which is hampered by problems related to the peculiar shape of plasmids, to the high viscosity of some process streams, and to the presence of an assortment of impurities that share a number of properties with the target molecule (Prazeres et al. 1999, 2001). This translates into purification yields around 50–60%, manufacturing costs that can top 15 €/mg (about 20 US$/mg) (50–80% of the total) and water consumption of more than 150 ton/kg (Freitas, Santos, and Prazeres 2006b).

13.2.2 Primary Isolation

The downstream processing is designed to recover plasmids and remove host impurities like genomic DNA, RNA, proteins, and lipopolysaccharides until a level of purity compatible with human use is met. The unit operations in the downstream processing train can be grouped in three stages: primary isolation, intermediate purification, and final purification (Prazeres et al. 1999, 2001). In the first stage, cells from the bioreactor are harvested (e.g., by microfiltration or centrifugation) and lysed to release plasmids. Mechanical lysis by sonication and homogenization constitutes an alternative, but the outcome is usually a low plasmid yield that hardly exceeds 50% (Carlson et al. 1995). Damage may also be significant, though this can be alleviated by shielding plasmid DNA with compaction agents (Murphy et al. 2006). Alkaline lysis, which is the most popular option for lysing *E. coli* and release plasmids, involves adding alkaline solutions to cell suspensions. Lysis is fast microscopically (60 s) but the high viscosity and non-Newtonian behavior make lysate handling hard (Levy et al. 1999). During the ensuing neutralization of the lysate (e.g., with potassium acetate), a mass of solids containing cell debris, denatured proteins, and nucleic acids is generated, which must be removed by low shear, solid-liquid unit operations (e.g., filtration). A fraction of these solids float with other sediments making separation difficult. The initial concentration of cells is a critical variable: experimental data show that plasmid yields can be significantly reduced when using large cell concentrations. Liquid entrainment in the solids and plasmid binding to cell membranes and genomic DNA likely account for yield drops.

13.2.3 Intermediate Purification

Intermediate purification acts upon clarified lysates, concentrating plasmid DNA, and reducing impurities. Though precipitation has been traditionally used for this purpose (Lander et al. 2002; Freitas, Santos, and Prazeres 2006a), tangential flow filtration (TFF; Eon-Duval et al. 2003) and aqueous two-phase systems (ATPS; Trindade et al. 2005) have emerged as reliable alternatives. For instance, studies with poly(ethylene glycol)/salt ATPS indicate that these systems can be used as an early step to capture plasmids directly from cell lysates, directing them toward the bottom phase and partitioning impurities toward the top phase (Trindade et al. 2005). However, for ATPS to outperform alternatives such as precipitation, improvements of the current selectivity are required to increase yields and purity degree. The inability of ATPS to deliver high concentration factors (>10) and the large amount of mass sepa-rating agents required when performing extraction directly from cell lysates are other drawbacks.

13.2.4 Final Purification

Final purification aims to remove the more recalcitrant impurities such as genomic DNA and the biologically less efficient plasmid DNA variants (open circular, linear, denatured). It is here that fixed bed chromatogra-phy plays a central role, especially in the form of modalities such as gel filtration, anion-exchange, hydrophobic interaction, and affinity (Diogo, Queiroz, and Prazeres 2005). The usefulness of chromatography in the con-text of plasmid purification further extends into the analytical department, where HPLC is a tool often used for process monitoring and quality control purposes. Sterile filtration with 0.22 micron filters is usually the terminal step in the downstream processing. Though this type of filtration is stan-dard for many bioproducts, plasmid DNA may cause fouling of filters and yield losses are expected due to the large size of plasmid molecules.

13.3 Plasmid Structure

13.3.1 Basics

The majority of the plasmid vectors used in gene therapy and DNA vaccina-tion are covalently closed, double-stranded DNA molecules. Each nucleotide in either of the two opposing DNA strands has a negatively charged phosphate group. This confers a polyanionic nature to plasmids, which at pH > 4 have a number of negative charges equal to two times the number of base pairs (bp) in the molecule (Sinden 1994; Neidle 1999). The double strandedness of plas-mids implies that the hydrophobic portion of the molecule (i.e., the A, T, G, and C bases) is shielded inside the right-handed double helix. Hence, plasmids are essentially hydrophilic molecules. Apart from their specific chemical nature (i.e., plasmids are made of nucleic acids), the mass, size, and shape are major distinctive characteristics of plasmids that clearly set them apart from proteins. Regarding mass and size, most plasmids designed for therapeutic applications feature between 3,000 bp and 10,000 bp. Given that the average molecular mass of a base pair is roughly 660 g/mol, this means that the typical plasmids will have masses of the order of 1,980 to 6,600 kDa.

One of the most striking characteristics of plasmids is their ability to adopt a higher order structure, which is characterized by an extra coiling of the DNA double helix in space (Vologodskii and Cozzarelli 1994). The majority of plasmids isolated from prokaryotes such as *Escherichia coli* have a similar degree of right-handed supercoiling, at which they adopt a branched shape (Boles, White, and Cozzarelli 1990). Although a fraction of a population of plasmid molecules can also exist as other topological forms (open circular, linear), the supercoiled plasmid topoisomer is the most common *in vivo* (Vologodskii and Cozzarelli 1994). Furthermore, supercoiled plasmid isoforms are often considered more effective at transferring gene expression than open circle and linear variants (Cupillard et al. 2005), even though this notion has been contradicted by a number of reports (see Bergan, Galbraith, and Sloane 2000).

Plasmid supercoiling can be described by a property known as the linking number, Lk, which is defined as the number of times the two strands of the double helix are intertwined (Vologodskii and Cozzarelli 1994). The linking number of an open circular plasmid, Lk_o, can be obtained easily by dividing the number of base pairs in the molecule by the helical repeat (10.6 bp/turn). Negatively supercoiled plasmids have a deficiency in the linking number (i.e., $Lk < Lk_o$), and the extent of supercoiling can thus be expressed in terms of a specific linking number difference, $\sigma = (Lk - Lk_o)/Lk_o$. The typical σ of plasmids isolated from *E. coli* falls within the $-0.07 - -0.05$ range (Vologodskii and Cozzarelli 1994).

13.3.2 Plasmid Dimensions

Supercoiled plasmid molecules with superhelix densities, σ, in the $-0.07 - -0.05$ range have a branched shape like the one illustrated in Figure 13.1. Two parameters can be used as a measure of the dimensions of such an irregular shape: the superhelix axis length, L, and the superhelix radius, r. In the figure, L corresponds to the sum of the dashed segments crossing the nodes and bisecting the area enclosed by the two DNA double strands between adjacent nodes, whereas r is the distance between the superhelix axis and the DNA double helix. Boles, White, and Cozzarelli have found that for 3,500 and 7,000 bp supercoiled plasmids, L is independent of the degree of supercoiling and constitutes about 41% of the total DNA length (1990). Since the total length of DNA in a plasmid can be calculated by multiplying its size in base pairs and

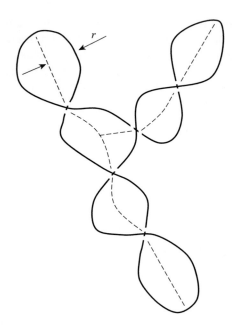

FIGURE 13.1
Schematic model of a negatively supercoiled plasmid DNA molecule. The thick line represents
the DNA double helix and the dashed line the superhelix axis, *L*. (Adapted from Vologodskii,
A. V., Levene, S. D., Klenin, K. V., Frank-Kamenetskii, M., and Cozzarelli, N. R., *J. Mol. Biol.*, 227,
1224–43, 1992.)

the average axial rise per base pair in the most common B-DNA aloform (3.4
Å/bp), L (Å) can be estimated as

$$L = 1.394 \text{ BP.} \tag{13.1}$$

   Given that typical plasmids used in gene therapy and DNA vaccines
have base pairs between 3,000 and 10,000, the superhelix axis will thus vary
between 4,200 Å and 14,000 Å. As for r, the same authors have shown that it
decreases hyperbolically with σ according to

$$\frac{1}{r} = 0.00153 - 0.2689\sigma, \tag{13.2}$$

with r in Å (Boles, White, and Cozzarelli 1990). Given the typical σ values
mentioned above, supercoiled plasmids isolated from *E. coli* will have superhelix diameters between 99 Å and 134 Å.
   Another parameter that is commonly used to infer the size of plasmids is
the average radius of gyration, R_G (Vologodskii and Cozzarelli 1994). This
quantity can be derived from a highly averaged set of experimental light
scattering data and is defined as the root mean square average distance of

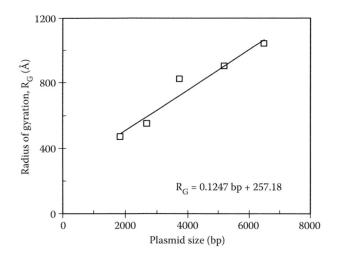

FIGURE 13.2
Radius of gyration as a function of plasmid mass/size. (Symbols–experimental data taken from Hammermann, M., Brun, N., Klenin, K., May, R., Toth, K., and Langowski, J., *Biophys. J.,* 75, 3057–63, 1998, 1868 bp, 468 Å; Chirico, G., and Baldini, G. *J. Chem. Phys.,* 104, 6009–19, 1996, 2686 bp, 550 Å; Fishman, D. M., and Patterson, G. D., *Biopolymers,* 38, 535–52, 1996, 3730 bp, 820 Å; Vologodskii, A. V., and Cozzarelli, N. R., *Annu. Rev. Biophys. Biomol. Struct.,* 23, 609–43, 1994, 5200 bp, 900 Å; and Voordouw, G., Kam, Z., Borochov, N., and Eisenberg, H., *Biophys. Chem.,* 8, 171–89, 1978, 6500 bp, 1040 Å.)

each point in the plasmid from the molecule's center of gravity (Fishman and Patterson 1996). Figure 13.2 displays experimental R_G values as a function of plasmid size and illustrates that as a first approach, R_G can be estimated by the linear correlation

$$R_G = 0.1247 \text{ bp} + 257.18, \tag{13.3}$$

with R_G in Å. By introducing the relationship between L and base pairs described above (Equation 13.1), an expression can also be derived to correlate R_G with L:

$$R_G = 0.0895 \text{ L} + 257.18, \tag{13.4}$$

with R_G in Å.

13.3.3 Plasmid Diffusion Coefficients

The typical 2–10 nm-wide globular protein (Tyn and Gusek 1990; He and Niemeyer 2003) tends to maintain its structure and shape within close boundaries. The micron-sized (>0.1 μm) interwound and flexible supercoiled plasmid, on the other hand, freely coils in a solution under the influence of Brownian motion, with branches forming and retracting rapidly (Boles, White, and Cozzarelli 1990; Olson 1999). Consequently, plasmids display

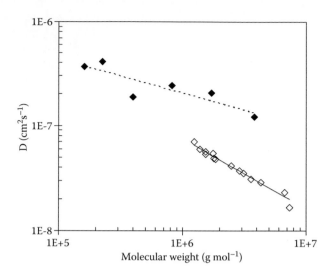

FIGURE 13.3
Plasmid diffusion coefficients as a function of plasmid size. Experimental diffusion coefficients of plasmids (◇) and proteins (◆) are shown together with predictions for plasmids obtained with the correlation in Equation 13.5 (solid line). Plasmid diffusion coefficients were taken from Prazeres (2008) and protein (γ-globulin, catalase, fibrinogen, R2-myoglobin, satellite tobacco necrosis virus, and pyruvate dehydrogenase) diffusion coefficients. (From He, L., and Niemeyer, B., *Biotechnol. Prog.*, 19, 544–48, 2003.)

an enormous variety of irregular shapes and are characterized by a highly dynamic behavior that determines, to some extent, their molecular diffusion in solution, within porous materials or in the extra-cellular space. The typical diffusion coefficient of a supercoiled plasmid DNA molecule in a dilute solution is of the order of 10^{-8} cm^2/s (Fishman and Patterson 1996), while that of a protein with an equivalent mass is one order of magnitude larger (Tyn and Gusek 1990; He and Niemeyer 2003). This difference in the magnitude of diffusion coefficients is exemplified in Figure 13.3, which displays a sample of experimental diffusion coefficients of plasmids (determined on the basis of light scattering) and globular proteins, as a function of the molecular weight (Prazeres 2008). The solid line shown in the figure corresponds to a recently proposed correlation that enables the prediction of the diffusion coefficient of supercoiled plasmid molecules in dilute solutions (<100 µg/ml) with an average error of 6.3% (Prazeres 2008). In this correlation, the size of the plasmid is used as the single input parameter:

$$D = \frac{AT}{\mu} bp^{-2/3},$$
(13.5)

where D (cm^{2}s^{-1}) is the diffusion coefficient, T (K) the temperature, µ (Pa.s) the solution viscosity and A is an empirical constant approximately equal to 3.31 × 10^{-11}. This correlation can be written in the form D ∝ M$^{-2/3}$, since the number of base pairs of a plasmid molecule can be readily converted to molecular weight, M, by assuming that the average molecular weight of a base pair is 660 g/mol. The order of magnitude of the diffusion coefficients of the larger RNA and genomic DNA fragments are also comparable to plasmids, even though supercoiling is nonexistent in those cases (Seils and Pecora 1990).

13.4 Plasmid Chromatography

13.4.1 The Challenges

With the exception of gel filtration, which always operates in a flowthrough mode, plasmid chromatography should in principle operate in a positive mode, for example, adequate matrices should be used to bind plasmid molecules and allow separation of impurities in the flowthrough fractions or upon elution under suitable buffer conditions. One of the key advantages associated with positive chromatography is the ability to elute the target plasmid in a pool that is concentrated relative to the feedstock. This is not possible when chromatographic operations are carried out in a negative mode, where elution of the target solute in the flowthrough under nonbinding conditions usually leads to dilution (Freitas, Santos, and Prazeres 2009). However, in order to take full advantage of a positive mode of operation, chromatographic matrices, whether of the anion-exchange, hydrophobic interaction, or affinity type, should selectively bind large amounts of plasmid and allow for fast separations to take place. Unfortunately, reality shows otherwise as will be described next.

Although it dominates the final purification stage, chromatography is faced with a number of limitations that by the most part are deeply rooted in the structural nature of the stationary phases and plasmid molecules involved (Prazeres et al. 1999; Diogo, Queiroz, and Prazeres 2005). Poor selectivity and coelution are attributed to similarities between plasmid DNA and *E. coli*-derived impurities: (i) lipopolysaccharide aggregates are similar in size and also negatively charged, (ii) RNA and genomic DNA are also polyanionic, and (iii) some RNA species and genomic DNA fragments that arise during the course of processing have molecular weights comparable to those of plasmids. These limitations can be partially alleviated by reducing the impurity load prior to chromatography with unit operations like precipitation (Lander et al. 2002; Freitas, Santos, and Prazeres 2006a), tangential flow filtration (Eon-Duval et al. 2003), and aqueous two-phase extraction (Trindade et al. 2005), as mentioned in Section 13.2.3 above. However, the more severe limitations encountered in plasmid chromatography are probably the slow internal diffusion and the low capacity (Prazeres et al. 1999; Diogo, Queiroz, and Prazeres 2005). These limitations can be ascribed to the awkward shape and large molecular weight of plasmid DNA molecules. The small diffusion coefficients that are characteristic of plasmids (Section 13.3.3) will typically lead to limitations in internal mass transfer, which result in broad peaks and low recovery (Huber 1998). Furthermore, the consequent need to use small flow rates and shallow gradients leads to long separation time (Huber 1998). This situation is clearly worsened by the fact that the pores of most stationary phases have not been engineered to handle such large macromolecules. These issues are discussed next.

13.4.2 Conventional Matrices

The vast majority of the chromatographic matrices available to bioprocess engineers were originally developed with the goal of purifying proteins. Most of those matrices feature particle sizes of 50 μm to 500 μm and have pores with diameters that seldom exceed the 300 Å (Tiainen et al. 2007). As an example let us consider Q-Sepharose (GE Healthcare, Uppsala, Sweden), a strong anion-exchanger with quaternary amine groups that has been used very often in plasmid purification (see Horn, Budahazi, and Marquet 1998; Prazeres, Schluep, and Cooney 1998). Q-Sepharose beads are made of highly cross-linked agarose (6%) and have a mean diameter of 90 μm. The material features a structure that combines rather large voids with dense matrix regions that contain extensively aggregated agarose double helices. In the case of the Q-Sepharose Fast Flow, the exclusion limit for globular proteins is 4×10^6 (GE Healthcare, Uppsala, Sweden). This corresponds to a diameter of approximately 332 Å* and thus assuming that spherical molecules with diameters higher than 200 Å are unlikely to enter the particle pores. In view of the unusual plasmid shape and dimensions described above, it is very clear that the geometry of conventional sorbents like Q-Sepharose is inadequate to handle plasmids, at least in an efficient way. The major problem is related to the fact that the majority of pores in a material like Q-Sepharose are smaller than plasmids ($d_p < L$) that is, the molecules are physically prevented from entering most conventional sorbents. As for the largest pores, and although diameters are comparable with the largest dimension of plasmids (i.e., the superhelix axis), approaching the mouth of a pore in a sideway manner is unlikely to permit access. This situation is illustrated in Figure 13.4a, which compares the dimensions of a typical 5,000 bp plasmid ($L = 6,900$ Å, see Section 13.3.2) with the largest pores (≈3,000 Å) than can be found in conventional ion-exchange adsorbents (DePhillips and Lenhoff 2001). In theory, and since the diameter of superhelices is of the order of 100 Å (see Section 13.3.2), plasmids could more easily enter these larger pores with an

---

* The radius of a globular protein can be related to its molecular weight according to: $r(\text{Å}) = 0.38 \, MW^{0.4}$ (Schnabel, Langer, and Breitenbach 1988).

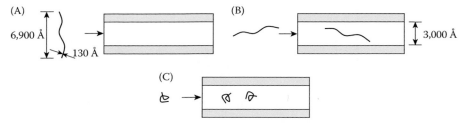

FIGURE 13.4
Schematic representation of the relative dimensions and accessibility of a typical 5,000 bp plasmid to the largest pore ($d_p \approx 3,000$ Å) in a conventional ion-exchange chromatographic matrix (DePhillips and Lenhoff 2001). The supercoiled plasmid is represented as a thick line with a length roughly proportional to the superhelix axis and a thickness proportional to the superhelix diameter. (A) Plasmids cannot enter the pore when facing the entrance side wise; (B) axial orientation of the plasmid facilitates entrance, but Brownian motion is likely to reorient the molecule leading to hindered diffusion; (C) divalent cations could be used to compact individual plasmid molecules into a globular structure and improve accessibility.

axial orientation (Figure 13.4b). However, Brownian motion and the constant coiling of plasmids would readily bring molecules into close proximity with pore walls. The resulting hindrance in molecular diffusion, together with the already very small diffusion coefficients (see Section 13.3.3), would make it unpractical for a plasmid molecule to diffuse in and out of such a chromatographic bead, at least in an acceptable timeframe.

The inability of conventional adsorbents to accommodate plasmid molecules translates directly into poor binding capacities that seldom exceed the 0.5–1.5 mg/ml (Table 13.1). In such cases, most of the plasmid molecules bind to an outer shell of the beads. In the case of 65 μm, 88 μm, and 111 μm Q-Sepharose XL particles and a 6,300 bp plasmid, for instance, the thickness of this shell is around 6.1 μm, 6.6 μm, and 7.3 μm, respectively (Ljunglöf et al. 1999). This means that a significant portion of the beads (59–66%) remains unused, as far as plasmid binding is concerned. The situation is perfectly illustrated in the top part of Figure 13.5, which shows confocal laser microscopy images of ordinary chromatographic beads (<200 nm pores) after being exposed to fluorescently labeled plasmid DNA (Stadler, Lemmens, and Nyhammar 2004). The depth discriminating properties of confocal microscopy makes it possible to slice the beads optically into the series of thin sections shown on the top right hand side of the figure. It is clear from these scanned images that plasmid molecules do not enter the core of the beads (Ljunglöf et al. 1999). Superficial adsorption of plasmids was confirmed also indirectly by the existence of an inverse correlation between the radius of anion-exchanger beads and static (Ferreira, Cabral, and Prazeres 2000) and dynamic (Theodossiou, Sondergaard, and Thomas 2001) plasmid capacity. Thus, an expeditious way to improve capacity (see Table 13.2 for other options as well) is to use beads with smaller sizes (Ljunglöf et al. 1999). However, the need to use high pressures when operating chromatographic columns

TABLE 13.1

Static and Dynamic[a] Binding Capacity of Plasmid DNA on Commercial Anion-Exchange Adsorbents

Stationary Phase	Bead Size (μm)	Pore Size (μm)	Capacity (mg/ml)	pDNA (kb)	Reference
Conventional adsorbents					
Q-Sepharose Fast Flow	90	<0.035	1.3	4.8	Ferreira et al. 2000
Q-Sepharose Fast Flow	90	<0.035	0.72[a]	5.9	Eon-Duval and Burke 2004
Q-Sepharose XL	90	<0.035	1.5[a]	6.9	Levy et al. 2000
Adsorbents with larger pores and improved topography					
Streamline QXL	200	n.a.	3	4.8	Ferreira et al. 2000
Q Hyper D	50	0.30	>5.3[a]	5.9	Eon-Duval and Burke 2004
Fractogel EMD DEAE	40–90	~0.80	2.45[a]	5.9	Eon-Duval and Burke 2004
Fractogel EMD DMAE	40–90	~0.80	4.5[a]	6.9	Levy et al. 2000
POROS 50 DEAE	50	<0.80	5.0[a]	6.9	Levy et al. 2000
POROS 50 HQ	50	<0.80	2.12[a]	5.9	Eon-Duval and Burke 2004
Superporous adsorbents					
Cytopore	230	30	~31	4.8	Ferreira, Prazeres, and Cabral 2003
Superporous agarose	60	4	3.0[a]	7.0	Tiainen et al. 2007
Superporous agarose	60	2	3.9[a]	7.0	Tiainen et al. 2007

[a] dynamic capacity. These values in the table not marked with an "a" represent slatic capacity.

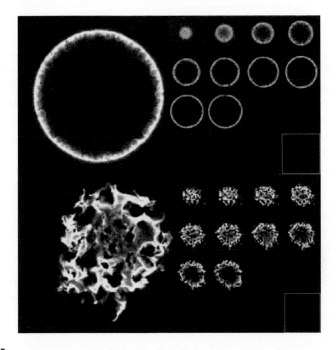

FIGURE 13.5
Confocal microscopy of chromatographic (series on top) and microcarrier (series at the bottom) beads that were incubated with fluorescently labeled plasmid DNA. After removal of unbound material by washing the beads, different sections of the beads were scanned layer by layer by a confocal laser beam, inducing fluorescence. The result clearly illustrates that, under given conditions, plasmid DNA does not deeply enter into the pores of conventional chromatographic beads, whereas it penetrates the microcarrier beads. (From Stadler, J., Lemmens, R., and Nyhammar, T., *J. Gene Med.*, 6, S54, 2004. With permission.)

TABLE 13.2

Strategies Used to Overcome the Major Limitations (Accessibility, Mass Transport, Capacity) Found in the Context of Plasmid Chromatography

Strategies	Comments
Compaction of plasmids	Mass separation agents are required that need to be removed afterward
Use unusual surface topographies	
Decrease bead size	The higher pressures required in columns packed with bead sizes <10–20 μm may not be practical from a bioprocess point of view
Operate in a negative mode	Dilution of plasmid material relative to the feed
Engineer superpores into matrices	Area available for binding decreases as superpore size increases

packed with small particles makes it impractical from a bioprocessing point of view to use beads smaller than 10 μm.

Improved binding capacities can be obtained by resorting to adsorbents that feature unusual surface topographies (Table 13.2). For example, plasmid molecules can be captured more densely in a bead with tentacle-like structures as the ones characteristic of materials like Fractogel EMD DEAE (Merck, Darmsatd) or Streamline QXL (GE Healthcare, Uppsala; Theodossiou, Sondergaard, and Thomas 2001). In this last case, the long dextran chains that protrude from the surface of 200 μm agarose beads are able to capture plasmid molecules efficiently, leading to capacities of 3–5 mg/ml (Table 13.1).

The size-diffusion barrier can be partially overcome by using multivalent cations or compaction agents that decrease repulsion between DNA phosphates and thus condense plasmid molecules from an elongated coiled state to a compacted globular state (Figure 13.4c, Table 13.2; Murphy et al. 1999). This strategy could (i) facilitate the access of plasmid to pores, (ii) allow a closer packing on the surface of the adsorbent, (iii) lead to a more homogeneous behavior during chromatography, and (iv) increase the diffusion coefficient up to the value displayed by a globular protein with an equivalent mass (see Figure 13.3). Ultimately, sharper peaks, higher yields and capacity, and increased selectivity could be obtained. For instance, the use of a 2 M ammonium sulphate buffer during size exclusion chromatography has been claimed to increase productivity (Lemmens et al. 2003; Stadler, Lemmens, and Nyhammar 2004). In another case, the use of poly(ethylene glycol) prior to anion-exchange chromatography of plasmid DNA with a Q-Sepharose adsorbent has claimed to increase recovery from 20% to 80% (Horn, Budahazi, and Marquet 1998).

Another possible solution to the diffusion-capacity conundrum is to design chromatography operations to run in a negative mode (Table 13.2). With such a strategy, plasmids are no longer required to enter in and out of chromatographic beads, but should simply travel with the flow through the extra particle spaces in the bed. The target impurities, on the other hand, should now diffuse into the beads and adequately bind to the available surfaces. This may be problematic in the case of the largest RNA and genomic DNA fragments that have sizes and diffusion coefficients on the order of magnitude of those found for plasmids. Hydrophobic interaction chromatography (HIC) has been explored in this context (Diogo et al. 2001; Diogo, Queiroz, and Prazeres 2005). By adequately manipulating conditions, like the ligand type and elution buffers, it is possible to promote the binding of RNA, genomic DNA, and lipopolysaccharides. As for plasmids, their low hydrophobicity prevents them from binding. Thus, plasmids can be purified with negative HIC by loading feed solutions with a high concentration of an adequate salt and subsequently performing step or gradient elution with low salt to remove bound impurities (Diogo et al. 2000, 2001).

13.4.3 Macroporous Matrices

13.4.3.1 Superporosity

The best way to overcome the capacity limitations described above is to engineer stationary phases to accommodate normal diffusion pores (300 Å) with very large pores, with diameters in the 0.2 μm to 30 μm range (Lyddiatt and O'Sullivan 1998; Table 13.2). Such large pores have been engineered into nonbeaded materials like the Mustang® chromatographic membranes from Pall Corporation (New York, USA) and the CIM® monolithic supports from Bia Separations (Klagenfurt, Austria). The improved plasmid binding capacities obtained with these anion-exchange monoliths and membranes (up to 10.0 mg/ml in either case) have been ascribed to the high accessibility made possible by the presence of 0.8 μm pores, in the case of Mustang membranes (Teeters et al. 2003), and to the highly interconnected network of 0.01–4 μm channels, in the case of CIM monoliths (Branovic et al. 2004). The application of these matrices to plasmid purification will not be examined in detail here since this review is focused on chromatographic materials of the classic bead type.

13.4.3.1.1 Perfusion Supports

Perfusion chromatography supports like the polystyrene-based POROS® beads (Applied Biosystems, Foster City, California) meet some of features that are desirable for plasmid chromatography applications, even though they were not developed explicitly with the purpose of enhancing plasmid binding capacities (Whitney et al. 1998). Rather, the introduction of 0.6–0.8 μm throughpores was designed to enable intrapore convective flow to take place and thus accelerate the distribution of solutes like proteins to the network of smaller (500–1,000 Å) diffusive pores. The major advantage of perfusion supports in the context of protein purification is the ability to perform separations in much shorter times (10–100 fold) due to the enhanced mass transfer that is brought about by intrapore convection. When used in plasmid chromatography, however, perfusion supports like POROS 50HQ make it possible to bind between 2 to 10 mg of plasmids per ml of adsorbent (Perspective Biosystems 1996; Levy et al. 2000; Eon-Duval and Burke 2004). This enhanced capacity has been attributed to the improved accessibility of plasmids to the bead interior, which is made possible by the larger throughpores (McCoy, Kalghatgi, and Afeyan 1996; Theodossiou, Sondergaard, and Thomas 2001). The dimensions of plasmid molecules relative to the diameter of the diffusion pores (500–1,000 Å) makes it clear that plasmids essentially bind to the walls of the large pores.

13.4.3.1.2 Cell Microcarriers

Cell microcarriers are another type of beaded material that feature extremely large pores. Such materials are typically suspended in culture media by gentle agitation and provide large surface areas for the attachment of cells.

This makes the large-scale culture of anchorage-dependent animal cells possible in a way that is efficient, practical, and economical (Economical microcarrier cell culture; product brochure 18113704AC, GE Healthcare Uppsala, Sweden). Cytopore® microcarriers (GE Healthcare, Uppsala, Sweden), for instance, have a sponge-like network structure in which almost 95% of the internal volume is available for cell growth. Cytopore® beads (230 μm) are made of hydrophilic and hydrated macroporous cross-linked cellulose with positively charged N,N-diethyl-aminoethyl groups. The material features superpores with diameters of the order of 30 μm and a surface area of 1.1 m²/g (Anon. 2005). Cell microcarriers are usually rather fragile and thus cannot be used in chromatographic fixed beds due to poor mechanical stability. Despite this inadequacy, plasmid binding experiments have been performed with Cytopore®, yielding static capacities of 1,240 mg/g (~31 mg/ml; Ferreira, Prazeres, and Cabral 2003). This extraordinary capacity, which is approximately one order of magnitude higher when compared with the capacity of conventional chromatographic supports, is associated with a deep penetration of plasmids into the superpores of Cytopore® beads. Confocal laser microscopy examination of Cytopore® beads demonstrates this very clearly, as seen in the series of successive slices of the bead shown in the bottom part of Figure 13.5 (Stadler, Lemmens, and Nyhammar 2004).

13.4.3.1.3 *Superporous Agarose*

Superporous agarose particles were developed specifically for chromatographic applications in the late 90s (Gustavsson and Larsson 1996; Gustavsson, Axelsson, and Larsson 1998). Like the POROS® materials, these supports combine the normal diffusion pores (~300 Å), which are characteristic of agarose materials with superpores of several micrometers in diameter (Gustavsson and Larsson 1996). The superporous structure is engineered into the agarose beads by using a double emulsification procedure (Gustavsson and Larsson 1996). An adequate manipulation of the conditions used during this process enables the preparation of superporous beads with different size distributions: 45–75 μm, 106–180 μm, 180–300 μm, and 300–500 μm have all been reported (Gustavsson, Axelsson, and Larsson 1998; Tiainen et al. 2007). Experimental data shows that the superpores in these particles are evenly distributed, accounting for around 40% of the bead volume. The exact superpore dimensions depend on the specific conditions used to prepare the beads. Average diameters of 2 μm, 4 μm, 20 μm, and 30 μm and a typical tortuosity around 1.2 have been reported (Gustavsson and Larsson 1996; Gustavsson, Axelsson, and Larsson 1998; Tiainen et al. 2007). The grafting of polyethyleneimine, phenyl, and protein A ligands onto the beads makes it possible to use these materials in ion-exchange (Gustavsson and Larsson, 1996), hydrophobic interaction (Gustavsson, Axelsson, and Larsson 1999), and affinity (Gustavsson et al. 1997) chromatography, respectively. The scanning electron microscopy photographs in Figure 13.6c and 13.6d show the network of regularly spaced superpores very well, which is characteristic of such superporous agarose beads (Tiainen et al. 2007). Homogeneous agarose beads with

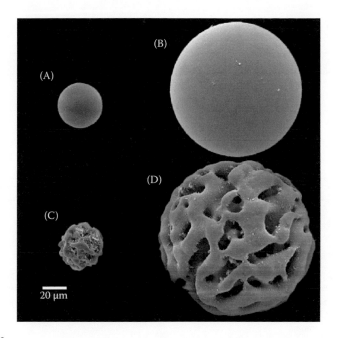

FIGURE 13.6
Scanning electron microscopy photographs of (A) a 45–75 μm homogenous agarose bead, (B) a 106–180 μm homogenous agarose bead, (C) a 45–75 μm agarose bead with 4 μm superpores, and (D) a 106–180 μm agarose bead with 20 μm superpores. The scale bar is 20 μm. (From Tiainen, P., Ljunglöf, A., Gustavsson, P.-E. and Larsson, P.-O., *J. Chromatogr. A*, 1138, 84–94, 2007. With permission.)

normal diffusion pores are shown also in Figure 13.6a and 13.6b for comparison. Superpores improve the access of plasmid and other very large molecules to the internal voids. Figure 13.7 illustrates this situation by comparing a model superpore with 5 μm diameter and 20 μm length with the typical 5,000 bp plasmid ($L = 6,900$ Å). Normal 300 Å diffusive pores are also schematized perpendicularly to the superpore. In this case, accessibility is clearly not an issue and thus higher binding capacities are expected. However, the experimental data available regarding the use of superporous agarose beads in plasmid chromatography shows that the gains in capacity have been very modest (3.0–3.9 mg/ml), at least when using beads with 2 μm or 4 μm superpores to bind a 7,000 bp plasmid (Tiainen et al. 2007). An explanation for this can be found in the fact that an increase in the diameter of superpores leads to a drop in the pore wall surface area available for plasmid binding (Tiainen et al. 2007).

13.4.3.1.4 *Superporous Methacrylate Beads*
Polymeric beads of ethylene dimethacrylate and glycidyl methacrylate with an average diameter in the 29–54 μm range have also been prepared, which combine 0.5–7.3 μm superpores (23% of the bead volume) with normal 20–1,000 Å diffusion pores (31% of the bead volume). The preparation of the

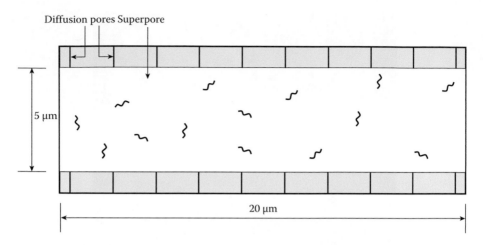

FIGURE 13.7
Schematic representation of the relative dimensions and accessibility of a typical 5,000 bp plasmid to a superpore with 5 μm diameter and 20 μm length. Normal 30 nm wide diffusion pores are also shown perpendicular to the superpore. The supercoiled plasmid is represented as a thick line (not proportional to plasmid diameter) with a length roughly proportional to the superhelix axis.

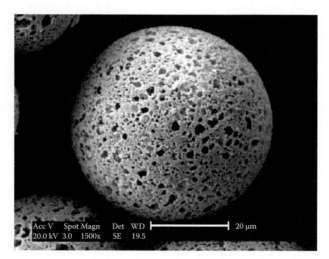

FIGURE 13.8
A scanning electron micrograph of a polymethacrylate biporous microsphere prepared by Sun and coworkers. (From Li, Y., Dong, X.-Y., and Sun, Y., *J. Appl. Polymer Sci.*, 104, 2205–11, 2007. With permission.)

beads is described in detail elsewhere (see Li, Dong, and Sun 2007). Specific areas of 20 m²/g have been reported for these particles (Li, Dong, and Sun 2005). As in the case of the previous superporous materials, the structure of the beads allows for intraparticle convective flow of the mobile phase. Scanning electron micrographs (Figure 13.8) of these materials indicates that the sphericity of a typical particle is close to 1. The figure also clearly shows that the matrix is full of micron-sized pores (Li, Dong, and Sun 2007). It is also interesting to notice that the topography and general aspect of this particle is quite different when compared with the superporous agarose particle shown in Figure 13.6d. These superporous methacrylate beads have been used in plasmid chromatography, but for operations run in a negative mode that rely on the binding of impurities and not of the target molecule (see Section 13.4.2). Further information is given in Section 13.4.3.3.

13.4.3.2 Diffusion and Convection in Superpores

Superporosity does not by itself improve diffusive mass transport of plasmids significantly, even though hindered diffusion due to pore wall proximity is less likely to occur. This is illustrated in Figure 13.9, which shows estimates of the internal diffusion time, t_D, of a 5,000 bp plasmid ($D = 3.37 \times 10^{-8}$ cm^2 s^{-1}, Equation 13.5) in hypothetical superporous particles with diameters d_p, in the 50 to 500 µm range. The diameter of superpores in such particles is assumed to be sufficient to allow unhindered diffusion to occur. The diffusion times were estimated according to (Gustavsson and Larsson 1999)

$$t_D = \frac{d_p^2}{4D}.$$

(13.6)

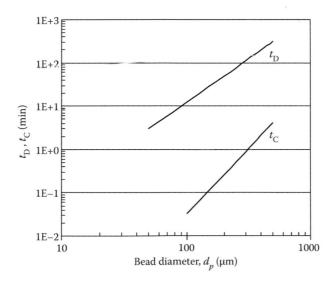

FIGURE 13.9
Characteristic diffusion and convection times of a 5,000 bp plasmid ($D = 3.4 \times 10^{-8}$ cm^{2}s^{-1}, Equation 13.5) in the 30 µm superpores of agarose beads with different diameters. An interstitial velocity of 1 cm/min and a bed porosity of 0.4 are further assumed. The characteristic times were estimated with Equations 13.6 and 13.9, as described in the text.

Figure 13.9 shows very clearly that diffusion times become very large for particles with diameters larger than 100 μm. Fortunately, the large dimensions of superpores allow intrapore convective flow to take place (Gustavsson and Larsson 1996; McCoy, Kalghatgi, and Afeyan 1996). Thus, liquid is able to flow right across the chromatographic particles in a packed bed via the superpores or flow pores, with a velocity that is similar in magnitude to the interstitial velocity of the mobile phase. The intraparticle convection times can then be estimated according to (Gustavsson and Larsson 1999):

$$t_C = \frac{3}{2}\frac{d_p}{v_{sp}}. \tag{13.7}$$

Gustavsson and coworkers have shown that for superporous agarose beads, the ratio of superpore fluid velocity, v_{sp}, relative to the interstitial velocity, v_i, in a column with an extra-particle porosity, ε_c, of 0.4 can be determined by the relationship (1999)

$$\frac{v_{sp}}{v_i} = 5.06\left(\frac{d_{sp}}{d_p}\right)^2, \tag{13.8}$$

where d_{sp} is the superpore diameter and d_p is the particle diameter. For instance, the fluid velocities in 30 μm superpores have been experimentally determined to be 3, 7, and 12% of the interstitial velocity in the cases of 300–500 μm, 180–300 μm, and 106–180 μm beads, respectively. Combining Equation 13.7 and 13.8 it is then possible to estimate the intraparticle convection times according to

$$t_C = 0.3\frac{d_p^3}{v_i d_{sp}^2}. \tag{13.9}$$

Figure 13.9 also shows estimates of this intraparticle convection time for a case in which a typical interstitial fluid velocity, v_i, of 1 cm/min and a typical superpore diameter, d_{sp}, of 30 μm are assumed. It is clear from the figure that in this case t_C values are around two orders of magnitude lower than t_D values. The significant reduction in the time required for plasmids to access the interior of particles, which is brought about by the existence of intraparticle convection inside the superpores, is obviously a function of the superpore diameter. This effect can be estimated by computing the ratio between the internal diffusion time (Equation 13.6) and the intraparticle convection time (Equation 13.9)

$$\lambda = 0.843\frac{v_i d_{sp}^3}{D d_p^2}. \tag{13.10}$$

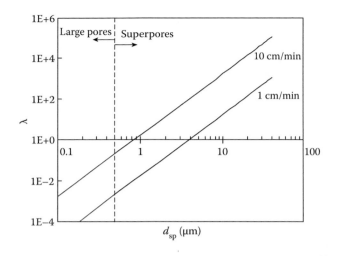

FIGURE 13.10
Relative magnitude of intraparticle convective and diffusive transport of a 5,000 bp plasmid in a 120 μm superporous agarose bead. The intraparticle Peclet number, λ, was calculated with Equation 13.10 and for two linear flow rates, for two interstitial velocities, 0.1 cm/min and 10 cm/min.

The dimensionless quantity λ, is known as the intraparticle Peclet number. Figure 13.10 shows estimates of λ for the case of the intraparticle transport of a 5,000 bp plasmid in 120 μm superporous particles packed in a bed with porosity equal to 0.4. The diameter of the pores–superpores is assumed to vary between 0.1 μm and 20 μm. The calculations are presented for two extreme interstitial velocities of 0.1 and 10 cm/min. The figure shows that intraparticle transport of a 5,000 bp plasmid becomes important (i.e., $\lambda > 1$) when the diameter of superpores is higher than 1 μm and 4 μm for interstitial velocities of 10 cm/min and 1 cm/min, respectively. In particles with lower pore dimensions, diffusive transport will dominate and chromatographic separations may be expected to take longer.

13.4.3.3 Plasmid Purification

13.4.3.3.1 Superporous Agarose

The ability of superporous agarose beads to purify plasmids directly from an RNase-treated alkaline lysate was reported recently (Tiainen et al. 2007). A 5 mm i.d. column was packed to a bed height of 3.7 cm with 45–75 μm agarose beads with 4 μm superpores and Q (quaternary amine) ligands. Following equilibration with a 0.4 M NaCl buffer, 10 ml of a clarified lysate with a conductivity matching that of the mobile phase were injected. The typical chromatogram (Figure 13.11) and supporting analysis of the collected fractions show that the majority of low molecular weight RNA species (97%)

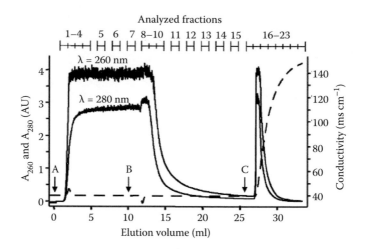

FIGURE 13.11

Anion-exchange chromatography purification of a 6,100 bp plasmid from an RNase-treated, clarified lysate using superporous agarose beads. Chromatography bead type: 45–75 μm agarose beads with 4 μm superpores and Q (quaternary amine) ligands. The dotted line shows the conductivity. Arrow A: injection of 10 ml, diluted (46%) clarified lysate. Arrow B: injection completed, washing starts. Arrow C: start of elution with 2M NaCl. (From Tiainen, P., Ljunglöf, A., Gustavsson, P.-E. and Larsson, P.-O., *J. Chromatogr. A*, 1138, 84–94, 2007. With permission.)

elute in the flowthrough. As for plasmid, it exited the column as a single peak from the column by stepping up the NaCl concentration in the elution buffer to 2M (see Figure 13.11). The authors further report that 59% of the adsorbed plasmid material was recovered (Tiainen et al. 2007).

13.4.3.3.2 Superporous Methacrylate Beads

Plasmids can be purified with the biporous polymeric beads developed by Sun and coworkers (Li, Dong, and Sun 2005, 2007). In the two cases described in the literature, the authors have opted to develop a chromatographic separation based on a negative mode of operation. In the first case the beads were coupled with phenyl ligands (Li, Dong, and Sun 2005) and a HIC operation was run in a similar fashion to the one described in previous reports (Diogo et al. 2000). Briefly, a 5 mm I.D. column was packed with the beads and equilibrated with a 1.0 M $(NH_4)_2SO_4$ buffer. Next, a given volume of a preconcentrated (with isopropanol precipitation) and prepurified (with 2.5 M $(NH_4)_2SO_4$ precipitation) lysate was injected, the column was washed with 1.0 M $(NH_4)_2SO_4$, and elution was performed with a $(NH_4)_2SO_4$-free buffer. Analysis of the successive fractions collected at the column outlet indicates that the separation pattern obtained with the superporous beads is qualitatively identical to the separation obtained with conventional phenyl-Sepharose beads (Diogo et al. 2001; Diogo, Queiroz, and Prazeres 2005): plasmid DNA remains unbound and elutes in the flowthrough at 1 M $(NH_4)_2SO_4$, whereas the strong hydrophobic interactions between RNA, protein, and lipopolysaccharide impurities with the phenyl ligands can only be broken with a $(NH_4)_2SO_4$-free buffer (Li, Dong, and Sun 2005). In a more recent application mercaptopyridine ligands were grafted into the superporous beads (Li, Dong, and Sun 2007). Although the ligand was now different, the chromatographic operation (feedstock, buffers, and elution strategy) and its outcome (plasmid in the flowthrough and impurities bound to the matrix) were virtually identical to the ones described for the phenyl bioporous sorbent (Li, Dong, and Sun 2007). The dynamic capacity of the phenyl and mercaptopyridine for *E. coli* RNA was estimated to be approximately equal to 2.5 mg/ml (Li, Dong, and Sun 2005) and 3.1 mg/ml (Li, Dung, and Sun 2007), respectively.

13.5 Discussion

The extremely large size and awkward shape of plasmid molecules are responsible for most of the problems faced by the conventional fixed bed chromatography used in the context of plasmid purification. Such problems are essentially related to mass transport—most pores in conventional beads (300 Å) are inaccessible to plasmids and small diffusion coefficients translate into separation inefficiencies. The capacity and internal mass transport limitations faced by conventional chromatographic materials can be partially overcome by engineering micrometer-sized superpores into materials like agarose or poly methacrylates. The highly interconnected networks of such superpores provide an appreciable internal surface area for plasmid binding. Furthermore, the dimensions of the pores are such that intraconvective transport of plasmid molecules become possible. The concomitant rapid mass transfer translates directly into faster separations and lower band broadening effects. However, and in spite of the significant progress made, capacities remain disappointingly low. Most likely, the problem relates to the fact that the larger the superpores, the lower the surface area for binding becomes. Thus, while increasing the dimension of pores in a first instance might increase access and improve plasmid binding capacities, after a certain critical superpore size, capacities are likely to drop due to the inherent decrease in the binding area available per ml of bead material. Although this critical superpore size probably depends on the material being crafted, this author speculates here that it probably falls in the 2 μm to 5 μm range.

Acknowledgment

This work was partially funded by FCT, the Portuguese Foundation for Science and Technology (PTDC/EQU-EQU/65492/06).

References

Anonymous 2005. Economical microcarrier cell culture. In product brochure 18113704AC (GE Healthcare, Uppsala, Sweden)

Bergan, D., Galbraith, T., and Sloane, D. L. 2000. Gene transfer *in vitro* and *in vivo* by cationic lipids is not significantly affected by levels of supercoiling of a reporter plasmid. *Pharm. Res.* 17: 967–73.

Boles, T. C., White, J. H., and Cozzarelli, N. R. 1990. Structure of plectonemically supercoiled DNA. *J. Mol. Biol.* 213:931–51.

Branovic, K., Forcic, D., Ivancic, J., Strancar, A., Barut, M., Kosutic Gulija, T., Zgorelec, R., and Mazuran, R. 2004. Application of short monolithic columns for fast purification of plasmid DNA. *J. Chromatogr. B.* 801:331–37.

Carlson, A., Signs, M., Liermann, L., Boor, R., and Jem, K. 1995. Mechanical disruption of *Escherichia coli* for plasmid recovery. *Biotechnol. Bioeng.* 48:303–15.

Carnes, A. E. 2005. Fermentation design for the manufacture of therapeutic pDNA. *BioProcess Int.* 3(9):36–42.

Chirico, G., and Baldini, G. 1996. Rotational diffusion and internal motions of circular DNA. I. Polarized photon correlation spectroscopy. *J. Chem. Phys.* 104:6009–19.

Cupillard, L., Juillard, V., Latour, S., Colombet, G., Cachet, N., Richard, S., Blanchard, S., and Fischer, L. 2005. Impact of plasmid supercoiling on the efficacy of a rabies DNA vaccine to protect cats. *Vaccine* 23:1910–16.

DePhillips, P., and Lenhoff, A. M. 2001. Determinants of protein retention characteristics on cation-exchange adsorbents. *J. Chromatogr. A* 933:57–72.

Diogo, M. M., Queiroz, J. A., Monteiro, G. A., Martins, S. A. M., Ferreira, G. N. M., and Prazeres, D. M. F. 2000. Purification of a cystic fibrosis plasmid vector for gene therapy using hydrophobic interaction chromatography. *Biotechnol. Bioeng.* 68:576–83.

Diogo, M. M., Queiroz, J. A., and Prazeres, D. M. F. 2005. Chromatography of plasmid DNA. *J. Chromatogr. A* 1069:3–22.

Diogo, M. M., Ribeiro, S., Queiroz, J. A., Monteiro, G. A., Tordo, N., Perrin, P., and Prazeres, D. M. F. 2001. Production, purification and analysis of an experimental DNA vaccine against rabies. *J. Gene Med.* 3:577–84.

Drape, R. J., Macklin, M. D., Barr, L. J., Jones, S., Haynes, J. R., and Dean, H. J. 2006. Epidermal DNA vaccine for influenza is immunogenic in humans. *Vaccine* 24:4475–81.

Edelstein, M. L., Abedi, M. R., and Wixon, J. 2007. Gene therapy clinical trials worldwide to 2007: An update. *J. Gene Med.* 9:833–42.

Eon-Duval, A., and Burke, G. 2004. Purification of pharmaceutical-grade plasmid DNA by anion-exchange chromatography in an RNase-free process. *J. Chromatogr. B* 804:327–35.

Eon-Duval, A., MacDuff, R. H., Fisher, C. A., Harris, M. J., and Brook, C. 2003. Removal of RNA impurities by tangential flow filtration in an RNase-free plasmid DNA purification process. *Anal. Biochem.* 316:66–73.

Ferreira, G. N. M., Cabral, J. M. S., and Prazeres, D. M. F. 2000. Studies on the batch adsorption of plasmid DNA onto anion-exchange chromatographic supports. *Biotechnol. Prog.* 16:416–24.

Ferreira, G. N. M., Prazeres, D. M. F., and Cabral, J. M. S. 2003. Processo cromatográfico para purificação de DNA plasmídico utilizando suportes superporosos. Portuguese patent PT 102,341. Instituto Superior Técnico.

Fishman, D. M., and Patterson, G. D. 1996. Light scattering studies of supercoiled and nicked DNA. *Biopolymers* 38:535–52.

Forde, G. M. 2005. Rapid-response vaccines: Does DNA offer a solution? *Nat. Biotechnol.* 23:1059–62.

Freitas, S. S., Santos, J. A., and Prazeres, D. M. F. 2006a. Optimization of isopropanol and ammonium sulfate precipitation steps in the purification of plasmid DNA. *Biotechnol. Prog.* 22:1179–86.

Freitas, S. S., Santos, J. A. L., and Prazeres, D. M. F. 2006b. Plasmid DNA production. In *Development of Sustainable Bioprocesses,* eds. E. Heinzle, A. Biwer, and C. Cooney, 271–85. West Sussex: John Wiley & Sons, Ltd.

Freitas, S., Santos, J.A.L., Prazeres, D.M.F. 2009. Plasmid purification by hydrophobic interaction chromatography using sodium citrate in the mobile phase", *Sep. Pur. Technol.* 65: 95–104.

Gurunathan, S., Klinman, D. M., and Seder, R. A. 2000. DNA vaccines: immunology, application, and optimization. *Annu. Rev. Immunol.* 18:927–74.

Gustavsson, P.-E., Axelsson, A., and Larsson, P.-O. 1998. Direct measurements of convective fluid velocities in superporous agarose beads. *J. Chromatogr. A* 795: 199–210.

Gustavsson, P.-E., Axelsson, A., and Larsson, P. O. 1999. Superporous agarose beads as a hydrophobic interaction chromatography support. *J. Chromatogr. A* 830:275–84.

Gustavsson, P.-E., and Larsson, P.-O. 1996. Superporous agarose, a new material for chromatography. *J. Chromatogr. A* 734:231–40.

Gustavsson, P.-E., and Larsson, P.-O. 1999. Continuous superporous agarose beds for chromatography and electrophoresis. *J. Chromatogr. A* 832:29–39.

Gustavsson, P.-E., Mosbach, K., Nilsson, K., and Larsson, P. O. 1997. Superporous agarose as an affinity chromatography support. *J. Chromatogr. A* 776:197–203.

Hammermann, M., Brun, N., Klenin, K., May, R., Toth, K., and Langowski, J. 1998. Salt-dependent DNA superhelix diameter studied by small angle neutron scattering measurements and Monte Carlo simulations. *Biophys. J.* 75:3057–63.

He, L., and Niemeyer, B. 2003. A novel correlation for protein diffusion coefficients based on molecular weight and radius of gyration. *Biotechnol. Prog.* 19:544–48.

Hoare, M., Levy, M. S., Bracewell, D. G., Doig, S. D., Kong, S., Titchener-Hooker, N., Ward, J. M., and Dunnill, P. 2005. Bioprocess engineering issues that would be faced in producing a DNA vaccine at up to 100 m³ fermentation scale for an influenza pandemic. *Biotechnol. Prog.* 21:1577–92.

Horn, N., Budahazi, G., and Marquet, M. 1998. Purification of plasmid DNA during column chromatography. US patent 5,707,812. Vical Incorporated, San Diego.

Huber, C. G. 1998. Micropellicular stationary phases for high-performance liquid chromatography of double-stranded DNA. *J. Chromatogr. A* 806:3–30.

Lander, R. J., Winters, M. A., Meacle, F. J., Buckland, B. C., and Lee, A. L. 2002. Fractional precipitation of plasmid DNA from lysate by CTAB. *Biotechnol. Bioeng.* 79:776–84.

Lemmens, R., Olsson, U., Nyhammar, T., and Stadler, J. 2003. Supercoiled plasmid DNA: selective purification by thiophilic/aromatic adsorption. *J. Chromatogr. B* 784:291–300.

Levy, M., Collins, I., Yim, S., Ward, J., Titchener-Hooker, N., Shamlou, P., and Dunnill, P. 1999. Effect of shear on plasmid DNA in solution. *Bioprocess Eng.* 20:7–13.

Levy, M. S., O'Kennedy, R. D., Ayazi-Shamlou, P., and Dunnill, P. 2000. Biochemical engineering approaches to the challenges of producing pure plasmid DNA. *Trends Biotechnol.* 18:296–305.

Li, Y., Dong, X.-Y., and Sun, Y. 2005. High-speed chromatographic purification of plasmid DNA with a customized biporous hydrophobic adsorbent. *Biochem. Eng. J.* 27:33–39.

Li, Y., Dong, X.-Y., and Sun, Y. 2007. Biporous polymeric microspheres coupled with mercaptopyridine for rapid chromatographic purification of plasmid DNA. *J. Appl. Polymer Sci.* 104:2205–11.

Ljunglöf, A., Bergvall, P., Bhikhabhai, R., and Hjorth, R. 1999. Direct visualisation of plasmid DNA in individual chromatography adsorbent particles by confocal scanning laser microscopy. *J. Chromatogr. A* 844:129–35.

Lyddiatt, A., and O'Sullivan, D. 1998. Biochemical recovery and purification of gene therapy vectors. *Curr. Opin. Biotechnol.* 9:177–85.

McCoy, M., Kalghatgi, K., and Afeyan, N. 1996. Perfusion chromatography: Characterization of column packings for chromatography of proteins. *J. Chromatogr. A* 743:221–29.

Merial. 2007. *USDA grants conditional approval for first therapeutic vaccine to treat cancer.* United Kingdom: Merial Limited.

Murphy, J. C., Cano, T., Fox, G. E., and Willson, R. C. 2006. Compaction agent protection of nucleic acids during mechanical lysis. *Biotechnol. Prog.* 22:519–22.

Murphy, J. C., Wibbenmeyer, J. A., Fox, G. E., and Willson, R. C. 1999. Purification of plasmid DNA using selective precipitation by compaction agents. *Nat. Biotechnol.* 17:822–23.

Neidle, S. 1999. *Oxford handbook of nucleic acid structure.* Oxford: Oxford Science Publications.

Novartis. 2005. *Novel Novartis vaccine to protect Canadian salmon farms from devastating viral disease.* Basel: Novartis Animal Health Inc.

Olson, W. K. 1999. DNA higher-order structures. In *Oxford handbook of nucleic acid structure*, ed. S. Neidle, 499–531. Oxford: Oxford Science Publications.

Perseptive-Biosystems. 1996. Rapid, preparative purification of plasmid DNA by anion exchange perfusion chromatography technology. *Biochemica* 1:9–11.

Powell, K. 2004. DNA vaccines-back in the saddle again? *Nat. Biotechnol.* 22:799–801.

Prazeres, D. M. F. 2008. Prediction of diffusion coefficients of plasmids. *Biotechnol. Bioeng.* 99:1040–44.

Prazeres, D. M. F., Ferreira, G. N. M., Monteiro, G. A., Cooney, C. L., and Cabral, J. M. S. 1999. Large-scale production of pharmaceutical-grade plasmid DNA for gene therapy: problems and bottlenecks. *Trends Biotechnol.* 17:169–74.

Prazeres, D. M. F., Monteiro, G. A., Ferreira, G. N. M., Diogo, M. M., Ribeiro, S. C., and Cabral, J. M. S. 2001. Purification of plasmids for gene therapy and DNA vaccination. In *Biotechnology Annual Review,* ed. M. R. El-Gewely, Vol. 7, 1–30. Netherlands: Elsevier.

Prazeres, D. M. F., Schluep, T., and Cooney, C. 1998. Preparative purification of supercoiled plasmid DNA using anion-exchange chromatography. *J. Chromatogr. A* 806:31–45.

Schnabel, R., Langer, P., and Breitenbach, S. 1988. Separation of protein mixtures by Bioran® porous glass membranes. *J. Membrane Sci.* 36:55–66.

Seils, S. S., and Pecora, R. 1990. A dynamic light scattering study of four DNA restriction fragments. *Macromolecules* 487–97.

Sinden, R. R. 1994. *DNA structure and function.* San Diego, CA: Academic Press.

Stadler, J., Lemmens, R., and Nyhammar, T. 2004. Plasmid DNA purification. *J. Gene Med.* 6:S54–S66.

Teeters, M. A., Conrardy, S. E., Thomas, B. L., Root, T. W., and Lightfoot, E. N. 2003. Adsorptive membrane chromatography for purification of plasmid DNA. *J. Chromatogr. A* 989:165–73.

Theodossiou, I., Sondergaard, M., and Thomas, O. R. 2001. Design of expanded bed supports for the recovery of plasmid DNA by anion exchange adsorption. *Bioseparation* 10:31–44.

Tiainen, P., Gustavsson, P.-E., Ljunglöf, A., and Larsson, P.-O. 2007. Superporous agarose anion exchangers for plasmid isolation. *J. Chromatogr. A* 1138:84–94.

Trindade, I. P., Diogo, M. M., Prazeres, D. M. F., and Marcos, J. C. 2005. Purification of plasmid DNA vectors by aqueous two-phase extraction and hydrophobic interaction chromatography. *J. Chromatogr. A* 1082:176–84.

Tyn, M., and Gusek, T. 1990. Prediction of diffusion coefficients of proteins. *Biotechnol. Bioeng.* 35:327–38.

Ulmer, J. B., Donnelly, J. J., Parker, S. E., Rhodes, G. H., Felgner, P. L., Dwarki, V. J., Gromkowski, S. H., Deck, R. R., DeWitt, C. M., and Friedman, A., et al. 1993. Heterologous protection against influenza by injection of DNA encoding a viral protein. *Science* 259:1745–49.

Vologodskii, A. V., and Cozzarelli, N. R. 1994. Conformational and thermodynamic properties of supercoiled DNA. *Annu. Rev. Biophys. Biomol. Struct.* 23:609–43.

Vologodskii, A. V., Levene, S. D., Klenin, K. V., Frank-Kamenetskii, M., and Cozzarelli, N. R. 1992. Conformational and thermodynamic properties of supercoiled DNA. *J. Mol. Biol.* 227:1224–43.

Voordouw, G., Kam, Z., Borochov, N., and Eisenberg, H. 1978. Isolation and physical studies of intact supercoiled, open circular and linear forms of CoIE1-plasmid DNA. *Biophys. Chem.* 8:171–89.

Whitney, D., McCoy, M., Gordon, N., and Afeyan, N. 1998. Characterization of large-pore polymeric supports for use in perfusion biochromatography. *J. Chromatogr. A* 807:165–84.

Wolff, J. A., Malone, R. W., Williams, P., Chong, W., Acsadi, G., Jani, A., and Felgner, P. L. 1990. Direct gene transfer into mouse muscle *in vivo*. *Science* 247:1465–68.

14

Cryogels as Matrices for Cell Separation and Cell Cultivation

Maria B. Dainiak, Ashok Kumar, Igor Yu. Galaev, and Bo Mattiasson

CONTENTS

14.1 Introduction

From the point of view of polymer chemistry, macroporous hydrogels prepared by cryotropic gelation have existed for some time. However, their potential for biotechnological and biomedical applications has been realized only recently (Lozinsky et al. 2003; Dainiak, Galaev, et al. 2007; Dainiak, Mattiasson, and Galaev 2007). In particular, it has been demonstrated that cryogels are highly suitable as matrices for cell separation and for cell culture applications due to the combination of elastic, three-dimensional (3-D), polymeric network and unique interconnective highly porous structure capable of absorbing and keeping large a amount of water or biologic fluids inside the pores as well as ensuring nonhindered mass transport of nutrients and metabolic products.

Cryogels are produced via gel formation in polymer systems at subzero temperatures. The essential feature of cryogelation is crystallization of the solvent, during which growing and merging ice crystals perform as a porogen. The dissolved monomers or polymer precursors get concentrated in a nonfrozen liquid microphase, while the whole sample looks like a frozen block. After melting the ice crystals, a system of large continuous interconnected pores is obtained. The resulting materials have porosity up to 90%, pore size of 10–200 μm, and excellent mechanical properties (Lozinsky et al. 2003; Dainiak, Galaev, et al. 2007; Plieva, Galaev, and Mattiasson 2007). They can withstand large deformations and can be easily compressed down to 20% of their original volume without getting mechanically destroyed (Dainiak, Kumar, et al. 2006).

Cryogels derivatized with affinity ligands have been developed for chromatographic cell separation. Efficient and inexpensive techniques for the isolation of target cells are necessary to provide defined cell populations for analysis and subsequent cultivation. There is a growing need for specific isolation and characterization of stem cells for cell-based therapy, for detection of low numbers of tumor cells in blood for disease prognosis, for isolation and assessing functionality of T lymphocytes, and so on (De Wynter et al. 1995; Beaujean 1997; Prestvik et al. 1997; Racila et al. 1998; Collins, Luebering, and Shaut 1998; Siewert et al. 2001). Separation and recovery of rare cell types, such as stem cells, antigen-specific B and T cells, and rare circulating tumor cells is a challenging task because target cells may occur at frequencies below one per million (Gross et al. 1995). For example, the frequency of B cells specific for any particular antigen is usually less than 1% (Oshiba et al. 1994). The percentage of hematopoietic stem cells (HSC) is less than 0.5% of peripheral blood cells and is 1.5% of bone marrow mononuclear cells (Beaujean 1997). Apart from biomedicine and immunology, another wide area of applications of cell separation techniques is the isolation and detection of pathogenic microorganisms and protozoan parasites in various samples in food, clinical and environmental microbiology, and parasitology (Islam and Lindberg 1992; Payne et al. 1992; Deng, Cliver, and Mariam 1997; Seesod et al. 1997).

An important feature of cryogelation technology is the possibility to modulate a wide range of material properties such as porosity, mechanical strength, biochemistry of pore wall surface, and degradability. Cryogels can be prepared in different formats (i.e., as monolithic columns, sheets, beads), from any hydrophilic polymer that forms a gel. While synthetic, mechanically robust cryogels prepared from such polymers as polyacrylamide are good chromatographic matrices, cryogels prepared from natural polymers or derivatized with appropriate bioligands are suitable as scaffolds for various cell culture applications. The latter include tissue engineering, stem cell expansion, production of therapeutic proteins, and cell-based assays. The environmentally friendly way of producing cryogels (with no organic solvents involved) and efficient control over the pore size and architecture make this approach favorable compared to other techniques, such as solvent casting-particulate leaching or phase separation combined with freeze-drying, which are used for production of macroporous scaffolds.

14.2 Affinity Cryogels for Cell Separation

The performance of cell separation is typically characterized by three parameters: throughput, purity, and recovery. Commercial applications of the existing affinity techniques for cell separation that is, immunoadsorption, fluorescence-activated cell sorting (FACS), and magnetic-activated cell sorting (MACS), have distinct advantages and disadvantages with respect to these parameters. The low yields have been a common problem of cell isolation procedures using immunoadsorbents due to the difficulties to recover bound cells without effecting cell viability and function (Haas, Schrader, and Szenberg 1974; Bell 1978). The FACS has some major limitations, such as losses in the yield, the requirement for special technical expertise, and a comparatively low throughput (Ibrahim and Van Den Engh 2003). Purity and recovery in MACS typically have large variances (Chalmers et al. 1998). Since both FACS and MACS modify cell membrane, neither technique is physiologically preferable for subsequent analysis or recultivation of the sorted cells (Seidl, Knuechel, and Kunz-Schughart 1999). As the demands placed on cell sorting technologies continue to increase, novel solutions to overcome these limitations have to be found. Application of novel monolithic matrices with properties different from those of conventional chromatographic materials is one of the possible solutions. Such matrices are now being increasingly used for liquid and gas chromatography.

The recent developments of monolithic adsorbents have shown tremendous potential in their application for the separation of nanoparticles. However, most of these matrices are produced from hydrophobic materials and have a limitation when dealing with separation processes of cell organelles or whole cells due to the nonspecific adsorption (Hentze and Antonietti 2002; Dainiak, Galaev, et al. 2007). Although the application of hydrophilic macroporous materials gives a possibility to overcome this limitation, the production of highly porous materials from hydrophilic polymers has not been addressed very much so far. The production of macroporous agarose gels by a double-emulsion procedure is one of the rare exceptions of successful production of hydrophilic macroporous material (Gustavsson and Larsson 1999), which is discussed in detail in the Chapter 7. However, the technology suffers from poor reproducibility and the necessity of intensive washing of the macroporous material to remove solvents and detergents used for double-emulsion preparation. The local concentration of the polymer in the pore walls is the same as when nonmacroporous agarose gel is formed. As hydrophilic polymers in aqueous environment bind a lot of water, the polymer phase in macroporous systems (walls of the macropores) has lower mechanical strength (rigidity) than the polymeric phase of macroporous materials formed by hydrophobic polymers. Thus macroporous materials produced from hydrophilic polymers by a double-emulsion procedure usually have poor mechanical characteristics and their application for chromatography is limited as these materials tend to collapse in chromatographic columns with increasing flow rates. Certainly, the mechanical properties of superporous agarose gels could be improved by chemical cross-linking of agarose chains.

Cryogelation process used for preparation of macroporous hydrogels involves compulsory displacement of polymeric precursors into nonfrozen microphase resulting in concentrated polymer gel in the pore walls (Figure 14.1), which ensures mechanical strength of these materials. The concentrated polymer gel swells poorly and ensures elasticity of cryogels even those produced from hydrophilic polymers such as poly(acrylamide) and agarose. In contrast to rather brittle polyacrylamide gels prepared at room temperatures, polyacrylamide cryogels can be easily compressed from four- to sixfold without getting mechanically damaged. Compressed cryogels re-swell almost instantaneously and restore their initial shape upon contact with liquid (Galaev et al. 2005; Dainiak, Kumar, et al. 2006). High elasticity makes it possible to compress and insert cryogel monoliths of an appropriate diameter into the wells of a standard 96-well plate (Galaev et al. 2005; Ahlqvist et al. 2006; Dainiak, Galaev, and Mattiasson 2007) or stack them as "building blocks" on top of each other inside a column when scale-up is needed (Dainiak, Galaev, et al. 2006; Dainiak et al. 2004; Figure 14.2). A 96-minicolumn plate with open-ended wells filled with cryogel monoliths

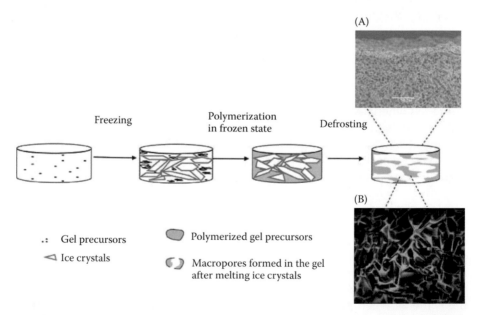

FIGURE 14.1
(See color insert following page 116.) Formation of cryogels. Insert graph: scanning electron micrograph of p(2-hydroxyethyl methacrylate cryogel, × 50 (A); porous structure of laminin-containing protein cryogel labeled with FITC and viewed by confocal laser scanning microscopy. (Reproduced from Dainiak, M. B., B. Mattiasson, and I. Yu. Galaev, *BioForum Europe*, 11, 28–29, 2007. With permission.)

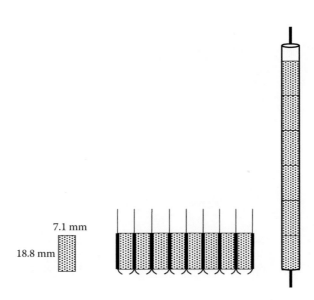

FIGURE 14.2
Schematic illustration of a "building block", cryogel monolith (18.8 mm × 7.1 mm diameter); 96-minicolumn plate with open-ended wells (7.0 mm diameter) filled with cryogel monoliths (only the first row of minicolumns out of 12 is presented) and a cryogel column (113 mm × 7.1 mm diameter) constructed using 6 cryogel monoliths. (Reproduced from Dainiak, M. B., I.Yu. Galaev, and B. Mattiasson, *J. Chromatogr. A*, 1123, 145–50, 2006. With permission.)

(18.8 mm × 7.1 mm) represents a system of 96 drainage-protected minicolumns and is ideal for parallel adsorption tests and screening for optimal separation conditions (Dainiak, Galaev, et al. 2006; Dainiak, Galaev, and Mattiasson 2007). Affinity cryogel monoliths in column format and in 96-minicolumn plate format have been used in chromatographic cell separations.

Large (up to 200 μm) interconnected pores in cryogels make it possible to apply these materials for chromatography of cells without a risk of their mechanical entrapment and clogging the column. Cells are rather fragile and sensitive to shear stress. Their diffusivity is negligible and only convective transport can be used. Despite the presence of large cells (up to 20 μm), a pulse of blood applied on a naked (underivatized) cryogel column is convectionally transported by the liquid flow through the column without substantial tailing (Figure 14.3A). The cells do not experience strong shear forces and are intact at the outlet of the cryogel column. There is no haemolysis in the cell-containing fractions (Figure 14.3B; Noppe et al. 2007).

Binding of cells to the cryogel matrix occurs if a suitable ligand has been immobilized on the matrix and is accessible to the cells passing it. Cells possessing the appropriate surface molecules will bind to the ligands (charges, hydrophobic moieties or affinity ligands) and will be removed from the solution. Various coupling chemistries can be used for the immobilization of ligands on polyacrylamide cryogel adsorbents. The ligand can be either coupled to a

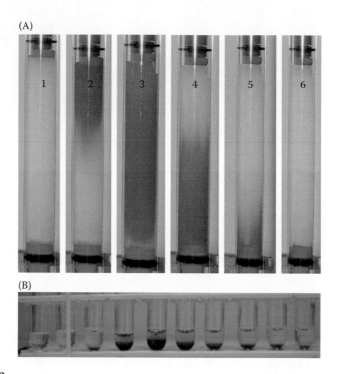

FIGURE 14.3
(See color insert following page 116.) Flow of the pulse of whole blood through an underivatized cryogel column. (A) One ml blood was applied to the naked cryogel column at a flow rate of 0.5 ml/min in isotonic buffer solution. Column 1: column before application; column 2–5: column during the run; column 6: column after the flow of blood sample. (B) Flow through fractions of the column were collected and left to stand. No red blood cells lysis was observed in the fractions. (Reproduced from Noppe, W., F. M. Plieva, K. Vanhoorelbeke, H. Deckmyn, M. Tuncel, A. Tuncel, I.Yu. Galaev, and B. Mattiasson, *J. Biotechnol.*, 131, 293–99, 2007. With permission.)

reactive group, for example epoxy group included in the matrix during its fabrication (Arvidsson et al. 2003; Dainiak et al. 2004; Dainiak et al. 2005; Dainiak, Galaev, and Mattiasson 2006), or coupled via a spacer (Kumar et al. 2003), or introduced by grafting polymeric chains onto cryogel surfaces followed by derivatization of the grafted polymer with required functionality (Savina, Galaev, and Mattiasson 2005; Savina, Mattiasson, and Galaev 2005; Savina, Mattiasson, and Galaev 2006). An alternative method is to copolymerize acrylamide with the molecules bearing the required functionality (Plieva, Bober, et al. 2006).

When cells are specifically bound to the column, an important question is whether the binding takes place in the whole volume of the column or the cells are accumulated only at some particular zones, for example only at the top. Ampicillin-resistant *Escherichia coli* cells were applied to an ion-exchange cryogel column at low ionic strength. Bacterial cells bear net negative charge on the surface and get bound to anion-exchanging cryogel (Arvidsson et al. 2002). The matrix with bound cells was taken out from the column, a central disc-shaped zone was cut out, and a few small pieces were taken from the central parts of the disks and placed on top of an agar plate containing ampicillin. After incubation overnight at 37°C, pronounced and approximately similar growth was observed around all pieces of the matrix exposed to the cells. This indicates cell binding inside the whole volume of the anion-exchange cryogel column. As a control, pieces of an unused column were placed on top of an agar plate without ampicillin and no growth occurred under the same conditions. Bound *E. coli* cells were also visualized using scanning electron microscopy (SEM; Figure 14.4). Clearly, bound cells are attached to the plain parts of the pore walls indicating specific interaction of cells rather than mechanical entrapment in dead flow zones (Arvidsson et al. 2002).

Different microbial cells are characterized by different cell surface properties. These differences can be exploited for separation of specific cells from mixed populations using cryogels derivatized with group-specific affinity ligands such as immobilized metal chelates. The screening using IMAC-(Me(II)-iminodiacetic)-cryogel monoliths in a 96-minicolumn plate format revealed significant cell type and growth phase-dependent differences in affinity to immobilized metal ions among three analyzed bacterial strains, wild type *E. coli*, recombinant *E. coli* cells displaying poly-His peptides (His-tagged *E. coli*) and haloalkalophilic *Bacillus halodurans* cells (Figure 14.5). Recombinant, as well as wild type *E. coli*, but not *B. halodurans* cells were shown to have affinity for Cu(II)-, Ni(II)-, and Zn(II)-IDA ligands in the following order: Cu(II) > Ni(II) > Zn(II) (Dainiak, Galaev, and

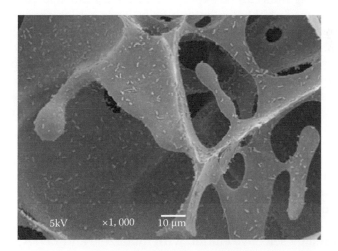

FIGURE 14.4
Scanning electron micrographs of anion-exchange cryogel with bound *E. coli* cells. The samples were fixed in 2.5% glutaraldehyde in 0.15 M sodium cacodylate buffer overnight, postfixed in 1% osmium tetroxide for 1 hour, dehydrated in ethanol and critical point dried. Dried samples were coated with gold/palladium (40/60) and examined using a JEOL JSM-5600LV scanning electron microscope. (Reproduced from Arvidsson, P., F. M. Plieva, I. N. Savina, V. I. Lozinsky, S. Fexby, L. Bulow, I.Yu. Galaev, and B. Mattiasson, *J. Chromatogr., A*, 977, 27–38, 2002. With permission.)

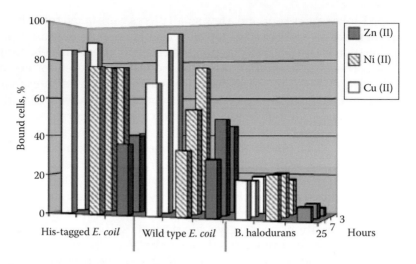

FIGURE 14.5
Binding of wild type *E. coli*, recombinant *E. coli* and *B. halodurans* cells to Cu(II)-, Ni(II)-, and Zn(II)-IDA-cryogel monoliths (0.5 ml, ligand density 23 μmol/ml adsorbent) as a function of cell fermentation time. Standard deviation of the experiment was five. Amount of applied cells was set at 100%. (Reproduced from Dainiak, M. B., I.Yu. Galaev, and B. Mattiasson, *Enzyme. Microb. Technol.*, 40, 688–95, 2007. With permission.)

Mattiasson 2007). The revealed poor affinity of *B. halodurans* to immobilized metal affinity ligands can be explained by the amino acid composition of the cell wall proteins of haloalkalophilic bacteria, which are enriched with acidic amino acids (Salamah 1992; Trotsenko and Khmelenina 2002). These residues play a role in binding Al(III), Ca(II), or Fe(III) ions, but practically do not contribute to binding Cu(II), Ni(II), and Zn(II) ions complexed with IDA chelator (Gaberc-Porekar and Menart 2001).

In contrast to *B. halodurans* cells, the surface of *E. coli* cells has accessible His, Arg, Lys, and Trp residues that are preferable for binding with Cu(II)-, Ni(II)-, and Zn(II)-IDA ligands. Aromatic side chains of Trp and Phe contribute to the binding as well when located in the vicinity of accessible His residues (Arnold 1991). However, His-tagged *E. coli* cells displayed a stronger binding to IMAC cryogels than wild type *E. coli*, which allowed separation from their mixture using a 5 ml Cu(II)-IDA cryogel column (Figure 14.6). Both types of cells were quantitatively captured from the feedstock containing equal amounts of cells of each type and recovered by selective elution with imidazole and EDTA, with yields of 80% and 77%, respectively. The peak obtained after EDTA elution was enriched eightfold with His-tagged *E. coli* cells as compared with the peak from imidazole elution, which contained mainly weakly bound wild type *E. coli* cells. *B. halodurans* cells were efficiently separated from a mixture with wild type *E. coli* cells, which were retained and recovered from the column with imidazole gradient (Figure 14.7; Dainiak et al. 2005).

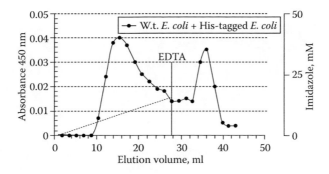

FIGURE 14.6
Chromatography profile of a mixture of wild type and His-tagged *E. coli* cells on a Cu(II)-IDA cryogel column (4.5 × 1.2 cm diameter). Experimental conditions: cell mixture containing equal amount of cells of both types (2 ml with OD_{450} of 1.0–1.2) was passed through the column equilibrated with 20 mM HEPES, 200 mM NaCl pH 7.0 at a flow rate of 1 ml/min. Bound cells were eluted with 0.0–0.05 M imidazole gradient (200 ml), which was interrupted and followed by 20 mM EDTA, 0.2 M NaCl pH 7.5. (Reproduced from Dainiak, M. B., F. M. Plieva, I.Yu. Galaev, R. Hatti-Kaul, and B. Mattiasson, *Biotechnol. Progr.*, 21, 644–49, 2005. With permission.)

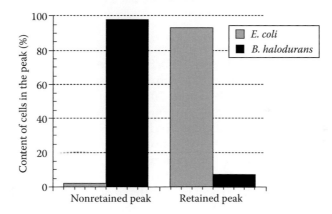

FIGURE 14.7
Content of wild-type *E. coli* and *B. halodurans* cells in the nonretained and retained peaks obtained after chromatography of the cell mixture on Cu(II)-IDA cryogel column (4.5 × 1.2 cm diameter). The total amount of cells in the peak was taken as 100%. (Reproduced from Dainiak, M. B., F. M. Plieva, I.Yu. Galaev, R. Hatti-Kaul, and B. Mattiasson, *Biotechnol. Progr.*, 21, 644–49, 2005. With permission.)

Due to the high heterogeneity of cell surface and the complexity of interactions of cells with affinity adsorbents, which is influenced by many factors (Wong, Leach, and Brown 2004; Dainiak, Galaev, and Mattiasson 2006), it may be difficult to predict theoretically the behavior of cells in a chromatographic process. Therefore, screening tests using cryogel-based 96-minicolumns may be very useful. For example, parallel chromatography on plain cryogel minicolumns and minicolumns with immobilized concanavalin A (18.8 mm × 7.1 mm) was successfully used for evaluation of

(A) (B)

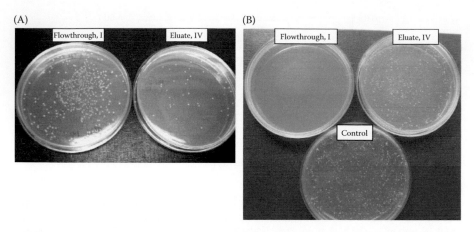

FIGURE 14.8
Analysis of *E. coli* (A) and *S. cerevisiae* (B) cell content in the flowthrough (OD_{600} 0.031) and in the eluate (OD_{600} 0.028). Samples were diluted 1:80 (A) and nondiluted (B) prior to streaking on the LB agar and malt extract agar, respectively. Control (B)-suspension of *S. cerevisiae* cells, OD_{600} 0.027. (Reproduced from Dainiak, M. B., I.Yu. Galaev, and B. Mattiasson, *J. Chromatogr. A*, 1123, 145–50, 2006. With permission.)

suitable conditions for separating *Saccharomyces cerevisiae* and *E. coli* cells (Dainiak, Galaev, and Mattiasson 2006). Such screening tests demonstrated that prolonged incubation of cells within the adsorbent promoted nonspecific adherence of *E. coli* cells where 70% of applied cells were captured on plain underivatized cryogels. Although incubation with ConA-cryogel monoliths leads to a more efficient binding of *S. cerevisiae* cells (Dainiak, Galaev, and Mattiasson 2006), it can not be used in the separation process as it will result in nonspecific retention of *E. coli* cells. Optimization of chromatographic separation was also carried out regarding the load of *S. cerevisiae* cells on ConA-cryogel monoliths in order to ensure quantitative capture of this type of cells during the separation procedure. Separation of a cell mixture containing equal amounts of cells of both types performed in a column format (113 mm × 7.1 mm) under the determined optimal conditions, resulted in a quantitative capture of applied *S. cerevisiae* cells, while *E. coli* passed through the column. The flow through fraction contained *E. coli* cells with nearly 100% purity, whereas the eluted fraction contained viable *S. cerevisiae* cells with 95% purity (Figure 14.8). The work clearly demonstrated that a nearly baseline chromatographic separation of different types of cells can be achieved under optimized conditions (Dainiak, Galaev, and Mattiasson 2006).

Along with optimization of chromatographic separation conditions, cryogel 96-minicolumn plates have been used for integrated binding-quantification of cells (Dainiak, Galaev, and Mattiasson 2006, 2007) and in enzyme-linked immunosorbent assay (ELISA) systems for direct analysis of bioparticles with surface-displayed antigens (Ahlqvist et al. 2006). Cryogels stay filled with the liquid due to the capillary forces inside the pores. This phenomenon combined with bulk diffusion conditions inside the large pores, allows cell metabolism and enzymatic reactions to proceed in the cryogels with bound cells. This, in turn, allows for direct biochemical and microbiological assays of bound cells, as was demonstrated by the example of *E. coli* cells bound to Cu(II)-IDA monoliths and *S. cerevisiae* cells bound to ConA monoliths (Dainiak, Galaev, and Mattiasson 2007). The former model was analyzed using tetrazolium salt XTT, which is reduced to colored formazan dye in the electron transport system in respiring cells, while the latter was assayed by monitoring pH changes of the medium, occurring due to production of protons during cell metabolism. In both models bound cells were viable and maintained ability to metabolize nutrients.

Chromatography on affinity cryogel columns has been successfully applied for separation of such large and fragile objects as animal cells. Affinity separation methods for isolation of rare cell populations most commonly use antibodies against differentially expressed cell-surface antigens. For instance, enrichment of HSC is based on the use of antibodies against surface CD34 and CD133 antigens. Antibodies are highly suitable as ligands for cell separation due to their great specificity and diversity. Typically, antibody labels are covalently linked to either a molecule (in FACS), a particle (in MACS), or a support matrix (immunoadsorbents in chromatography; Dainiak, Kumar, et al. 2007). Isolation of cell populations in which surface antigens are labeled with specific IgG on cryogel columns with immobilized protein A offers a generic approach to the separation of specific cells from a variety of biomedical samples. Protein A is a protein from *S. aureus,* which binds to the Fc portion of the IgG from a wide range of species. Cryogel matrix with covalently immobilized protein A can be used as an efficient adsorbent for cells that have been coated with a specific antibody (IgG type) and can separate them from cells that lack the surface antigen against which the antibody is directed. The feasibility of this approach was demonstrated by an example of fractionations of T- and B-lymphocytes.

Fractionating lymphocytes is of great importance because of the increased need for specific transfusion of lymphocytes and also if one wishes to describe the roles of lymphocyte subpopulations in immunological processes. After treating lymphocytes with goat antihuman IgG(H + L), the IgG-positive B-lymphocytes were bound to cryogel matrix with covalently attached protein A through the Fc region of antibodies. Nonbound T-lymphocytes passed through the column. More than 90% of the B-lymphocytes were retained in the column while the cells in the breakthrough fraction were enriched in T-lymphocytes (81%; Figure 14.9). The viability of the T-lymphocytes was greater than 90%. About 60–70% of the bound B-lymphocytes were recovered by the addition of a displacer, human or dog IgG, which competes for binding to protein A. The recovered B-cells maintained their viability (Kumar et al. 2003). It is noteworthy that the coupling chemistry used for immobilization of protein A is an important parameter that affects performance of

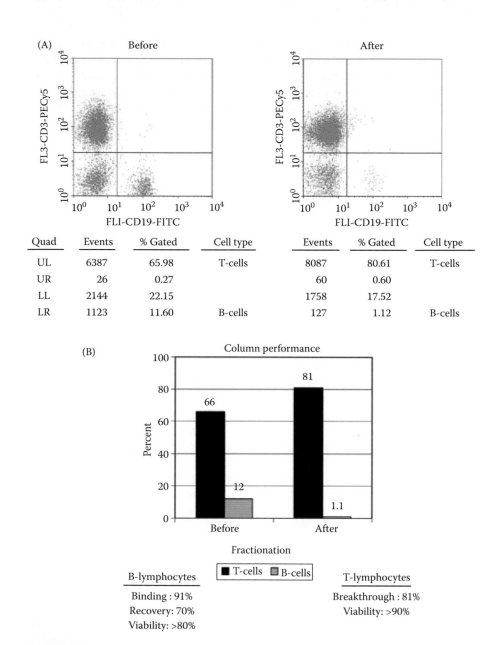

FIGURE 14.9
Composition of human peripheral blood lymphocytes before and after passage through the supermacroporous monolithic cryogel-protein A column. Lymphocytes (1 ml, 3.0 × 10⁷ cells/ml) were treated with goat antihuman IgG(H + L) and applied to a 2-ml cryogel-protein A column. (A) Flow cytometric analysis. (B) Column performance. The scatter gates were set on the lymphocyte fraction. The cells bound on the column were released with 2 ml of dog IgG (30 mg/ml). (Reproduced from Kumar, A., F. M. Plieva, I.Yu. Galaev, and B. Mattiasson, *J. Immunol. Methods*, 283, 185–94, 2003. With permission.)

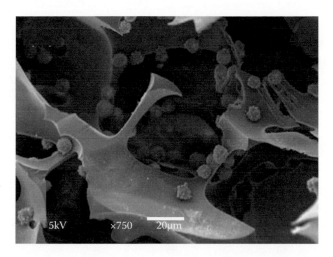

FIGURE 14.10
Scanning electron micrograph of CD34+ human acute myeloid leukemia cells (KG-1) cells bound to protein A-cryogel adsorbent. Magnification × 750. (Reproduced from Kumar, A., A. Rodriguez-Caballero, F. M. Plieva, I.Yu. Galaev, K. S. Nandakumar, M. Kamihira, R. Holmdahl, A. Orfao, and B. Mattiasson, *J. Mol. Rec.*, 18, 84–93, 2005. With permission.)

the affinity adsorbent. Among three derivatization protocols that is, direct coupling to epoxy groups without a spacer, via a long spacer (11 carbon atoms), and via a shorter spacer (7 carbon atoms), the latter proved to be optimal in terms of selectivity and binding capacity. A longer spacer of 11 carbon atoms promoted nonspecific binding, while cryogels with protein A immobilized directly to the matrix had a lower binding of B-cells, probably due to poor accessibility of the ligand to the passing cells.

As mentioned earlier, chromatography using the protein A cryogel column can be applied in general to cell separation systems provided IgG antibodies against specific cell surface markers are available. A similar approach has been employed for capture (Figure 14.10) and release of human acute myeloid leukemia KG-1 cells expressing the CD34 surface antigen (Kumar et al. 2005). As CD34 surface antigen is recognized as an important marker for HSC, the system may be a good model for the separation of CD34+ cells from bone marrow or peripheral blood. Capture of CD34+ KG-1 cells on monolithic cryogels or cryogel beads (diameter 200–500 μm) derivatized using different methods was also shown to be dependent on ligand coupling chemistry and matrix architecture. KG-1 cells labeled with anti-CD34 antibodies were efficiently bound (95%) to monolithic protein A cryogel columns while only about 76% of the cell binding was achieved in the case of protein A cryogel beads (Table 14.1). Surface architecture is an important aspect: in monolithic columns concave surface of the pores presumably is more advantageous for multivalent interactions with the cells than the convex surface of beads. A spacer arm of seven carbon atoms improved the selective binding of KG-1 cells to

TABLE 14.1

Recovery of Antibody-Labeled CD34 Human Acute Myeloid Leukemia (KG-1) Cells from Protein A-Cryogel Monoliths and Protein A-PVA Beads

	Protein A-PVA Beads			Protein A-Cryogel Monoliths		
	Recovery (%)	Viability (~%)	Viable Cells Recovered (%)	Recovery (%)	Viability (~%)	Viable Cells Recovered (%)
Elution with IgG	—	—	—	40	90	**36**
Elution with squeezing	—	—	—	75	85	**64**
Elution with IgG + squeezing	—	—	—	85	80	**68**
Elution with vortexing	60	60	**36**	—	—	—
Elution with vortexing + IgG	80	70	**56**	—	—	—

monolithic columns while introduction of a spacer between the ligand and the surface of the bead had no effect on cell binding.

Alternatively, nonlabeled KG-1 cells were applied to cryogel beads with immobilized anti-CD34 antibodies. Affinity binding was lower (50%) than in the case of adsorption of labeled cells on protein A beads probably due to poor orientation or partial inactivation of the IgG molecules when coupled directly to the matrix. One of the most common methods for optimization in orienting IgG molecules is by their coupling via Fc fragment to immobilized protein A. This technique leaves the variable heavy and light chain regions (Fab fragments) of immobilized IgG available for binding to antigen epitopes (Babacan et al. 2000). The efficiency of cell binding to affinity cryogel beads was improved (66%) when IgG were immobilized via protein A.

Monolithic protein A cryogel columns can be used both in positive and in negative selection. The latter is preferable since labeling of the target cells with IgG is not desirable. Cell chromatography on affinity cryogels offers high selectivity due to very low nonspecific adsorption, and high yield due to a very efficient method in recovery of bound cells (i.e., by mechanical compression of the cryogel matrix), which is discussed in the next section.

14.3 Recovery of Bound Cells by Mechanical Compression of a Cryogel

The adsorption of cells to surfaces in affinity-based separation have a polyvalent nature. The difficulty in disrupting polyvalent interactions has been one of the main problems in designing affinity techniques for cell separation.

According to theoretical studies, it is unlikely that reasonable concentrations of a soluble monovalent competitor (i.e., a biospecific eluent) can displace the binding equilibrium in situations where the number of interactions is >10 (Hubble 1997). Under typical chromatographic conditions (10^{10}–10^{12} ligands and receptors per cm^2, and 10^{-10}–10^{-8} cm^2 contact area) the number of specific binding interactions is 1–10,000 (Cao, Eisenthal, and Hubble 2002). Thus, in most cases an external force affecting the entire bioparticle, or the matrix in an integral way, is required to simultaneously disrupt multiple bonds and detach specifically adsorbed cells (Bell 1978).

Various strategies have been applied in an attempt to solve the problem of strong multipoint attachment of cells to affinity adsorbents. In membrane-based chromatography, bound B-cells were released by transmembrane diffusion of hydrochloric acid (pH 1) into a flow of neutralizing normal saline (Madrusov et al. 1995). Other approaches used for cell detachment are the passage of air-liquid interfaces (Cao, Eisenthal, and Hubble 2002), use of ligands immobilized through cleavable bonds (Bonnafous et al. 1983), and application of flow-induced or mechanical shear forces (Ming, Whish, and Hubble 1998; Ujam et al. 2003).

Elasticity of cryogels allows for efficient binding and release of cells (Figure 14.11; Dainiak, Kumar, et al. 2006; Galaev et al. 2007). The efficiency in the release of bound particles by compression depends on the elastic modulus of the cryogel, size of the particles, and density of the receptors on the surface of the particle. The four-fold difference in elasticity had no effect on the efficiency of the release of bound microgels of cross-linked poly(N-isopropyl acrylamide-*co*-N-vinylimidazole; 0.4 µm) and IgG-labeled inclusion bodies (~1 µm) from corresponding affinity cryogels. However, the release of larger particles like *E. coli* cells (rods 1×3 µm; Figure 14.12) and especially yeast cells (8 µm) was highly dependent on the cryogel elasticity. The more elastic the cryogel, the more efficient the detachment of larger particles upon the deformation. The lower the density of the receptors on the particle's surface, the higher the recovery from affinity cryogels (Figure 14.13).

The phenomenon of particle detachment upon elastic deformation is of a generic nature and was demonstrated for a variety of bioparticles of different sizes and for synthetic particles, for different ligand-receptor pairs (IgG-protein A, sugar-ConA, metal ion-chelating ligand), and when the deformation was caused by either external forces (mechanical deformation) or internal forces (the shrinkage of thermosensitive, macroporous hydrogel upon an increase in temperature). The main mechanisms involved in the compression-induced detachment of bound particles from the surface are the physical dislodging of cells by microscopic deformation of the surface carrying affinity ligands and the removal of dislodged particles by the flow of liquid squeezed out. The presence of specific eluent contributes to the detachment by decreasing the equilibrium number of bonds and preventing readsorption of detached particles on their way out of the column. It

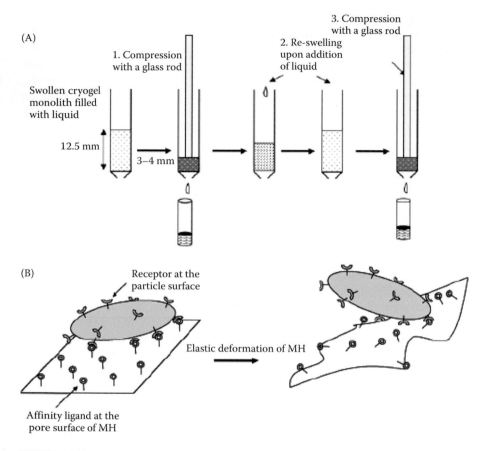

FIGURE 14.11
Schematic presentation of the (A) procedure used for the release of captured bioparticles by mechanical compression of monolithic affinity cryogel and (B) of the mechanism of detachment of bound cells induced by cryogel deformation. (Reproduced from Dainiak, M. B., I.Yu. Galaev, and B. Mattiasson, *J. Chromatogr. A*, 1123, 145–50, 2006; and Galaev, I.Yu., M. B. Dainiak, F. M. Plieva, and B. Mattiasson, *Langmuir*, 23, 35–40, 2007. With permission.)

is worthwhile to note that matrix deformation has no effect on the release of affinity-bound macromolecules, even when these macromolecules have more than one binding site like recombinant His$_6$-tagged tetrameric lactate dehydrogenase from *Bacillus stearothermophilus*.

Because the detachment conditions are very mild, the viability of bound and subsequently released cells is retained. The efficiency of detachment and viability of fragile mammalian cells eluted by compression was compared to that of cells detached by shear force when vortexing the beads with bound cells. The KG-1 cells labeled with anti-CD34 antibodies were bound either to protein A-cryogel monolithic column or protein A-poly(vinyl alcohol; PVA) cryogel beads. The porous structure and elastic properties of PVA beads are rather

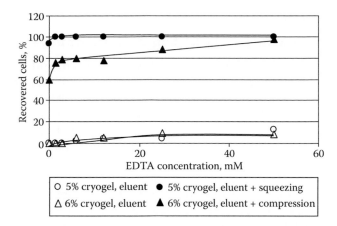

FIGURE 14.12
Release of bound recombinant *E. coli* cells by conventional elution and by mechanical compression of Ni(II)-IDA-cryogel monoliths equilibrated with different concentrations of EDTA. The amount of bound cells was assumed to be 100%. (Reproduced from Dainiak, M. B., I.Yu. Galaev, and B. Mattiasson, *J. Chromatogr. A*, 1123, 145–50, 2006. With permission.)

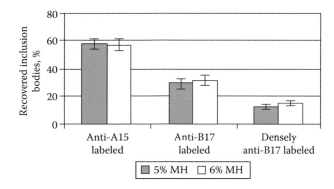

FIGURE 14.13
Release of bound inclusion bodies labeled with anti-A15-IgG and anti-B17-IgG by mechanical compression of protein A-cryogel monoliths. Anti-B17-IgG have higher affinity to the antigens on the surface of inclusion bodies and are present at higher density on the surface of IgG-labeled inclusion bodies than anti-A15-IgG (Ahlqvist et al. 2006). Different dilutions of anti-B17-IgG were used in order to ensure different densities of these IgG on the particle surface. (Reproduced from Galaev, I.Yu., M. B. Dainiak, F. M. Plieva, and B. Mattiasson, *Langmuir*, 23, 35–40, 2007. With permission.)

similar to those of cryogel monoliths; however, the pore size (0.1–1 μm) of PVA beads is too small for KG-1 cells to penetrate inside the pores, so the cells are bound only to the outer surface and are fully exposed to the shear force during vortexing. Compression of cryogel monoliths significantly increased the recovery of cells (Table 14.1). Moreover, both the efficiency of cell detachment by compression and the viability of the detached cells were significantly higher than when vortexing beads with cells bound to the bead surface.

14.4 Cryogel Scaffolds: Tissue Engineering Applications

The purpose of tissue engineering is to create tissues in culture for use as replacements for damaged body parts. It is generally based on culturing cells on 3-D scaffolds under conditions necessary for them to develop into functional tissue. After the implantation of tissue-biomaterial construct into the patient, the biomaterial gradually absorbs, ensuring that only the natural tissue remains in the body. Alternatively, scaffolds are implanted into the injured tissue to stimulate the bodies own cells to promote local tissue repair (*in situ* tissue regeneration). Scaffolds provide the physical and chemical cues to guide cell differentiation and assembly into 3-D tissues and play a crucial role in tissue engineering. Generally, they are porous, mechanically stable, degradable structures fabricated from either natural materials (e.g., fibrin, gelatin, collagen), or synthetic polymers (e.g., polylactide coglycolide; Griffith and Naughton 2002). Synthetic polymers are often modified with bioactive components in order to mimic the extracellular matrix (ECM) environment *in vivo* and stimulate specific cellular responses at the molecular level.

Cryogels possess all the properties of a successful 3-D biomimetic scaffold (Bloch et al. 2005; Plieva, Oknianska, et al. 2006; Dainiak, Mattiasson, and Galaev 2007; Savina et al. 2007; Bolgen et al. 2007). Their high porosity and optimally designed pore size provide structural space for cell accommodation (Figure 14.14) and migration and enable the exchange of nutrients between the scaffold and the environment. Cryogel scaffolds have been prepared from a wide range of synthetic and natural materials, including poly(2-hydroxyethyl methacrylate; pHEMA), poly(ethylene glycol; PEG), gelatin, chitosan, albumin/chitosan, dextran, alginate, and agarose (Dainiak, Mattiasson, and Galaev 2007). In order to achieve biomolecular recognition of synthetic cryogels by cells, they are subjected to surface or bulk modification with bioactive molecules such as ECM proteins (Savina et al. 2009; Dainiak et al. 2008). The type or level of modification, which is optimal for cell-surface interaction, varies depending on the specific cell types and intrinsic properties of biomaterials (Shin, Jo, and Mikos 2003). Bulk modification of synthetic scaffolds is typically performed by copolymerization or attachment of functional groups to the polymer chains before scaffold fabrication and can be achieved through physical (Stile and Healy 2001), chemical (Elbert and Hubbell 2001; Mao et al. 2003; Dainiak et al. 2008), photochemical (Borkenhagen et al. 1999), and ionic cross-linking (Rowley, Madlambayan, and Mooney 1999). Recognition sites obtained by bulk modification are present not only on the surface but also in the bulk of the materials. Such modification with enzymatically degradable sequences can render biomaterials biodegradable by specific proteases (West and Hubbell 1999; Halstenberg et al. 2002). Surface modification is carried out after scaffold fabrication

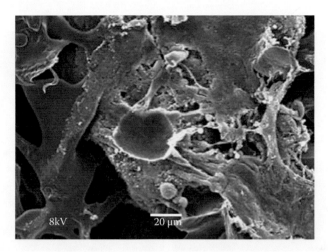

FIGURE 14.14
Scanning electron micrograph of human osteoblasts (× 650) cultured for three days on collagen-coated cryogel scaffold. (Reproduced from Dainiak, M. B., I.Yu. Galaev, and B. Mattiasson, *Enzyme. Microb. Technol.*, 40, 688–95, 2007. With permission.)

and is usually achieved by covalent binding of peptides or proteins on biomaterial surfaces using different coupling techniques (Shin, Jo, and Mikos 2003). To maintain the biological activity, immobilized proteins should have intact recognition sites available for interaction with cells. The mobility and accessibility of the ligand strongly influence the adhesion and proliferation of cells (Thid et al. 2007).

Analysis of poly(2-hydroxyethyl methacrylate; pHEMA) cryogels modified with bovine collagen type I (CG) and fibrinogen (FG) from human plasma demonstrated significant variations in spatial distribution and surface density of the incorporated proteins depending on the method used for modification and protein nature. Bulk modification was performed by cross-linking cryocopolymerisation of HEMA and poly(ethylene glycol) diacrylate in the presence of proteins (CG/pHEMA and FG/pHEMA cryogels). One can expect that the protein introduced during gel formation will build up the cryogel pore walls together with the polymer. Such incorporation can take place partially due to the involvement of proteins in chain termination during free radical copolymerisation or by physical entrapment during the formation of polymeric network. Surface modification was performed by covalent immobilization of the proteins to the active groups (epoxy groups introduced by copolymerization of allyl glycidyl ether with HEMA, these cryogels are designated as epHEMA) present on cryogel after fabrication (CG-epHEMA and FG-epHEMA cryogels).

The concentration of incorporated proteins in protein/pHEMA and protein-epHEMA cryogels was 80–85 and 130–140 µg per 1 ml cryogel, respectively. As was demonstrated by immunostaining and confocal laser scanning microscopy, bulk modification resulted in spreading of CG in the

(A)   (B)

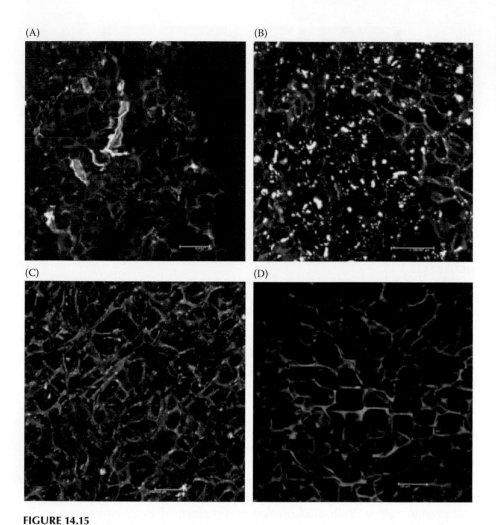

(C)   (D)

FIGURE 14.15
(See color insert following page 116.) Confocal laser scanning microscopy images of (A,B) spatial distribution of collagen and fibrinogen in bulk-modified and (C,D) surface-modified pHEMA cryogel scaffolds. (Reproduced from Savina, I. N., M. B. Dainiak, H. Jungvid, S. V. Mikhalovsky, and I.Yu. Galaev, *J. Biomater. Sci. Polymer Edn.*, 20, 1781–95, 2009, DOI:10.1163/156856208X386390. With permission.)

polymer matrix and spot-like distribution of FG (Figure 14.15A,B). However, surface modification resulted in spot-like distribution of CG and homogeneous spreading of FG, which evenly coated the surface (Figure 14.15C,D). Such different behavior of CG and FG could be explained by different solubility of the proteins in the reaction media and their interaction with the surface in the case of surface modification, and by solubility–compatibility of proteins with HEMA solution in the case of bulk modification. The basic conditions (pH 9.0) used for surface modification of epHEMA cryogels enhanced solubility of FG, while CG, which has poor solubility in aqueous solutions especially at high pH, was partially in the aggregated form. On the other hand, bulk modification compatibility of CG with HEMA was better and CG spread out in polymer solution building into polymer walls of cryogels. Separated spot-like distribution of FG molecules along the cryogel matrix with embedded protein could be a result of poor compatibility of FG with HEMA solution. HEMA is hydrophilic. FG molecules spread out on hydrophobic materials, as opposed to hydrophilic ones (Fuss, Palmaz, and Sprague 2001). Moreover, pH of the reaction mixture used for cryopolymerization was close to or lower than the isoelectric point of FG leading to the decreased solubility. Thus, FG was embedded in an aggregated form during production of bulk-functionalized cryogels (Savina et al. forthcoming).

The proliferation rate of fibroblasts was higher on cryogels with even distribution of the ligands (i.e., on FG-epHEMA and CG/pHEMA). After 30 days of culture, fibroblasts formed several monolayers and deposited ECM filling the pores of these cryogels. The best result in terms of cell proliferation was obtained on FG-epHEMA. Distinct cells are not seen in the layers formed on FG-epHEMA presumably due to flattened morphology and coverage with ECM secreted by the cells (Figure 14.16). On the surface, the ligands displayed these scaffolds in native conformation, while in bulk-modified CG/pHEMA cryogels, most of the proteins were buried inside the polymer matrix as was demonstrated by immunostaining with antibodies against native proteins. The method used for modification of cryogels with bioligands strongly affects spatial distribution, density, and conformation of the ligand on the surface of the scaffold that in turn affects cell-surface interactions (Savina et al. 2009).

Along with biocoated synthetic cryogels, protein cryogels have been developed in which a bulk protein is copolymerized with appropriate ECM protein (e.g., laminin or fibronectin; Figure 14.17). The biodegradation rate of such protein-based cryogels is modulated by changing the concentration of the cross-linker and can be decreased 100-fold by a 10-fold increase in degree of cross-linking (*in vitro* studies; M. Dainiak unpublished data). Synthetic biodegradable cryogels have been produced using biodegradable cross-linkers like N,N′-bis(methacryloyl)-L-cystine (Andac et al. 2008) or biodegradable macromers such as (lactate)-hydroxyethyl methacrylated dextran and lactate-HEMA (Bolgen et al. 2007). Tissue response to the latter was evaluated in rats after dorsal subcutaneous implantation, iliac intramuscular implantation, auricular implantation or in a calvarial defect model (Bolgen et al. in press). The cryogel scaffolds integrated with the surrounding tissue and the formation of a new tissue was accompanied with significant ingrowth of connective tissue cells and new blood vessels into the cryogel (Figure 14.18). The tissue responses were significantly lower in auricular and calvarial implantations when compared to the subcutaneous and the intramuscular implantations. The degradation of the scaffold was slower in bone compared to that in soft tissues. The preliminary bone regeneration in critical size cranial bone

(A)

(B)

(C)

(D)

(E)

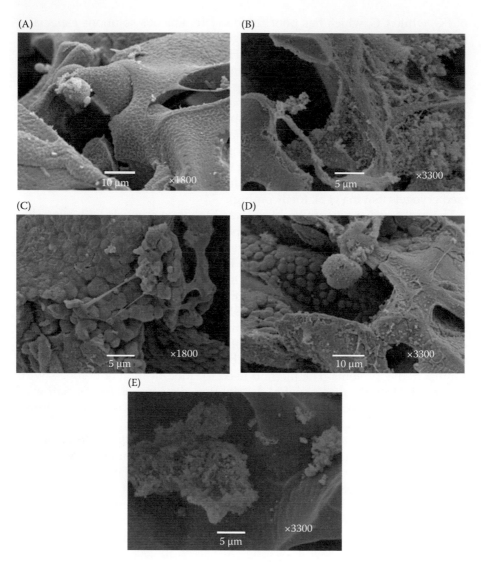

FIGURE 14.16
Scanning electron micrographs of human embryonic fibroblasts cultured for 30 days on (A, B) bulk-modified, (C, D) surface-modified and (E) plain pHEMA cryogel scaffolds. pHEMA cryogel scaffolds. (Reproduced from Savina, I. N., M. B. Dainiak, H. Jungvid, S. V. Mikhalovsky, and I.Yu. Galaev, *J. Biomater. Sci. Polymer Edn.*, 20, 1781–95, 2009, DOI:10.1163/156856208X386390. With permission.)

defects model is promising. A significant ingrowth of connective tissue cells and new blood vessels into the cryogel scaffold guided new bone formation.

One of many potential applications of cryogel scaffolds is the transplantation of insulin-producing cells for treatment of insulin-dependent diabetic patients. Agarose cryogel scaffolds with grafted gelatin were shown to support

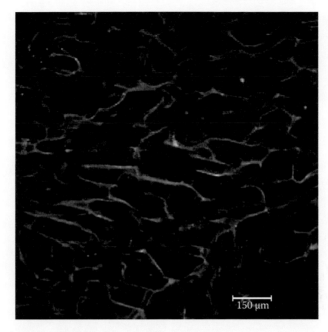

FIGURE 14.17
(See color insert following page 116.) Distribution of fibrinogen in fibrinogen-containing protein-based cryogel scaffolds viewed by confocal laser scanning microscopy. Samples were immunostained with monoclonal antifibrinogen mouse antibodies and with secondary FITC-labeled antimouse antibodies.

proliferation and insulin secretion by insulinoma cells (INS-1E) (Bloch et al. 2005). Moreover, the results of initial studies showed that subcutaneous implantation of cryogel sponges created from agarose and gelatin induced a strong promoting effect on neovascularization in transplanted mice. The angiogenesis-inducing property of cryogel scaffolds is very important as adequate oxygenation is critical to cell function and survival, in particular to pancreatic islets.

Biopolymer cryogels have great potential as biomaterials for healing burns and healing chronic ulcers. Hundreds of thousands of patients seek medical attention for burn injuries each year. Skin substitution with skin replacement materials is a lifesaving measure in the treatment of acute burns. Despite the huge need for tissue-engineered biological dressings most of the marketed cellular replacements have suffered economical setbacks due to a number of deficiencies that include complicated logistics of delivery, poor mechanical properties, short shelf life, high costs, and a variable outcome in the wound environment (Hutmacher and Vanscheidt 2002; Ehrenreich and Ruszczak 2006; Anthony et al. 2006). Cryogel-based biological dressing is being designed as a template for regeneration of neodermis and should seal the wound, stop infection, prevent scar-inducing contraction, and promote vascularization. The research on tailoring cryogels to meet these requirements is in progress. Preliminary *in vitro* tests with a protein-based cryogel

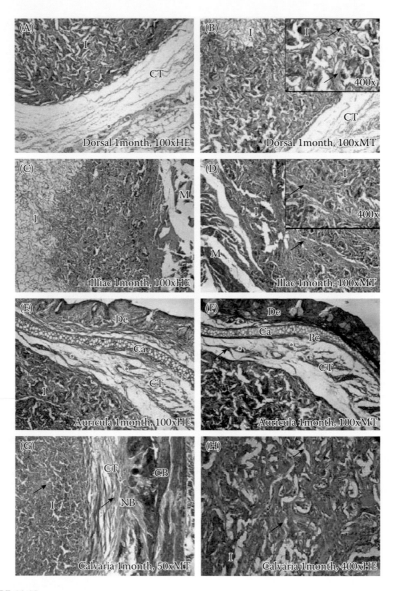

FIGURE 14.18

(See color insert following page 116.) Representative histological data at one month: the particles of cryogel scaffold are surrounded by a moderate capsule characterized by infiltration of mononuclear phagocytic cells, macrophages, lymphocytes, fibroblasts, and some polymorphonuclear leukocytes with foreign body giant cells in hard and soft tissues at the implantation sites. Collagen fibers and the blood vessels are in close relation with the implant. Auricular cartilage appear healthy. HE: haematoxylin eosin, MT: Masson's trichrome, I: implant, CT: connective tissue, FC: fibrous callus, CB: compact bone, NB: new trabecular bone, Ca: cartilage, Pc: pericondrium, De: dermis, Sk: skin, M: muscles, (*): blood vessels, Arrow: collagen fibers. (Reproduced from Bolgen, N., I. Vargel, P. Korkusuz, E. Guzel, F. Plieva, I.Yu. Galaev, B. Mattiasson, and E. Piskin, *J. Biomed. Mater. Res. Part A*, 2009, DOI: 10.1002/jbm.a.32193. With permission.)

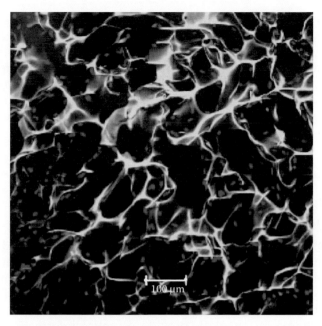

FIGURE 14.19
(See color insert following page 116.) Confocal laser scanning microscopy of primary dermal fibroblasts SKF375 cultured on protein-based cryogel dermal regeneration template for 6 days. Snapshot of the bottom of the scaffold. Cells were stained with 4′,6-diamidino-2-phenylindole (DAPI), cryogel was stained with fluorescein isothiocyanate (FITC).

dermal regeneration template has demonstrated its high biocompatibility, degradability, and ability to promote efficient infiltration by dermal fibroblasts (Figure 14.19).

Currently available skin grafts all need a patient's own epidermis to help full recovery. However, harvesting sufficient keratinocytes is a major obstacle in producing high quality skin equivalents. Stem cells hold great promise for treating damaged tissue where the source of cells for repair is extremely limited or not readily accessible (Thomson et al. 1998). The difficulty to expand stem cells to high densities in porous matrices is one of the bottlenecks among many unresolved difficulties in stem cell research. One can foresee the use of cryogel bioreactors for cell expansion.

The spongy elastic and macroporous nature of cryogels is quite close to native cartilage architecture. The water content in cryogels is about 80–95% while the polymer content is up to 10–20%, which is similar to that of natural cartilage tissue. Cartilage is a predominately avascular, aneural, and alymphatic tissue with a limited ability of self-repair. In this respect cartilage is an ideal system for tissue engineering. The recent work to use a novel cryogelation technology to construct mechanically stable, macroporous 3-D scaffolds with gradient porosity from agarose-gelatin and chitosan-gelatin cryogels have shown a great promise for cartilage tissue engineering (Tripathy, Kathuria, and Kumar 2008; Kathuria et al. 2008). Different hybrid cryogel scaffolds from natural polymers like chitosan, agarose, alginate, and gelatin have been synthesized and demonstrated a good mechanical integrity and biocompatibility. Initial experiments demonstrated proliferation of chondrocytes on these scaffolds with the resulting construct possessing some of the essential characteristics required to make up a good *in vitro* neo-cartilage tissue (unpublished results).

14.5 Cryogels in a High Throughput Screening Format for Cell-Based Assays

Cryogel scaffolds in a 96-minicolumn plate represent a novel 3-D culture system for high throughput toxicity assays. 3-D tissue culture models have an invaluable role in tumor biology, and the possibility of 3-D modeling in a high throughput screening mode offers great prospects for facilitating the research in this area.

In vitro cytotoxicity tests can provide essential information about the potential effects of therapeutic agents on specific cell properties, and provide a more relevant basis for molecular studies than the laboratory animal models. However, *in vitro* experiments can truly become more predictive of *in vivo* systems only if they are designed to mimic a complex 3-D cell environment in a living organism (Zhang 2004). Relative to 2-D *in vitro* adhesions, 3-D matrix interactions display enhanced cell biological activities and narrowed integrin usage and are more biologically relevant to living organisms (Cukierman et al. 2001). It is also important to recognize that cells respond to external stimuli through complex pathways, and are finely tuned by cell-cell and cell-ECM interactions. *In vivo* cell-ECM interactions regulate proliferation, differentiation, motility, and apoptosis (Weaver and Roskelley 1997; Lukashev and Werb 1998; Santini, Rainaldi, and Indovina 2000). Although malignant cells have often been assumed to be anchorage-independent, they remain sensitive to regulatory signals originating from the ECM (Varner and Cheresh 1996; Keely, Parise, and Juliano 1998; Henriet et al. 2000). Therefore, scaffold modification with ECM components is an important factor to be considered when designing a realistic experimental model for toxicity tests. Along with biochemical modifications, surface characteristics of the underlying material such as hydrophobicity, charge, and mechanical compliance are also important surface cues that substantially affect cellular responses (Pelham and Wang 1997; Allen et al. 2003; Wong, Leach, and Brown 2004) and therefore may affect the outcome of cell-based assays. For instance, substrate stiffness modulates cell motility and spreading (Pelham and Wang 1997) and through alterations in cell shape is able to modulate gene expression (Maniotis, Chen, and Ingber 1997). The behavior of some cells on soft materials is characteristic of important phenotypes; for instance, cell growth on soft agar gels is used to identify cancer cells (Discher, Janmey, and Wang 2005).

Most of the existing high throughput toxicity assays do not provide an environment in which three-dimensionality, elasticity, surface chemistry, and relevant biochemical modifications are taken into account. At present high throughput screening is carried out mostly in multiwell culture plates in which cells grow in a 2-D monolayer environment on rigid flat surfaces (Sundberg 2000; Falconnet et al. 2006).

Cryogels prepared in a 96-minicolumn plate format and functionalized with ECM proteins or with a mimetic cell-recognition motif RGD (Arg-Gly-Asp) provide a more realistic *in vitro* model for cell-based assays. The agmatine-based mimetic of RGD peptide (RGDm) was synthesized via reaction of agmatine with itaconic anhydride (Dainiak et al. 2008). About half of the 24 integrins have been shown to bind to ECM molecules in a RGD dependent manner and surface functionalization with this peptide has found wide applications in medicine and academic studies (Hersel, Dahmen, and Kessler 2003). However, RGD-containing peptides and their derivatives suitable for covalent coupling to the matrices are expensive. Therefore, an attempt has been made to develop cost-effective structural analogue of RGD for functionalization of cryogels used in high throughput toxicity tests.

Adhesion, proliferation, and chemosensitivity of HCT116 human colon cancer cells, human embryonic fibroblasts, and KG-1 human acute myeloid leukemia cells on collagen- and RGDm-containing native underivatized cryogels has been studied. Functionalization of cryogels with CG and RGDm monomer induced cell-cell interactions and formation of multicellular aggregates in cancer cell cultures that was demonstrated by scanning electron microscopy (Figure 14.20). Cell clusters were also observed in the medium displaced from the cryogels during the medium exchange process. Formed cell aggregates were compact and contained deposited ECM. In multicellular aggregates the quantity of ECM proteins is greater making it more likely that integrins can bind ECM components and transduce antiapoptotic stimuli (Santini, Rainaldi, and Indovina 2000). The ability of malignant cells to adhere to each other is associated with the early stages of cancer metastasis. Presumably, it is the combination of surface cues including elasticity of the matrix that triggers cell aggregation on cryogels. Elasticity may play an important role in this process. A decrease in substrate stiffness results in weakening of cell-surface interactions (Pelham and Wang 1997). The survival signals, which come from cell-surface adhesion, must be substituted by those arising from cell-cell contact.

Significant differences were revealed with respect to adhesion and morphology of fibroblasts cultured for three days on functionalized and on plain cryogels. Cells adhered to a collagen-coated surface displayed spreading

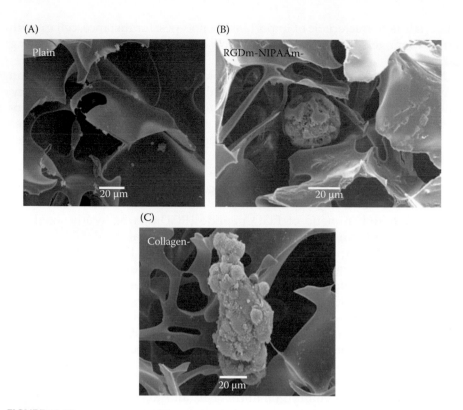

FIGURE 14.20
Scanning electron micrographs of HCT116 cells cultured for 6 days on (A) plain, (B) modified with RGDm plus poly(N-isopropylacrylamide), and (C) modified with collagen polyacrylamide cryogels. Magnification × 850. (Reproduced from Dainiak, M. B., I. Savina, I. Musolino, A. Kumar, B. Mattiasson, and I.Yu. Galaev, *Biotechnol. Progr.*, 24, 1373–83, 2008. With permission.)

with the outgrowth of a number of cell extensions (Figure 14.21). Spreading was also observed on RGDm-cryogels. Fibroblasts cultured on plain cryogels had a completely round shape and did not form contacts and adhesions with the matrix.

HCT116 cells were treated with different concentrations of *cis*-diaminedichloroplatinum (II; cisplatin). It is a commonly used anticancer agent that is assumed to induce apoptosis and clonogenic outgrowth by DNA damage-induced signalling (Berndtsson et al. 2006). KG-1 cells were treated with cytosine 1-β-D arabino furanoside (Ara-C). Ara-C must be metabolized to a triphosphate (Ara-CTP) derivative to exert its cytotoxic effect, which occurs due to interference with DNA polymerases and termination of DNA chain elongation (Chabner 1990). HCT116 cultured on collagen- and RGDm-cryogels were more resistant to the treatment with the drugs during the first 18–24 hours of incubation than single cells grown on unmodified cryogels (Figure 14.22A,B,C). HCT116 cells grown as 2-D cultures in conventional

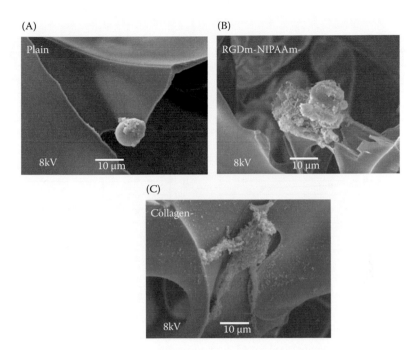

FIGURE 14.21
Scanning electron micrographs of fibroblasts cultured for 3 days on (A) plain, (B) modified with RGDm plus poly(N-isopropylacrylamide), and (C) modified with collagen polyacrylamide cryo-gels. Magnification × 1500 and × 1900 (C). (Reproduced from Dainiak, M. B., I. Savina, I. Musolino, A. Kumar, B. Mattiasson, and I.Yu. Galaev, *Biotechnol. Progr.*, 24, 1373–83, 2008. With permission.)

96-well tissue culture plates were 1.5–3.5-fold more sensitive to the treatment with 70 µM cisplatin than cells in 3-D cultures in functionalized cryogels (Figure 14.22D; Dainiak et al. 2008).

Presumably, the formation of aggregates by HCT116 and KG-1 cells cultured on collagen- and RGDm-containing cryogels explains the enhanced drug resistance of the cells grown as 3-D cultures on these matrices. It is well documented that investigations with multicellular tumor spheroids yield much useful information that is often in contradiction to that obtained with 2-D cultures, but is closer to that derived from *in vivo* studies (Santini, Rainaldi, and Indovina 2000; Kim 2005). Similarly to the situation within intact human tumors, 3-D cultures contain cells with different phenotypes such as proliferating, nonproliferating, and necrotic cells. Such cellular heterogeneity within 3-D culture models is far more realistic than the cellular homogeneity found in monolayer culture. It was also shown that integrin expression observed in spheroids closely resembled the expression pattern found in *in vivo* tumors and that cell-cell contact and the spheroid microenvironment can modulate integrin expression (Waleh et al. 1994). Most importantly, *in vitro* 3-D culture systems have been shown to recapitulate the drug sensitivity patterns of tumor cells *in vivo* (Kim 2005). The

(A)

(B)

(C)

(D)

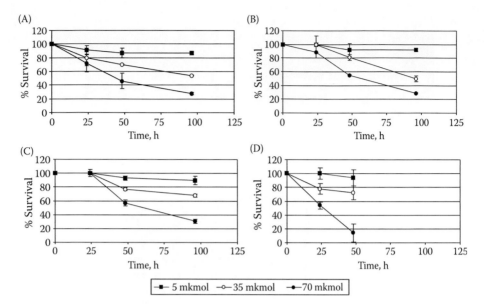

-■- 5 mkmol   -○- 35 mkmol   -●- 70 mkmol

FIGURE 14.22

Treatment of HCT116 cells cultured on plain (A), modified with collagen (B), and modified with RGDm plus poly(N-isopropylacrylamide) (C) polyacryalmide cryogels, and on polystyrene tissue culture plates (D) with different concentrations of cisplatin. Cells were cultured on cryogels and tissue culture plates for 4 and 1 days, respectively prior to exposure to 5, 35, and 70 μM of *cis*-diamminedichloroplatinum (II) (cisplatin) in the medium. After the incubation with the drug for 24, 48, and 96 hours, the density of viable cells in MHs was determined by XTT-based assay. Fresh medium containing drug was added to cryogels not subjected to the analysis after 24 and 48 h from the start of drug treatment. Medium not containing drug was added to control MHs. Number of cells in control MHs or wells was taken as 100%. Percent survival is the percent of viable cells remaining in MHs and wells after the drug treatment. (Reproduced from Dainiak, M. B., I. Savina, I. Musolino, A. Kumar, B. Mattiasson, and I.Yu. Galaev, *Biotechnol. Progr.*, 24, 1373–83, 2008. With permission.)

possibility of the formation of multicellular aggregates inside the pores of functionalized cryogels may not only provide the opportunity to mimic the architectural intricacies of physiological cell-cell interactions in a high throughput screening format but also to avoid an important limitation of the currently used 3-D cultures. This limitation is an external stress due to unlimited space available in spheroid cultures grown in spinner flasks or by using a liquid overlay method (Helmlinger et al. 1995).

The possibility of culturing fibroblasts on cryogels in a high throughput screening format may represent a basis for development of coculture systems for studying paracrine and growth factor interactions between tumor and host cells. Such systems are playing an important role in toxicity testing due to their ability to recreate important intracellular networks *in vitro* (Bhogal et al. 2005). For instance, fibroblasts were shown to increase proliferation and decreased apoptosis of human acute myelogenous leukemia cells (Ryningen et al. 2005).

14.6 Extracorporeal Medical Devices

There is a growing interest in developing hybrid artificial devices and they are still in the early stage of development. These bioartificial devices combine biological systems with some nonbiological forms such as biomaterial that can maintain their activity outside the body. These can then be used as a substitute or as an assisting device for the damaged organ system. One of the most active areas of investigation involve hepatic-assisted devices. Currently there are nine bioartificial devices that are being tested clinically and most of them utilize a hollow fiber bioreactor. A similar functional bioartificial liver device is under development using cryogel as the matrix for cell immobilization and further purification of a patient's plasma. A cryogel bioreactor offers as much as 10 times greater surface area than a conventional hollow fiber bioreactor combined with unrestricted diffusion and convective mass transfer. Cryogel bioreactors made up of a combination of synthetic and natural polymers have proved to be favorable for growth of hepatocytes. Further use of this device, for purifying patient plasma, demonstrated the potential of cryogel as bioartificial liver devices (unpublished data). Similar approaches of using monolithic, cryogel-based extracorporeal haemoperfusion devices is under active research development.

14.7 Cryogel Bioreactors for Production of Therapeutic Proteins

Mammalian cells are the source of a variety of high-value bioproducts, usually fragile molecules that require highly specialized and mild processing conditions and may need to be separated from a complex mixture of molecules, including cell debris. Many products of culture bioprocesses are inhibitory to the cells producing them and require continuous on-line monitoring and *in situ* product recovery methods to remove them. One of the promising strategies for achieving a better operational economy is the reduction in the number of steps by integrating the primary capture of the product from the cell suspension with the initial purification (Larsson et al. 1989). Such integrated downstream processing also implies the application of the separation technologies capable of processing particulate-containing solutions. In animal cell culture processes, significant development has been achieved by the introduction of hollow fiber bioreactors, which can be integrated with the product recovery processes (Luther et al. 1992).

An integrated cell cultivation and protein product separation process was developed using a cryogel support matrix (Kumar et al. 2006). Human fibrosarcoma HT1080 and human colon cancer HCT116 cell lines grown on pAAm-gelatin cryogel were used to secrete urokinase into the culture medium. The cells were attached to the matrix within 4–6 hr postseeding and grew as a tissue sheet inside the cryogel matrix. Continuous urokinase secretion into the circulating medium was monitored as a parameter of growth and viability of cells inside the bioreactor. No morphological changes were observed in the cells eluted from the gelatin-cryogel support and recultured in tissue culture flasks.

Urokinase [EC 3.4.99.26], a single polypeptide of about 55 kDa found in urine, blood, and other body fluids, catalyzes the conversion of plasminogen to plasmin by cleaving the Arg-Val linkage in the Pro-Gly-Arg-Val sequence of the former. As the resulting plasmin dissolves, clots of fibrin in blood vessels, urokinase is intravenously administered for the treatment of thromboembolic diseases. Secretion of urokinase by cultured renal cells suggested tissue culture as suitable means for the production of the enzyme (Gohji et al. 1997). The production of urokinase is controlled by negative feedback and stops when its level in the medium reaches a critical concentration. However, when the enzyme is removed continuously by replacing the medium, the cells maintain enzyme production at a constant rate for a period of several weeks. Solid microcarrier-based perfusion bioreactors were reported previously for the production of urokinase (Jo et al. 1998). These, however, lead to the formation of large heterogeneous cell aggregates that decrease the viable cell density and subsequently the urokinase production levels.

The gelatin-cryogel bioreactor was further connected to pAAm cryogel column carrying Cu(II)-iminodiacetic acid (Cu(II)-IDA)-ligands, which had been optimized for the capture of urokinase from the conditioned medium of the cell lines (Figure 14.23). Such affinity cryogel columns provide the possibility to process unclarified crude feeds in a chromatographic mode (Arvidsson et al. 2003; Dainiak et al. 2004). Thus an automated system was built, which integrated the features of a hollow fiber reactor with a chromatographic protein separation system. The urokinase was continuously captured by the Cu(II)-IDA-cryogel column and periodically recovered through elution cycles. The urokinase activity increased from 250 PU/mg in the culture fluid to 2,310 PU/mg after recovery from the capture column, which gave about ninefold purification of the enzyme. Increased productivity was achieved by operating integrated bioreactor system continuously for 32 days under product inhibition-free conditions during which no backpressure or culture contamination was observed (Figure 14.24). A total 152,600 Plough units of urokinase activity was recovered from 500 ml culture medium using 38 capture columns over a period of 32 days.

A comparative study for the practical applicability of Cu(II)-IDA cryogel and that of Cu(II)-IDA Sepharose columns for the direct capture of urokinase from culture broth of HT1080 cell line in an integrated setup revealed that cryogel capture column showed better operational efficiency as compared to that of the Sepharose column. The Sepharose-based column was clogged in time by the cells and cell debris released from the cell culture device, whereas in the cryogel matrix these cells pass easily through the capture column and presented no problem.

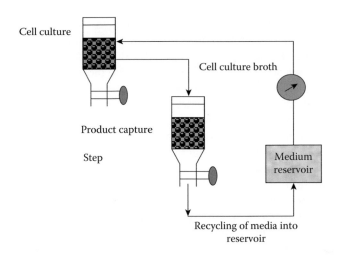

FIGURE 14.23

Integrated set-up of urokinase production and capture by human kidney cell line HT1080 using cryogels. Cell culture device: cells are growing and proliferating on gelatin-pAAm cryogel column (bed volume 5 ml); Product capture step: capture of urokinase on 5 ml Cu(II)-IDA-pAAm cryogel column; Medium reservoir: 500 ml circulating continuously through the columns at a flow rate of 0.2 ml/min. (Reproduced from Kumar, A., V. Bansal, K. S. Nandakumar, I.Yu. Galaev, P. K. Roychoudhury, R. Holmdahl, and B. Mattiasson, *Biotechnol. Bioeng.*, 93, 636–46, 2006. With permission.)

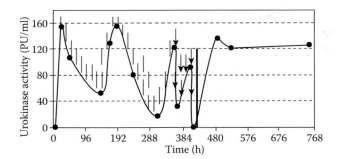

FIGURE 14.24

Production and capture of urokinase secreted by human kidney cell line HT1080. The cells were grown on gelatin-pAAm cryogel matrix for 32 days. The urokinase produced was captured on Cu(II)-IDA-pAAm cryogel column (bed volume, 5 ml). Short vertical lines indicate the replacement of the capture column with a fresh one. Arrows indicate connecting two capture columns (each with a bed volume of 5 ml) in series. On day 18 the medium reservoir was changed with new medium (indicated by thick vertical line). The samples were withdrawn regularly from the production column and the capture columns for the determination of urokinase activity and of any cells detached from the cryogel matrix. (Reproduced from Kumar, A., V. Bansal, K. S. Nandakumar, I.Yu. Galaev, P. K. Roychoudhury, R. Holmdahl, and B. Mattiasson, *Biotechnol. Bioeng.*, 93, 636–46, 2006. With permission.)

Thus, cryogels provided a scaffold for the growth and proliferation of anchorage-dependent cell lines and also a suitable support matrix for the preparation of an affinity chromatographic column that could efficiently capture urokinase without developing backpressure. The bioreactor could be run for periods as long as 32 days without contamination or other operational problems. This development is particularly significant with regard to production of urokinase, an enzyme of immense therapeutic significance, and hence, is a pursuit central to the modern biotechnology and pharmaceutical industry.

After the successful use of cryogel bioreactors for the production of therapeutic protein-like urokinase, the challenging task was to use these reactors for the production of monoclonal antibodies (mAb) from hybridoma cells. Cryogels have been used in disposable and economic bioreactor devices for production of an antibody against type II collagen using hybridoma cell line M2139 (Nilsang et al. 2007). Hybridoma cells were immobilized in the porous bed matrix of polyacrylamide-gelatin cryogel (10 ml bed volume). The cells were attached to the matrix within 48 hr after which they grew as a confluent sheet over the matrix for a period of 55 days (Figure 14.25). Cells were in lag phase for 15 days after which they secreted antibodies into the circulating

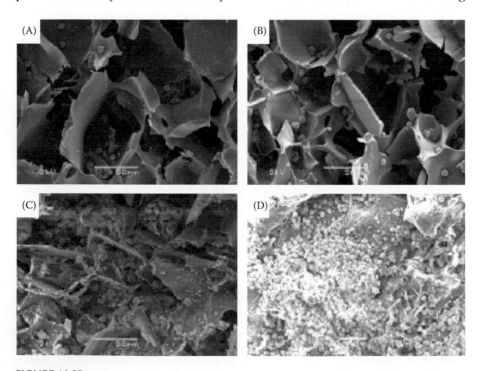

FIGURE 14.25
Cultivation of hybridoma cell line M2139 on supermacroporous cryogel matrix. Scanning electron micrograph pictures of the cells grown inside the matrix on (A) day 1, (B) day 7, (C) day 36, and (D) day 55. (Reproduced from Nilsang, S., K. S. Nandakumar, I.Yu. Galaev, S. K. Rakshit, R. Holmdahl, B. Mattiasson, and A. Kumar, *Biotechnol. Progr.*, 23, 932–29, 2007. With permission.)

medium. To evaluate the growth of cells inside the bioreactor, their metabolic profile was monitored during the total period of the bioreactor run. The yield of the mAb after purification was 67.5 mg L^{-1}, which was three times greater than the mAb yield obtained from the T-flask batch cultivation (Figure 14.26). The cells in the cryogel column were active even after the change of medium reservoir and had a stable mAb production for an extended period of time. The results from ELISA as well as *in vivo* arthritis experiments demonstrate that the antibodies secreted by cells grown on the cryogel column did not differ from antibodies purified from the cells grown in commercial CL-1000 culture flasks. Thus, supermacroporous cryogels can be useful as a supporting material for hybridoma cell culture and continuous production of antibodies.

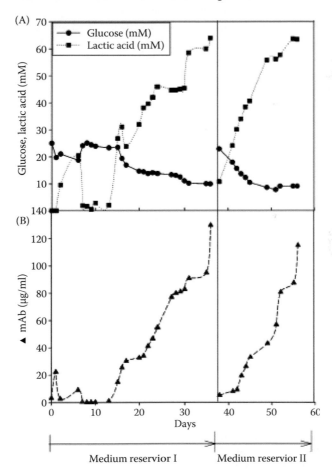

FIGURE 14.26
Glucose consumption, lactate, and mAb production from cryogel column in medium reservoirs I and II: (A) glucose consumption and lactate production; (B) mAb production. (Reproduced from Nilsang, S., K. S. Nandakumar, I.Yu. Galaev, S. K. Rakshit, R. Holmdahl, B. Mattiasson, and A. Kumar, *Biotechnol. Progr.*, 23, 932–29, 2007. With permission.)

14.8 Conclusion

Solid amounts of accumulated data clearly demonstrate the great potential of cryogel matrices in two major areas of biotechnology and biomedicine: cell separation and cell culture. A combination of unique macroporous structure, mechanical strength and elasticity, and the versatility of the cryogelation technology provide a successful application of cryogels as adsorbents for cell chromatography, as scaffolds and supports for tissue engineering, cell-based assays, cell expansion, and bioreactors. Various issues such as antigenicity, biocompatibility, rate of degradability, and so forth can be addressed by modulating the cryogel composition, concentration of polymeric precursors, freezing conditions, and surface modification of cryogels. Further *in vitro* and *in vivo* studies of cryogel scaffolds are needed and are being carried out in order to facilitate cryogels entering the clinical trials phase.

References

Ahlqvist, J., M. B. Dainiak, A. Kumar, E. G. Hörnsten, I. Yu. Galaev, and B. Mattiasson. 2006. Monitoring the production of inclusion bodies during fermentation and ELISA analysis of intact inclusion bodies using cryogel minicolumn plates. *Anal. Biochem.* 354:229–37.

Allen, L. T., E. J. Fox, I. Blute, Z. D. Kelly, Y. Rochev, A. K. Keenan, K. A. Dawson, and W. M. Gallagher. 2003. Interaction of soft condensed materials with living cells: Phenotype/transcriptome correlations for the hydrophobic effect. *Proc. Natl. Acad. Sci. USA* 100:6331–36.

Andac, M., F. M. Plieva, A. Denizli, I. Yu. Galaev, and B. Mattiasson. 2008. Poly(hydroxyethyl methacrylate) based macroporous hydrogels with disulfide cross-linker. *Macromol. Chem. Phys.* 209:577–84.

Anthony, E. T., M. Syed, S. Myers, G. Moir, and H. Navsaria. 2006. The development of novel dermal matrices for cutaneous wound repair. *Drug Discovery Today* 3:81–86.

Arnold, F. H. 1991. Metal-affinity separations: A new dimension in protein processing. *Bio/Technology* 9:150–55.

Arvidsson, P., F. M. Plieva, V. I. Lozinsky, I. Yu. Galaev, and B. Mattiasson. 2003. Direct chromatographic capture of enzyme from crude homogenate using immobilized metal affinity chromatography on a continuous supermacroporous adsorbent. *J. Chromatogr. A* 986:275–90.

Arvidsson, P., F. M. Plieva, I. N. Savina, V. I. Lozinsky, S. Fexby, L. Bülow, I. Yu. Galaev, and B. Mattiasson. 2002. Chromatography of microbial cells using continuous supermacroporous affinity and ion-exchange columns. *J. Chromatogr. A* 977:27–38.

Babacan, S., P. Pivarnik, S. Letcher, and A. G. Rand. 2000. Evaluation of antibody immobilization methods for piezoelectric biosensor application. *Biosens. Bioelectron.* 15:615–21.

Beaujean, F. 1997. Methods of CD34+ cell separation: Comparative analysis. *Transfus. Sci.* 18:251–61.

Bell, G. I. 1978. Model for specific adhesion of cells. *Science* 200:618–27.

Berndtsson, M., M. Hägg, T. Panaretakis, A. M. Havelka, M. C. Shoshan, and S. Linder. 2006. Acute apoptosis by cisplatin requires induction of reactive oxygen species but is not associated with damage to nuclear DNA. *Int. J. Cancer* 120:175–80.

Bhogal, N., C. Grindon, R. Combes, and M. Balls. 2005. Toxicity testing: Creating a revolution based on new technologies. *Trends Biotechnol.* 23:299–307.

Bloch, I., V. I. Lozinsky, I.Yu. Galaev, K. Yavriyanz, M. Vorobeychik, D. Azarov, L. G. Damshkaln, B. Mattiasson, and P. Vardi. 2005. Functional activity of insulinoma cells (INS-1E) and pancreatic islets cultured in agarose cryogel sponges. *J. Biomed. Mater. Res.* 75A:802–9.

Bolgen, N., F. Plieva, I.Yu. Galaev, B. Mattiasson, and E. Piskin. 2007. Cryogelation for preparation of novel biodegradable tissue-engineering scaffolds. *J. Biomater. Sci. Polym. Edn.* 8:1165–79.

Bolgen, N., I. Vargel, P. Korkusuz, E. Guzel, F. Plieva, I.Yu. Galaev, B. Mattiasson, and E. Piskin. 2009. Tissue responses to novel tissue engineering biodegradable cryogel-scaffolds: an animal model. *J. Biomed. Mater. Res. Part A.* DOI: 10.1002/jbm.a.32193.

Bonnafous, J. C., J. Dornand, J. Favero, M. Sizes, E. Boschetti, and J. C. Mani. 1983. Cell affinity chromatography with ligands immobilized through cleavable mercury-sulfur bonds. *J. Immunol. Methods* 11:93–107.

Borkenhagen, M., J-F. Clemence, H. Sigrist, and P. Aebischer. 1999. Three dimensional extracellular matrix engineering in the nervous system. *J. Biomed. Mater. Res.* 40:392–400.

Cao, X., R. Eisenthal, and J. Hubble. 2002. Detachment strategies for affinity-adsorbed cells. *Enzyme. Microb. Technol.* 31:153–60.

Chabner, B. A. 1990. Cytidine analogues. In *Cancer chemotherapy: Principles and practice*, ed. J. M. Collins. Philadelphia: J.B. Lippincott.

Chalmers, J. J., M. Zborowski, L. Sun, and L. Moore. 1998. Flow through, immuno-magnetic cell separation. *Biotechnol. Prog.* 14:141–48.

Collins, D. P., B. J. Luebering, and D. M. Shaut. 1998. T-lymphocyte functionality assessed by analysis of cytokine receptor expression, intracellular cytokine expression, and femtomolar detection of cytokine secretion by quantitative flow cytometry. *Cytometry* 33:249–55.

Cukierman, E., R. Pankov, D. R. Stevens, and K. M. Yamada. 2001. Taking cell-matrix adhesions to the third dimension. *Science* 294:1708–12.

Dainiak, M. B., I.Yu. Galaev, A. Kumar, F. M. Plieva, and B. Mattiasson. 2007. Chromatography of living cells using supermacroporous hydrogels, cryogels. *Adv. Biochem. Engin. Biotechnol.* 106:101–27.

Dainiak, M. B., I.Yu. Galaev, and B. Mattiasson. 2006. Affinity cryogel monoliths for screening for optimal separation conditions and chromatographic separation of cells. *J. Chromatogr. A* 1123:145–50.

Dainiak, M. B., I.Yu. Galaev, and B. Mattiasson. 2007. Macroporous monolithic hydrogels in a 96-minicolumn plate format for cell surface-analysis and integrated binding/quantification of cells. *Enzyme. Microb. Technol.* 40:688–95.

Dainiak, M. B., A. Kumar, I.Yu. Galaev, and B. Mattiasson. 2006. Detachment of affinity-captured bioparticles by elastic deformation of a macroporous hydrogel. *Proc. Natl. Acad. Sci. USA* 103:849–54.

Dainiak, M. B., A. Kumar, I.Yu. Galaev, and B. Mattiasson. 2007. Methods in cell separations. *Adv. Biochem. Engin. Biotechnol.* 106:1–18.

Dainiak, M. B., A. Kumar, F. M. Plieva, I.Yu. Galaev, and B. Mattiasson. 2004. Integrated isolation of antibody fragments from microbial cell culture fluids using super-macroporous cryogels. *J. Chromatogr. A* 1045:93–98.

Dainiak, M. B., B. Mattiasson, and I.Yu. Galaev. 2007. Macroporous hydrogels, cryo-gels: Novel biomaterials for tissue engineering. *BIOforum Europe* 12:18–19.

Dainiak, M. B., F. M. Plieva, I.Yu. Galaev, R. Hatti-Kaul, and B. Mattiasson. 2005. Cell chromatography: Separation of different microbial cells using IMAC superma-croporous monolithic columns. *Biotechnol. Progr.* 21:644–49.

Dainiak, M. B., I. Savina, I. Musolino, A. Kumar, B. Mattiasson, and I.Yu. Galaev. 2008. Biomimetic macroporous hydrogel scaffolds in a high throughput screening for-mat for cell-based assays. *Biotechnol. Progr.* 24:1373–83.

Deng, M. Q., D. O. Cliver, and T. W. Mariam. 1997. Immunomagnetic capture PCR to detect viable *Cryptosporidium parvum* oocysts from environmental samples. *Appl. Environ. Microbiol.* 63:3134–48.

De Wynter, E. W., L. H. Coutinho, X. Pei, J. C. W. Marsh, J. Hows, T. Luft, and N. G. Testa. 1995. Comparison of purity and enrichment of CD34+ cells from bone marrow, umbilical cord and peripheral blood (primed for apheresis) using five separation systems. *Stem Cells* 13:524–32.

Discher, D. E., P. Janmey, and Y.-L. Wang. 2005. Tissue cells feel and respond to the stiffness of their substrate. *Science* 310:1139–43.

Ehrenreich, M., and Z. Ruszczak. 2006. Update on tissue-engineered biological dres-sings. *Tissue Engin.* 12:2407–24.

Elbert, D. L., and J. A. Hubbell. 2001. Conjugate addition reactions combined with free-radical cross-linking for the design of materials for tissue engineering. *Biomacromolecules* 2:430–31.

Falconnet, D., G. Csucs, H. M. Grandin, and M. Textor. 2006. Surface engineering approaches to micropattern surfaces for cell-based assays. *Biomaterials* 27:3044–63.

Fuss, C., J. C. Palmaz, and E. A. Sprague. 2001. Fibrinogen: Structure, function, and surface interactions. *J. Vasc. Interv. Radiol.* 12:677–82.

Gaberc-Porekar, V., and V. Menart. 2001. Perspectives of immobilized-metal affinity chromatography. *J. Biochem. Biophys. Methods* 49:335–60.

Galaev, I.Yu., M. B. Dainiak, F. M. Plieva, R. Hatti-Kaul, and B. Mattiasson. 2005. High throughput screening of particulate-containing samples using supermacro-porous elastic monoliths in microtiter (multiwell) plate format. *J. Chromatogr. A* 1065:169–75.

Galaev, I.Yu., M. B. Dainiak, F. M. Plieva, and B. Mattiasson. 2007. Effect of matrix elas-ticity on affinity binding and release of bioparticles. Elution of bound cells by temperature-induced shrinkage of the macroporous hydrogel. *Langmuir* 23:35–40.

Gohji, K., M. Nakajima, D. Boyd, C. P. N. Dinney, C. D. Bucana, S. Kitazana, S. Kamidono, and I. J. Fidler. 1997. Organ-site dependence for the production of urokinase-type plasminogen activator and metastasis by human renal cell carci-noma cells. *Am. J. Pathol.* 151:1655–61.

Griffith, L. G., and G. Naughton. 2002. Tissue engineering-current challenges and expanding opportunities. *Science* 295:1009–14.

Gross, H. J., B. Verwer, D. Houck, R. A. Hoffman, and D. Recktenwald. 1995. Model study detecting breast cancer cells in peripheral blood mononuclear cells at fre-quencies as low as 10^{-7}. *Proc. Natl. Acad. Sci. USA* 92:537–41.

Gustavsson, P.-E., and P.-O. Larsson. 1999. Continuous superporous agarose beds for chromatography and electrophoresis. *J. Chromatogr. A* 832:29–39.

Haas, W., J. W. Schrader, and A. Szenberg. 1974. A new, simple method for the preparation of lymphocytes bearing specific receptors. *Eur. J. Immunol.* 4:565–70.

Halstenberg, S., A. Panitch, S. Rizzi, H. Hall, and J. A. Hubbel. 2002. Biologically engineered protein-graft-poly(ethylene glycol)hydrogels: A cell plasmindegradable biosynthetic material for tissue repair. *Biomacromolecules* 3:710–23.

Helmlinger, G., P. A. Netti, H. C. Lichtenbeld, R. J. Melder, and R. K. Jain. 1995. Solid stress inhibits the growth of multicellular tumor spheroids. *Nat. Biotechnol.* 15:778–83.

Henriet, P., Z.-D. Zhong, P. C. Brooks, K. I. Weinberg, and Y. A. DeClerck. 2000. Contact with fibrillar collagen inhibits melanoma cell proliferation by up-regulating p27^{KIP1}. *Proc. Natl. Acad. USA* 97:10026–31.

Hentze, H.-P., and M. Antonietti. 2002. Porous polymers and resins for biotechnological and biomedical applications. *Rev. Mol. Biotechnol.* 90:27–53.

Hersel, U., C. Dahmen, and H. Kessler. 2003. RGD modified polymers: Biomaterials for stimulated cell adhesion and beyond. *Biomaterials* 24:4385–4415.

Hubble, J. 1997. Multivalent antibody mediated interactions. *Immunol. Today* 18:305–306.

Hutmacher, D. W., and W. Vanscheidt. 2002. Matrices for tissue-engineered skin. *Drugs Today* 38:113–33.

Ibrahim, S. F., and G. Van Den Engh. 2003. High-speed cell sorting: Fundamentals and recent advances. *Curr. Opin. Biotechnol.* 14:5–12.

Islam, D., and A. A. Lindberg. 1992. Detection of *Shigella dysenteriae* type 1 and *Shigella flexneri* in feces by immunomagnetic isolation and polymerase chain reaction. *J. Clin. Microbiol.* 30:2801–6.

Jo, E. C., J. W. Yun, K. H. Jung, S. I. Chung, and J. H. Kim. 1998. Performance study of perfusion cultures for the production of single-chain urokinase-type plasminogen activator (scu-PA) in a 2.5 l spin-filter bioreactor. *Bioprocess Bioeng.* 19:363–72.

Kathuria, N., A. Tripathy, K. K. Kar, and A. Kumar. 2008. Synthesis and characterisation of elastic and macroporous chitosan-gelatin cryogels for tissue engineering. *Acta Biomaterialia* 5:406–18.

Keely, P., L. Parise, and R. Juliano. 1998. Integrins and GTPases in tumor cell growth, motility and invasion. *Trends Cell Biol.* 8:101–6.

Kim, J. B. 2005. Three-dimensional tissue culture models in cancer biology. *Seminars Cancer Biol.* 15:365–77.

Kumar, A., V. Bansal, K. S. Nandakumar, I.Yu. Galaev, P. K. Roychoudhury, R. Holmdahl, and B. Mattiasson. 2006. Integrated bioprocess for the production and isolation of urokinase from animal cell culture using supermacroporous cryogel matrices. *Biotechnol. Bioeng.* 93:636–46.

Kumar, A., F. M. Plieva, I.Yu. Galaev, and B. Mattiasson. 2003. Affinity fractionation of lymphocytes using supermacroporous monolithic cryogel. *J. Immunol. Methods* 283:185–94.

Kumar, A., A. Rodriguez-Caballero, F. M. Plieva, I.Yu. Galaev, K. S. Nandakumar, M. Kamihira, R. Holmdahl, A. Orfao, and B. Mattiasson. 2005. Affinity binding of cells to cryogel adsorbents with immobilized specific ligands: Effect of ligand coupling and matrix architecture. *J. Mol. Rec.* 18:84–93.

Larsson, M., V. Arasaratnam, and B. Mattiasson. 1989. Integration ofbioconversion and downstream processing. Starch hydrolysis in aqueous two-phase syste. *Biotechnol. Bioeng.* 33, 758–66.

Lozinsky, V. I., I.Yu. Galaev, F. M. Plieva, I. N. Savina, H. Jungvid, and B. Mattiasson. 2003. Polymeric cryogels as promising materials of biotechnological interest. *Trends Biotechnol.* 21:445–51.

Lukashev, M. E., and Z. Werb. 1998. ECM signalling: Orchestrating cell behaviour and misbehaviour. *Trends Cell Biol.* 8:437–41.

Luther, H., S. Hirsch, E. Schuster, and E. Weber. 1992. Hollow fibre modules as membrane reactor in biocatalysis. *Acta Biotechnol.* 12:133–42.

Madrusov, E., A. Houng, E. Klein, and E. F. Leonard. 1995. Membrane-based cell affinity chromatography to retrieve viable cells. *Biotechnol. Prog.* 11:208–13.

Maniotis, A. J., C. S. Chen, and D. E. Ingber. 1997. Demonstration of mechanical connections between integrins, cytoskeletal filaments, and nucleoplasm that stabilize nuclear structure. *Proc. Natl. Acad. Sci. USA* 94:849–54.

Mao, J. S., H. F. Liu, Y. J. Yin, and K. D. Yao. 2003. The properties of chitosan-gelatin membranes and scaffolds modified with hyaluronic acid by different methods. *Biomaterials* 24:1621–29.

Ming, F., W. J. D. Whish, and J. Hubble. 1998. Estimation of parameters for cell-surface interactions: Maximum binding force and detachment constant. *Enzyme. Microb. Technol.* 22:94–99.

Nilsang, S., K. S. Nandakumar, I.Yu. Galaev, S. K. Rakshit, R. Holmdahl, B. Mattiasson, and A. Kumar. 2007. Monoclonal antibody production using a new supermacroporous cryogel perfusion reactor. *Biotechnol. Progr.* 23:932–29.

Noppe, W., F. M. Plieva, K. Vanhoorelbeke, H. Deckmyn, M. Tuncel, A. Tuncel, I.Yu. Galaev, and B. Mattiasson. 2007. Macroporous monolithic gels, cryogels, with immobilized phages from phage-display library as a new platform for fast development of affinity adsorbent capable of target capture from crude feeds. *J. Biotechnol.* 131:293–99.

Oshiba, A., H. Renz, J. Yata, and E. W. Gelfand. 1994. Isolation and characterization of human antigen-specific B lymphocytes. *Clin. Immunol. Immunopathol.* 72:342–49.

Payne, M. J., S. Campbell, R. A. Patchett, and R. G. Kroll. 1992. The use of immobilized lectins in the separation of *Staphylococcus aureus, Escherichia coli, Listeria* and *Salmonella* spp. from pure cultures and foods. *J. Appl. Bacteriol.* 73:41–52.

Pelham, R. J., and Y. Wang. 1997. Cell locomotion and focal adhesions are regulated by substrate flexibility. *Proc. Natl. Acad. Sci. USA* 94:13661–65.

Plieva, F., B. Bober, M. Dainiak, I.Yu. Galaev, and B. Mattiasson. 2006. Macroporous polyacrylamide monolithic gels with immobilized metal affinity ligand: The effect of porous structure and ligand coupling chemistry on protein binding. *J. Mol. Recognit.* 19:305–12.

Plieva, F. M., I.Yu. Galaev, and B. Mattiasson. 2007. Macroporous gels prepared at subzero temperatures as novel materials for chromatography of particulate-containing fluids and cell culture applications. *J. Sep. Sci.* 30:1657–71.

Plieva, F., A. Oknianska, E. Degerman, I.Yu. Galaev, and B. Mattiasson. 2006. Novel supermacroporous dextran gels. *J. Biomater. Sci. Polymer Ed.* 17:1075–92.

Prestvik, W. S., A. Berge, P. C. Mork, P. M. Stenstad, and J. Ugelstad. 1997. Preparation and application of monosized magnetic particles in selective cell separation. In *Scientific and clinical applications of magnetic carriers*, eds. U. Häfeli, W. Schutt, J. Teller, and M. Zborowski. New York, London: Plenum Press.

Racila, E., D. Euhus, A. J. Weiss, C. Rao, J. McConnell, L. Terstappen, and J. Uhr. 1998. Detection and characterization of carcinoma cells in the blood. *Proc. Natl. Acad. Sci. USA* 95:4589–94.

Rowley, J. A., G. Madlambayan, and D. J. Mooney. 1999. Alginate hydrogels as synthetic extracellular matrix materials. *Biomaterials* 20:45–53.

Ryningen, A., L. Wergeland, N. Glenjen, B. T. Gjertsen, and O. Bruserud. 2005. *In vitro* crosstalk between fibroblasts and native human acute myelogenous leukemia (AML) blasts via local cytokine networks results in increased proliferation and decreased apoptosis of AML cells as well as increased levels of proangiogenic interleukin 8. *Leukemia Res.* 29:185–96.

Salamah, A. A. 1992. Effect of medium salinity on amino acid and phospholipid composition of two halophilic *Bacillus* species from Saudi Arabia. *J. Univ. Kuwait, Science (CAN 118:143092)* 19 (2):231–38.

Santini, M. T., G. Rainaldi, and P. L. Indovina. 2000. Apoptosis, cell adhesion and the extracellular matrix in the three-dimensional growth of multicellular tumor spheroids. *Critical Rev. Oncol./Hematol.* 36:75–87.

Savina, I. N., V. Cnudde, S. D'Hollander, L. VanHoorebeke, B. Mattiasson, I.Yu. Galaev, and F. DuPrez. 2007. Cryogels from poly(2-hydroxyethyl methacrylate): Macroporous, interconnected materials with potential as cell scaffolds. *Soft Matter* 3:1176–84.

Savina, I. S., M. B. Dainiak, H. Jungvid, S. V. Mikhalovsky and I. Yu. Galaev. 2009. Biomimetic macroporous hydrogels: Protein ligand distribution and cell response to the ligand architecture in the scaffold. *J. Biomater. Sci.* 20:1781–95.

Savina, I. N., I. Yu. Galaev, and B. Mattiasson. 2005. Anion-exchange supermacroporous monolithic matrices with grafted polymer brushes of N,N-dimethyl-aminoethyl-methacrylate. *J. Chromatogr. A* 1092:199–205.

Savina, I. N., B. Mattiasson, and I.Yu. Galaev. 2005. Graft polymerization of acrylic acid onto macroporous polyacrylamide gel (cryogel) initiated by potassium diperiodatocuprate. *Polymer* 46:9596–603.

Savina, I. N., B. Mattiasson, and I.Yu. Galaev. 2006. Graft polymerization of vinyl monomers inside macroporous polyacrylamide gel, cryogel, in aqueous and aqueous-organic media initiated by diperiodatocuprate(III) complexes. *J. Polymer Sci. Part A: Polym. Chem.* 44 (6):1952–63.

Seesod, N., P. Nopparat, A. Hedrum, A. Holder, S. Thaithong, M. Uhlen, and J. Lundeberg. 1997. An integrated system using immunomagnetic separation, polymerase chain reaction, and colorimetric detection for diagnosis of *Plasmodium falciparum. Am. J. Trop. Med. Hyg.* 56:322–28.

Seidl, J., R. Knuechel, and L. A. Kunz-Schughart. 1999. Evaluation of membrane physiology following fluorescence activated or magnetic cell separation. *Cytometry* 36:102–11.

Shin, H., S. Jo, and A. G. Mikos. 2003. Biomimetic materials for tissue engineering. *Biomaterials* 24:4353–64.

Siewert, C., M. Herber, N. Hunzelmann, O. Fodstad, S. Miltenyi, M. Assenmacher, and J. Schmitz. 2001. Rapid enrichment and detection of melanoma cells from peripheral blood mononuclear cells by a new assay combining immunomagnetic cell sorting and immunocytochemical staining. *Recent Results Cancer Res.* 158:51–60.

Stile, R. A., and K. E. Healy. 2001. Thermo-responsive peptide-modified hydrogels for tissue regeneration. *Biomacromolecules* 2:185–94.

Sundberg, S. A. 2000. High-throughput and ultra-high-throughput screening: Solution- and cell-based approaches. *Curr. Opin. Biotechnol.* 11:47–53.

Thid, D., M. Bally, K. Holm, S. Chessari, S. Tosatti, M. Textor, and J. Gold. 2007. Issues of ligand accessibility and mobility in initial cell attachment. *Langmuir* 23:11693–704.

Thomson, J. A., J. Itskovitz-Eldor, S. S. Shapiro, M. A. Waknitz, J. J. Swiergiel, V. S. Marshall, and J. M. Jones. 1998. Embryonic stem cell lines derived from human blastocysts. *Science* 282:1145–47.

Tripathy, A., N. Kathuria, and A. Kumar. 2008. Elastic and macroporous agarose-gelatin cryogels with isotropic and anisotropic porosity for tissue engineering. *J. Biomed. Mater. Res. Part A* 90A:680–94.

Trotsenko, Yu. A., and V. N. Khmelenina. 2002. The biology and osmoadaptation of haloalkaliphilic methanotrophs. *Microbiology* 71 (2):123–32.

Ujam, L. B., R. H. Clemmitt, S. A. Clarke, R. A. Brooks, N. Rushton, and H. A. Chase. 2003. Isolation of monocytes from human peripheral blood using immuno-affinity expanded-bed adsorption. *Biotechnol. Bioeng.* 83:554–66.

Varner, J. A., and D. A. Cheresh. 1996. Integrins and cancer. *Curr. Opin. Cell Biol.* 8:724–30.

Waleh, N. S., J. Gallo, T. D. Grant, B. J. Murphy, R. H. Kramer, and R. M. Sutherland. 1994. Selective down-regulation of integrin receptors in spheroids of squamous cell carcinoma. *Cancer Res.* 54:838–43.

Weaver, V. M., and C. D. Roskelley. 1997. Extracellular matrix: The central regulator of cell and tissue homeostasis. *Trends Cell Biol.* 7:40–42.

West, J. L., and J. A. Hubbell. 1999. Polymeric biomaterials with degradation sites for proteases involved in cell migration. *Macromolecules* 32:241–44.

Wong, J. Y., J. B. Leach, and X. Q. Brown. 2004. Balance of chemistry, topography, and mechanics at the cell-biomaterial interface: Issues and challenges for assessing the role of substrate mechanics on cell response. *Surf. Sci.* 570:119–33.

Zhang, S. 2004. Beyond the Petri dish. *Nature Biotechnol.* 22:151–52.

15

Macroporous Polymeric Scaffolds for Tissue Engineering Applications

Ashok Kumar, Era Jain, and Akshay Srivastava

CONTENTS

15.1 Introduction

A paradigm shift is taking place in medicine from using synthetic implants and tissue grafts to a tissue engineering approach that uses degradable porous material scaffolds integrated with biological cells or molecules to regenerate tissues. This new paradigm requires scaffolds that balance temporary mechanical function with mass transport to aid biological delivery and tissue regeneration [1]. Currently there are three approaches used in tissue engineering, the first one requires delivery of the appropriate cells at the defect site. The second and third approach requires a biodegradable scaffold with or without cells to be implanted at the defect site, which

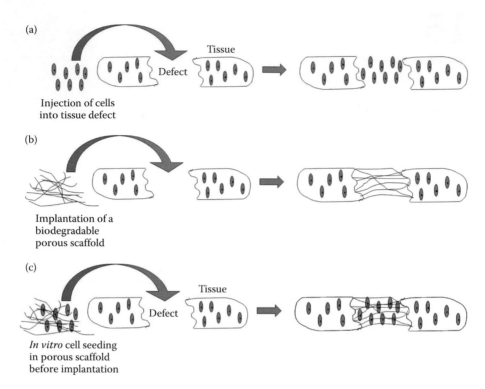

FIGURE 15.1
Schematic representation of the three main generic tissue engineering approaches that can be used to regenerate a tissue defect. (a) Cell self-assembly, (b) acellular scaffold, and (c) cell seeded polymeric scaffold.

eventually helps in healing the defect (Figure 15.1). Thus making macroporous three-dimensional polymer scaffolds an important part in tissue engineering. Successful tissue regeneration by both the second and third approaches requires strong interaction among three components: the cells that restore tissue, a scaffold to hold the cells as they create tissues, and signaling moieties that direct the cells to form the tissue. Interactions at the cell–polymer interface are of fundamental importance in tissue engineering. Functional substitutes of damaged tissues result through complex interactions of living cells, bioactive molecules, and three-dimensional porous scaffolds. The 3-D porous scaffolds support cell attachment, proliferation, and differentiation. The physicochemical nature of the scaffold surface strongly influences the number of cells that attach and their course of differentiation and growth.

There are a few basic requirements that have been widely accepted for designing polymer scaffolds: (a) A scaffold has to have high porosity and proper pore size. (2) A high surface area is needed. (3) Biodegradability is generally required and a proper degradation rate is needed to match the rate of neotissue formation. (4) The scaffold must have the required mechanical integrity to maintain the predesigned tissue structure. (5) The scaffold should not be toxic to the cells (i.e., biocompatible). (6) The scaffold should positively interact with cells, including enhanced cell adhesion, growth, migration, and induce differentiation to maintain required function. Many techniques have been investigated in recent years to form highly porous biodegradable scaffolds suitable for use in tissue engineering [2]. Many of these methods are able to form foams with high porosity to encourage cell attachment. Of these, the methods of fiber bonding, solvent casting/particulate leaching, gas foaming/particulate leaching, and liquid–liquid phase separation produce materials with large, interconnected pores to facilitate cell seeding and migration. The fiber bonding, solvent casting/particulate leaching, and gas foaming/particulate leaching methods produce materials that exhibit good biocompatibility, making these techniques especially promising for future use in tissue-engineered cell-polymer constructs.

More specifically, the regeneration of specific tissues aided by synthetic or natural materials has been shown to be dependent on the porosity and pore size of the supporting three-dimensional structure [3]. A large surface area favors cell attachment and growth, whereas a large pore volume is required to accommodate and subsequently deliver a cell mass sufficient for tissue repair. Highly porous biomaterials are also desirable for the easy diffusion of nutrients to and waste products from the implant and for vascularization, which are major requirements for the regeneration of highly metabolic organs such as the liver and pancreas. The surface area/volume ratio of porous materials depends on the density and average diameter of the pores. Nevertheless, the diameter of cells in suspension dictates the minimum pore size, which varies from one cell type to another. Depending on the envisioned applications, pore size must be carefully controlled. The effect of implant pore size on tissue regeneration is emphasized by experiments demonstrating optimum pore size of 5 μm for neovascularization, 5–15 μm for fibroblast ingrowth, close to 20 μm for the ingrowth of hepatocytes, 20–125 μm for regeneration of adult mammalian skin, 40–100 μm for osteoid ingrowth [5] and 100–350 μm for regeneration of bone [5]. Fibrovascular tissues appear to require pores sizes greater than 500 μm for rapid vascularization and for the survival of transplanted cells [6]. Another important consideration is the continuity of the pores within a synthetic matrix. Mass transport and cell migration will be inhibited if the pores are not interconnected even if the matrix porosity is high [7]. Mass transport is one of the most significant challenges in tissue engineering. Large-scale cell transplantation in open structures is presently limited by inadequate nutrient delivery. Cells located more than approximately 200 μm from a blood supply are either metabolically inactive or necrotic due to low oxygen tension [8]. It is for this reason that hodrocytes, with its very low metabolic activity, has been one of the few cell types successfully engineered into large tissue structures, namely cartilage. A further concern is the changes in the effective pore structure over time *in vivo*. If the matrices are biodegradable, as in the case with polylactic acid (PLA) and polyglycolic acid (PGA) matrices, the average pore size will increase and bottlenecks in the continuity of the pore structure will open [9]. If the matrix does not degrade, its effective pore size may be reduced by *in vivo* events such as the invasion of fibrous tissue into the pores and the nonspecific adsorption of proteins onto the material's surface [8]. Besides pore size and porosity, the shape and tortuosity can affect tissue ingrowth [10]. Strong cell adhesion and spreading often favor proliferation while rounded cell morphology is required for cell-specific function [11]. Thus, a polymer scaffold must act as a suitable substrate to maintain differentiated functions without hindering proliferation [12].

A compilation of various techniques being used for fabrication of 3-D porous scaffolds is given in Table 15.1. However, almost all currently available techniques require the use of organic solvents, which could reduce the ability of cells to form new tissues *in vivo*. Thus, long processing times to fully remove these solvents are necessary. To overcome this problem, other combinations of materials and pore-forming techniques must be explored to create constructs that can be fabricated during surgery and tailored for specific applications. It is only when these clinical design criteria have been addressed that tissue engineered constructs will see widespread use to aid patients suffering from various types of organ and tissue failure. Recently, a new type of polymeric scaffold technique has emerged with great potential, known as "cryogelation." The monolithic supermacroporous cryogels with pore size up to 100 μm have been developed [33–35]. These matrices are produced by polymerization of the hydrophilic monomers or by crosslinking of hydrophilic polymers at subzero temperatures when most of the solvent that is water is frozen while the dissolved substances (monomers or polymers) are concentrated in small nonfrozen regions. The reaction proceeds in these nonfrozen regions while the crystals of frozen solvent perform like porogen. After melting of the ice crystals, a system of large interconnected pores is formed. The pore size depends on the initial concentration of reagents in solution and the freezing conditions. A matrix with a system of large interconnected pores allows application of cell suspension directly on a column. These cryogels are spongy and elastic and form attractive matrices for cell culture. Cryogels composed of a mixture of natural and synthetic polymers have been tested for tissue engineering application [27–32]. Thermoresponsive cryogels composed of polyvinylcaprolactam have also been applied for cell culture and are further being developed as potential thermoresponsive scaffolds for tissue engineering [21]. The fabrication of macroporous materials using the above techniques have been described in detail in other chapters. A compilation of different types of polymers being used for various tissue engineering applications is given in Table 15.2. This chapter focuses on the applications of 3-D macroporous materials in tissue engineering.

TABLE 15.1

Methods Used to Process Biomaterials into Tissue Engineering Scaffolds

Fabrication Technique	Requirement for Materials	Reproducibility	Scaffold Architecture	Biomaterials	Problems	Reference
Impregnate sintering	Withstand high temperature	Sensitive to sintering	Pore size: 200–1000 μm; porosity: 50% foam dependent	HA, TCP	Brittle	16–18
Solvent casting and particulate leaching	Soluble in cell nontoxic solvent	User and materials dependent	Pore size: 50–1000 μm; porosity: 30–90%	PLA, PLGA, collagen and so on	Solvent toxicity Particulate remnant	13, 14
Phase separation/ emulsion in combination with freezing drying/ critical point drying	Soluble in cell nontoxic solvent	Emulsion formation sensitive to stirring	Pore size, 200 μm; Porosity: 70–95%	PLGA, PLA, PLLA and collagen	Solvent toxicity Pore size difficult to control	15
Fiber knitting/ nonwoven/ bonding	Fiber	Machine control Solvent sensitive	Interconnected channels, 20–100 μm in diameter	PVA, PLA, PLGA	Lack of rigidity	16, 17
Solid free form	Low melting point and thermoplastic	Computer control	Interconnected channels Complex shape and structure ≈150 μm Customer based	PEG, PLA, PLGA Collagen, starch, HA, TCP	Costly	18–26
Cryogel	Soluble in water or solvents	User and material dependent	Interconnected pore capillaries with pore size upto 100 μm and porosity of 90%	Chitosan, NiPAAm, agarose, gelatin, polyacrylonitrile, polyvinylcaprolactam, polyacrylamide, alginate, dextran, etc.		27–32

Source: Reproduced from Liu C., Xia Z., Czernuszka J.T., *Trans. IChemE, Part A, Chem. Eng. Res. Design.*, 85 (A7), 1051–64, 2007.

TABLE 15.2

Common Polymeric Macroporous Material Studied for Tissue
Engineering Applications

Material	Application	Reference
Natural materials		
Collagen	Cartilage	40, 64
	Skin	203, 221
	Neural	218, 219, 246, 251
	Bone	122, 123, 153
Silk	Cartilage	91, 92
	Bone	136, 138–140
Fibrin	Cartilage	44, 62
	Skin	215, 220
Hyaluronic acid	Cartilage	80, 97, 101
Chitosan	Cartilage	71, 76, 84
	Bone	123, 125–130
	Skin	192–193, 195–200
	Neural	244, 245
Alginate	Cartilage	71, 72, 82
	Bone	165
	Neural	265–268
Gelatin	Cartilage	89
	Skin	200, 213, 214
Synthetic material		
Poly ethylene glycol (PEG)	Cartilage	48, 49, 52
PLGA	Cartilage	94, 96, 98, 141
	Bone	161–163
	Neural	238–240
PLLA	Cartilage	72, 98
	Bone	153, 166
	Skin	181, 206
	Neural	260, 222, 243
PCL	Cartilage	95, 144
	Bone	159, 160
Poly(HEMA)	Neural	228, 230

15.2 3-D Macroporous Scaffolds for Tissue Engineering

15.2.1 Cartilage Tissue Engineering

For cartilage and bone tissues, a suitable scaffold provides porosity, initial mechanical stability, and supports even cell distribution. Research is
focused on developing bioresorbable scaffolds that exhibit optimal physical

properties (e.g., macroporosity) coupled with excellent biocompatibility [36]. Scaffolds act as shape and guidance templates for *in vitro* and *in vivo* tissue development [37]. Natural polymeric gels, such as hyaluronic acid (HA), collagen, alginate [38], and chitosan have been used successfully [39]. These scaffolds permit 3-D immobilization of cells and maintain the differentiated phenotype of chondrocytes [40]. However, their mechanical behavior is insufficient for tissue transplantation and so solid bioresorbable fiber scaffolds or other porous structures are used to achieve initial biomechanical stability [41]. Synthetic biodegradable poly(α-hydroxy esters) such as PLA, PGA, and copolymer poly(lactide-*co*-glycolic) acid (PLGA) [42] have been used extensively in this context. Both types of materials increase proteoglycan synthesis compared with collagen scaffolds [43]. Injectable *in situ* cross-linkable polymeric preparations that entrap cells have been designed [44] and techniques that combine the advantages of both porous fiber structures and gels are being explored as suitable alternatives to either gel or fiber scaffolds [45] (Figure 15.2). Research is also focused on developing smart scaffolds that incorporate inflammatory inhibitors or antibiotics. Slow and controlled release of these bioactive molecules provides sufficient time to the new cartilage to adapt and mature in a hostile *in vivo* situation or to prevent early infection after surgery [46].

To date, a wide range of natural and synthetic materials have been investigated as scaffolding for cartilage repair. Natural polymers that have been

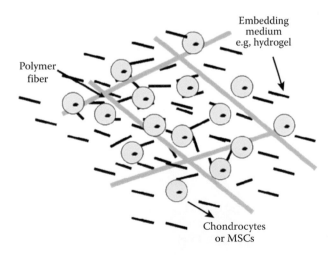

FIGURE 15.2
Schematic drawing showing the strategy of developing tissue engineered cartilage constructs using fibers and embedding substances. Embedding substances offer 3-D immobilization and uniform distribution of cells in the fibre mesh. (From Risbud M.V., Sittinger M., *Trends Biotechnol.*, 20, 351–56, 2002. With permission.)

explored as bioactive scaffolds for cartilage engineering include: alginate, agarose, fibrin, HA, collagen, gelatin, chitosan, chondroitin sulfate, and cellulose [47].

On the other hand, synthetic polymers are more controllable and predictable, where chemical and physical properties of a polymer can be modified to alter mechanical and degradation characteristics. Synthetic polymers currently explored for cartilage repair include: poly(α-hydroxy esters), polyethylene glycol (PEG), poly(N-isopropyl acrylamide) (polyNiPAAm), poly(propylene fumarates), and polyurethanes. However, unless suitably modified, synthetic polymers do not benefit from direct cell-scaffold interactions, which can play a role in adhesion, cell signaling, directed degradation, and matrix remodeling. In addition, degradation by-products may be toxic or elicit an inflammatory response. Finally, scaffold architecture also plays a major role in dictating cellular behavior. The primary focus has been on polymeric materials, in forms of hydrogels, sponges, and fibrous meshes [47] (Figure 15.3).

Macroporous hydrogels support the nutrients transport, and can homogenously suspend cells in a 3-D environment, where encapsulated cells typically retain a rounded morphology that may induce a chondrocyte phenotype. The PEG is a relatively inert polymer and supports chondrogenesis when cross-linked into hydrogels [48]. Further modifications of PEG, including the addition of hydrolyzable units and bioactive peptides have improved cartilage tissue growth [49]. For instance, degradable lactic acid units have been added to PEG hydrogels to increase cell proliferation and extracellular matrix (ECM) deposition [48–51]. Recently, Lee and colleagues covalently incorporated a CMP into PEG hydrogels [52]. CMP is known to associate with type I collagen and other ECM fibers, forming physical cross-links that can then be manipulated by cells. This study showed that PEG hydrogels conjugated with CMP limited the diffusion of exogenous type I collagen and increased ECM production by encapsulated chondrocytes. PEG has also been combined with methacrylated poly(glycerol succinic acid) dendrimers [53]. Another polymer used for cartilage tissue regeneration is HA, a linear polysaccharide found natively in cartilage. It functions as a core molecule for

FIGURE 15.3
Different formats of macroporous scaffold used for tissue engineering. (Adapted from Chung C., Burdick J.A., *Adv. Drug Deliv. Rev.*, 60, 243–62, 2008. With permission.)

the binding of keratin sulfate and chondroitin sulfate in forming aggrecan in cartilage and degrades primarily by hyaluronidases found throughout the body.

HA plays a role in cellular processes like cell proliferation, morphogenesis, inflammation, and wound repair [54] and may function as a bioactive scaffold, where cell surface receptors for HA (CD44, CD54, and CD168) allow for cell/scaffold interactions. Macroporous scaffolds of modified HA with methacrylate groups [55,56] were prepared via photopolymerization and by varying the molecular weight and concentration of the modified HA, a wide range of physical properties can be obtained [57]. Increased macromer concentrations significantly increased the network compressive modulus and degradation time while decreasing the swelling ratio and cell viability [57]. These variations in scaffold properties also affected neocartilage formation by auricular chondrocytes *in vivo* [58]. In a recent *in vivo* rabbit defect model, Liu et al. investigated the quality of repair using macroporous HA-gelatin hydrogels seeded with mesenchymal stem cells (MSCs). Defects with MSCs alone exhibited hyaline-like cartilage on the peripheral defect area and fibrous repair in the middle, whereas defects filled with a macroporous scaffold and MSCs resulted in elastic, firm, translucent cartilage with zonal architecture and good integration with the surrounding cartilage [59]. Chondroitin sulfate, another major constituent of cartilage, can also be photopolymerized with similar modifications to produce macroporous hydrogels that exhibit viscoelastic behavior [60]. These chondroitin sulfate-based macroporous hydrogels support viable chondrocytes and can be degraded in the presence of chondroitinase ABC. Furthermore, chondroitin sulfate can be copolymerized with PEG to increase the hydrogel pore size and provide bioactive cues for encapsulated cells [60].

Fibrin glue is a natural polymer formed from the polymerization of fibrinogen with thrombin, and it elicits good biocompatibility as a wound adhesive and can facilitate cell–matrix interaction via integrin binding [61]. It is attractive as a natural scaffold because it can be made from autologous blood. However, one drawback is that macroporous gels tend to shrink *in vivo*. Recently, a long-term stable fibrin gel was developed that is transparent and stable for 3 weeks [62]. This gel exhibits a broad linear viscoelastic region, withstands loads of 0.0001–10 kPa, and supports chondrocyte proliferation and cartilaginous ECM production while retaining its size and shape. Studies in nude mice have shown the suitability of using fibrin glue as a biomaterial, where degradation and polymerization time can be controlled by fibrinogen and thrombin concentrations, respectively [44]. Fibrin glue has also been combined with other polymers like polyurethane and improved cell seeding viability and distribution, and increased the expression of aggrecan and type II collagen [63]. Type I and type II collagen scaffolds have inherent biological cues that allow chondrocytes to interact and remodel the hydrogel. A type I collagen gel seeded with autologous chondrocytes has been used to treat full thickness defects in rabbits with newly regenerated cartilaginous tissue formation seen after 6 months and tissue organization after 12 months [64]. Gelatin, which is derived from collagen, is also biocompatible and can be modified to cross-link with visible light and support chondrocytes, though some potential diffusion limitations may exist [65]. Also, macroporous gelatin/alginate gels promoted chondrocyte proliferation, a rounded morphology, and expression of hyaline matrix molecules with increased spatial deposition of proteoglycans and constant expression of type II collagen [66].

Alginate is a polyanionic polymer found in brown algae, and can be cross-linked with bivalent cations to form stable macroporous ionically cross-linked gels. Alginate hydrogels have been used to expand chondrocytes and induce stem cell differentiation [67–69]. Recently, investigators have modified alginate gels with synthetic adhesion peptides [70] or combined alginate with other materials to make hybrid scaffolds [71,72]. The RGD-functionalized alginate gels have been shown to affect articular chondrocyte attachment and morphology and chondrogenesis [73]. Also with increasing cross-linking density and substrate stiffness, chondrocytes grown on alginate gels exhibited a more flattened morphology with stress fibers observed via phalloidin staining [73]. Despite its advantages for studying *in vitro* chondrogenesis, limitations to alginate gels include low mechanical properties and low degradation rates. Agarose is a linear polysaccharide derived from red alagae and has been widely used to study chondrocyte response to deformational loading since it is able to transmit applied mechanical forces to cells during compression [74]. Chitosan and its hybrid macroporous hydrogels support normal chondrocyte phenotypes in 2-D [75,76] and 3-D cultures [77]. Some synthetic copolymers have also been investigated as thermoreversible hydrogels, with gelation occurring above their lower critical solution temperature (LCST). These include Poly(N-isopropyl acrylamide-*co*-acrylic acid) (p(NiPAAm-co-AAC)) and poly(propylene fumarate-*co*-ethylene glycol) (p(PF-co-EG)), which are capable of retaining chondrocyte phenotype and viability [78,79]. Within these gels, cells remained responsive to coencapsulated soluble factors like HA and TGF-β3, which can lead to increased expression and synthesis of cartilage-specific ECM proteoglycans [80].

Sponges are macroporous porous scaffolds whose properties are dependent on pore size, porosity, and interconnectivity. To date, numerous materials have been used to fabricate sponge scaffolds, including poly(α-hydroxy esters) [81], alginate [82], polyglactin/polydioxanone [83], chitosan [84,85], silk fibroin [86], HA [87], collagen [88], and gelatin [89]. A novel biodegradable elastomer scaffold from poly(1,8-octanediol citrate) (POC) has been fabricated by salt leaching and supported the growth of chondrocytes *in vitro* [85]. In addition, resorbable polyglactin/polydioxanone scaffolds have been used in full thickness equine defects and showed good cartilage repair with integration into surrounding tissue [83]. Chitosan can also be formed into sponges via freeze-drying and lyophilization [84]. Chitosan has also been hybridized with gelatin, which serves as a substrate for cell adhesion [85]. This chitosan/gelatin scaffold was used for elastic cartilage repair and neocartilage exhibited type II collagen, elastic fibers, and glycosoaminoglycan (GAG) production, with total GAG content ~90% of that found in native auricular cartilage [85].

Sponges can be formed from silk fibroin by a solvent casting/salt leaching method that supports both chondrocytes [91] and stem cells [92]. Compared to fast degrading collagen scaffolds, silk scaffolds supported greater proliferation and chondrogenesis of MSCs [93]. Collagen and collagen hybrid sponges have also been formed that support chondrocyte growth and phenotype retention [88]. The use of collagen microsponges in the porous openings of PLGA fibers [94] and sponges [79] has yielded new hybrid scaffolds with improved properties. Furthermore, type II collagen-GAG scaffolds with varying cross-linking densities can mediate cell behavior. In a study by Vickers et al., chondrocytes seeded on type II collagen-GAG scaffolds with low cross-linking densities experienced cell-mediated contraction, an increase in cell number, enhanced chondrogenesis, and increased degradation rates [88]. Recently, electrospinning has generated much interest to produce biomaterials with nanoscale polymer fibers that mimic collagen fibrils in cartilage ECM [95]. Fibers are generated as the surface charge of the polymer droplet overcomes its surface tension in an applied electric field, causing an instability that creates jets of polymer that can then be collected as solvent evaporates. Advantages of using electrospun scaffolds include high surface area to volume ratios and fully interconnected pores, and the ability to create aligned fibers. By collecting the nanofibers on a rotating mandrel, aligned fibrous scaffolds can be fabricated, and can mimic the anisotropic morphology of some tissues. These nanofibrous scaffolds support chondrocytes and stem cells [95]. The most commonly used meshes are made of poly(α-hydroxy esters). These meshes have been used since the early 1990s for cartilage regeneration and include PLA, PGA and their copolymers (PLGA). Shin et al. showed that immobilization of type II collagen on PLLA/PLGA scaffolds increased chondrocyte proliferation and GAG deposition while decreasing inflammatory responses by preventing host tissue infiltration and capsule formation [96]. Immobilization of HA to the surface of PLGA scaffolds enhanced chondrocyte attachment and substantially increased GAG and collagen synthesis [97]. Furthermore, MSCs seeded on PLGA scaffolds resulted in smooth, shiny white hyaline-like tissue after 12 weeks of *in vivo* culture in a rabbit defect [98].

Poly(caprolactone) (PCL) is another member of the poly(α-hydroxy ester) family with slower degradation kinetics. Recently, PCL has been electrospun to form nanofibrous scaffolds capable of supporting proliferating chondrocytes that produce proteoglycan-rich matrices [95]. Furthermore, these scaffolds can also support chondrogenesis of MSCs comparable to cell pellet controls [99]. Several natural materials have also been processed as fibrous scaffolds, including cellulose [100] and HA derivatives [101–103]. Nonwoven cellulose II fabrics coated with calcium phosphate supported better cell adhesion than unmodified fabrics, where calcium leaching from the scaffold has the potential to mimic the cartilage microenvironment in the vicinity of subchondral bone [100]. Nonwoven HA esters (HYAFF® derivatives) are semisynthetic, resorbable meshes that support cell adhesion, proliferation, and production of cartilage-specific ECM *in vitro* [101,102] and *in vivo* [103]. In a pilot study by Radice et al., HYAFF®11 elicited no inflammatory response and completely degraded within 4 months of implantation [104]. In a clinical setting, Hyalograft® C (a graft composed of autologous chondrocytes grown on a HYAFF®11 scaffolds) has been used to treat a number of human articular cartilage defects [104–106]. Hyalograft® C repaired cartilage showed significant improvements over preoperation assessments with cartilage regeneration even in joints with progressed osteoarthritis [105].

15.2.2 Bone Tissue Engineering

Bone is a dynamic, highly vascularized tissue with a unique capacity to heal and remodel without leaving a scar [107]. These properties, together with its capacity to rapidly mobilize mineral stores on metabolic demand, make it the ultimate smart material [108]. Owing to these properties of bone, generating bone *in vivo* or *in vitro* has been a challenging task but till date no ideal or satisfactory bone substitute could be developed. Although material science technology has resulted in clear improvements in the field of bone substitution medicine, no adequate bone substitute has been developed and hence large bone defects/injuries still represent a major challenge for orthopedic and reconstructive surgeons. The three key elements for generating bone tissue, namely osteogenic progenitor cells, osteoinductive growth factors, and osteoconductive matrices are involved in this process [109–111].

To be used as a bone tissue replacement a scaffold must have some of the essential properties. Scaffolds must possess an open and interconnected porous structure. The accepted pore size range lies in between 200–900 μm [109,112]. Studies have been done to define the optimal pore size for bone tissue ingrowth, the results of these are compiled in Table 15.3 [112,113,114]. Though some researchers have also reported a pore size of 1.2 to 2 μm to be favorable for tissue ingrowth and vascularization, due to high surface to volume ratio [115]. Surface properties, both chemical and topographical, are an important aspect and regulate cell adhesion and migration [108]. In general the scaffold should be osteoconductive and should allow growth of osteoblast cells. It has been previously shown [108,116] that a more rough surface will be able to imprison the fibrin matrix, better than a smooth surface, and hence facilitate the migration of osteogenic cells to the materials surface [108]. *In vitro* the scaffolds should have sufficient mechanical strength to withstand the hydrostatic pressures and to maintain the spaces required for cell ingrowth and matrix production. While in the case of *in vivo* or implantable bone, constructs should have mechanical strength equivalent to

TABLE 15.3

Studies Defining Optimal Pore Size for Bone Regeneration

Reference	Scaffold Pore Size (μm)	Porosity	Mineralize Tissue Ingrowth/ Comments
[113]	Type I: 2–6 μm	33.5%	No tissue ingrowth
	Type II: 15–40 μm	46.2%	No bone ingrowth, fibrous tissue ingrowth
	Type III: 30–100 μm 80% pores <100 μm	46.9%	50 μm of bone ingrowth osteoid and fibrous tissue ingrowth
	Type IV 50–100μm 63% pores <100μm	46.9%	20 μm bone ingrowth by 11 weeks and 500 μm of ingrowth by 22 weeks, osteoid and fibrous tissue ingrowth
	Type V: 60–100 μm 37% <100 μm	48%	600 μm bone ingrowth by 11 weeks and 1,500 μm of ingrowth by 22 weeks, osteoid and fibrous tissue ingrowth
[114]	≤100 μm	35.3%	Not statistically different from treated controls
	≤200 μm	51.0%	Not statistically different from treated controls
	≤350 μm	73.9%	Statistically significant more bone than all other groups

Source: From Yang S., Leong K.F., Du Z., Chua C.K., *Tissue Eng.*, 7, 679–89, 2001. With permission.

natural bone to achieve early mobilization of the injured site [117]. A variety of biomaterials have been proposed for constructing bone substitutes. The range of these materials lies from hard materials like metals and ceramics [118–121] to soft biodegradable polymers [108,112,117]. In ceramics, porous scaffolds composed of hydroxyapatite or β-tricalcium phosphate have mainly been explored [121]. However, hydroxyapatite and calcium phosphate are not themselves osteoinductive and are resorbed relatively slowly. They are brittle and present a low mechanical stability, which prevent their use in the regeneration of large bone defects [108].

As an alternative, several polymeric materials have been investigated for bone tissue engineering. These generally fall into two categories natural and synthetic, both being biodegradable in nature. The main advantages of natural materials are their low immunogenic potential, the potential bioactive behavior and the capability of interacting with the host tissue, chemical versatility, and in some cases their source, as in starch and chitosan, which is almost unlimited.

Natural and synthetic polymers both have been used in various forms but 3-D macroporous scaffolds have found greater potential as bone tissue scaffold. Collagen being one of the main components of the ECM and is an obvious choice for bone tissue engineering. Collagen/hydroxyapatite composite scaffold fabricated by solid liquid phase separation have shown to be a potential material for bone regeneration *in vitro* [122]. It has also been reported that collagen-based porous scaffolds reinforced by chitosan fibers show higher mechanical strength and faster osteoblasts growth rate than the unreinforced fibers [123]. A recent report suggests that coculture of MSCs with endothelial progenitor cells on collagen fiber mesh enhances tissue-engineered bone formation *in vivo* [124]. Chitosan-based macroporus scaffolds both alone or in combination with Ca-phosphate allowed for potential osteoblast growth and bone mineral deposition owing to their structural similarity to natural ECM components [125–127]. Due to the cationic nature of chitosan it has also been used for attachment and delivery of several growth factors in a controlled manner for bone healing [128,129]. Chitosan surfaces chemically modified with imidazole promote mineralization, induced bone formation and filled critical size bone defects with the apposition of trabecular bone [130]. When applied in rat calvarial defect, chitosan sponges incorporating platelet derived growth factor (PDGF) induced new bone formation [131].

Poly(hydroxyalkonates) (PHAs) are a new class of biodegradable polyesters that have found their usage in bone tissue engineering recently. In particular PHB (polyhydroxybutyrate) has demonstrated production of a consistent favorable bone tissue adaptation response with no evidence of an undesirable chronic inflammatory response after implantation periods of up to 12 months. Bone is formed close to the material and subsequently becomes highly organized, with up to 80% of the implant surface lying in direct apposition to new bone. The materials showed no evidence of extensive structural breakdown *in vivo* during the implantation period of the study [132,133]. The copolymers of hydroxybutyrate with hydroxyvaleric acid (PHBV) are less crystalline, more flexible, and more readily processable than PHB itself, which has aroused interest in them as scaffold material for bone [134]. The PHBV is now being studied intensely as a tissue engineering substrate [135]. Another natural polymer that has shown interesting results with human mesenchymal cells is silk [136–138]. Porous biodegradable silk scaffolds and human bone marrow derived mesenchymal stem cells (hMSCs) were used to engineer bone-like tissue *in vitro*. The silk fiber surface was modified by using adhesion peptide RGD. The 1.2-mm long, interconnected, and organized bone-like trabeculae with cuboid cells on the silk-RGD scaffolds were formed while such features were absent on collagen based scaffold. The results suggest that RGD-silk scaffolds are particularly suitable for autologous bone tissue engineering, presumably because of their stable macroporous structure, tailorable mechanical properties matching those of native bone, and slow degradation [139]. Also silk fibroin fiber scaffolds containing bone morphogenetic protein 2 and nanoparticles of hydroxyapatite prepared via electrospinning had been tested for there potential *in vitro* bone formation from hMSCs [140].

In spite of several advantages that the natural polymers provide, synthetic polymers allow a better control of physicochemical properties, and, in principle, of delivery kinetics for specific molecules and cells than natural polymers. The PLGA and PLA/PLLA are the most widely used, biodegradable polymeric scaffold material for bone tissue engineering. The polymeric 3-D porous scaffolds composed of these polymers have been fabricated using various methodologies like solvent casting, particulate leaching, rapid prototyping technique, and so on. All of these methods used for 3-D fabrication of scaffolds have their own limitations. PLGA scaffolds fabricated by solvent casting and particulate leaching are limited by the pore shape (cubic crystal) and remnants of soluble particles in the polymer matrix. The PLGA scaffolds fabricated to date have mechanical strength in the range of trabecular/spongy bones. In fact, most of the porous materials prepared by solvent casting and particulate leaching method are limited to thickness ranging from 0.5 to 2 mm [141]. In addition, their limited interpore connectivity is disadvantageous for uniform cell seeding and tissue growth.

An improvement to make scaffolds for bone tissue engineering via particulate leaching methodology is the use of paraffin spheres as porogen. Homogenous foam like scaffolds with spherical pore shape and well-controlled interpore connectivity [142], can be generated by this route (Figure 15.4). The control of porosity and the pore size can be achieved by changing the concentration of the polymer solution, the number of the casting steps, and the size of the paraffin spheres. The main advantage of this method is that it can ensure the creation of a totally interconnected pore network in the polymer scaffold,

(a)   (b)

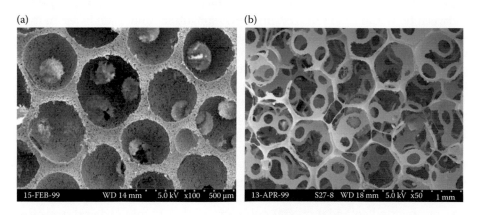

FIGURE 15.4
SEM micrographs of poly(α-hydroxy acids) scaffolds. (a) PLLA foams prepared with paraffin spheres with a size range of 250–350 µm. (b) PLGA foams prepared with paraffin spheres with a size range of 420–500 µm. (Adopted from Ma P.X., Choi J.W., *Tissue Eng.*, 7, 23–33. 2001. With permission.)

and, the paraffin sphere assembly can be dissolved in some organic solvents but not water. Therefore, certain water-soluble polymers can be involved in fabricating such scaffolds. However, what is the ideal pore size and interpore connectivity of such scaffolds for bone tissue engineering is yet to be investigated. A combination of phase-separation process with a novel sugar sphere template leaching had been used by Wei and Ma for producing macroporous PLLA scaffolds with nanofibrous pore walls. The PLLA matrices were bioactive and deposition of bone-like apatite was seen when incubated in simulated body fluid (SBF). These properties of the matrices were enhanced when combined with nanosized hydroxyapatite particles [143]. A PCL scaffold has been fabricated using selective laser sintering (SLS), a rapid prototyping technique. PCL scaffolds fabricated by SLS with porous architecture had compressive modulus and yield strength values ranged from 52 to 67 MPa and 2.0 to 3.2 Mpa, respectively, lying within the lower range of properties reported for human trabecular bone.

Histological evaluation and microcomputed tomography analysis of implanted scaffolds show that bone can be generated *in vivo*. The technology was further used to design and fabricate prototype mandibular condyle scaffold based on an actual pig condyle, which might prove useful for designing of scaffolds of required shape [144]. Three-dimensional hollow root form scaffolds of poly-L-lactic acid/polyglycolic acid composites (50/50, 65/35, and 75/25 ratios), using the solvent casting/compression molding/particulate leaching technique have also been utilized for making roots for alveolar bone regeneration. Salt particle with two different sizes were used, 150–180 and 180–300 μm, to effect porogenesis and it was observed that tissue reaction favored larger sized pores [145]. Porous biodegradable elastomeric polyurethane scaffolds combined with the patient's own bone marrow could be potential bone substitutes. The elastomeric bone substitute prevents shear forces at the interface between bone and rigid (e.g., ceramic bone substitutes and establishes an intimate contact with the native bone ends), thus facilitating the proliferation of osteogenic cells and bone regeneration. In previous studies biodegradable polyurethane sponges of varying hydrophilicity implanted in mono- and tricortical defects in the ilium of healthy sheep and in bicortical defects in the ilium of estrogen-deficient sheep induced regeneration of new cancellous bone [146,147]. Polyurethane sponges impregnated with autogenous bone marrow induced regeneration of critical-size defects in the sheep tibiae [148].

Several reports signify the effects of scaffold architecture on the cell behavior and mineral deposition. Also it can be seen that the mechanical strength is greatly affected by the technique used for the fabrication of scaffolds. Recently Cuddihy and Kotov have generated a novel PLGA scaffold having inverted colloidal crystal geometry with a highly ordered arrangement of identical spherical cavities. Colloidal crystals were constructed with soda lime beads of 100, 200, and 330 μm diameters. The scaffolds demonstrated high mechanical properties for PLGA alone (>50 MPa), *in vitro* biocompatibility, and maintenance of osteoblast phenotype (Figure 15.5) [149].

(a) (b) (c)

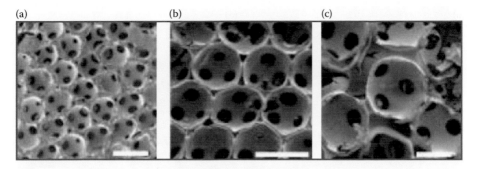

FIGURE 15.5
Scanning electron microscopy images of poly(lactic-*co*-glycolic acid) inverted colloidal crystals resulting from colloidal crystals of microspheres of diameter (a) 100 mm, (b) 200 mm, and (c) 330 mm. Scale bar 200 µm. (Adapted from Cuddihy M.J., Kotov N.A, *Tissue Eng. A.*, 14, 1639–49, 2008. With permission.)

Recently a cryogel scaffold made of cross-linked 2-hydroxyethyl methacrylate (HEMA)–lactate–dextran with interconnected macropores was studied in bioreactors at different regimes (static, perfusion, and compression–perfusion) using a osteoblast cell line. The osteoblasts secreted a significant amount of extracellular matrices in the scaffolds, and the cells subjected to dynamic culture exhibited highly interconnective morphology. These matrices might find application in healing of nonload bearing bone defects [150].

A matrix designed to biomimic the natural nanoscale features of bone can enhance new bone growth and, thus, provide a better material for bone tissue regeneration applications. Attempts are being now made to combine nanoscale materials with porous matrices and hydrogel. It has been observed that the nanofibrous structure not only helps in improving the mechanical strength but also influence cell behavior [151]. To mimic the nanofibrous architecture, a 3-D interconnected fibrous network of PLLA with a fiber diameter ranging from 50 to 500 nm by liquid–liquid phase separation technique. (Figure 15.6) [152]. Typically, the nanoscale fibrous matrices were fabricated with five steps: polymer dissolution, phase separation and gelation, solvent extraction, freezing, and then freeze-drying under vacuum. These synthetic analogs of natural extracellular matrices combine the advantages of the synthetic biodegradable polymers and the nanoscale architecture similar to the natural ECM. They were found to selectively enhance protein adsorption and promote osteoblastic cell adhesion [152].

Osteoblasts cultured on the nanofibrous scaffolds exhibited higher alkaline phosphatase activity and an earlier and enhanced expression of the osteoblast phenotype versus solid-walled scaffolds. Most notable were the increases in runx2 protein and in bone sialoprotein mRNA in cells cultured on nanofibrous scaffolds versus solid-walled scaffolds [151]. Similar results have also been reported for nanofibrous scaffolds of PLGA fabricated by reverse solid free form fabrication. Fiber diameters in these scaffolds were

(a)　　　　　　　　　　　　　　　　(b)

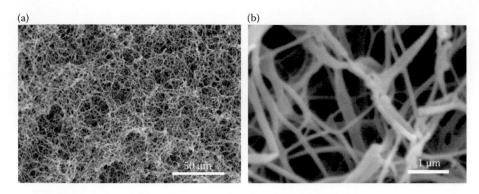

FIGURE 15.6

SEM micrographs of a PLLA fibrous matrix prepared from 2.5% (wt/v) PLLA/THF solution at a gelation temperature of 8 °C. (Adapted from Ma P.X., Zhang R.Y., *J. Biomed. Mater. Res. A.*, 46, 60–72, 1999.)

50–500 nm, similar to type I collagen. After six weeks in culture, bone-like tissue was evident within the nanofibrous scaffolds [153]. A nanofibrous tri-phasic scaffold was electrospun from a mixture of PCL, type-I collagen, and hydroxyapatite nanoparticles (nano-HA) with a mixture dry weight ratio of 50/30/20, respectively. Fibers with an average diameter of 180 ± 50 nm, are formed, which coincides well with the collagen fiber bundle diameter of the native ECM of bone. Young's modulus of the produced material was in the range 0.50–3.9 GPa h increasing in the order PCL < PCL/nano-HA < collagen < triphasic < collagen/nano-HA. The modulus data emphasize the importance of collagen and its interaction with other components in affect-ing mechanical properties of osteoconductive scaffolds [154].

In another study by a different group, electrospun nanofibrous biocom-posite scaffold using polycaprolactone/nanohydroxyapatite/collagen (PCL/nHA/Col) was fabricated to provide mechanical support and to direct the growth of human fetal osteoblasts (hFOB). The fiber diameters were around 189 ± 0.026 to 579 ± 272 nm and pore sizes up to 2–35 µm. Resulting nano-fibrous scaffolds were highly porous (>80%) structures and provided a sufficient open pore structure. The unique nanoscale biocomposite system had inherent surface functionalization for hFOB adhesion, migration, pro-liferation, and mineralization to form a bone tissue for the regeneration of bone defects [155]. An interesting research report explores the use of self-assembling helical rosette nanotubes (HRNs) embedded in a hydrogel. HRNs are formed by chemically immobilizing two DNA base pairs. They are soft nanomaterial that biomimics natural nanostructured components of bone (such as collagen) since they are 3.5 nm in diameter and self-assemble into a helical structure in aqueous solutions. HRNs undergo a phase transition from a liquid to a viscous gel when heated to slightly above body tempera-ture. When embedded in a polyhydroxymethylmethacrylate (polyHEMA) hydrogel, they led to enhanced osteoblast adhesion [156].

A variety of composite materials are being developed for bone tissue engineering owing to one or other shortcomings of a single material [157]. Synthetic polymeric scaffolds have been combined with bioactive ceramics to generate composite having favorable properties of both. The greatest advantage of combining both is conferring of bioactive property to the synthetic polymer network, which is achieved by inclusion of bioactive particles or coatings. The degree of bioactivity is adjustable by the volume fraction, size, shape, and arrangement of inclusions [157,158]. It has been shown that increased volume fraction and higher surface area to volume ratio of inclusions favor higher bioactivity, hence in some applications the incorporation of fibers instead of particles is favored [159,160]. Addition of bioactive phases to bioresorbable polymers can also alter the polymer degradation behavior, by allowing rapid exchange of protons in water for alkali in the glass or ceramic. This mechanism is suggested to provide a pH buffering effect at the polymer surface, modifying the acidic polymer degradation [158].

Poly (D, L-lactic acid) (PDLLA)/Bioglass composites exhibit high bioactivity, and the potential of these scaffolds in bone and soft-tissue engineering has been demonstrated *in vitro* with optimized concentrations of 45S5 Bioglass® added to PDLLA or PLGA matrices [161,162]. It has also been shown that the composite foams support the migration, adhesion, spreading, and viability of MG-63 cells (osteosarcoma cell line) [161]. PLGA/bioactive glass [163] microspheres were synthesized by emulsification and were heated in molds to fabricate porous 3-D scaffolds. The scaffolds demonstrated bioactivity and were able to promote osteogenesis of marrow stromal cells. Other composite scaffolds of bioresorbable PLLA and ceramic fillers, HA, or β-tricalcium phosphate, demonstrates viscoelastic behavior with anisotropy in pore direction as well as good mechanical strength [164]. Composite of alginate and HA synthesized by freeze-drying has also been tested for bone tissue engineering by [165]. The average pore diameter of the porous composite was around 150 μm and the size of the pores was variable in different portions of the same scaffold. *In vitro* studies showed adhesion of osteoblast-like cell line more to the composite scaffold than the alginate alone. Bioactive coating over porous scaffolds is also a form making composite materials for bone tissue engineering. Bioceramic-coated porous scaffolds have been produced either as foams [166], fibrous bodies [167], or meshes [168,169]. Recently, 3-D PDLLA foams containing both TiO_2 nanoparticles and Bioglass® additions have been synthesized by TIPS (thermally induced phase separation). These foams demonstrate enhanced bioactivity and surface nanotopography [170].

15.2.3 Skin Tissue Engineering

Skin is an important tissue engineering target for reconstructive surgery of burn victims, but more and more to assist in the healing of diabetes related ulcers. Skin tissue engineering has developed into the most advanced clinical application in the field to date. Several commercial products are now available to regenerate different layers of the skin in patients suffering from burns and chronic wounds. Different variety of scaffolds made up of biodegradable synthetic and natural polymers are being synthesized for making artificial products. The aim of skin tissue engineering is to develop a bilayer of cells that closely resembles the natural architecture of the skin and can help in restoring the functions of lost skin in case of burns and wounds [171,172].

The process of wound repair involves the timed and balanced activity of inflammatory, vascular, connective tissue, and epithelial cells. All of these components need an ECM to facilitate the healing process. To minimize scar formation and to accelerate healing time, different techniques of skin substitution have been introduced in the last decades [173]. The clinical use of cultured skin substitutes for wound closure has reduced the amount of donor skin required by more than 10 times compared with conventional skin grafts and has reduced the number of surgeries required to harvest donor skin while at the same time decreasing the time of recovery of severely burn-injured patients [174,175].

There are three different strategies used for wound healing. The first one employs cells without matrix to restore normal functioning of skin. The second approach is to use synthetic and natural polymeric scaffolds similar to natural ECM without cells to assist in wound healing. This strategy depends upon employment of cells from the wound area and further proliferation of these cells to replace the lost skin. A third one is a combination of the above two strategies, which is the cells are grown within a three dimensional polymeric matrix that is then used as an artificial substitute for restoring normal skin functioning. For example, bilayered skin substitutes composed of keratinocytes and fibroblasts incorporate into biologic collagen matrices or synthetic bioabsorbable scaffolds before implantation and serve to enhance or promote tissue repair processes. Currently, the available skin substitutes can be classified as (a) temporary dressings—material designed to be placed on a fresh wound (superficial or partial thickness) and left until healed, (b) semipermanent—material remaining attached to the excised wound, and eventually replaced by autogeneous skin grafts, and (c) permanent incorporation of an epidermal analog, dermal analog, or both as a permanent replacement. The biologic components in these substituents may be either autogenous, allogeneic, or xenogeneic [171,174].

Most of the artificial skin substitute available commercially are based on collagen [173,176]. One of the main issues in using collagen is the risk of immune reactions that may be caused due to the source of collagen. However there are only a few reports that have actually looked into this aspect. Studies on two forms of commercially available artificial skin (Integra® Dermal Regeneration Template or Apligraf® Graftskin) shows that they do not possess immunogenic potential to the human host [176–178]. The skin products available commercially may be divided as acellular and cellular [179]. A classification of commercially available tissue engineered skin products is given in Table 15.4 [180]

TABLE 15.4

Classification of Commercially Available Tissue Engineered Products (Skin)

	Acellular	Cellular						
		Allogenic			Autologous			
		Fibroblast	Keratinocytes	Cadaver skin	Fibroblast	Keratinocyte	Hair follicle	Autograft from skin biopsy
Epidermal substitute						Laserskin Myskin	Epidex	Epicel
Dermal substitutes	Oasis Integra Permacol EZ-Derm Matriderm	Dermagraf Transcyte ICX-SKN		Alloderm				
Bilayer		Orcel Apligraf			Permaderm (permanent skin substitute)			
Hydrogel	E-Matrix							

Source: From van Winterswijk P.J., Nout E., *Wounds.*, 19, 277–84, 2007.

Acellular products contain, as the name implies, no cells and consist of a matrix that functions by binding to the host, allowing matrix-cell interactions. Due to its porous nature, the matrix allows host cells to infiltrate. Today, a matrix can contain virus vectors or plasmids, which can transcribe and translate the in-built DNA leading to secretion of specific growth factors [180,181]. These growth factors carry out their specific functions by stimulating host cells to enhance wound healing. In specific situations it is even possible to manufacture matrices containing plasmids that can release hormones. The matrix contains ECM components such as collagen, HA, and fibronectin, thus ensuring biocompatibility. Examples of these products are E-Matrix™ (Encelle Inc, Greenville, North Carolina), OASIS® (Healthpoint Ltd, Fort Worth, Texas) and (Integra LifeSciences, Plainsboro, New Jersey), Permacol® (Tissue Science Laboratories Inc, Andover, Maryland), Matriderm® (Dr. Suwelack Skin & Health Care AG, Germany), and EZDerm® (Brennen Medical, St. Paul, Minnesota) [175,182,183].

MATRIDERM® is a native, structurally intact matrix of collagen and elastin. These ECM proteins serve as a scaffold for reconstitution of the skin and the modulation of scar tissue. MATRIDERM® is used for dermal regeneration in cases of full skin defects (Figure 15.7). Its use is indicated in plastic surgery and in surgery for burn injuries combined with autologous skin grafts. The matrix serves as a support structure for the ingrowth of cells and vessels. Its elastin component improves the stability and elasticity of the regenerating tissue. As the healing process advances, fibroblasts produce their own ECM, and MATRIDERM® is resorbed. MATRIDERM® also remains stable and elastic in wet conditions without any chemical cross-linking agents.

INTEGRA® Dermal Regeneration Template is a bilayer membrane system for skin replacement. The dermal replacement layer is made of a porous matrix

(a)   (b)

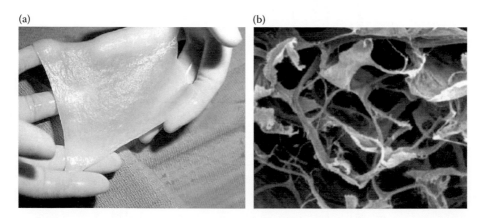

FIGURE 15.7
Illustrations of some of the commercially available wound dressings: (a) rehydrated MATRIDERM® adapts smoothly to any surfaces and shows considerable tear resistance. (b) native, structurally intact collagen elastin matrix (MATRIDERM®. With permission.)

of fibers made by cross-linked bovine tendon collagen and a glycosamino-glycan (chondroitin-6-sulfate) that is manufactured with a controlled poros-ity and defined degradation rate. The temporary epidermal substitute layer is made of synthetic polysiloxane polymer (silicone) and functions to con-trol moisture loss from the wound. The collagen dermal replacement layer serves as a matrix for the infiltration of fibroblasts, macrophages, lympho-cytes, and capillaries derived from the wound bed. As healing progresses an endogenous collagen matrix is deposited by fibroblasts; simultaneously, the dermal layer of INTEGRA® Dermal Regeneration Template is degraded. Upon adequate vascularization of the dermal layer and availability of donor autograft tissue, the temporary silicone layer is removed and a thin, meshed layer of epidermal autograft is placed over the neodermis. Cells from the epi-dermal autograft grow and form a confluent stratum corneum, thereby clos-ing the wound reconstituting a functional dermis and epidermis [175]. This approach requires cells to be cultured in a 3-D polymeric matrix which is finally used as an implant forming an epidermal layer skin substitute [179].

Autologous cells are used in these products to minimize the risk of rejection. Autologous keratinocytes are derived from progenitor cells from dermal sheets in the outer root surrounding hair follicles [184] or from epithelial cells obtained via a biopsy of the recipient's skin. A permanent autologous epidermal skin graft is applied, which functions as a reliable barrier and promotes the formation of granulation tissue. Examples of tissue-engineered epidermal substitutes are: Epicel® (Genzyme Biosurgery, Cambridge, Massachusetts), Laserskin® (Fidia Advanced Biopolymers, Abano Terme, Italy), Myskin™ (Celltran Ltd, Sheffield, United Kingdom), and EpiDex™ (Modex Therapeutics, Switzerland). Laserskin autograft is an epidermal substitute. There are orderly arrays of laser-perforated microholes for the ingrowth and proliferation of keratinocytes [185]. As kerati-nocytes are directly cultivated on Laserskin, the graft can easily be peeled off from the skin. Composite Laserskin grafts are good human skin substitutes in terms of durability, biocompatibility, high seeding efficacy for keratinocytes, high graft take rate, and low infection rate [186].

Besides epidermal substitutes, tissue-engineered dermal substitutes have also been developed. The dermis is composed of loose connective tissue con-taining collagen and fibrils, anchoring the dermis to the epidermis [187]. The upper layer of the dermis, the papillary layer, is a cell-rich layer contain-ing fibroblasts and macrophages allowing dermis-epidermis interactions. These interactions trigger synthesis of ECM components and stimulation of differentiation and growth of keratinocytes. Inclusion of living fibroblasts leads to active release of cytokines. However, until now, only allogenic tissue-engineered dermal wound products have been developed, and unfor-tunately always carry the risk of (chronic) graft rejection. Examples of these products include: Dermagraft® (Advanced BioHealing, La Jolla, California), Alloderm® (LifeCell Inc, Branchburg, New Jersey), TransCyte® (Advanced BioHealing, La Jolla, California), and ICX-SKN® (Intercytex, Cambridge, United Kingdom) [175,182,183].

Dermagraft is a cryopreserved human fibroblast derived dermal substitute. Fibroblasts incorporated with Dermagraft secreted vascular endothelial growth factor (VEGF), platelet-derived growth factor (PDGF) insulin-like growth factor I (IGF-I), colony stimulating factors (CSF), interleukins (IL), tumor necrosis factor (TNF), and transforming growth factor (TGF-β) [179]. Alloderm is a dermal collagen matrix derived from banked human skin that is specially treated to remove most cellular components. Alloderm has been successfully used in the resurfacing of full-thickness burn wounds in combination with an ultra thin autograft that replaces the epidermis [188]. TransCyte is a laboratory-grown ECM of allogeneic human dermal fibroblasts. Noordenbos et al. [189], reported on the safety and efficacy of TransCyte for treatment of partial-thickness burns. It has been indicated for use as a temporary covering for excised burns prior to autografting or burns that do not require autografting (partial-thickness burns) and its physical properties (pliability) allow it to easily conform to the contours of the face. Wounds treated with TransCyte healed faster, with no infection and less hypertrophic scarring, compared to wounds treated with silver sulfadiazine. More recently, ICX-SKN, an autosynthesized human collagen-based ECM with human dermal fibroblasts vascularized during healing of acute surgical wounds and showed integration and persistence during the healing process [190].

Currently, (allogeneic) bilayered products are available (e.g., Apligraf® Organogenesis, Canton, Massachusetts and OrCel® OrCel International, New York, New York). Apligraf, a living bilayered skin substitute, is capable of regenerating tissue in response to an injury. This is apparent because of the good interaction between Apligraf and the wound, thus making it suitable for wide application [171]. OrCel is a bilayered cellular matrix in which normal human allogeneic skin cells (epidermal keratinocytes and dermal fibroblasts) are cultured in two separate layers into a type I bovine collagen sponge. Like Apligraf, OrCel is a bilayer dressing resembling normal skin. OrCel was developed as a tissue-engineered biological dressing. OrCel delivers ECM components and growth factors and creates an environment conducive to wound healing [175,183].

15.3 Skin Substitutes Under Development

Though a number of skin replacements products are being marketed or are under clinical trials, there are still challenges that need to be resolved. The main obstacles are improving safety, finding a substitute for split thickness grafts, improving angiogenesis in replacement tissue, and improving ease of use [183]. In this section we specifically discuss the use of macroporous polymeric scaffolds, which are being developed for wound healing in burns and other associated problems. The matrixes being developed utilize both natural as well as synthetic polymers and a composite of aforementioned. Most of the matrices being developed using natural polymers are based mainly on collagen, chitosan, gelatin, and HA or their combinations with different materials.

15.4 Acellular Approaches

Chitin and chitosan have been a major interest in the area of wound healing owing to its structural similarity to HA, an extra cellular matrix (ECM) component. Therefore, chitin should possess the characteristics favorable for promoting rapid dermal regeneration and accelerated wound healing suitable for applications extending from simple wound coverings to sophisticated artificial skin matrixes. There have been several reports supporting the above hypothesis, which demonstrates that both chitin and chitosan act as accelerators of wound healing by several mechanisms. They have been used in several forms such as films, solutions, powders, and sponges [191–195]. A bilayer of chitosan film and sponge has also been investigated as a scaffold for neofetal dermal fibroblasts. The chitosan sponge was synthesized using different porogens by freezing at –28°C and lyophilization to obtain one of the bilayer chitosan materials. Cells grew and proliferated in an extended shape on the bottom of some large pores but in spherical form on the rough wall of large pores or at the edges of small pores. During the cell culture of four weeks, no contraction of the bilayer chitosan materials was observed unlike the collagen sponge as a scaffold of tissue engineering [196]. An asymmetric chitosan sponge synthesized by immersion—precipitation phase separation method have shown potential as wound dressing material. By this fabrication method, asymmetric membranes comprising different thicknesses of skin layer and different porosity of sponge sublayer could be prepared. They consist of an upper skin microporous layer while the bottom layer is made of macroporous chitosan sponge. The asymmetric chitosan membrane showed excellent oxygen permeability, controlled evaporative water loss and promoted fluid drainage ability, but could inhibit exogenous microorganism invasion due to its dense top layer and inherent antimicrobial property of chitosan [197]. This same bilayered chitosan membrane was subsequently loaded with silver sulfadiazine (AgSD). The release of sulfadiazine showed a burst effect, while silver displayed a sustained longer term release. The combination release was found to be effective in controlling *Pseudomonas aeruginosa* and *Staphyloccocus aureus* populations in cell culture for up to one week [198].

Liu and colleagues have synthesized a porous matrix for artificial skin by freeze-drying using a combination of chitosan, gelatin, and HA [199]. The matrix had heterogeneous porosity, which is the upper portion in contact with air, had larger pores in the range of 65 to 80 μm while the lower portion had pores in the range of 10 to 20 μm. A coculture of keratinocytes and a fibroblast over the matrix at air water interface for a period of two weeks demonstrates the presence of skin-like architecture. In a comparative study macroporous sponges based on a combination of four different natural polymers viz gelatin-alginate, gelatin-hyaluronic acid, chitosan-hyaluronic acid were tested *in vivo* with or without AgSD. Gelatin-based sponges showed increased porosity due to the addition of alginate or hyaluronate, resulting in an increased water uptake ability that ranged 10–40 times the weight of the original sponge. On applying these wound dressings to full thickness dorsal skin defect of Wistar rat indicated that the AgSD-impregnated gelatin-based sponges containing sodium alginate or sodium hyaluronate made the skin defect almost re-epitheliaze at the 12th postoperative day with a minor infection and good integrity of dermis. The combination gelatin-hyaluronic acid with AgSD was found to show the best wound healing properties as a wound dressing resulting from histological findings and computerized morphometric analysis of epidermal healing [200].

15.5 Cellular Approaches

Skin substitutes under investigation generally aim at making a bilayer of skin containing both the fibroblast and keratinocyte in a three dimensional matrix. A dermal substitute is made keeping the fact that it should guide cells moving into the repair area, to serve as a scaffold for cells such as fibroblasts, and to help synthesize ECM components [201,202]. The dermal element may vary from an interface for healing (temporary substitute) to incorporation of dermal elements (permanent substitutes). It is responsible for restoring normal tissue architecture and for the prevention of scars [203].

In particular, collagen-based materials in the form of a sponge are frequently used as three-dimensional insoluble scaffolds for the preparation of artificial dermis or skin [202,203]. PCL with collagen biocomposite membranes support human dermal fibroblasts and keratinocytes in tissue-engineered skin in regenerative medicine [204]. Nanofiber matrices of PCL, collagen, and PCL coated–collagen support the attachment and proliferation of human dermal fibroblasts and might have potential in tissue engineering as a dermal substitute for skin regeneration [205]. A biodegradable hybrid mesh of PLGA and collagen was constructed by forming web-like collagen microsponges in the openings of a PLGA knitted mesh (Figure 15.8). The growth of human fibroblast cells over the hybrid mesh was found to be more homogenous than that on the PLGA mesh alone. Fibroblasts cultured in the hybrid mesh implanted in the back of a nude mouse resulted in formation of dermal tissues after two weeks and became epithelialized after four weeks (Figure 15.9) [206].

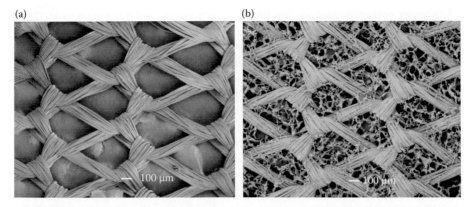

FIGURE 15.8
SEM photomicrographs of (a) PLGA knitted mesh and (b) PLGA–collagen hybrid mesh at original magnification ×90. (Adapted from Chen G., Sato T., Ohgushi H., Ushid T., Tateishi T., Tanak J., *Biomaterials*, 26, 2559–66, 2005. With permission.)

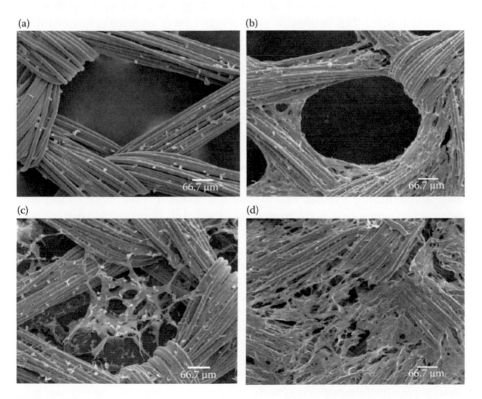

FIGURE 15.9
SEM photomicrographs of fibroblasts cultured in (a,b) PLGA knitted mesh and (c,d) PLGA–collagen hybrid mesh after culture for (a,c) 30 min and (b,d) 5 days at original magnification ×200. (Adapted from Chen G., Sato T., Ohgushi H., Ushid T., Tateishi T., Tanak J., *Biomaterials*, 26, 2559–66, 2005. With permission.)

Chitosan collagen composites have shown to be favorable for growth of skin stem cells [207]. A collagen-chitosan, tissue-engineered skin sponge was developed to serve as a scaffold for the reconstruction of a tissue-engineered skin *in vitro*. Human fibroblast when cultured in this sponge for a period of 27 days expressed biosynthetic activities similar to the ones exhibited *in vivo*. The newly synthesized matrix was highly differentiated, as shown by the presence of a dense network of quarter-staggered collagen fibrils (42 nm ± 6 nm in diameter) surrounding the cells. The size and the shape of these fibrils demonstrated that the newly synthesized procollagen was fully processed in collagen by removal of their N- and C-terminal propeptides [208]. This sponge was used to develop artificial skin *in vitro* by long-term culture of human fibroblast and keratinocytes [209]. This material when implanted in nude mice lead to nerve regeneration over a period of 120 days [210]. Research reports on porous collagen-chitosan sponges cross-linked using glutaraldehyde suggest that chitosan helps in improving the biostability by acting as a cross-linking bridge. *In vivo* studied animal tests further revealed that the scaffold could sufficiently support and accelerate the fibroblasts infiltration from the surrounding tissue [211]. Berthod and colleagues have reported elastic system remodeling in the mice model using collagen–chitosan–chondroitin sulfate sponges. Reconstructed skin made of a collagen sponge seeded with human fibroblasts and keratinocytes were grown *in vitro* for 20 days. On implantation into nude mice after 40 to 90 days, fibril bundles were arranged in a basket weave pattern, which is nearly similar to ECM in normal dermis [209]. Studies on hybrid scaffolds of collagen with β-chitin have also been reported to favor cell growth and their application as wound dressing [212]. Gelatin, a denatured form of collagen, has also been widely used in skin tissue engineering [200,213]. A porous scaffold of gelatin and β-glucan was used to prepare stratified artificial skin *in vitro* by coculturing keratinocytes and dermal fibroblast. *In vivo* testing revealed that the cellular construct is superior to acellular construct in promoting early wound healing [214]. Recently, Mazlyzam and colleagues have created a living bilayer of human skin equivalent using fibrin matrix derived form human plasma. The bilayer had fibrin–fibroblast and fibrin–keratinocyte layers. Transplantation of the bilayer into athymic mice revealed an expression of basal Keratin 14 gene in the epidermal layer and collagen type I gene in the dermal layer. The utilization of culture-expanded human skin cells and fibrin matrix from human blood allowed a fully autologous human skin equivalent construction [215].

15.5.1 Neural Tissue Engineering

The complexity of the nervous system allows for information to be received and transmitted through the body. As a result the brain, spinal cord, and peripheral nerve tissue pose unique challenges when designing scaffolds for repair, replacements, and regeneration of injured or diseased tissue. The key design parameters required for neural tissue engineering are establishing a permissive interface with the host tissue, mobilizing the host's cells within the polymer matrix, and stimulating tissue formation. Various strategies which have been used in attempts to repair injured areas of the adult mammalian CNS (central nervous system) [216,217] and will be discussed in the following text.

There are four types of macroporous polymeric scaffold polymers for nerve tissue engineering: (i) Natural biological materials: they may be active and inactive materials. The former one is derived from autologous nerve, skeletal muscle, vessels, membrane tube, and so on; the latter one includes chitosan, gelatin, collagen [218,219], and acellular extracellular matrix (AECM). (ii) Synthetic materials: there are nondegradable/unassimilable abiological materials (such as nylon, silicone tube, and urethane), degradable polymer materials in the second generation (such as PLA, PGA) [220] and polyphosphoester [PPE]), and chemically modified materials. (iii) Compound materials: the compound materials are optimally made of two or more different materials. (iv) Bio-derived materials (i.e., proteins and peptides) and nanometer materials [221]. Recently, hybrid materials of collagen and PLA are widely researched and used for nerve tissue engineering. High-directional, macroporous PLA scaffold materials established by using lyophilization technique have proved to effectively promote regeneration of peripheral nerve and central nervous axon [222–224].

Cell transplantation has been the most widely investigated strategy, aimed at restoring a tissue defect by exogeneous cell replacement. In contrast, tissue engineering is an emerging field of research and aims at restoring a tissue defect by inducing endogenous tissue regeneration and manipulating the cascades of cellular events during the healing response. Tissue engineering uses polymer devices with controlled macro- and microstructures and chemical properties to achieve organ regeneration. Ideally, a biohybrid tissue is formed, which becomes a permanent part of the host organ by acting as the functional analog of the original tissue or organ and that continues to support tissue functions. The use of porous hydrogel to assist tissue repair in CNS, axonal regeneration in the brain [225–230] and the spinal cord has been investigated. Hydrogels that are formed with a three-dimensional network of hydrophilic copolymers are well tolerated by living tissues and may serve as a substrate for tissue formation [225].

The interest in hydrogel as a biomaterial for soft tissue replacement resides in their ability to retain water within the polymer network and also in the possibility to prepare macroporous matrices that have a controlled porous structure and a specific surface area (e.g., the area delimited by the surface of the polymer network) large enough for cell attachment, growth, and organization and deposition of extracellular matrices. The swollen hydrogel allows for the maintenance of a chemical balance with the surrounding tissue because the water held within the swollen polymer network determines transport properties and allows the exchange between water in the hydrogel and ions and metabolites of tissue fluids. In addition, viscoelastic behavior, low interfacial tension with biological fluids, and structural stability make porous hydrogels suitable for implantation in soft tissue [231–233]. Hence, hydrogels are a big class of macroporous polymeric material that display a 3-D porous structure and have recently been explored to serve as templates for neural tissue regeneration and organization. The rationale is based on the observation that the success of axonal regeneration depends to a large extent on the existence of a preestablished cellular matrix that serves to guide and support axonal growth [234]. The use of hydrogels for cell transplantation provides an attractive challenge for neurotransplantation when considering the clinical application of neural grafting technology. The ultimate goal here is to repair damage and restore function in lesioned or diseased brain areas by the replacement of lost neurons with similar tissue, usually of embryonic origin. The incorporation of dissociated neuronal cells into a highly porous matrix and the implantation of this cellular hydrogel in the target tissue area represent an attractive approach for optimizing graft survival and integration. Indeed, while the polymer matrix provides a scaffold to guide ingrowth of the host tissue, it maintains the transplanted cells in a 3-D organization and promotes interactions with the host tissue. In the following section the polymeric macroporous scaffolds studied for nerve tissue engineering has been discussed and classified into macroporous material used for peripheral nerve regeneration, CNS, and spinal cord regeneration.

15.6 Peripheral Nerve Regeneration

The two basic components for structure supporting axonal migration, natural and synthetic materials are available. Pursuing the logical way of designing the scaffold for peripheral nerve repair, several groups have implanted natural and synthetic biomaterials, cells, microfibers, nanofibers, chondroitinase ABC digested autografts, and Schwann cells seeded in Matrigel to enhance regeneration across peripheral nerve gaps [235]. An analysis of these various approaches reveals that four essential components of grafts are typically manipulated to enhance regeneration across peripheral nerve gaps. These components are the growth permissive substrates (hydrogels or nano/micro fibers), neurostimulatory ECM proteins or peptides (typically LN-1 or LN-1 fragments), trophic factors (bFGF, NGF, or BDNF), and glial cells or other support (Schwann cells or stem cells) [235] (Figure 15.10).

Hydrogels based on cross-linked poly(2-hydroxyethyl methacrylate) (PHEMA) and poly(N-2-hydroxypropyl methacrylamide) (HPMA) implanted into the rat cortex showed astrocytes and NF160-positive axons grew similarly into both types of hydrogels. No cell types other than astrocytes were found in the PHEMA hydrogels. In the PHPMA hydrogels, a massive ingrowth of connective tissue elements was found [236]. The ability of DRG (dorsal root

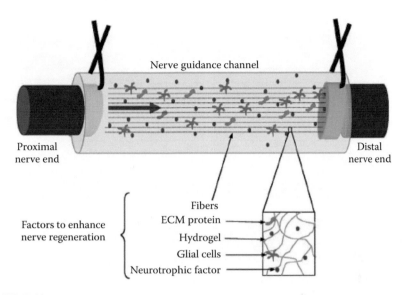

FIGURE 15.10
A schematic illustration of the components of grafts that influence peripheral nerve regeneration. The components include scaffolds (hydrogel or fibers), ECM proteins, glial or other cells, and neurotrophic factors. The spatial distribution of one or more of these components determines the degree of anisotropy of the graft. (Adapted from Bellamkonda R.V., *Biomaterials*, 27, 3515–18, 2006. With permission.)

ganglia)-derived neurons to survive and attach onto macroporous PLA foams was assessed *in vitro*. The foams were fabricated using a thermally induced polymer-solvent phase separation. It produces two types of pore structures, namely oriented and interconnected pores. The capacity of highly oriented foams to support *in vivo* peripheral nerve regeneration was studied in rats. A sciatic nerve gap of 5 mm length was bridged with a polymer implant showing macrotubes of 100 µm diameter. Four weeks postoperatively, the implant was well integrated into the surrounding tissue. Its shape and porous structure are maintained during regeneration and resist collapse during and after implantation. A massive invasion of Schwann cells and nerve fibers is observed, but preferentially at the outer surface of the implant. The scaffolds 1 [Figure 15.11(a–b)] and 2 [Figure 15.11(c–d)], which consist of pores with uniform pore size distribution, were more favorable to cell attachment. Numerous cells covered the surface of these polymer scaffolds; some of them had formed aggregates [Figure 15.11(a,b), in the case of foam 1]. Cell spreading and extensive neurite outgrowth were also observed. Although some neurite extension can be seen on foam 3 [Figure 15.12(a,b)], only a few isolated cells were seen on foam 4, which were localized either in closed macropores or in small cavities, as shown by Figure 15.12(c,d), respectively [237].

To enhance the efficacy of peripheral nerve regeneration, PLLA was manufactured into porous biodegradable conduits using a combined

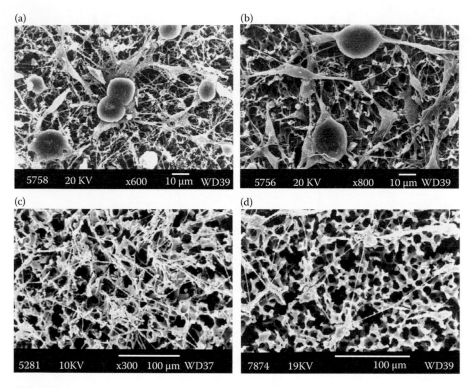

FIGURE 15.11
Scanning electron micrographs of neuronal cells seeded on the surface of scaffolds with inter-connected pores: (a–b) foam 1 (resomer L210 with 5% dioxane/water having interconnected pore) and (c–d) foam 2 (resomer R206 with 10% dioxane/water having interconnected pore) for 48 hr. Cell density on the surface is very important, particularly for foam 1. Cells of spherical or ellipsoidal shape form a network of neurites. Aggregates of few neuronal cells are seen that adhere to the polymer surface. (Adapted from Maquet V., Martin D., Malgrange B., Franzen R., Schoenen J., Moonen G., Jerome R., *J. Biomed. Mater. Res.*, 52, 639–51, A 2000.)

solvent casting, extrusion, and particulate leaching technique [238–240]. The efficacy of extruded biodegradable polymer conduits was increased by placing Schwann cells within an empty PLLA conduit containing collagen, to make them bioactive for peripheral nerve regeneration. The conduits were then implanted into a 12 mm right sciatic nerve defect in rats. The study provides promising results for further investigation to reach equivalent nerve regeneration when compared to autografts. Not only the macroporosity but also the design of macroporous material affects the nerve tissue regeneration [241].

The coil-reinforced composite tubes provide a new design strategy to create tubes with mechanical integrity from low modulus materials, such as hydrogels that match the modulus of soft tissues. These strategies could be applied to other device designs and indicate that nerve autografts can be replaced with synthetic alternatives. The coil-reinforced tubes of poly(2-hydroxyethyl

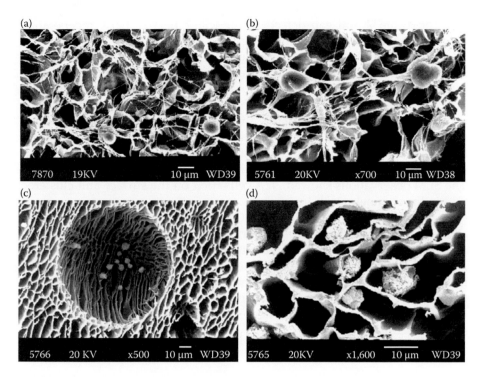

FIGURE 15.12
Scanning electron micrographs of neuronal cells seeded on the surface of scaffolds with oriented channels: (a–b) foam 3 (resomer L210 with 5% dioxane having oriented pore) and (c–d) foam 4 (resomer R206 with 5% dioxane having oriented pore) for 48 hr. The number of cells is small on 4, some isolated cells being seen within the small cavities. The cell surface coverage on foam 3 is higher compared to foam 4. Neurite extension can also be observed on foam 3. (Adapted from Maquet V., Martin D., Malgrange B., Franzen R., Schoenen J., Moonen G., Jerome R., *J. Biomed. Mater. Res. A.*, 52, 639–51, 2000.)

methacrylate-*co*-methyl methacrylate) used as nerve guidance channels demonstrated equivalence to nerve autografts (the current gold standard) in a 10 mm rat transected peripheral nerve injury model, as assessed by several validated outcome measures of nerve regeneration [241]. A new method to fabricate porous, biodegradable conduits using a combined injection molding, thermally induced phase transition technique was developed, which produced conduits with dimensionally tolerated, longitudinally aligned channels [242]. The geometry of the channels was designed to approximate the architecture of peripheral nerves and to support the monolayer adherence of physiologically relevant numbers of Schwann cells. *In vivo* studies were conducted using the nerve conduits described above in which adherent Schwann cells were introduced or a model neurotrophin was encapsulated in the porous conduit walls and was slowly released over a period of weeks to months. In the first study, seven-channel conduits were seeded with Schwann cells and the adherent Schwann cell/conduit composites were implanted into 7 mm gap defects in rat sciatic nerves. Early regeneration results compared favorably with autografts [243].

Chitosan is a potential biomaterial for nerve repair. Previous studies have shown that fetal mouse cerebral cortex cells could grow well on chitosan films [244]. Porous multichannel chitosan conduits were fabricated using a novel phase-separation technique with an axial temperature gradient. These conduits contained longitudinally aligned channels for guiding the outgrowing nerve fibers, an outer membrane for preventing ingrowth of fibrous tissue into the nerve gap and allowing fluid transport. Neuroblastoma cells (N2A, mouse) were used to evaluate the nerve cell affinity of the materials. The SEM observations showed N2A cells tiled on the surface and clustered within the channels [245].

The commercially available NeuraGen® Nerve Guide is an absorbable collagen tube designed to be an interface between the nerve and the surrounding tissue and to create a conduit for axonal growth across a nerve gap. Although the axons of severed peripheral nerves regenerate spontaneously, they will not establish functional connections unless the nerve stumps are surgically reconnected. The NeuraGen Nerve Guide offers a rapid method for rejoining severed peripheral nerves, in contrast to conventional microsurgical techniques [246].

15.7 Central Nervous System (CNS)

The adult mammalian CNS possesses a limited capacity for endogenous repair, as shown by astrocytic differentiation and proliferation, mesenchymal-epithelial interactions, deposition of extracellular matrices, angiogenesis, and axonogenesis [247–249]. Tissue engineering approaches used to repair CNS require polymer devices with controlled macro- and microstructures and chemical properties to achieve organ regeneration. Matrices of collagen [250,251] containing neuroactive agents [252] or cell grafts [253, 254] and synthetic polymers including nitrocellulose membranes, tubes of poly(acrylonitrile-*co*-vinylchloride), of polycarbonate, of nitrocellulose, and of polylactide implants [255] have been used for tissue repair of CNS lesions. Hydrogels of PHPMA have been synthesized with basic properties for stimulating tissue remodeling and organization during natural healing of neural tissues [256]. To demonstrate that the response of host CNS tissue was the result of the influence of the hydrogel on the wound healing process, a large tissue defect was surgically created. In normal conditions, this response results in the formation of a scar that seals off the lesion and impedes axon regeneration. The purpose of implanting the PHPMA hydrogel was to provide a stable three-dimensional substrate that facilitated tissue replacement, encouraged axon regrowth, and reduced scar formation. Tissue infiltration into the gels occurred and was assessed using immunocytochemical and ultrastructural techniques. In addition, tissue ingrowth was correlated with *in vivo* physiological changes in diffusion parameters within the PHPMA hydrogel.

This study confirms previous studies that the PHPMA hydrogel can effectively bridge tissue defects in the CNS, with recruitment of cells and blood vessels from host tissue contacting the hydrogel surface, and axonal regrowth into and through the polymer matrix [236]. A hyaluronic acid (HA)–poly-D-lysine (PDL) copolymer hydrogel with an open porous structure and viscoelastic properties similar to neural tissue has been developed for brain tissue engineering [257]. The incorporation of PDL peptides into the HA–PDL hydrogel allowed for the modulation of neuronal cell adhesion and neural network formation. Recently, a biodegradable and injectable scaffold had been developed that can be used in conjunction with cell replacement therapies. Thermally responsive and biodegradable 2-D (thin film made by placing 15 µl of polymer solution in a well plate) and 3-D films (thick film made by placing 150 µl of polymer solution in a well plate were prepared). The 3-D film composed of hydrogel of chitosan/glycerophosphate salt were used to repair damaged neural pathways within the brain [258]. In later studies [259], experiments were performed on 8-week-old Sprague-Dawley female rats. A block of cortical tissue (2.0 mm × 3.0 mm × 2.0 mm) was removed mechanically by excavator spoon and a lesion cavity was made in the left frontal region of the cortex. A piece of hydrogel sized to the dimensions and shape of the cavity was placed into the lesion site (Figure 15.13) [259]. The results showed that the HA hydrogels had mechanical properties and rheological behavior similar to that of the brain tissue. After being implanted into the lesion of the cortex, the porous hydrogels created a scaffold, which could support cell

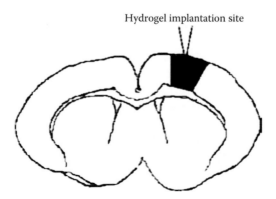

FIGURE 15.13
The scheme showing HA hydrogels implantation site as a shaded area in the right hemisphere on a coronal section. (Adapted from Hou S., Xu Q., Tian W., Cui F., Cai Q., Ma J., Lee I., *J. Neurosci. Methods*, 148, 60–70, 2005. With permission.)

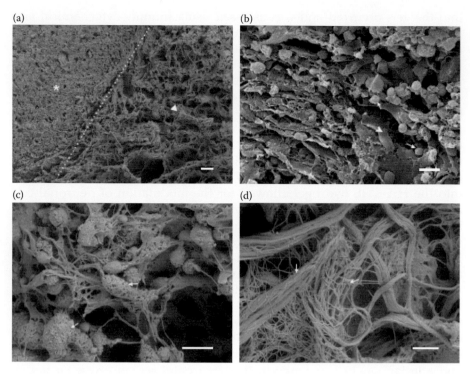

FIGURE 15.14
SEM examination of the implantation site 6 weeks showing the structure of hydrogels filled in the defects of the cortex on a coronal section. A low magnification assay reveals hydrogels (on the right of dotted line, arrowhead) have merged with the normal tissue (on the left of dotted line, asterisk) of the brain and re-established the anatomical continuity of the tissue (a). In the internal structure of the porous hydrogels, the three-dimensional network (arrowhead) is invaded by ample cells (arrows) in the group of modified (b) and unmodified hydrogels (c). Bundles of like collagen (arrows) can be seen between invaded cells in the grafts under high power (d). Scale bar = 40 μm (a); 10μm (b and c); 2μm (d). (From Hou S., Xu Q., Tian W., Cui F., Cai Q., Ma J., Lee I., *J. Neurosci. Methods*, 148, 60–70, 2005. With permission.)

infiltration and angiogenesis and simultaneously inhibit the formation of glial scar. In addition, HA hydrogels modified with laminin could promote neurite extension (Figure 15.14).

It seems possible that the tissue engineering technique may pave the way to repair injuries in the CNS as suggested by the results. To improve cell adhesion and neurite outgrowth in some natural polymeric scaffold, PDL was immobilized onto chitosan via azidoaniline photocoupling. It was shown that immobilized PDL improved cell survival up to an optimum concentration of 0.1%, however, further increases resulted a drop in cell number and neurite outgrowth. This was attributed to a higher cell interaction with PDL within a 3-D hydrogel compared to the corresponding 2-D surface. The results showed that thermally responsive chitosan/glycerophosphate hydrogels provide a suitable 3-D scaffolding environment for neural tissue engineering [258]. Furthermore, it was also shown that porous HA hydrogels, unmodified or modified with the molecule laminin, may induce neural tissue formation after implantation into a lesion of the adult CNS [259]. The results showed that the HA hydrogels synthesized had mechanical properties and rheological behavior similar to the brain tissue. After being implanted into the lesion of the cortex, the porous hydrogels created a scaffold, which could support cell infiltration and angiogenesis, and simultaneously inhibit the formation of a glial scar. In addition, HA hydrogels modified with laminin could promote neurite extension.

Some recent experimental demonstration again showed that the three-dimensional distribution and growth of cells within the porous scaffold are of clinical significance for nerve tissue engineering. An attempt was made to develop porous polymeric nanofibrous scaffolds using a biodegradable PLLA for *in vitro* culture of nerve stem cells (NSCs) [260]. The processing of PLLA scaffolds has been carried out by a liquid–liquid phase separation method. The prepared scaffold was highly porous and fibrous with diameters down to nanometer scale. The nanostructured PLLA scaffolds mimic natural ECM and thus are employed as biodegradable cell carriers in nerve tissue engineering. The *in vitro* performance of NSCs seeded on nanofibrous scaffolds showed that the NSCs could differentiate on the nanostructured scaffold and the scaffold acted as a positive cue to support neurite outgrowth. In conclusion the hydrogel and nanofibrous scaffolds are potential cell carriers in nerve tissue engineering.

15.8 Spinal Cord

The core approach of tissue engineering for spinal cord repair and regeneration consists of the provision of an interactive environment between cells, scaffolds, and bioactive molecules to promote tissue repair. To achieve this goal, the ex vivo engineered cell-scaffold constructs could be transplanted to the site of injury. Alternatively, the repair is achieved by delivery of scaffold-free cells or acellular scaffolds to the damaged tissue. As the lack of ECM at the lesion site that directs and organizes the wound healing cells is one of the mechanisms that interferes with regenerative process after spinal cord injury (SCI), different studies have been conducted to investigate the potential of bioscaffold grafts to promote regeneration in the injured spinal cord, and to provide a bridge through which the regenerating axons can be properly guided from one end of the injury to the other end [261].

Macroporous scaffolds were applied either alone or, to increase their healing effects, in combination with different growth factors or cellular components. In spinal cord injuries, collagen has been used to fill the gap and the present evidence shows that it supports axonal regeneration. Application of cross-linked collagen and collagen filaments [262,263] have been studied in animal models of spinal cord injury. They increased regenerative activity in the spinal cord and improved the functional disability. The results of implantation of a collagen tube in the injured spinal cords of rats were also promising showing that regenerating spinal axons regrow into the ventral root through this tube [264]. When an alginate sponge was implanted in the spinal cord of rats, it promoted axonal elongation, and the axons establish electrophysiologically functional projections and lead to functional improvements [265,266]. Also, interestingly, it was found that the axons that entered the sponge from the rostral and caudal stumps were able to leave the sponge from the opposite side and establish functional synapses with local neurons [267]. When compared with collagen, alginate reduced glial scar formation at the construct-tissue interface [268]. Also, the number of axons entered the alginate sponge was significantly higher than that entering collagen collagen sponge. Recently, an alginate-based anisotropic capillary hydrogel was implanted into the cervical spinal cord injury of rats and the increased ingrowth of longitudinally directed regenerating axons into this implant was demonstrated [269].

For provision of a better 3-D construct, macroporous scaffolds (foams) made of PLA containing block copolymer of poly(ethylene oxide) with PLA were tested for injuries in spinal cords. The foams were molded into small diameter rods and 14–20 rods were assembled using acidic fibroblast growth factor (aFGF) containing fibrin glue and used to bridge the transected rat spinal cord. The construct was invaded by blood vessels and axons from proximal and distal spinal stumps, and axonal regrowth preferentially occurred along the main pore direction [270,271]. In another experiment, the same foam was made with the same diameter as rat spinal cord, treated with the neuroprotective brain-derived neurotrophic factor (BDNF), and embedded in fibrin glue containing aFGF. Apart from easier handling, this construct possessed a good flexibility and was able to support formation of blood vessels and migration of astrocytes, Schwann cells, and axons.

The BDNF led to the ingrowth of more regenerating axons to the implant, mainly at the rostral part. But the implants did not improve functional performance [271]. Synthetic hydrogels, such as PHPMA hydrogel (NeuroGel™) [272] and poly(2-hydroxyethyl methacrylate-*co*-methyl methacrylate) (PHEMA-MMA) [273] consist of cross-linked networks of hydrophilic copolymers that swell in water and provide three-dimensional substrates for cell attachment and growth. After implantation of NeuroGel into the transected cat spinal cord, the NeuroGel was infiltrated by blood vessels, glial cells, and regenerating descending supraspinal axons of the ventral funiculus and afferent fibers of the dorsal column, and most of regenerating axons were myelinated, mainly by Schwann cells. The regenerating axons were able to leave the implant both rostrally and caudally. The animals showed variable degrees of locomotor improvements [274]. NeuroGel was also implanted in the post-traumatic lesion cavity in a rat model of chronic compression-produced injury of spinal cord. The hydrogel was invaded by blood vessels and glial cells. Also, ingrowth of regenerating axons was observed from the rostral stump into the NeuroGel. The axons were associated with well-organized myelin sheets and Schwann cells. Functional recovery was also observed [275]. In another interesting study, the cell-adhesive sequence Arg-Gly-Asp (RGD) of the central-binding domain of the ECM glycoprotein fibronectin was incorporated into the NeuroGel (PHPMA-RGD hydrogel). This core tripeptide sequence plays a central role in the adhesion-mediated cell migration required for tissue construction during development and repair. The PHPMA-RGD hydrogel was implanted in the transected cord of rats and led to angiogenesis and axonal growth (Figure 15.15). It was shown that axons enter the construct from the rostral cord and leave it into the caudal stump. The axons were myelinated by Schwann cells, and supraspinal axons and synaptic connections were observed in the reconstructed cord segment. The rats showed some degrees of functional improvements [276,277].

In another experiment, PHEMA soaked in BDNF solution was implanted in hemisected rat spinal cords. BDNF did not have any effect on the scarring and angiogenesis, but it promoted axonal regeneration [278]. Axonal regeneration into the implant is also improved when PHEMA-MMA channels are filled with the matrices such as collagen and Matrigel [279].

Recently, a freeze-dried agarose scaffold with uniaxial linear pores extending through its full length was manufactured and its biocompatibility and ability to function as a depot for growth factors was confirmed by *in vitro* studies [280]. They can retain their guidance capabilities within the spinal cord for at least one month. Implantation of BDNF-incorporated scaffolds in a rat model of spinal cord injury, led to organized and linear axonal growth into the agarose. The implant was also penetrated with Schwann cells, blood vessels, and macrophages. It has also been shown that adult neural progenitor cells harvested from rats cervical spine can be mounted on an alginate-based anisotropic capillary hydrogel and this construct supports axonal regeneration *in vitro* [269]. In another experiment, neurospheres prepared from fetal rat hippocampus were injected into the alginate sponge and implanted in the injured spinal cord of rats. Alginate increased the survival of neurospheres after transplantation and supported their migration, differentiation, and integration to the host spinal cord [281]. Microencapsulation of fibroblasts producing BDNF in alginate-poly-L-ornithine is another method for application of alginate in treatment of spinal cord injury. Another kind of cell-scaffold construct includes a two-component scaffold made of a blend of 50:50 PLGA and a block copolymer of PLGA with-polylysine. The scaffold's inner portion emulated the gray matter via a porous polymer layer and its outer portion emulated the white matter with long, axially oriented pores for axonal guidance and radial porosity to allow fluid transport while inhibiting ingrowth of scar tissue. The inner layer was seeded with a clonal multipotent neural precursor cell line originally derived from the external

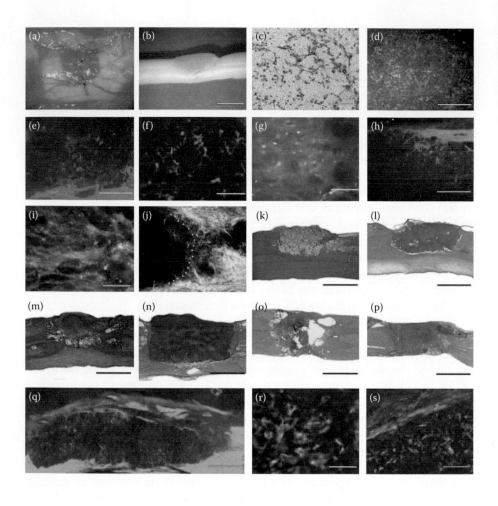

germinal layer of neonatal mouse cerebellum. Implantation of this construct into the hemisection adult rat model of spinal cord injury led to a long-term functional improvement accompanied by reduction of epidural and glial scar formation and growing of regenerating corticospinal tract fibers through the construct, from the injury epicenter to the caudal cord [282].

15.8.1 Extracorporeal Artificial Organs (EAO)

Extracorporeal artificial organs (EAO) may be defined as those life-supporting devices that substitute or replace an organ function temporarily or intermittently and require the online processing of blood outside the patient's body. Several assisting devices fall into this category including kidney machines used for dialysis, hemofiltration, and lung machines, for example [283]. A somewhat new area of interest in EAO is the development

FIGURE 15.15   (Opposite)
(a) HPMA-RGD hydrogel implanted in a hemisection lesion. Laminectomy was performed at the Th8 level. The dura was cut and a 2 mm segment in the right half of the spinal cord was dissected out. The hydrogel was trimmed to adapt to the cavity. The dura was sutured to avoid scarring from the extradural region. (b) The hydrogel implanted inside a hemisection cavity one month after implantation. It adhered well to the spinal cord (scalebar = 2mm). (c) Detail of the central part of the positively charged HEMA hydrogel. The pores of the scaffold are infiltrated with cells (CV staining, scalebar = 100 μm). (d) Regenerating blood vessels infiltrating the pores the HPMA-RGD hydrogel implant two weeks after implantation (RECA staining, scalebar = 250μm) (e, f) Two weeks after the HPMA-RGD implantation inside a hemisection. Neurofilaments were found mostly in the peripheral parts of the hydrogel, while four weeks after implantation the central part of the hydrogel was infiltrated as well (NF160 staining, scalebarG = 250μm, scalebarH = 100μm). (g) Regenerating neurofilaments on the HPMA-RGD hydrogel-tissue interface (GAP-43 staining, scalebar = 50μm). (h) Schwann cells grew from the spinal root entry zone into the HPMA-RGD hydrogel (p75 staining, scalebar = 250μm). (i) Within the implanted hydrogel scaffold Schwann cells grew along the regenerating neurofilaments (NF160-p75 staining, scalebar = 100μm). (j) Astrocytes only rarely crossed the border zones (white dotted line) of the HPMA-RGD hydrogel (GFAP staining, scalebar = 250μm). (k–l) Negatively (c) and positively (d) charged HEMA hydrogels bridging a hemisection cavity. Fewer cells grew inside the negatively charged hydrogel compared to the positively one. (m) Biodegradable HPMA hydrogel with a degradation time of 7 days. The obliterated hydrogel is surrounded by a zone where the polymer has already degraded and newly formed tissue has appeared (black bar) (HE staining, scalebar = 1mm). (n) Biodegradable HPMA hydrogel with a degradation time of 34 days. The hydrogel is minimally degraded with a large central part consisting of amorphous matter infiltrated with connective tissue and capillaries (HE staining, scalebar = 1 mm). (o, p) Positively charged HEMA hydrogels were implanted either immediately (o) or after a one week delay (p) after the spinal cord transection at the Th8 level. It is apparent that delayed implantation was associated with a reduction in the size of pseudocysts (HE staining, scalebar = 1mm) (q) HPMA-RGD hydrogel implanted inside a balloon compression spinal cord injury lesion 5 weeks after the injury. The hydrogel formed a bridge across the cavity and provided a scaffold for regenerating neurofilaments (NF160 staining, scalebar = 1mm). (r, s) HPMA-RGD hydrogel seeded with GFP-positive mesenchymal stem cells and implanted into a hemisection cavity. Two weeks after implantation the stem cells were found not only in the hydrogel (r) but they also infiltrated the spinal cord tissue surrounding the lesion (s) (GFP, scalebar = 25 μm). (From Hejčl A., Lesný P., Přádný M., Michálek J., Jendelová P., Stulík J., Syková E., *Physiol. Res.*, 57, S121–S132, 2008.)

of hybrid/bioartificial organ technologies. These bioartificial organs combine cells of desired type with nonbiological materials such as polymers or membranes. Bioartificial liver and pancreatic assisted devices are currently being investigated for use.

15.9 Bioartificial Liver Devices

Liver is a vital organ of the body that performs the function of detoxification and metabolism. Owing to its importance in maintaining the body functions, liver failure is a serious threat and is the wide spread cause of death due to unavailability of any proper medical cure except liver transplants, which are not readily available [284,285]. In such cases a temporary liver assist device, which can replace its function for sometime, may help the liver to regenerate or wait until a donor is available. To this effect a number of bioartificial liver devices has been developed and a number of them are in clinical trials. Among them are the extracorporeal liver assist device, the bioartificial liver support system, the liver support system, the modular extracorporeal liver support (The Academic Medical Center bioartificial liver, AMC-BAL), the Hepat-Assist, and the radial flow bioreactor (RFB). Most of these are based on hollow fiber technology and are in phase I and II clinical trials. Others like the University of California–Los Angeles bioartificial liver (UCLA-BAL) and polyurethane foam bioartificial liver (PUF-BAL) are based on either packed bed technology, flat membrane technology, or encapsulation technology and are in preclinical trial phase. Though most of the BAL devices being developed utilize hollow fibers and semipermeable membranes, some of the new BALs under investigation have also utilized macroporous scaffolds for hepatocyte immobilization and proliferation [284–289]. A further brief description of the macroporous polymeric matrices being used for bioartificial liver development will be given.

A variety of polymeric matrices is being explored for this purpose with several modifications. Adhesion, proliferation, and maintenance of functionality of hepatocytes were shown to be extremely dependent on surface properties of polymeric materials [290,291]. Attachment of monomeric carbohydrate sugar moieties to the porous scaffolds have been found to be favorable in terms of the introduction to specific hepatocyte recognition signal interaction within the matrix leading to greater cell attachment and viability. Improvement in spheroid formation and long-term maintenance of liver specific functions on the sugar modified scaffolds has also been reported. Both synthetic and natural polymers have been modified with galactose and used as a 3-D scaffold for liver cells [292]. Porous scaffolds of alginate and galactosylated chitosan sponges [293] and fructose conjugated chitosan sponges [294] have been prepared by lyophilization for liver tissue engineering.

Synthetic biodegradable polymers like PLLA and PLGA have also been investigated for liver tissue engineering. Ranucci and Moghe demonstrated that polymer substrate topography actively regulates the multicellular organization and liver-specific functions of cultured hepatocytes [295]. Porous foams of amorphous PLGA with a wide range of controlled pore size distributions (~1 to 100 μm) were used as culture model surfaces. Foams with intermediate void sizes (~17μm), promoted both 2-D and 3-D reorganization kinetics [295]. PLLA sponges fabricated by a salt leaching method were used for hepatocyte immobilization and preconditioned in a pulsatile flow bioreactor for a period of six days. This led to the formation of tightly packed cellular aggregates with a smooth boundary. The spheroids were homogenously distributed, occupying the pores within the entire volume of the 3-D porous structure of the PLLA polymer. Both the number and diameter of the spheroids increased during the first two days and then remained constant until day six. Pretransplant conditioning in a bioreactor and spheroid formation of hepatocytes polymer may represent a functionally active and easy transplantable neotissue and may serve as an *in vivo* substitute for lost liver function [296].

A 3-D scaffold based on chitosan gelatin has been fabricated by combining solid free-form fabrication, microreplication, and lyophilization techniques. It has a predefined multilevel internal architecture and highly porous structure. The scaffold possesses multilevel organized internal morphologies including vascular systems (portal vein, artery, and hepatic vein) and parenchymal component (hepatocyte chamber). These organized structures enable orderly arrangement of hepatocyte and hepatic nonparenchymal cells and coculture in the same 3-D scaffold to guide liver regeneration in a controlled manner. Cell culture experiment *in vitro* showed that hepatocytes perform better in the well-defined chitosan/gelatin scaffold than in porous scaffold [297].

15.10 Conclusion

Macroporous scaffolds are an integrated part needed to produce a engineered artificial tissue. It is important to maintain cells in a proper shape within a 3-D environment to keep them functionally active. A number of biomaterials are available for this purpose, both synthetic and natural. Natural polymers are more likely to be ideal and successful as they mimic the naturally occurring substances that constitute tissues. Newer strategies for scaffolding have lead to an improvement in the scaffold architecture ultimately affecting cell matrix interaction. The utility of each technique for engineering of specific tissues will ultimately depend on several design criteria including mechanical stability, chemical composition, degradation, cellular organization, and nutrient requirements. The hierarchical structures of the scaffolds have important influence on the cellular behavior of cells. Constant research is being carried on to understand cell physiology and cell matrix interactions and factors affecting them. However reviewing the experimental studies, it can be concluded that the ideal scaffold and matrix material for tissue engineering is still to be generated. The field of tissue engineering is definitely moving ahead with a promise to deliver products that will improve health conditions and life expectancy.

Acknowledgments

The work was supported by grants from the Department of Biotechnology (DBT) and the Department of Science and Technology (DST), Government of Indian Organizations. **Era Jain and Akshay Srivastava** would like to thank IITK and UGC respectively, for granting fellowship during the PhD program.

References

1. Hollister S.J. Porous scaffold design for tissue engineering. *Nat. Mater.* 2005; 4:518–24.
2. Liu C., Xia Z., Czernuszka J.T. Design and development of three-dimensional scaffolds for tissue engineering. *Trans. IChemE, Part A, Chem. Eng. Res. Design.* 2007; 85 (A7):1051–64.
3. Cima L.G., Vacanti J.P., Vacanti C., Ingber D., Mooney D., Langer R. Tissue engineering by cell transplantation using degradable polymer substrates. *J. Biomech. Eng.* 1991; 113:143–51.
4. Whang K., Healy E., Elenz D.R., Nam E.K., Tsai D.C., Thomas C.H., Nuber G.W., Glorieux F.H., Travers R., Sprague S.M. Engineering bone regeneration with bioabsorbable scaffolds with novel microarchitecture. *Tissue Eng.* 1999; 5:35–51.
5. Klawitter J.J., Hulbert S.F. Application of porous ceramics for the attachment of load-bearing internal orthopedic applications. *J. Biomed. Mater. Res. Symp.* 1971; 2:161–229.
6. Wake M.C., Patrick C.W., Mikos A.G. Pore morphology effects in the fibrovascular tissue growth in porous polymer substrates. *Cell Transplant.* 1994; 3:339–43.
7. Mooney D.J., Baldwin D.F, Suh N.P. Vacanti J.P., Langer R. Novel approach to fabricate porous sponges of poly(D,L-lacto-*co*-glycolic acid) without the use of organic solvents. *Biomaterials* 1996; 17:1417–22.
8. Colton C.K. Implantable biohybrid artificial organs. *Cell Transplant* 1995; 4:415–36.
9. Cohen S., Yoshioka T., Lucarelli M., Hwang L.H., Langer R. Controlled delivery systems for proteins based on poly(lactic/glycolic acid) microspheres. *Pharm. Res.* 1991; 8:713–20.
10. Cima L.G., Cima M.J. Tissue regeneration matrices by solid free form fabrication techniques. U.S. patent 1996, #5,518,680.
11. Mooney D., Hansen L., Vacanti J., Langer R., Farmer S., Ingber D. Switching from differentiation to growth in hepatocytes: Control by extracellular matrix. *J. Cell Physiol.* 1992; 51:497–505.
12. Lu L., Mikos A.G. The importance of new processing techniques in tissue engineering. *MRS Bulletin* 1996; 21:28–32.

13. Liu D.M. Fabrication of hydroxyapatite ceramic with controlled porosity. *J. Mater. Sci.: Mater. Medicine* 1997; 8:227–32.
14. Meenan B.J., Boyd A., Love E., Akay M. Evaluation of the quality of calcium phosphate bioceramics via a study of the effects of thermal processing. *Bioceramics* 2000; 192-1: 15–18.
15. Meenen N.M., Osborn J.F., Dallek M., Donath K. Hydroxyapatite-ceramic for juxta-articular implantation, *J. Mater. Sci.: Mater. Medicine* 1992; 3:345–51.
16. Chen G.P., Ushida T., Tateishi T. Development of biodegradable porous scaffolds for tissue engineering. *Mater. Sci. Eng. C.* 2001; 17:63–69.
17. Miko A.G., Thorsen A.J., Czerwonka L.A., Bao Y., Langer R. Preparation and characterization of poly(L-lactic acid) foams. *Polymer.* 1994; 35:1068–77.
18. Zhao F., Yin Y.J., Lu W.W., Leong J.C., Zhang W., Zhang J.Y., Zhang M.F., Yao K.D. Preparation and histological evaluation of biomimetic three-dimensional hydroxyapatite/chitosan-gelatin network composite scaffolds. *Biomaterials* 2002; 23:3227–34.
19. Cooper J.A., Lu H.H., Ko F.K., Freeman J.W., Laurencin C.T. Fiber-based tissue-engineered scaffold for ligament replacement: Design considerations and *in vitro* evaluation. *Biomaterials* 2005; 26:1523–32.
20. Ouyang H.W., Goh J.C., Thambyah A., Teoh S.H., Lee, E.H. Knitted poly-lactide-co-glycolide scaffold loaded with bone marrow stromal cells in repair and regeneration of rabbit Achilles tendon. *Tissue Eng.* 2003; 9:422–31.
21. Chu T-M.G., Orton D.G., Hollister S.J., Feinberg S.E., Halloran J.W. Mechanical and *in vivo* performance of hydroxyapatite implants with controlled architectures. *Biomaterials* 2002; 23:1283–93.
22. Das S., Hollister S.J., Flanagan C., Adewunmi A., Bark K., Chen C., Ramaswamy K., Rose D., Widjaja E. Freeform fabrication of nylon-6 tissue engineering scaffolds. *Rapid Prototyping J.* 2003; 9:43–49.
23. Hoque M.E., Hutmacher D.W., Feng W., Li S., Huang M.H., Vert M., Wong Y.S. Fabrication using a rapid prototyping system and *in vitro* characterization of PEG-PCL-PLA scaffolds for tissue engineering. *J. Biomater. Sci.-Polymer Ed.* 2005; 16:1595–1610.
24. Khalil S., Nam J., Sun W. Multi-nozzle deposition for construction of 3D biopolymer tissue scaffolds. *Rapid Prototyping J.* 2005; 11:9–17.
25. Taboas J.M., Maddox R.D., Krebsbach P.H., Hollister S.J. Indirect solid free form fabrication of local and global porous, biomimetic and composite 3D polymer-ceramic scaffolds. *Biomaterials* 2003; 24:181–94.
26. Woodfield T.B.F., Malda J., de Wijn J., Peters F., Riesle J., van Blitterswijk C.A. Design of porous scaffolds for cartilage tissue engineering using a three-dimensional fiber-deposition technique. *Biomaterials* 2004; 25:4149–61.
27. Plieva F., Oknianska A., Degerman E., Galaev I.Yu., Mattiasson B. Novel super-macroporous dextran gels. *J. Biomater. Sci. Polymer Edn.* 2006; 17:1075–92.
28. Jain E., Srivastava A., Kumar A. Macroporous interpenetrating cryogel network of poly(acrylonitrile) and gelatin for biomedical applications. *J. Mater. Sci.: Mater. Med.* 2008; DOI 10.1007/s10856-008-3504-4.
29. Tripathi A., Kathuria N., Kumar A. Elastic and macroporous agarose-gelatin cryogels with isotropic and anisotropic porosity for tissue engineering. *J. Biomed. Mater. Res. A* 2008 DOI 10.1002/jbm.a.32127.

450													*Macroporous Polymers*

30. Kathuria N., Tripathi A., Kar K.K., Kumar A. Synthesis and characterization of elastic and macroporous chitosan-gelatin cryogel for tissue engineering. *Acta Biomaterialia* 2008; 5:406–18.
31. Jain E., Kumar A. Designing supermacroporous cryogels based on polyacrylonitrile and a polyacrylamide–chitosan semi-interpenetrating network. *J. Biomater. Sci. Polymer Edn.* 2009; 20(7–8):877–902.
32. Srivastava A., Kumar A. Synthesis and characterization of temperature responsive biocompatible poly(N-vinylcaprolactam) cryogel: A step towards designing a novel cell scaffold. *J. Biomater. Sci. Polymer Edn.* 2009; 20(10): 1393–1415.
33. Kumar A., Plieva F.M., Galaev I.Yu., Mattiasson B. Affinity fractionation of lymphocytes using a monolithic cryogel. *J. Immunol. Methods* 2003; 283:185–94.
34. Arvidsson P., Plieva F.M., Lozinsky V.I., Galaev I.Yu., Mattiasson B. Direct chromatographic capture of enzyme from crude homogenate using immobilized metal affinity chromatography on a continuous supermacroporous adsorbent. *J. Chromatogr. A* 2003; 986:275–90.
35. Arvidsson P., Plieva F.M., Savina I.N., Lozinsky V.I., Fexby S., Bülow L., Galaev I.Yu., Mattiasson B. Chromatography of microbial cells using continuous supermacroporous affinity and ion exchange columns. *J. Chromatogr. A.* 2002; 977:27–38.
36. Tuli R., Li W-J, Tuan R.S. Current state of cartilage tissue engineering. *Arthritis. Res. Ther.* 2003; 5:235–38.
37. Honda M., Yada T., Ueda M., Kimata K. Cartilage formation by cultured chondrocytes in a new scaffold made of poly(L-lactide-epsilon-caprolactone) sponge. *J. Oral Maxillofac. Surg.* 2000; 58:767–75.
38. Atala A., Kim W., Paigek. T., Vacanti C.A., Retik A.B. Endoscopic treatment of vesicoureteral reflux with a chondrocyte-alginate suspension. *J. Urol.* 1994; 152:641–43.
39. Risbud M., Ringe J., Bhonde R., Sittinger M. *In vitro* expression of cartilage-specific markers by chondrocytes on a biocompatible hydrogel: implications for engineering cartilage tissue. *Cell Transplant.* 2001; 10:755–63.
40. Roche S., Ronzière M-C., Herbagem D., Freyria A-M. Native and DPPA crosslinked collagen sponges seeded with fetal bovine epiphyseal chondrocytes used for cartilage tissue engineering. *Biomaterials* 2001; 22:9–18.
41. Rodriguez A., Cao Y.L., Ibarra C., Pap S., Vacanti M., Eavey R.D., Vacanti C.A. Characteristics of cartilage engineered from human pediatric auricular cartilage. *Plast. Reconstr. Surg.* 1999; 103:1111–19.
42. Ma P.X., Schloo B., Mooney D., Langer R. Development of biomechanical properties and morphogenesis of *in vitro* tissue engineered cartilage. *J. Biomed. Mater. Res.* 1995; 29:1587–95.
43. Grande D.A., Halberstadt C., Naughton G., Schwartz R., Manji R. Evaluation of matrix scaffolds for tissue engineering of articular cartilage grafts. *J. Biomed. Mater. Res.* 1997; 34:211–20.
44. Silverman R.P., Passaretti D., Huang W., Randolph M.A., Yaremchuk M.J. Injectable tissue engineered cartilage using a fibrin glue polymer. *Plast. Reconstr. Surg.* 1999; 103:1809–18.

45. Sittinger M., Bujia J., Minuth W.W., Hammer C., Burmester G.R. Engineering of cartilage tissue using bioresorbable polymer carriers in perfusion culture. *Biomaterials* 1994; 15:451–56.
46. Risbud M.V., Sittinger M. Tissue engineering: Advances in *in vitro* cartilage generation. *Trends Biotechnol.* 2002; 20:351–56.
47. Chung C., Burdick J.A. Engineering cartilage tissue. *Adv. Drug Deliv. Rev.* 2008; 60:243–62.
48. Bryant S.J., Anseth K.S. Hydrogel properties influence ECM production by chondrocytes photoencapsulated in poly(ethylene glycol) hydrogels. *J. Biomed. Mater. Res.* 2002; 59:63–72.
49. Bryant S.J., Anseth K.S. Controlling the spatial distribution of ECM components in degradable PEG hydrogels for tissue engineering cartilage. *J. Biomed. Mater. Res. Part A* 2003; 64A:70–79.
50. Bryant S.J., Durand K.L., Anseth K.S. Manipulations in hydrogel chemistry control photoencapsulated chondrocyte behavior and their extracellular matrix production. *J. Biomed. Mater. Res. A.* 2003; 67:1430–36.
51. Rice M.A., Anseth K.S. Encapsulating chondrocytes in copolymer gels: Bimodal degradation kinetics influence cell phenotype and extracellular matrix development. *J. Biomed. Mater. Res. A* 2004; 70:560–68.
52. Lee H.J, Lee J.S., Chansakul T., Yu C., Elisseeff J.H., Yu S.M. Collagen mimetic peptide-conjugated photopolymerizable PEG hydrogel. *Biomaterials* 2006; 27:5268–76.
53. Sontjens S.H.M., Nettles D.L., Carnahan M.A., Setton L.A., Grinstaff M.W. Biodendrimer-based hydrogel scaffolds for cartilage tissue repair. *Biomacromolecules* 2006; 7:310–16.
54. Chen W.Y., Abatangelo G. Functions of hyaluronan in wound repair. *Wound Repair Regen.* 1999; 7:79–89.
55. Nettles D.L., Vail T.P., Morgan M.T., Grinstaff M.W., Setton L.A. Photo-crosslinkable hyaluronan as a scaffold for articular cartilage repair. *Ann. Biomed. Eng.* 2004; 32:391–97.
56. Smeds K.A., Pfister-Serres A., Miki D., Dastgheib K., Inoue M., Hatchell D.L., Grinstaff M.W. Photocrosslinkable polysaccharides for *in situ* hydrogel formation. *J. Biomed. Mater. Res.* 2001; 54:115–21.
57. Burdick J.A., Chung C., Jia X.Q., Randolph M.A., Langer R. Controlled degradation and mechanical behavior of photopolymerized hyaluronic acid networks. *Biomacromolecules* 2005; 6:386–91.
58. Chung C., Mesa J., Randolph M.A., Yaremchuk M., Burdick J.A. Influence of gel properties on neocartilage formation by auricular chondrocytes photoencapsulated in hyaluronic acid networks. *J. Biomed. Mater. Res. A* 2006; 77:518–25.
59. Liu Y.C., Shu X.Z., Prestwich G.D. Osteochondral defect repair with autologous bone marrow-derived mesenchymal stem cells in an injectable, *in situ*, cross-linked synthetic extracellular matrix. *Tissue Eng.* 2006; 12:3405–16.
60. Li Q., Williams C.G., Sun D.D., Wang J., Leong K, Elisseeff J.H. Photocrosslinkable polysaccharides based on chondroitin sulfate. *J. Biomed. Mater. Res. A* 2004; 68:28–33.
61. Mosesson M.W. Fibrinogen and fibrin structure and functions. *J. Thromb. Haemost.* 2007; 3:1894–1904.

62. Eyrich D., Brandl F., Appel B., Wiese H., Maier G., Wenzel M., Staudenmaier R., Goepferich A., Blunk T. Long-term stable fibrin gels for cartilage engineering. *Biomaterials* 2007; 28:55–65.
63. Lee C.R., Grad S., Gorna K., Gogolewski S., Goessl A., Alini M. Fibrinpolyurethane composites for articular cartilage tissue engineering: A preliminary analysis. *Tissue Eng.* 2005; 11:1562–73.
64. Franceschi L.D., Grigolo B, Roseti L., Facchini A., Fini M., Giavaresi G., Tschon M., Giardino R. Transplantation of chondrocytes seeded on collagen-based scaffold in cartilage defects in rabbits. *J. Biomed. Mater. Res. A.* 2005; 75:612–22.
65. Hoshikawa A., Nakayama Y., Matsuda T., Oda H., Nakamura K., Mabuchi K. Encapsulation of chondrocytes in photopolymerizable styrenated gelatin for cartilage tissue engineering. *Tissue Eng.* 2006; 12:2333–41.
66. Schagemann J.C., Mrosek E.H., Landers R., Kurz H., Erggelet C. Morphology and function of ovine articular cartilage chondrocytes in 3-D hydrogel culture. *Cells Tissues Organs* 2006; 182:89–97.
67. Awad H.A., Wickham M.Q., Leddy H.A., Gimble J.M., Guilak F. Chondrogenic differentiation of adipose-derived adult stem cells in agarose, alginate, and gelatin scaffolds, *Biomaterials* 2004; 25:3211–22.
68. Erickson G.R., Gimble J.M., Franklin D.M., Rice H.E., Awad H., Guilak F. Chondrogenic potential of adipose tissue-derived stromal cells *in vitro* and *in vivo*. *Biochem. Biophys. Res.* 2002; 290:763–69.
69. Lin Y.F., Luo E., Chen X.Z., Liu L., Qiao J., Yan Z.B., Li Z.Y., Tang W., Zheng X.H., Tian W.D. Molecular and cellular characterization during chondrogenic differentiation of adipose tissue-derived stromal cells *in vitro* and cartilage formation *in vivo*. *J. Cell Mol. Med.* 2005; 9:929–39.
70. Connelly J.T., Garcia A.J., Levenston M.E. Inhibition of *in vitro* chondrogenesis in RGD-modified three-dimensional alginate gels. *Biomaterials* 2007; 28:1071–83.
71. Iwasaki N., Yamane S.T., Majima T., Kasahara Y., Minami A., Harada K., Nonaka S., et al. Feasibility of polysaccharide hybrid materials for scaffolds in cartilage tissue engineering: evaluation of chondrocyte adhesion to polyion complex fibers prepared from alginate and chitosan. *Biomacromolecules* 2004; 5:828–33.
72. Wayne J.S., McDowell C.L., Shields K.J., Tuan R.S. *In vivo* response of polylactic acid-alginate scaffolds and bone marrow-derived cells for cartilage tissue engineering. *Tissue Eng.* 2005; 11:953–63.
73. Genes N.G., Rowley J.A., Mooney D.J., Bonassar L.J. Effect of substrate mechanics on chondrocyte adhesion to modified alginate surfaces. *Arch. Biochem. Biophys.* 2004; 422:161–67.
74. Mauck R.L., Soltz M.A., Wang C.C.B., Wong D.D., Chao P.H.G., Valhmu W.B., Hung C.T., Ateshian G.A. Functional tissue engineering of articular cartilage through dynamic loading of chondrocyte-seeded agarose gels. *J. Biomech. Eng-T Asme* 2000; 122:252–60.
75. Sechriest V.F., Miao Y.J., Niyibizi C., Westerhausen-Larson A., Matthew H.W., Evans C.H., Fu F.H., Suh J.K. GAG-augmented polysaccharide hydrogel: A novel biocompatible and biodegradable material to support chondrogenesis. *J. Biomed. Mater. Res. A* 1999; 49:534–41.

76. Subramanian A., Lin H.Y. Crosslinked chitosan: Its physical properties and the effects of matrix stiffness on chondrocyte cell morphology and proliferation. *J. Biomed. Mater. Res. A.* 2005; 75: 742–53.
77. Chen J.P., Cheng T.H. Thermo-responsive chitosan-graft-poly(N-isopropylacrylamide) injectable hydrogel for cultivation of chondrocytes and meniscus cells. *Macromol. Biosci.* 2006; 6:1026–39.
78. Au A., Ha J., Polotsky A., Krzyminski K., Gutowska A., Hungerford D.S., Frondoza C.G. Thermally reversible polymer gel for chondrocyte culture. *J. Biomed. Mater. Res. A* 2003; 67:1310–19.
79. Fisher J.P., Jo S., Mikos A.G., Reddi A.H. Thermoreversible hydrogel scaffolds for articular cartilage engineering. *J. Biomed. Mater. Res. A* 2004; 71:268–74.
80. Na K., Kim S., Woo D.G., Sun B.K., Yang H.N., Chung H.M., Park K.H. Synergistic effect of TGF beta-3 on chondrogenic differentiation of rabbit chondrocytes in thermo-reversible hydrogel constructs blended with hyaluronic acid by *in vivo* test. *J. Biotechnol.* 2007; 128:412–22.
81. Chen G.P., Sato T., Ushida T., Ochiai N., Tateishi T. Tissue engineering of cartilage using a hybrid scaffold of synthetic polymer and collagen. *Tissue Eng.* 2004; 10:323–30.
82. Miralles G., Baudoin R., Dumas D., Baptiste D., Hubert P., Stoltz J.F., Dellacherie E., Mainard D., Netter P., Payan E. Sodium alginate sponges with or without sodium hyaluronate: *in vitro* engineering of cartilage. *J. Biomed. Mater. Res.* 2001; 57:268–78.
83. Barnewitz D., Endres M., Kruger I., Becker A., Zimmermann M., Wilke I., Ringe J., Sittinger M., Kaps C. Treatment of articular cartilage defects in horses with polymer-based cartilage tissue engineering grafts. *Biomaterials* 2006; 27:2882–89.
84. Nettles D.L., Elder S.H., Gilbert J.A. Potential use of chitosan as a cell scaffold material for cartilage tissue engineering. *Tissue Eng.* 2002; 8:1009–16.
85. Xia W.Y., Liu W., Cui L., Liu Y.C., Zhong W., Liu D.L., Wu J.J., Chua K.H., Cao Y.L. Tissue engineering of cartilage with the use of chitosan:Gelatin complex scaffolds. *J. Biomed. Mater. Res. B-Appl. Biomat.* 2004; 71:373–80.
86. Wang Y.Z., Kim H.J., Vunjak-Novakovic G., Kaplan D.L. Stem cell based tissue engineering with silk biomaterials. *Biomaterials* 2006; 27:6064–82.
87. Solchaga L.A., Temenoff J.S., Gao J.Z., Mikos A.G., Caplan A.I., Goldberg V.M. Repair of osteochondral defects with hyaluronan- and polyester-based scaffolds. *Osteoarthritis Cartilage* 2005; 13:297–309.
88. Vickers S.M., Squitieri L.S., Spector M. Effects of cross-linking type II collagen-GAG scaffolds on chondrogenesis *in vitro*: Dynamic pore reduction promotes cartilage formation. *Tissue Eng.* 2006; 12:1345–55.
89. Goodstone N.J., Cartwright A., Ashton B. Effects of high molecular weight hyaluronan on chondrocytes cultured within a resorbable gelatin sponge. *Tissue Eng.* 2004; 10:621–31.
90. Kang Y., Yang J., Khan S., Anissian L., Ameer G.A. A new biodegradable polyester elastomer for cartilage tissue engineering. *J. Biomed. Mater. Res. A* 2006; 77:331–39.
91. Wang Y.Z., Blasioli D.J., Kim H.J., Kim H.S., Kaplan D.L. Cartilage tissue engineering with silk scaffolds and human articular chondrocytes. *Biomaterials* 2006; 27:4434–42.

92. Wang Y.Z., Kim U.J., Blasioli D.J., Kim H.J, Kaplan D.L. *In vitro* cartilage tissue engineering with 3D porous aqueous-derived silk scaffolds and mesenchymal stem cells. *Biomaterials* 2005; 26:7082–94.
93. Hofmann S., Knecht S., Langer R., Kaplan D.L., Vunjak-Novakovic G., Merkle H.P., Meinel L. Cartilage-like tissue engineering using silk scaffolds and mesenchymal stem cells. *Tissue Eng.* 2006; 12:2729–38.
94. Chen G.P., Sato T., Ushida T., Hirochika R., Shirasaki Y., Ochiai N., Tateishi T. The use of a novel PLGA fiber/collagen composite web as a scaffold for engineering of articular cartilage tissue with adjustable thickness. *J. Biomed. Mater. Res. A* 2003; 67:1170–80.
95. Li W.J., Danielson K.G., Alexander P.G., Tuan R.S. Biological response of chondrocytes cultured in three-dimensional nanofibrous poly(ε-caprolactone) scaffolds. *J. Biomed. Mater. Res. A* 2003; 67:1105–14.
96. Hsu S.H., Chang S.H., Yen H.J., Whu S.W., Tsai C.L., Chen D.C. Evaluation of biodegradable polyesters modified by type II collagen and Arg-Gly-Asp as tissue engineering scaffolding materials for cartilage regeneration. *Artificial Organs* 2006; 30:42–55.
97. Yoo H.S., Lee E.A., Yoon J.J., Park T.G. Hyaluronic acid modified biodegradable scaffolds for cartilage tissue engineering. *Biomaterials* 2005; 26:1925–33.
98. Uematsu K., Hattori K., Ishimoto Y., Yamauchi J, Habata T., Takakura Y., Ohgushi H., Fukuchi T., Sato M. Cartilage regeneration using mesenchymal stem cells and a three-dimensional poly-lactic-glycolic acid (PLGA) scaffold. *Biomaterials* 2005; 26:4273–79.
99. Li W.J., Tuli R., Okafor C., Derfoul A., Danielson K.G., Hall D.J., Tuan R.S. A three-dimensional nanofibrous scaffold for cartilage tissue engineering using human mesenchymal stem cells. *Biomaterials* 2005; 26:599–609.
100. Muller F.A., Muller L., Hofmann I., Greil P., Wenzel M.M., Staudenmaier R. Cellulose-based scaffold materials for cartilage tissue engineering. *Biomaterials* 2006; 27:3955–63.
101. Girotto D., Urbani S., Brun P., Renier D., Barbucci R., Abatangelo G. Tissue-specific gene expression in chondrocytes grown on three dimensional hyaluronic acid scaffolds. *Biomaterials* 2003; 24:3265–75.
102. Grigolo B., Lisignoli G., Piacentini A., Fiorini M., Gobbi P., Mazzotti G., Duca M., Pavesio A., Facchini A. Evidence for redifferentiation of human chondrocytes grown on a hyaluronan-based biomaterial (HYAFF (R) 11): Molecular, immunohistochemical and ultrastructural analysis. *Biomaterials* 2002; 23:1187–95.
103. Radice M., Brun P., Cortivo R., Scapinelli R., Battaliard C., Abatangelo G. Hyaluronan-based biopolymers as delivery vehicles for bone-marrow derived mesenchymal progenitors. *J. Biomed. Mater. Res. A* 2000; 50:101–9.
104. Gobbi A., Kon E., Berruto M., Francisco R., Filardo G., Marcacci M. Patellofemoral full-thickness chondral defects treated with hyalograft-C: A clinical, arthroscopic, and histologic review. *Am. J. Sports Med.* 2006; 34:1763–73.
105. Hollander A.P., Dickinson S.C., Sims T.J., Brun P., Cortivo R., Kon E., Marcacci M., et al. Maturation of tissue engineered cartilage implanted in injured and osteoarthritic human knees. *Tissue Eng.* 2006; 12:1787–98.
106. Marcacci M., Berruto M., Brocchetta D., Delcogliano A., Ghinelli D., Gobbi A., Kon E., et al. Articular cartilage engineering with Hyalograft (R) C: 3-year clinical results. *Clin. Orthop. Relat.* 2005; 435:96–105.

107. Sommerfeldt D.W., Rubin C.T. Biology of bone and how it orchestrates the form and function of the skeleton. *Eur. Spine J.* 2001; 10:S86–S95.
108. Salgado A.J., Coutinho O.P., Reis R.L. Bone tissue engineering: State of the art and future trends. *Macromol. Biosci.* 2004; 4:743–65.
109. Schieker M., Seitz H., Drosse I., Seitz S., Mutschler W. Biomaterials as scaffold for bone tissue engineering. *Eur. J. Trauma* 2006; 32:114–24.
110. Orban J.M., Marra K.G., Hollinger J.O. Composition options for tissue-engineered bone. *Tissue Eng.* 2002; 8:529–39.
111. Rose R.A.J., Oreffo R.O.C. Bone tissue engineering: Hope vs hype felicity. *Biochem. Biophy. Res. Commun.* 2002; 292:1–7.
112. Yang S., Leong K.F., Du Z., Chua C.K. The design of scaffolds for use in tissue engineering. Part I. Traditional factors. *Tissue Eng.* 2001; 7:679–89.
113. Klawitter J.J., and Hulbert S.F. Application of porous ceramics for the attachment of load-bearing internal orthopedic applications. *J. Biomed. Mater. Res. Symp.* 1971; 5:161–229.
114. Whang K., Healy E., Elenz D.R., Nam E.K., Tsai D.C., Thomas C.H., Nuber G.W., Glorieux F.H., Travers R., Sprague S.M. Engineering bone regeneration with bioabsorbable scaffolds with novel microarchitecture. *Tissue Eng.* 1999; 5:35–51.
115. Holly C.E., Schoichet M.S., Davies J.E. Engineering three-dimensional bone tissue *in vitro* using biodegradable scaffolds: Investigating initial cell-seeding density and culture period. *J. Biomed. Mater. Res.* 2000; 51:376–82.
116. Albrektsson T., Johansson C. Osteoinduction, osteoconduction and osseointegration. *Eur. Spine J.* 2001; 10:S96–S101.
117. Hutmacher D.W. Scaffolds in tissue engineering bone and cartilage. *Biomaterials* 2000; 21:2529–43.
118. Grynpas M.D., Pilliar R.M., Kandel R.A., Renlund R., Filiaggi M., Dumitriu M. Porous calcium polyphosphate scaffolds for bone substitute applications *in vivo* studies. *Biomaterials* 2002; 23:2063–70.
119. Dong J., Kojima H., Uemura T., Kikuchi M., Tateishi T., Tanaka J. *In vivo* evaluation of a novel porous hydroxyapatite to sustain osteogenesis of transplanted bone marrow-derived osteoblastic cells. *J. Biomed. Mater. Res.* 2001; 57:208–16.
120. Yoshikawa T., Ogushi H., Nakajima H., Yamada E., Ichijima K., Tamai S., Ohta T. *In vivo* osteogenic durability of cultured bone in porous ceramics: A novel method for autogenous bone graft substitution. *Transplantation* 2000; 69:128–34.
121. LeGeros R.Z. Properties of osteoconductive biomaterials: Calcium phosphates. *Clin. Orthop. Relat. Res.* 2002; 395:81–98.
122. Liu L., Zhang L., Ren B., Wang F., Zhang Q. Preparation and characterization of collagen-hydroxyapatite composite used for bone tissue engineering scaffold. *Artificial Cells, Blood Substitutes, and Biotech.* 2003; 31:435–48.
123. Li X., Feng Q., Jiao Y., Cui F. Collagen-based scaffolds reinforced by chitosan fibers for bone tissue engineering. *Polymer Internat.* 2005; 54:1034–40.
124. Usami K., Mizuno H., Okada K., Narita Y., Aoki M., Kondo T., Mizuno D., et al. Composite implantation of mesenchymal stem cells with endothelial progenitor cells enhances tissue-engineered bone formation. *J. Biomed. Mater. Res. A* 2008 DOI 10.1002/jbm.a.32142.
125. Seol Y.J., Lee J.Y., Park Y.J., Lee Y.M., Young-Ku, Rhyu I.C., Lee S.J., Han S.B., Chung C.P. Chitosan sponges as tissue engineering scaffolds for bone formation. *Biotechnol. Lett.* 2004; 26:1037–41.

126. Zhang Y., Zhang M. Three-dimensional macroporous calcium phosphate bioceramics with nested chitosan sponges for load bearing bone implants. *J. Biomed. Mater. Res.* 2002; 61:1–8.
127. Zhang Y., Ni M., Zhang M., Ratner B. Calcium phosphate chitosan composite scaffolds for bone tissue engineering. *Tissue Eng.* 2003; 9:337–45.
128. Park Y.J., Lee Y.M., Park S.N., Sheen S.Y., Chung C.P., Lee S.J. Platelet derived growth factor releasing chitosan sponge for periodontal bone regeneration. *Biomaterials* 2000; 21:153–59.
129. Lee J.Y., Nam S.H., Im S.Y., Park Y.J., Lee Y.M., Seol Y.J., Chung C.P., Lee S.J. Enhanced bone formation by controlled growth factor delivery from chitosan-based biomaterials. *J. Control. Release* 2002; 78:187–97.
130. Muzzarelli R.A., Mattioli-Belmonte M., Tietz C., Biagini R., Ferioli G., Brunelli M.A., Fini M., Giardino R., Ilari P., Biagini G. Stimulatory effect on bone formation exerted by a modified chitosan. *Biomaterials* 1994; 15:1075–81.
131. Lee Y.M., Park Y.J., Lee S.J., Ku Y., Han S.B., Klokkevold P.R., Chung C.P. The bone regenerative effect of platelet-derived growth factor-BB delivered with a chitosan/tricalcium phosphate sponge carrier. *J. Periodontol.* 2000; 71:418–24.
132. Doyle C., Tanner E.T., Bonfield W. *In vitro* and *in vivo* evaluation of polyhydroxybutyrate and of polyhydroxybutyrate reinforced with hydroxyapatite. *Biomaterials* 1991; 12:841–47.
133. Rezwan K., Chen Q.Z., Blaker J.J., Boccaccini A. R. Biodegradable and bioactive porous polymer/inorganic composite scaffolds for bone tissue engineering. *Biomaterials* 2006; 27:3413–31.
134. Gassner F., Owen A.J. Some properties of poly(3-hydroxybutyrate-*co*-3-hydroxyvalerate) blends. *Polym. Int.* 1996; 39:215–19.
135. Kose G.T., Kenar H., Hasırcı N., Hasırcı V. Macroporous poly(3-hydroxybutyrate-*co*-3-hydroxyvalerate) matrices for bone tissue engineering. *Biomaterials* 2003; 24:1949–58.
136. Altman G.H., Horan R.L., Lu H.H., Moreau J., Martin I., Richmond J.C., Kaplan D.L. Silk matrix for tissue engineered anterior cruciate ligaments. *Biomaterials* 2002; 23:4131–41.
137. Kaplan D.L. Spiderless spider webs. *Nat. Biotechnol.* 2002; 20:239–40.
138. Sofia S., McCarthy M.B, Gronowicz G., Kaplan D.L. Functionalized silk-based biomaterials for bone formation. *J. Biomed. Mater. Res. A* 2001; 54:139–48.
139. Meinel L., Karageorgiou V., Hofmann S., Fajardo R., Snyder B., Li C., Zichner L., Langer R., Vunjak-Novakovic G., Kaplan D.L. Engineering bone-like tissue *in vitro* using human bone marrow stem cells and silk scaffolds. *J. Biomed. Mater. Res. A* 2004; 71:25–34.
140. Li C., Vepari C., Jin H-J, Kim H.J., Kaplan D.L. Electrospun silk-BMP-2 scaffolds for bone tissue engineering. *Biomaterials* 2006; 27:3115–24.
141. Liao C.J., Chen C.F., Chen J.H., Chiang S.F., Lin Y.J., Chang K.Y. Fabrication of porous biodegradable polymer scaffolds using a solvent merging/particulate leaching method. *J. Biomed. Mater. Res. A* 2002; 59:676–81.
142. Ma P.X., Choi J.W. Biodegradable polymer scaffolds with well-defined interconnected spherical pore network. *Tissue Eng.* 2001; 7:23–33.
143. Wei G., Ma P.X. Macroporous and nanofibrous polymer scaffolds and polymer/bone-like apatite composite scaffolds generated by sugar spheres. *J. Biomed. Mater. Res. A* 2006; 78:306–15.

144. Williams J.M., Adewunmi A., Schek R.M., Flanagan C.L., Krebsbach P.H., Feinberg S.E., Hollister S.J., Das S. Bone tissue engineering using polycaprolactone scaffolds fabricated via selective laser sintering. *Biomaterials* 2005; 26:4817–27.
145. Marei M.K., Nouh S.R., Fata M.M., Faramawy A.M. Fabrication of polymer root form scaffolds to be utilized for alveolar bone regeneration. *Tissue Eng.* 2003; 9:713–31.
146. Gogolewski S., Gorna K., Rahn B., Wieling R. Biodegradable polyurethane cancellous bone graft substitute promotes bone regeneration in the iliac crest defects. In: Transaction, 27th Society for Biomaterials Annual Meeting, Saint Paul, MN, USA. 2001; 24:573.
147. Gogolewski S., Gorna K., Turner A.S. Regeneration of bicortical defects in the iliac crest of estrogen deficient sheep using new biodegradable polyurethane cancellous bone graft substitutes. A pilot study. Presented at the Proceedings of the 48th Annual Meeting, Orthopaedic Research Society, Dallas, TX. Abstract 0740. 2002; 10–13.
148. Gogolewski S., Gorna K., Wieling R. Biodegradable polyurethane cancellous bone graft substitutes in the treatment of critical-size segmental diaphyseal defects in the sheep tibiae. Presented at the Proceedings of the 49th Orthopaedic Research Society Meeting, New Orleans, LA. 2003.
149. Cuddihy M.J., Kotov N.A. Poly(lactic-*co*-glycolic acid) bone scaffolds with inverted colloidal crystal geometry. *Tissue Eng.* A 2008; 14:1639–49.
150. Bolgen N., Yang Y., Korkusuz P., Guzel E., El Haj A. J, Piskin E. Three-dimensional ingrowth of bone cells within biodegradable cryogel scaffolds in bioreactors at different regimes. *Tissue Eng.* A 2008, 14:1743–50.
151. Woo K.Mi., Jun Ji-Hae, Chend V.J., Seo J., Baek J-H, Ryoo H-Mo, Kim G.-S., Somerman M.J., Ma P.X. Nano-fibrous scaffolding promotes osteoblast differentiation and biomineralization. *Biomaterials* 2007; 28:335–43.
152. Ma P.X., Zhang R.Y. Synthetic nano-scale fibrous extracellular matrix. *J. Biomed. Mater. Res.* A 1999; 46:60–72.
153. Chen V.J., Smith L.A., Ma P.X. "Collagen-inspired nano-fibrous poly(l-lactic acid) scaffolds for bone tissue engineering created from reverse solid freeform fabrication." MRS Proceeding, 2004, Volume 823, W11.13.1.
154. Catledge S.A., Clem W.C., Shrikishen N., Chowdhury S., Stanishevsky A.V., Koopman M., Vohra Y.K. An electrospun triphasic nanofibrous scaffold for bone tissue engineering. *Biomed. Mater.* 2007; 2:142–50.
155. Venugopal J., Vadgama P., Sampath Kumar T.S., Ramakrishna S. Biocomposite nanofibres and osteoblasts for bone tissue engineering. *Nanotechnology* 2007; 18:Note(s):055101.1–055101.8.
156. Zhang L., Ramsaywack S., Fenniri H., Webster T.J. Enhanced osteoblast adhesion on self-assembled nanostructured hydrogel scaffolds. *Tissue Eng.* A 2008; 14:1353–64.
157. Rezwan K., Chen Q.Z., Blaker J.J., Boccaccini A.R. Biodegradable and bioactive porous polymer/inorganic composite scaffolds for bone tissue engineering. *Biomaterials* 2006; 27:3413–31.
158. Maquet V., Boccaccini A.R., Pravata L., Notingher I., Jerome R. Porous poly(α-hydroxyacid)/Bioglass® composite scaffolds for bone tissue engineering. I: Preparation and *in vitro* characterisation. *Biomaterials* 2004; 25:4185–94.

159. Jiang G., Evans M.E., Jones I., Rudd C.D., Scotchford C.A., Walker G.S. Preparation of poly(ε-caprolactone)/continuous bioglass fibre composite using monomer transfer moulding for bone implant. *Biomaterials* 2005; 26:2281–88.
160. Jaakkola T., Rich J., Tirri T., Narhi T., Jokinen M., Seppala J., Yli-Urpo A. *In vitro* Ca-P precipitation on biodegradable thermoplastic composite of poly(ε-caprolactone-*co*-lactide) and bioactive glass(S53P4). *Biomaterials* 2004; 25:575–81.
161. Blaker J.J., Gough J.E., Maquet V., Notingher I., Boccaccini A.R. *In vitro* evaluation of novel bioactive composites based on Bioglass®-filled polylactide foams for bone tissue engineering scaffolds. *J. Biomed. Mater. Res. A* 2003; 67:1401–11.
162. Verrier S., Blaker J.J., Maquet V., Hench L.L., Boccaccini A.R. PDLLA/Bioglass(R) composites for soft-tissue and hard-tissue engineering: An *in vitro* cell biology assessment. *Biomaterials* 2004; 25:3013–21.
163. Yao J., Radin S., Leboy P.S., Ducheyne P. The effect of bioactive glass content on synthesis and bioactivity of composite poly(lactic-*co*-glycolic acid)/bioactive glass substrate for tissue engineering. *Biomaterials* 2005; 26:1935–43.
164. Mathieua L.M., Mueller T.L., Bourban P-E, Pioletti D.P., Muller R., Manson J-A.E. Architecture and properties of anisotropic polymer composite scaffolds for bone tissue engineering. *Biomaterials* 2006; 27:905–16.
165. Lin H-R, Yeh Y-J. Porous alginate/hydroxyapatite composite scaffolds for bone tissue engineering: Preparation, characterization, and *in vitro* studies. *J. Biomed. Mater. Res. B: Appl. Biomater.* 2004; 71:52–65.
166. Roether J.A., Boccaccini A.R., Hench L.L., Maquet V., Gautier S., Jerome R. Development and *in vitro* characterisation of novel bioresorbable and bioactive composite materials based on polylactide foams and Bioglass® for tissue engineering applications. *Biomaterials* 2002; 23:387, 1–8.
167. Boccaccini A.R., Stamboulis A.G., Rashid A., Roether J. Composite surgical sutures with bioactive glass coating. *J. Biomed. Mater. Res. B: Appl. Biomater.* 2003; 67:618–26.
168. Day R.M., Boccaccini A.R., Shurey S., Roether J.A., Forbes A., Hench L.L., Gabe S.M. Assessment of polyglycolic acid mesh and bioactive glass for soft-tissue engineering scaffolds. *Biomaterials* 2004; 25:5857–66.
169. Stamboulis A.G., Boccaccini A.R., Hench L.L. Novel biodegradable polymer/bioactive glass composites for tissue engineering applications. *Adv. Eng. Mater.* 2002; 4:105–9.
170. Boccaccini A.R., Blaker J.J, Maquet V., Chung W., Jerome R., Nazhat S.N. PDLLA foams with TiO_2 nanoparticles and PDLLA/TiO_2-bioglass foam composites for tissue engineering scaffolds. *J. Mater. Sci.* 2006; 41:3999–4008.
171. Demling R.H., Desanti L., Orgill D.P. Biosynthetic skin substitutes: Purpose, properties and clinical indications, use of skin substitutes. burnsurgery.org (2000) http://www.burnsurgery.org.
172. Priya S.G., Jungvid H., Kumar A. Skin tissue engineering for tissue repair and regeneration. *Tissue Eng. B*, 2008; 14:105–18.
173. Ruszczak Z.B. Modern aspects of wound healing: An update. *Dermatol. Surg.* 2000; 26:219–29.
174. Horch R.E., Kopp J., Kneser U., Beier J., Bach A.D. Tissue engineering of cultured skin substitutes. *J. Cell. Mol. Med.* 2005; 9:592–608.
175. van Winterswijk P.J., Nout E. Tissue engineering and wound healing: An overview of the past, present, and future. *Wounds.* 2007; 19:277–84.

176. Auger F.A., Rouabhia M., Goulet F., Berthod F., Moulin V., Germain L. Tissue-engineered human skin substitutes developed from collagen populated hydrated gels: Clinical and fundamental applications. *Med. Biol. Eng. Comput.* 1998; 36:801–12.

177. Yannas I.V. Studies on the biological activity of the dermal regeneration template. *Wound Repair Regen.* 1998; 6518–524.

178. Muhart M., McFalls S., Kirsner R.S., Elgart G.W., Kerdel F., Sabolinski M.L., Hardin-Young J., Eaglstein W.H. Behaviour of tissue-engineered skin: A comparison of living skin equivalent, autograft, and occlusive dressing in human donor sites. *Arch. Dermatol.* 1999; 135:913–18.

179. Jimenez P.A., Jimenez S.E. Tissue and cellular approaches to wound repair. *The American J. Surg* 2004; 187:(Suppl to May) 56S–64S.

180. Bonadio J., Smiley E., Patil P., Goldstein S. Localized, direct plasmid gene delivery *in vivo*: Prolonged therapy results in reproducible tissue regeneration. *Nat. Med.* 1999; 5:753–59.

181. Luu Y.K., Kim K., Hsiao B.S., Chu B., Hadjiargyrou M. Development of a nanostructured DNA delivery scaffold via electrospinning of PLGA and PLA-PEG block copolymers. *J. Control Release* 2003; 89:341–53.

182. Pham C., Greenwood J., Cleland H., Woodruff P., Maddern G., Bioengineered skin substitutes for the management of burns: A systematic review. *Burns* 2007; 33:946–57.

183. MacNeil S. Progress and opportunities for tissue engineered skin. *Nature* 2007; 445:874–80

184. Jahoda C.A., Reynolds A.J. Hair follicle dermal sheath cells: Unsung participants in wound healing. *Lancet* 2001; 358:1445–48.

185. Chan E.S., Lam P.K., Liew C.T., Lau H.C., Yen R.S., King W.W. A new technique to resurface wounds with composite biocompatible epidermal graft and artificial skin. *J. Trauma* 2001; 50:358–62.

186. Lam P.K., Chan E.S., To E.W., Lau H.C., Yen S.C., King W.W. Development and evaluation of a new composite Laserskin graft. *J. Trauma* 1999; 47:918–22.

187. Gartner L.P., Hiatt J.L. Integument. In *Color Textbook of Histology*. Philadelphia:WB Saunders; 1997.

188. Tsai C.C., Lin S.D., Lai C.S., Lin T.M. The use of composite acellular allodermis-ultrathin autograft on joint area in major burn patients—one year follow-up. *Kaohsiung J. Med. Sci.* 1999; 15:651–58.

189. Noordenbos J., Dore C., Hansbrough J.F. Safety and efficacy of TransCyte for the treatment of partial-thickness burns. *J. Burn Care Rehabil.* 1999; 20:275–81.

190. Boyd M., Flasza M., Johnson P.A., Roberts J., Kemp P. Integration and persistence of an investigational human living skin equivalent (ICX-SKN) in human surgical wounds. *Regen. Med.* 2007; 2:363–70.

191. Cho Y.W., Cho Y.N., Chung S.H., Yoo G., Ko S.W. Water-soluble chitin as a wound healing accelerator. *Biomaterials* 1999; 20:2139–45.

192. Ueno H., Yamada H., Tanaka I., Kaba N., Matsuura M., Okumura M., Kadosawa T. Fujinaga T. Accelerating effects of chitosan for healing at early phase of experimental open wound in dogs. *Biomaterials* 1999; 20:1407–14.

193. Mizuno K., Yamamura K., Yano K., Osada T., Saeki S., Takimoto N., Sakura T., Nimura Y. Effect of chitosan film containing basic fibroblast growth factor on wound healing in genetically diabetic mice. *J. Biomed. Mater. Res. A* 2003; 64:177–81.

194. Yusof N.L.M., Lim L.Y., Khor E. Preparation and characterization of chitin beads as a wound dressing precursor. *J. Biomed. Mater. Res.* 2001; 54:59–68.
195. Ono K., Saito Y., Yura H., Ishikawa K., Kurita A., Akaike T., Ishihara M. Photocrosslinkable chitosan as a biological adhesive. *J. Biomed. Mater. Res.* 2000; 49:289–95.
196. Ma J., Wang H., He B., Chen J. A preliminary *in vitro* study on the fabrication and tissue engineering applications of a novel chitosan bilayer material as a scaffold of human neofetal dermal "fibroblasts." *Biomaterials* 2001; 22:331–36.
197. Mi F-L., Shyu S-S., Wu Y-B., Lee S-T., Shyong J-Y., Huang R-N. Fabrication and characterization of a sponge-like asymmetric chitosan membrane as a wound dressing. *Biomaterials* 2001; 22:165–73.
198. Mi F-L., Wu Y-B., Shyu S-S., Schoung J-Y., Huang Y-B., Tsai Y-H., Hao J-Y. Control of wound infections using a bilayer chitosan wound dressing with sustainable antibiotic delivery. *J. Biomed. Mater. Res. A* 2002; 59:438–49.
199. Liu H., Yin Y., Yao K. Construction of chitosan–gelatin–hyaluronic acid artificial skin *in vitro*. *J. Biomater. App.* 2007; 21:413–30.
200. Choi Y.S., Lee S.B., Hong S.R., Lee Y.M., Song K.W., Park M.H. Studies on gelatin-based sponges. Part III: A comparative study of cross-linked gelatin/alginate, gelatin/hyaluronate and chitosan/hyaluronate sponges and their application as a wound dressing in full-thickness skin defect of rat. *J. Mater. Sci.: Mater. in Med.* 2001; 12:67–73.
201. Chevallay B., Abdul-Malak N., Herbage D. Mouse fibroblasts in long-term culture within collagen three-dimensional scaffolds: Influence of crosslinking with diphenylphosphorylazide on matrix reorganization, growth, and biosynthetic and proteolytic activities. *J. Biomed. Mater. Res. A* 2000; 49:448–59.
202. Tanaka M., Nakakita N., Kuroyanagi Y. Allogeneic cultured dermal substitute composed of spongy collagen containing fibroblasts: Evaluation in animal test. *J. Biomater. Sci.-Polym. E* 1999; 10:433–53.
203. Ruszczak Z. Effect of collagen matrices on dermal wound healing. *Adv. Drug Delivery Reviews* 2003; 55:1595–1611.
204. Dai. N-T., Williamson M.R., Khammo N., Adams E.F., Coombes A.G.A. Composite cell support membranes based on collagen and polycaprolactone for tissue engineering of skin. *Biomaterials* 2004; 25:4263–71.
205. Venugopal J., Ramakrishna S. Biocompatible nanofiber matrices for the engineering of a dermal substitute for skin regeneration. *Tissue Eng.* 2005; 11:847–54.
206. Chen G., Sato T., Ohgushi H., Ushida T., Tateishi T., Tanaka J. Culturing of skin fibroblasts in a thin PLGA–collagen hybrid mesh. *Biomaterials* 2005; 26:2559–66.
207. Shi C., Cheng T., Su Y., Mai Y. Significance of dermis-derived multipotent stem cells in wound healing and skin equivalent construction. Joint International Tissue Engineering Society (TESI) and European Tissue Engineering Society (ETES) Meeting, Lausanne October 10–13, 2004, Switzerland.
208. Berthod F., Sahuc F., Hayek D., Damour O., Collombel C. Deposition of collagen fibril bundles by long-term culture of fibroblasts in a collagen sponge, *J. Biomed. Mater. Res.* 1996; 32:87–93.
209. Berthod F., Germain L., Li H., Xu W., Damour O., Auger F.A. Collagen fibril network and elastic system remodeling in a reconstructed skin transplanted on nude mice. *Matrix Biol.* 2001; 20:463–73.

210. Gingras M., Paradis I., Berthod F. Nerve regeneration in a collagen–chitosan tissue-engineered skin transplanted on nude mice. *Biomaterials* 2003; 24:1653–61.
211. Ma L., Gao C., Mao Z., Zhou J., Shen J., Hu X., Han C. Collagen/chitosan porous scaffolds with improved biostability for skin tissue engineering. *Biomaterials* 2003; 24:4833–41.
212. Lee S.B., Kim Y.H., Chong M.S., Lee Y.M. Preparation and characteristics of hybrid scaffolds composed of β-chitin and collagen. *Biomaterials* 2004; 25:2309–17.
213. Hong S.R., Lee S.J., Shim J.W., Choi Y.S., Lee Y.M., Song K.W., Park M.H., Nam Y.S., Lee S.I. Study on gelatin-containing artificial skin IV: A comparative study on the effect of antibiotic and EGF on cell proliferation during epidermal healing. *Biomaterials* 2001; 22:2777–83.
214. Lee S.B., Jeon H.W., Lee Y.W., Lee Y.M., Song K.W., Park M.H., Nam Y.S., Ahn H.C. Bio-artificial skin composed of gelatin and (1-3),(1-6)-β-glucan. *Biomaterials* 2003; 24:2503–11.
215. Mazlyzam A.L., Aminuddin B.S., Fuzina N.H., Norhayati M.M., Fauziah O., Isag M.R., Saim L., Ruszymah B.H.I. Reconstruction of living bilayer human skin equivalent utilizing human fibrin as a scaffold. *Burns* 2007; 33:355–63.
216. Stichel C.C., Muller, W.H. Experimental strategies to promote axonal regeneration after traumatic central nervous system injury. *Progress Neurobiol.* 1998; 56:119–48.
217. Olson L. Regeneration in the adult central nervous system: experimental repair strategies. *Nat. Med.* 1997; 3:1329–35.
218. Phillips J.B., Bunting S.C., Hall S.M., Brown R.A. Neural tissue engineering: A self-organizing collagen guidance conduit. *Tissue Eng.* 2005; 11:1611–17.
219. Stang F., Fansa H., Wolf G., Keilhoff G. Collagen nerve conduits-assessment of biocompatibility and axonal regeneration. *Biomed. Mater. Eng.* 2005; 15:3–12.
220. Pu Y., Guo Q.S., Wang A.M., Wu S.Y., Xing S.X., Zhang Z.R. Repair of acutely injured spinal cord through constructing tissue-engineered neural complex in adult rats. *Chin. J. Traumatol.* 2007; 10:171–76.
221. Ellis-Behnke R.G., Liang Y.X., You S.W., Tay D.K., Zhang S., So K.F., Schneider G.E. Nano neuro knitting: Peptide nanofiber scaffold for brain repair and axon regeneration with functional return of vision. *Proc. Natl. Acad. Sci. USA* 2006; 103:5054–59.
222. Hurtado A., Moon L.D., Maquet V., Blits B., Jérôme R., Oudega M. Poly(D,L-lactic acid) macroporous guidance scaffolds seeded with Schwann cells genetically modified to secrete a bi-functional neurotrophin implanted in the completely transected adult rat thoracic spinal cord. *Biomaterials* 2006; 27:430–42.
223. Almirall A., Larrecq G., Delgado J.A., Martínez S., Planell J.A., Ginebra M.P. Fabrication of low temperature macroporous hydroxyapatite scaffolds by foaming and hydrolysis of an α-TCP paste. *Biomaterials* 2004; 25:3671–80.
224. Itoh S., Takakuta K., Kawabata S., Aso Yu, Kasai K., Itoh H., Shinomiya K. Evaluation of cross-linking procedures of collagen tubes used in peripheral nerve repair. *Biomaterials* 2002; 23:4475–81.
225. Woerly S., Marchand R., Lavalle'e C. Intracerebral implantation of synthetic polymer/biopolymer matrix: A new perspective for brain repair. *Biomaterials* 1990; 11:97–107.

226. Woerly S. Porous hydrogels for neural tissue engineering. In *Porous Materials for Tissue Engineering*, eds. D.-M. Liu, V. Dixit. Switzerland: Trans. Tech. Publications, 1997, 53.

227. Woerly S., Ulbrich K., Chytry V., Smetane K., Petrovicky P., Rihova B., Morassuti D.J. Synthetic polymer matrices for neural cell transplantation. *Cell Transpl.* 1993; 2:229–39.

228. Plant, G.W., Harvey, A.R., Chirila T.V. Axonal growth within poly(2-hydroxyethyl methacrylate) sponges infiltrated with Schwann cells and implanted into the lesioned rat optic tract. *Brain Res.* 1995; 671:119–30.

229. Plant G.W., Woerly S., Harvey A.R. Hydrogels containing peptide or amino-sugar sequences implanted into the rat brain: Influence on cellular migration and axonal growth. *Exp. Neurol.* 1997; 143:287–99.

230. Plant G.W., Chirila T.V., Harvey A.R. Implantation of collagen IV/poly(2-hydroxyethyl methacrylate) hydrogels containing Schwann cells into the lesioned rat optic tract. *Cell Transpl.* 1998; 7:381–91.

231. Ratner B.D., Hoffman A.S. Synthetic hydrogels for biomedical applications. In *Hydrogels for Medical and Related Applications*, ed. J.D. Andrade. ACS Symp. Ser. 31, Washington DC, 1976, 1.

232. Park H., Park K. Hydrogels in Bioapplications. In *Hydrogels and Biodegradable Polymers for Bioapplications*, eds. R.M. Ottenbrite, S.J. Huang, K. Park. ACS Symp. Series 627, Washington DC, 1996, 2.

233. Dumitriu S., Dumitriu-Medvichi C. Hydrogel and general properties of biomaterials. In *Polymeric Biomaterials*, ed. S. Dumitriu. New York: Marcel Dekker, Inc., 1994, p. 3.

234. Bunge R.P. Hopkins J.M. The role of peripheral and central neuroglia in neural regeneration in vertebrates. *Sem. Neurosci.* 1990; 12:509–18.

235. Bellamkonda R.V. Peripheral nerve regeneration: An opinion on channels, scaffolds and anisotropy. *Biomaterials* 2006; 27:3515–18.

236. Lesny P., Croos J.De, Přádný M,. Vacík J, Michálek J, Woerly S, Sykova E. Polymer hydrogels usable for nervous tissue repair. *J. Chem. Neuroanatomy* 2002; 23:243–47.

237. Maquet V., Martin D., Malgrange B., Franzen R., Schoenen J., Moonen G., Jerome R. Peripheral nerve regeneration using bioresorbable macroporous polylactide scaffolds. *J. Biomed. Mater. Res. A* 2000; 52:639–51.

238. Evans G.R.D., Brandt K., Widmer M., Gurlek A., Savel T., Gupta P., Lohman R., et al. Tissue engineered conduits: The use of biodegradable poly-dl-lactic-co-gylcolic acid PLGA scaffolds in peripheral nerve regeneration. In *Biological matrices and tissue reconstruction*, eds. G.E. Stark, R. Horch E. Tanczos. Berlin: Springer, 1998. 187–92.

239. Widmer M.S., Gupta P.K., Lu L., Meszlenyi R., Evans G.R.D., Brandt K., Savel T., Gurlek A., Patrick Jr C.W., Mikos A.G. Manufacture of porous biodegradable polymer conduits by an extrusion process for guided tissue regeneration. *Biomaterials* 1998; 19:1945–55.

240. Saad B., Casotti M., Huber Th., Schmutz P., Welti M., Uhlschmid G.K., Neuenschwander P., Suter, U.W. *In vitro* evaluation of the biofunctionality of osteoblasts cultured on DegraPol-foam. *J. Biomater. Sci.*, Polymer Edition. 2000; 11(8): 787–800(14).

241. Katayama Y., Montenegro R., Freiera T., Midhab R., Belkasb J.S., Shoichet M.S. Coil-reinforced hydrogel tubes promote nerve regeneration equivalent to that of nerve autografts. *Biomaterials* 2006; 27:505–18.

242. Sundback C., Hadlock T., Cheney M., Vacanti J.P. Manufacture of porous polymer nerve conduits by a novel low-pressure injection molding process. *Biomaterials* 2003; 24:819–30.

243. Hadlock T., Sundback C., Hunter D., Cheney M., Vacanti J.P. A polymer foam conduit seeded with Schwann cells promotes guided peripheral nerve regeneration. *Tissue Eng.* 2000; 6:119–27.

244. Haipeng G., Yinghui Z., Jianchun L., Yandao G., Nanming Z., Xiufang Z. Studies on nerve cell affinity of chitosan-derived materials. *J. Biomed. Mater. Res.* 2000; 52:285–95.

245. Qiang A.O., Wang A., Cao W., Zhao C., Gong Y., Zhao N., Zhang X. Fabrication and characterization of chitosan nerve conduits with microtubular architectures. *Tsinghua Sci. Tech.* 2005; 10:435–38.

246. Li S.T., Archibald S.J., Krarup C., Madison R. Peripheral nerve repair with collagen conduits. *Clinical Materials* 1992; 9:195–200.

247. Prehn R.T. Regeneration versus neoplastic growth. *Carcinogenesis* 1997; 18:1439–44.

248. Guth L., Barrett C.P., Donati E.J., Anderson F.D., Smith M.V., Lifson M. Essentiality of a specific cellular terrain for growth of axons into a spinal cord lesion. *Exp. Neurol.* 1985; 88:1–12.

249. Beattie M.S., Bresnahan J.C., Komon J., Tovar C.A., Van Meter M., Anderson D.K., Faden A.I.,et al. Endogeneous repair after spinal cord contusion injuries in the rat. *Exp. Neurol.* 1997; 148:453–63.

250. Marchand R., Woerly S. Transected spinal cords grafted with *in situ* self-assembled collagen matrices. *Neuroscience* 1990; 36:45–60.

251. Spilker M.H., Yannas I.V., Hsu H-P., Norregaard T.V, Kostyk S.K., Spector M. The effect of collagen-based implants on early healing of the adult rat spinal cord. *Tissue Eng.* 1997; 3:309–17.

252. Goldsmith H.S., de la Torre J.C. Axonal regeneration after spinal cord transection and reconstruction. *Brain Res.* 1992; 589:217–24.

253. Bernstein J.J., Goldberg W.J. Transplantation of cultured fetal spinal cord grafts, grown on a histocompatible substrate, into adult spinal cord. *Brain Res.* 1986; 377:403–8.

254. Paino C.L., Fernandez-Valle C., Bates M.L., Bunge M.B. Regrowth of axons in lesioned adult rat spinal cord: Promotion by implants of cultured. Schwann cells. *J. Neurocytol.* 1994; 23:433–52.

255. Schreyer D.J., Jones E.G. Growth of corticospinal axons on prosthetic substrates introduced into the spinal cord of neonatal rats. *Brain Res.* 1987; 432:291–99.

256. Woerly S., Petrov P., Sykova E., Roitbak T., Simonova Z., Harvey A.R. Neural tissue formation within porous hydrogels implanted in brain and spinal cord lesions: Ultrastructural, immunohistochemical, and diffusion studies. *Tissue Eng.* 1999; 5:467–88.

257. Tian W.M., Hou S.P., Zhang C.L., Xu Q.Y., Lee I.S., Li H.D., Spector M., Cui F.Z. Hyaluronic acid–poly-D-lysine-based three-dimensional hydrogel for traumatic brain injury. *Tissue Eng.* 2005; 11:513–25.

258. Crompton K.E., Goud J.D., Bellamkonda R.V., Gengenbach T.R., Finkelsteine D.I., Hornef M.K., Forsythea J.S. Polylysine-functionalised thermoresponsive chitosan hydrogel for neural tissue engineering. *Biomaterials* 2007; 28:441–49.

259. Hou S., Xu Q., Tian W., Cui F., Cai Q., Ma J., Lee I. The repair of brain lesion by implantation of hyaluronic acid hydrogels modified with laminin. *J. Neurosci. Methods* 2005; 148:60–70.

260. Yang F., Murugan R., Ramakrishna S., Wang X., Mac Y.X., Wang S. Fabrication of nano-structured porous PLLA scaffold intended for nerve tissue engineering. *Biomaterials* 2004; 25:1891–1900.

261. Samadikuchaksaraei A. An overview of tissue engineering approaches for management of spinal cord injuries. *J. NeuroEng. Rehabilitation* 2007; 4:15–30.

262. Yoshii S., Oka M., Shima M., Akagi M., Taniguchi A. Bridging a spinal cord defect using collagen filament. *Spine* 2003; 28:2346–51.

263. Yoshii S., Oka M., Shima M., Taniguchi A., Taki Y., Akagi M. Restoration of function after spinal cord transection using a collagen bridge. *J. Biomed. Mater. Res. A* 2004; 70:569–75.

264. Liu S., Said G., Tadie M. Regrowth of the rostral spinal axons into the caudal ventral roots through a collagen tube implanted into hemisected adult rat spinal cord. *Neurosurgery* 2001; 49:143–50.

265. Kataoka K., Suzuki Y., Kitada M., Ohnishi K., Suzuki K., Tanihara M., Ide C., Endo K., Nishimura Y. Alginate, a bioresorbable material derived from brown seaweed, enhances elongation of amputated axons of spinal cord in infant rats. *J. Biomed. Mater. Res.* 2001; 54:373–84.

266. Suzuki K., Suzuki Y., Ohnishi K., Endo K., Tanihara M, Nishimura Y. Regeneration of transected spinal cord in young adult rats using freeze-dried alginate gel. *Neuroreport* 1999; 10:2891–94.

267. Suzuki Y., Kitaura M., Wu S., Kataoka K., Suzuki K., Endo K., Nishimura Y., Ide C. Electrophysiological and horseradish peroxidase tracing studies of nerve regeneration through alginate-filled gap in adult rat spinal cord. *Neurosci. Lett.* 2002; 318:121–24.

268. Kataoka K., Suzuki Y., Kitada M., Hashimoto T., Chou H., Bai H., Ohta M., Wu S., Suzuki K., Ide C. Alginate enhances elongation of early regenerating axons in spinal cord of young rats. *Tissue Eng.* 2004; 10:493–504.

269. Prang P., Muller R., Eljaouhari A., Heckmann K., Kunz W., Weber T., Faber C., Vroemen M, Bogdahn U., Weidner N. The promotion of oriented axonal regrowth in the injured spinal cord by alginate-based anisotropic capillary hydrogels. *Biomaterials* 2006; 27:3560–69.

270. Blacher S., Maquet V., Schils F., Martin D., Schoenen J., Moonen G., Jerome R., Pirard J.P. Image analysis of the axonal ingrowth into poly(D,L-lactide) porous scaffolds in relation to the 3-D porous structure. *Biomaterials* 2003; 24:1033–40.

271. Patist C.M., Mulder M.B., Gautier S.E., Maquet V., Jerome R., Oudega M. Freeze-dried poly(D,L-lactic acid) macroporous guidance scaffolds impregnated with brain-derived neurotrophic factor in the transected adult rat thoracic spinal cord. *Biomaterials* 2004; 25:1569–82.

272. Woerly S. Restorative surgery of the central nervous system by means of tissue engineering using NeuroGel implants. *Neurosurg. Rev.* 2000; 23:59–77.

273. Tsai E.C., Dalton P.D., Shoichet M.S., Tator C.H. Synthetic hydrogel guidance channels facilitate regeneration of adult rat brainstem motor axons after complete spinal cord transection. *J. Neurotrauma* 2004; 21:789–804.

274. Woerly S., Doan V.D., Sosa N., de Vellis J., Espinosa A. Reconstruction of the transected cat spinal cord following NeuroGel implantation: Axonal tracing, immunohistochemical and ultrastructural studies. *Int. J. Dev. Neurosci.* 2001; 19:63–83.
275. Woerly S., Doan V.D., Evans-Martin F., Paramore C.G., Peduzzi J.D. Spinal cord reconstruction using NeuroGel implants and functional recovery after chronic injury. *J. Neurosci. Res.* 2001; 66:1187–97.
276. Woerly S., Pinet E., de Robertis L., Van Diep D., Bousmina M. Spinal cord repair with PHPMA hydrogel containing RGD peptides (NeuroGel). *Biomaterials* 2001; 22:1095–11.
277. Hejčl A., Lesný P., Přádný M., Michálek J., Jendelová P., Stulík J., Syková E. Biocompatible hydrogels in spinal cord injury repair. *Physiol. Res.* 2008; 57:S121–S132.
278. Bakshi A., Fisher O., Dagci T., Himes B.T., Fischer I., Lowman A. Mechanically engineered hydrogel scaffolds for axonal growth and angiogenesis after transplantation in spinal cord injury. *J. Neurosurg. Spine* 2004; 1:322–29.
279. Tsai E.C., Dalton P.D., Shoichet M.S., Tator C.H. Matrix inclusion within synthetic hydrogel guidance channels improves specific supraspinal and local axonal regeneration after complete spinal cord transection. *Biomaterials* 2006; 27:519–33.
280. Stokols S., Tuszynski M.H. The fabrication and characterization of linearly oriented nerve guidance scaffolds for spinal cord injury. *Biomaterials* 2004; 25:5839–46.
281. Wu S., Suzuki Y., Kitada M., Kitaura M., Kataoka K., Takahashi J., Ide C., Nishimura Y. Migration, integration, and differentiation of hippocampus-derived neurosphere cells after transplantation into injured rat spinal cord. *Neurosci. Lett.* 2001; 312:173–76.
282. Teng Y.D., Lavik E.B., Qu X., Park K.I., Ourednik J., Zurakowski D., Langer R., Snyder E.Y. Functional recovery following traumatic spinal cord injury mediated by a unique polymer scaffold seeded with neural stem cells. *Proc. Natl. Acad. Sci. USA* 2002; 99:3024–29.
283. Malchesky P.S. Extracorporeal artificial organsratner B.D. In *Biomaterials Science: An Introduction to Materials and Medicine,* eds. A.S. Hoffman, J.S. Frederick, J.E. Lemons, 2nd ed. London: Elsevier Academic Press, 514–26.
284. Strain A.J., Neuberger J.M. A bioartificial liver and state of the art. *Science* 2002; 295:1005–8.
285. Court F.G., Wemyss-Holden S.A., Denniso A.R., Maddern A.J. Bioartificial liver support devices: Historical perspectives. *ANZ J. Surg.* 2003; 73:739–48.
286. Tzanakakis E.S., Hess D.J., Sielaff T.D., Hu W.-S. Extracorporeal tissue engineered liver-assist devices. *Annu. Rev. Biomed. Eng.* 2000; 02:607–32.
287. Chamuleau R.A.F.M., Poyck P.P.C., van de Kerkhove M.-P. Bioartificial liver: Its pros and cons. *Therapeutic Apheresis and Dialysis* 2006; 10:168–74.
288. Park J.-K., Lee D.-H. Bioartificial liver systems: Current status and future perspective. *J. Biosci. Bioeng.* 2005; 99:311–19.
289. Legallais C., David B., Doré E. Bioartificial livers (BAL): Current technological aspects and future developments. *J. Membrane Sci.* 2001; 181:81–95.
290. Gerlach J., Koppel K., Schauwecker H.H., Tauber R., Muller C., Bucherl E.S. Use of hepatocytes in adhesion and suspension culture for liver support bioreactors. *Artif. Organs* 1989; 12:788–92.

291. Webb K., Hlady V., Tresco P.A. Relationships among cell attachment, spreading, cytoskeletal organisation and migration rate for anchorage-dependent cells on model surfaces. *J. Biomed. Mater. Res.* 2000; 49:362–68.

292. Cho C.S., Seo S.J., Park I.K., Kim S.H., Kim T.H., Hoshiba T., Harada I., Akaike T. Galactose-carrying polymers as extracellular matrices for liver tissue engineering. *Biomaterials* 2006; 27:576–85.

293. Yang J., Chung T.W., Nagaoka M., Goto M., Cho C.-S., Akaike T. Hepatocyte-specific porous polymer-scaffolds of alginate/galactosylated chitosan sponge for liver-tissue engineering. *Biotechnol. Lett.* 2001; 23:1385–89.

294. Li J., Pan J., Zhang L., Yu Y. Culture of hepatocytes on fructose-modified chitosan scaffolds. *Biomaterials* 2003; 24:2317–22.

295. Ranucci C.S., Moghe P.V. Polymer substrate topography actively regulates the multicellular organization and liver-specific functions of cultured hepatocytes. *Tissue Eng.* 1999; 5:407–20.

296. Török É., Pollok J.-M., Mab P.X.C., Vogela M., Dandric J., Petersen M.R., Burda P.M. Hepatic tissue engineering on 3-dimensional biodegradable polymers within a pulsatile flow bioreactor. *Dig. Surg.* 2001; 18:196–203.

297. Jiankang H., Dichen L., Yaxiong L., Bo Y., Bingheng L., Qin L. Fabrication and characterization of chitosan/gelatin porous scaffolds with predefined internal microstructures. *Polymer* 2007; 48:4578–88.

16

Polymeric Scaffolds for Regenerative Medicine

Paul A. De Bank, Matthew D. Jones, and Marianne J. Ellis

CONTENTS

16.1 Introduction

Regenerative medicine is concerned with aiding the body's natural healing process. Tissue engineering is a branch of regenerative medicine in which cell biology is combined with engineering principles to replicate the tissue *in vitro*. Tissue engineering is therefore used to develop tissue constructs within the body to aid healing, or to model the tissue to gain a better understanding of its function. There is a vast range of materials used in tissue engineering but the most widely used are natural and synthetic polymers. Synthetic polymers have the advantage that their properties can be tailored from the molecular structure to the scaffold architecture, which makes them highly versatile. This chapter will give an overview of the synthetic polymers currently used in tissue engineering and will focus on polyesters, specifically alpha-hydroxy acids, which are the most common and widely used since their introduction into the medical field as sutures in 1960s and fixators [1]. The chapter will describe the polymerization of polyesters by ring opening polymerization (ROP), different scaffold fabrication techniques, surface modification of polyester scaffolds to improve cell culture properties, and finally selecting a suitable bioreactor for a given scaffold. Examples are drawn from all types of tissue engineering, the majority from bone tissue engineering. The reader is also directed to a number of recent reviews that may be of interest on biomaterials for tissue engineering [2–6], surface modification [7], and bioreactor design [8–11] as well as references throughout the chapter.

16.1.1 Biomaterials in Tissue Engineering

Biomaterials can be classified into four groups as shown in Figure 16.1. Ceramics are particularly well suited to bone tissue engineering [6,12,13]; a well-known metal scaffold was used to regenerate a jaw [14]. The scaffold was preformed by using a computer tomography scan of the patient's defect, and a computer-aided design (CAD)-operated milling machine to prepare a mold for a titanium mesh scaffold. Hydrogels are ideal for injectable scaffolds [15] and polymers are used for most applications depending on how they are prepared. A review of bone tissue engineering summarizes the available polymers used in this field [16] and their characteristics, a list of which is shown in Table 16.1, and the earlier review by Middleton & Tipton [1] details the poly(α-hydroxy esters).

16.2 Polymerization

16.2.1 Ring Opening Polymerization (ROP)

ROP has been extensively utilized for the controlled production (in terms of rate of polymerization, molecular weight, polydispersity, and polymer

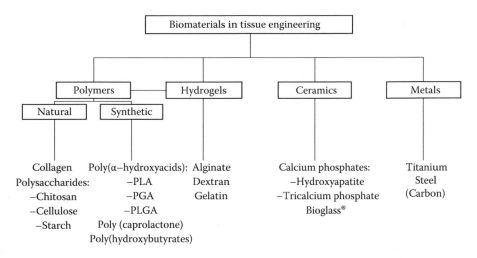

FIGURE 16.1
Classification of biomaterials and examples of each class.

TABLE 16.1

Examples of Natural and Synthetic Polymers

Natural	Synthetic
• Collagen	• Poly(α-hydroxy acids)
• Chitosan	• Poly(ε-caprolactone)
• Fibrin	• Poly(hydroxybutyrate)
	• Poly(phosphazenes)
	• Poly(anhydrides)

Note: Natural and synthetic refer to the polymerization process.
Source: From Salgado, A. J., Coutinho, O. P., and Reis, R. L., *Macromolecular Bioscience*, 4 (8), 743–65, 2004. With permission.

stereochemistry) of polyesters [17–19]. Monomers typically encountered in such polymerizations are lactide (racemic (PLA) and L-stereochemical forms (PLLA)), ε-caprolactone, and glycolide, see Figure 16.2.

One of the most common initiators for the ROP of cyclic esters is tin(II) 2-ethylhexanoate (stannous octoate) [20]. This initiator has been studied in great detail and is currently used in the industry for bulk production of Polyactic acid (PLA) and its copolymers. Due to the toxic nature of tin and the fact that this initiator only produces atactic (see Section 16.2.3) PLA, there is a tremendous drive to prepare initiators using biologically benign metals that, significantly, offer a high degree of control over the stereochemistry and molecular weight of the polymer. The most important mechanisms for the ROP of lactide are (i) a coordination-insertion mechanism used for metal complex initiation and (ii) an activated monomer process for organo/cationic initiators [17]. Here, the focus is on the coordination-insertion mechanism, which has been extensively examined in recent years by experimental and computational chemists [21–26].

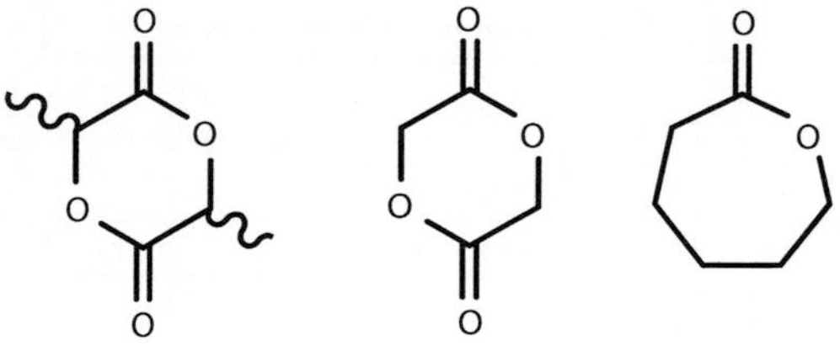

FIGURE 16.2
Typical monomers used in the preparation of polyesters. 1 = lactide (3,6-dimethyl-1,4-dioxane-2,5-dione , 2 = glycolide (1,4-dioxane-2,5-dione) and 3 = ε-caprolactone (6-hexanolactone).

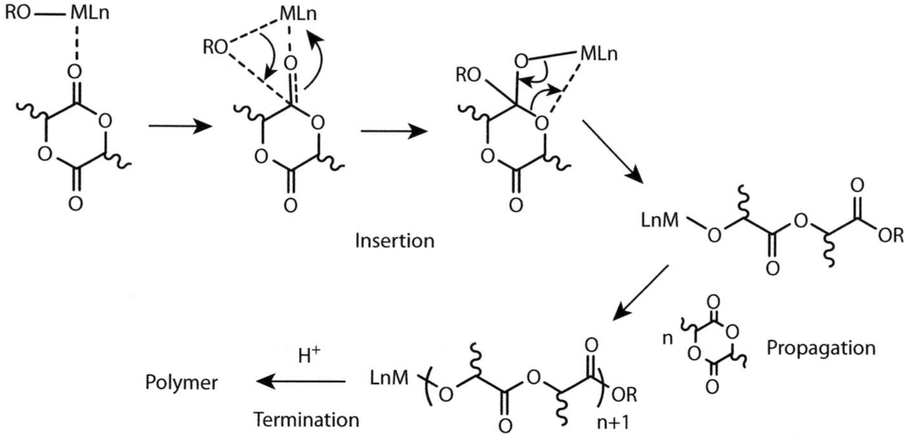

FIGURE 16.3
Coordination insertion mechanism for the polymerization of lactide.

16.2.2 The Coordination-Insertion Mechanism

This coordination-insertion mechanism is shown in Figure 16.3 for a simple metal alkoxide catalyst. The initial step involves coordination of the cyclic ester to the metal center, a nucleophilic labile ligand or coinitiator then is inserted into the activated carbonyl group resulting in O-acyl bond cleavage. This then produces another metal alkoxide, which acts as an initiator for subsequent monomer additions. This coordination-insertion mechanism proceeds until all of the monomer is consumed. The polymerization is terminated by the addition of a proton source (for example, methanol), which quenches the polymerization. This mechanism is supported via polymer end-group analysis obtained from MALDI-TOF (matrix-assisted laser desorption/ionization time of flight) mass spectrometry and ^{1}H NMR (nuclear magnetic resonance) spectroscopy. There are two main competing undesired

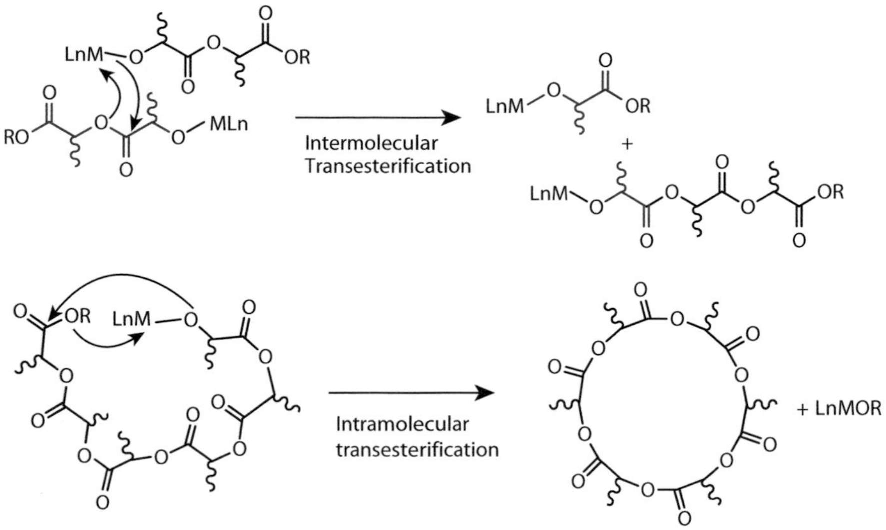

FIGURE 16.4
Possible transesterification competing reactions.

reactions to consider: (i) intramolecular transesterification or "backbiting" (this results in the formation of cyclic moieties) and (ii) intermolecular transesterification, both of which are illustrated in Figure 16.4 [17]. The degree of transesterification can be determined via (i) GPC (gel permeation chromatography) where an increase in PDI (polydispersity index) is observed and (ii) MALDI-TOF mass spectrometry for intramolecular transesterification cyclic polymers that are detected and for intermolecular transesterification polymers with an odd number of lactic acid monomer units that are observed. For PLA it is also possible to use ^{13}C- or ^{1}H-NMR spectroscopy to study transesterification, the observation of a resonance originating from the *iis* tetrad is indicative of transesterification [27].

The choice of initiator is critical and can dramatically affect the properties of the resulting polymer. For controlled polymerization, the initiators must be capable of producing stereoregular polymers with a narrow molecular weight distribution and predictable final molecular weights with well-defined end groups.

16.2.3 The Importance and Control of Stereochemistry

One of the most significant factors for the ROP of racemic lactide is the need for stereocontrol in the polymerization. Racemic lactide is a 50:50 mixture of the L and D forms of the monomer, consequently, as shown in Figure 16.5, there are three common tacticities observed: (i) atactic, (ii) isotactic stereoblock, and (iii) heterotactic [28]. The stereochemistry can significantly alter the properties

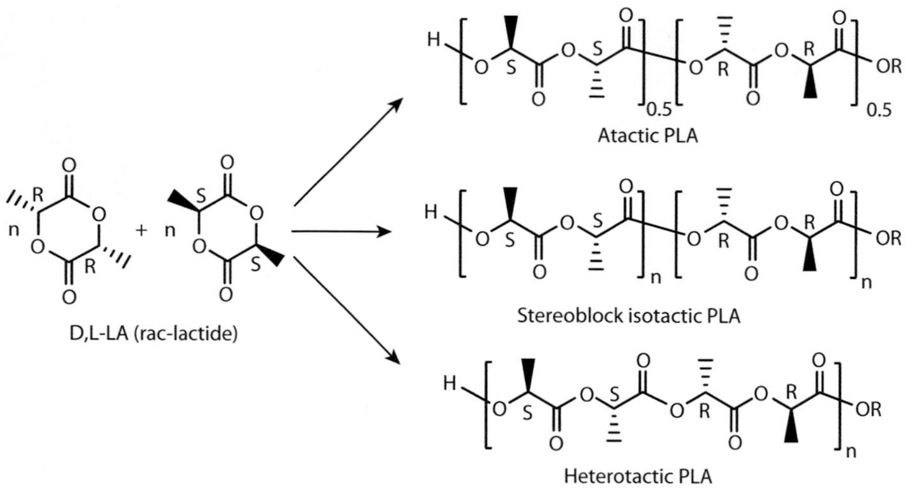

FIGURE 16.5
Possible polymers produced for the ROP of rac-lactide.

of the final polymer, for example, isotactic PLA is highly crystalline (high melting point) whereas atactic and heterotactic PLA are both amorphous.

The resulting stereoselectivity can be quantified by analysis of the methine region of the ^{1}H homonuclear decoupled NMR spectrum and comparing to those predicted by Bernouillan statistics [29]. For atactic PLA five tetrads (*sis, iis, sii, iii, isi*) are observed in a 1:1:1:3:2 ratio. From this analysis of the homonuclear decoupled NMR spectrum the parameters P_r (the probability of heterotactic enchainment) or P_m (the probability of isotactic enchainment) can be determined. For heterotactic PLA an enhancement of the *isi* and *sis* tetrads are observed and for isotactic PLA an enhancement of the *iii* tetrad is detected.

16.2.4 Selection of the Initiator

Initiators commonly encountered are those based on alkaline earth metal, Group 3, Group 4, Al(III), Zn(II), lanthanides, and Sn(II) metal centers [21,28,30–36]. Remarkable advances have been made in this area over the last decade. Three catalysts will be highlighted that were prepared by the groups of Coates, Gibson, and Feijen. Coates prepared a Zn(II) β-diiminate complex (Figure 16.6) that produced almost exclusively heterotactic PLA (Conditions: Temp. = 20°C, [monomer]:[initiator] 200:1, conversion = 95% after 20 minutes using CH_2Cl_2 as the solvent, M_n = 37,900 gmol^{-1}, P_r = 0.9, PDI = 1.10) [28,37]. This zinc alkoxide was one of the first single-site, living homogeneous catalysts for the stereoselective polymerization of racemic lactide. Gibson has shown that the ligand employed can have a remarkable effect on the

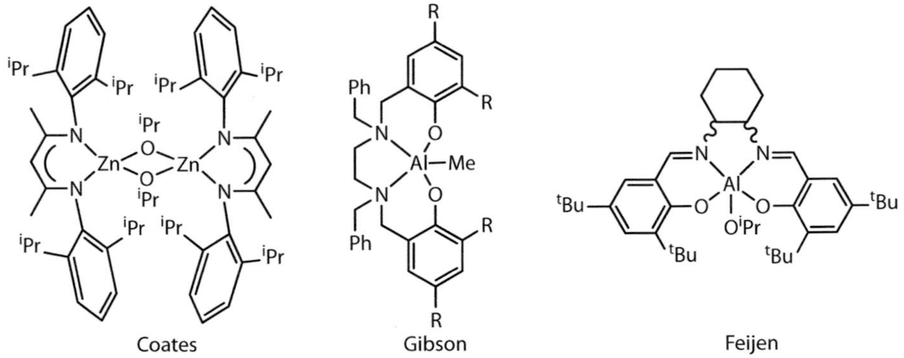

FIGURE 16.6
Highly selective initiators for the ROP of rac-lactide.

properties of the resultant polymer. During the course of their work a series of Al(III) complexes bearing salan ligands were prepared, Figure 16.6 [23]. For example changing the substituent (R) on the salan-type ligand from a Cl to an H moiety switches the tacticity from $P_r = 0.96$ to $P_m = 0.79$ [23]. One of the most significant breakthroughs came when Feijen demonstrated, for the first time, that it is possible to produce stereoselective PLA in the melt [34]. Again an Al(III) initiator was employed, a conversion of 95% with a [monomer]:[initiator] 200:1, $M_n = 24,900$ gmol^{-1}, $P_m = 0.88$, PDI = 1.37 after 48 hours. The so-called melt conditions (temperature in excess of 130°C and, most important, in the absence of solvent) are those preferred by the industry for the production of PLA.

Recently, Davidson and colleagues have prepared a series of C_3 symmetric complexes based on group 4 metal centers for the stereoselective production of PLA in the melt at 130°C, Figure 16.7 [31].

Initiators based on Zr(IV) and Hf(IV) were extremely active and selective for the polymerization. It is significant that almost pure heterotactic PLA was quantitatively produced after 30 minutes. Whereas when using the Ti(IV) initiator atactic PLA was isolated. Recent computational studies on magnesium-alkoxide catalyzed ROP of cyclic esters have highlighted the importance of chelation of the lactate units of the growing polymer chain in determining the stereoselectivity of the subsequent inserted monomer [22]. Zirconium and hafnium complexes are prone to higher coordination numbers than titanium complexes. Therefore it is possible that chelation of the polymer chain occurs in a greater extent, which may in-turn favor either coordination of the S,S or R,R monomer.

Even though impressive advances have been made in the metal-catalyzed initiation of lactide, there is a drive to produce PLA with no metal contaminant. There are two main methods to achieve this goal: (i) organo-initiators such as N-heterocyclic carbenes [38,39]; and (ii) immobilizing homogeneous

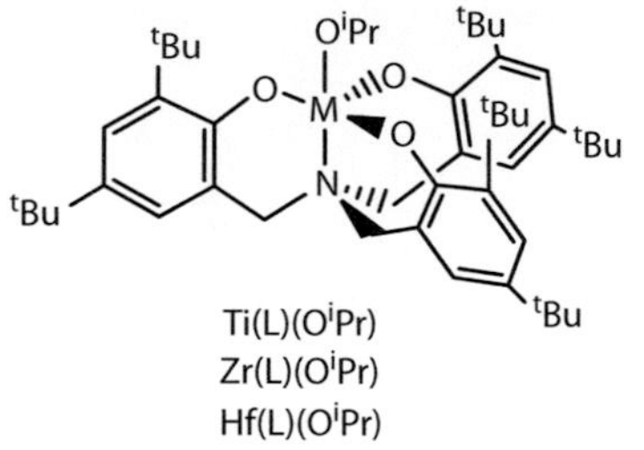

Metal	Time	Con.	M_n	PDI	P_r
Ti	0.5	50	37100	1.38	0.5
Zr	0.5	75	45900	1.23	0.92
Hf	0.5	95	71150	1.19	0.88

Ti(L)(O^{i}Pr)
Zr(L)(O^{i}Pr)
Hf(L)(O^{i}Pr)

FIGURE 16.7
Group 4 initiators for the ROP of rac-lactide. Conditions of polymerization Temp. = 130°C, [monomer]:[initiator] 300:1.

catalysts [40]. Jones and colleagues have used the second approach to heterogenize a simple titanium initiator to the walls of silica and have utilized this and an initiator for the ROP of racemic lactide. Using this heterogeneous catalyst, the polymer produced contained 15 ppm of titanium, this is compared to 1,110 ppm for the homogeneous analogue [40]. These results imply that heterogeneous catalysts could play an important role in ROP catalysts for the production of PLA for biomedical applications, where there needs to be minimal metal residue present.

16.3 Scaffold Fabrication

16.3.1 The Ideal Scaffold

There are a number of prerequisites, listed below, for an ideal scaffold for tissue engineering and regenerative medicine applications and these are largely dependent on the target tissue.

The scaffold should be

- Biocompatible: The scaffold as a whole and its degradation products should not elicit an inflammatory response by the recipient.
- Biodegradable: Ideally at the same rate as tissue formation, so a functional, polymer-free tissue is formed.
- Cytocompatible: The material must enable the adhesion and proliferation of cells and should be conducive, if necessary, to the differentiation of the seeded cells.
- Mechanically similar to the target tissue: (e.g., bone or soft tissues).
- Allow controlled release: (e.g., growth factors).

- Architecturally similar to the tissue being generated.
- Able to promote angiogenesis and innervation.

There is currently a wide range of materials used all of which have varying degrees of the ideal properties. Selection of material, architecture, and fabrication technique should be made by considering which of these properties are essential or desirable. Polymers are the most versatile biomaterials when considering the range of fabrication techniques and resulting architecture (Table 16.2).

16.3.2 Scaffold Fabrication Techniques

Phase inversion and temperature or pressure are the two generalized categories for the fabrication techniques used for making scaffolds. They are generalized because there is some crossover between the groups, for instance temperature is an important parameter in phase inversion processes. However, these classifications are convenient to describe the major mechanism of formation. The two categories are further described below. Yang et al. published two reviews on scaffold fabrication techniques; the first an overview [41] and the second specifically on rapid prototyping [42], and Hutmacher published a review for bone and cartilage engineering although many of the techniques described can be applied to other tissues [43].

16.3.2.1 Phase Inversion

Phase inversion is a very popular option for scaffold formation; it is cheap, quick, easy, and gives good repeatable structures if the conditions are kept the same on each run. The resulting pore morphology can be controlled by the conditions used and the scaffold architecture can be selected by the technique used, as shown in Table 16.2. For example immersion precipitation, humidified solvent casting and electrospinning are shown in Figure 16.8.

16.3.2.1.1 Solvent Evaporation

Solvent evaporation, where the polymer is dissolved in a volatile solvent and left exposed to the air allowing the solvent to evaporate, is the simplest and most common. Two-dimensional films can be cast onto glass cover slips in this way [44,45]. Three-dimensional structures can also be formed using solvent evaporation by simply pouring the solution alone into a mold [46] or combined with particle leaching. Solvent casting in a humidified environment can drastically affect the film structure. Solvent casting in air will form smooth surfaces but humidification can result in a honeycomb structure if the casting dope concentration, temperature, and solvent are appropriate for the chosen polymer (Figure 16.8).

TABLE 16.2

The Available Scaffold Fabrication Techniques and Resulting Architecture of Scaffolds That Have Been Used in Tissue Engineering

Technique	Material				Architecture						
	Polymer	Ceramic	Metal	Hydrogel	Films	Porous Sheets	Porous Sponges/Blocks	Channeled Sponges/Blocks	Fibers	Hollow Fibers/Tubes	Microspheres/Particles
Phase Inversion											
TIPS	✓				✓	✓	✓				
Double emulsion	✓										✓
Solvent casting	✓				✓	✓	✓				
(Dry-)wet spinning	✓	✓				✓			✓	✓	
Particle leaching	✓			✓		✓	✓				
Dip coating	✓			✓	✓			✓	✓	✓	
Temperature and Pressure											
Supercritical fluid	✓					✓		✓	✓	✓	
Extruding	✓	✓	✓	✓			✓	✓	✓	✓	✓
Sintering	✓	✓	✓		✓	✓	✓	✓	✓	✓	✓
Rapid prototyping	✓	✓	✓	✓				✓	✓		
Melt molding	✓	✓	✓			✓			✓		

16.3.2.1.2 Immersion Precipitation

Another phase inversion process, immersion precipitation, is used to fabricate scaffolds. Instead of a volatile solvent being allowed to evaporate, a third component, the nonsolvent, is used. The nonsolvent is miscible with the solvent but not the polymer so on the addition of nonsolvent to polymer-solvent solution, the polymer precipitates to form a porous structure. Structures formed include membranes [47], fibers [48–50], and microspheres with diameters in the range of 200–900 μm [51].

The polymer is dissolved in the solvent to form a solution, called the casting or spinning dope and the resulting structure is the membrane. The solvent moves into the nonsolvent and the nonsolvent into the polymer-solvent solution, the relative rates of each flux will directly affect the membrane thickness. Once phase inversion is complete, a solid-phase membrane will be present in a solution rich in nonsolvent. The resulting pore structure of the membrane depends on the mechanism of membrane formation, which in turn depends on the interactions of the three components of the ternary system.

While membrane design evolves around modifying the structure to suit the application, the exact mechanisms of formation are still not fully understood. Park et al. (1999) [52] and Xu and Qusay (2004) [53] reference a number of publications that describe the theories of membrane formation by phase inversion. The porous structure of membranes is generally believed to arise from liquid–liquid demixing, by either nucleation and growth, or spinodal decomposition, or a combination of both. The rate of the demixing mechanism leads to either an asymmetric cross section with finger-like pores (Figure 16.8) or a sponge-like structure [54].

Liquid–liquid demixing can be described with the aid of a phase diagram that relates solution composition, temperature, and solution state. A phase

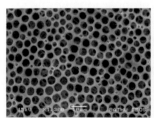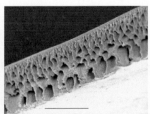

FIGURE 16.8
Examples of poly(D,L-lactide-*co*-glycolide) (PLGA) scaffolds prepared using phase inversion (left) immersion precipitation (20% w/w 75:25 in N-Methyl-2-pyrrolidone, water nonsolvent, 20°C), scale bar 100 μm; (middle) honeycomb scaffold fabricated in a controlled humidified environment (1 mg/ml 50:50 in CHCl₃, 26°C, 90% humidity), scale bar 10 μm; (right) a nonwoven sheet prepared by electrospinning (10% 75:25 PLGA in 3:1 chloroform:methanol, flow rate = 1.5 ml/h, fabrication time = for 15 min, voltage = 10 kV, spinning gap = 10 cm. All scaffolds were prepared in the tissue engineering group laboratories at the University of Bath.

diagram can be 2-D if the system contains two components or 3-D if the system contains three components and detailed explanations are given in membrane textbooks such as that by Mulder [54]. Below is an overview of the process.

Liquid–liquid demixing of a solution will occur when the binodal is reached. The binodal can be reached either by a drop in temperature, or by the addition of another component, the nonsolvent. Above a critical temperature T_c, a solution containing any concentration of polymer and solvent will be completely miscible, such as solution A at temperature T_1. If the temperature of solution A is lowered from T_1 to T_2, the binodal will eventually be reached and the solution will demix into two liquid phases, one rich and one poor in polymer. This process is called liquid–liquid demixing. The temperature at which the binodal is reached is dependent of the polymer concentration in the solvent. Similarly, if a ternary system is used, the polymer, solvent, and nonsolvent will be miscible in any combination of concentrations above the critical temperature. When the binodal is reached, either by a drop in temperature or change in concentration combination, liquid–liquid demixing occurs and two liquid phases are formed, again one rich and one poor in polymer. The composition of these two phases at temperature T_2 are linked by a tie line.

For concentrations Φ where $\Phi^1 < \Phi < \Phi^2$, where Φ^1 and Φ^2 lie on the spinodal, the solution is thermodynamically unstable. If the solution is thermodynamically unstable, it will demix spontaneously into very small interconnected regions with a fluctuating composition. The amplitude of these fluctuations increases with time and results in a lacy structured membrane; this is spinodal decomposition and solution A will undergo this type of liquid–liquid demixing at T_2.

For concentrations Φ where $\Phi^I < \Phi < \Phi^1$ and $\Phi^2 < \Phi < \Phi^{II}$, the solution is metastable. A metastable solution is stable with respect to small fluctuations and does not spontaneously demix, but demixing occurs when a stable nucleus has formed. This will occur near compositions Φ^I and Φ^{II}. The nucleus grows by downhill diffusion, this is nucleation and growth, the final structure will depend on the initial polymer-solvent concentration. A dilute polymer solution will form a latex-type structure whereas an open porous structure is likely to form with a high polymer concentration [54].

The structure is also governed by whether the liquid–liquid demixing is delayed or instantaneous. Instantaneous demixing leads to a porous top layer whereas delayed mixing leads to a dense top layer. Exposure to air will induce a separate demixing mechanism at the surface, and while this does not affect the type of demixing once the nonsolvent is added, the morphology will be altered because of the formation of the top layer. Instantaneous demixing often results in membranes containing macrovoids [54,55]. Macrovoids are large teardrop-shaped pores extending from the top layer through a large section of the membrane wall. Macrovoids are formed due to a stable polymer solution underneath the nuclei so no new nuclei are formed allowing the top nuclei to continue to grow down. Li et al. (2004) [55] suggest that macrovoids are a transition stage between a sponge-like and finger-like structure, the transition being dependent on membrane thickness [55]. Although macrovoids are generally not favorable as they can be weaknesses in the structure [54], they pose no resistance to mass transfer and may be beneficial in some applications.

16.3.2.1.3 Particle Leaching

Particle leaching involves adding salt or glucose to the polymer-solvent solution then washing the particles out with water after the solvent has been removed. This creates macropores the size of the particle, for example 150 μm to 710 μm when sieved sodium chloride is used [56], and 300 μm to 1.5 mm using glucose crystals [57]. El-Amin et al. (2003) [46] produced porous foams from solvent casting and molding with dimensions 4 mm thick and 15 mm in diameter. Combining solvent casting and particulate leaching can produce thicker scaffolds due to the larger pores obtained than when using solvent casting alone.

16.3.2.1.4 Electrospinning

Nanofibers can be formed using electrospinning in which a polymer-solvent solution stream is ejected from a charged capillary toward a grounded collector plate, the solvent evaporates as the stream moves down so forming a charged nonwoven mesh [58] (Figure 16.8).

16.3.2.1.5 Other Techniques

Other techniques include compression molding [59] and photo curing [60] both *in vitro* or *in situ*. Goldstein et al. [59] produced PLGA scaffolds 6 mm thick using a combination of solvent casting, compression molding, and solvent leaching.

16.3.2.1.6 Phase Inversion Parameters

Just how versatile phase inversion is can be seen from the number of variables that affect the resulting scaffold microstructure:

Phase inversion scaffold properties depend on

- Stereoisomer
- Polymer composition
- Mn
- PDI
- Tacticity
- Chain ends
- Monomer sequence
- Polymer concentration
- Solvent
- Nonsolvent
- Temperature

Dry and wet spinning:

- Air gap
- Take-up rate
- Bore flow rate
- Needle diameter

Salt leaching:

- Porogen material
- porogen size
- porogen concentration

Dip coating:

- Mandrel diameter and length
- Number of dips
- Time of mandrel immerison

Polymer properties depend on

- Metal
- Ligand
- Initiator (usually an alcohol)
- Time

- Stereoisomer
- Polymer composition
- Temperature
- Monomer:initiator

It is clear there are a lot of options to tailor make a polymer scaffold using phase inversion. The effect of varying just the solvent is shown in Figure 16.9.

16.3.2.2 Temperature and Pressure

The use of temperature and pressure has seen the advent of a commercialized drug delivery construct (Critical Pharmaceuticals Ltd., United Kingdom), the preparation method is also used to make scaffolds and recently it has been shown that cells are able to survive the fabrication process so it is possible to carry out fabrication and seeding in one step [61,62]. Rapid prototyping uses computer-aided design [63] and fabrication coupled with medical images has huge potential in scaffold fabrication, allowing scaffolds to be formed to fit individual defects. Another technique using temperature and pressure is injection molding, used to produce blocks with multiple conduits [64].

16.3.3 Scaffold Architecture

Whatever the chosen architecture and fabrication technique, a balance between mechanical integrity and porosity to allow angiogenesis and inner-vation that is suitable for the specific application must be met, as well as the other ideal properties described in Section 16.3.1. Scaffold architectures

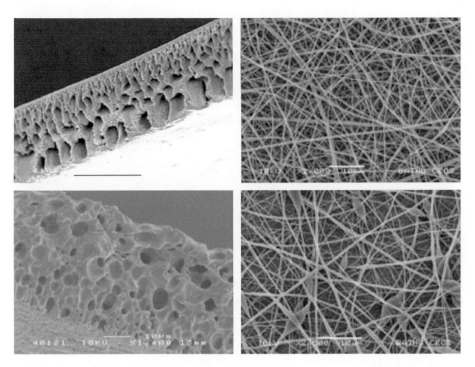

FIGURE 16.9

Examples of PLGA scaffolds prepared by membrane casting (left top and bottom) and electrospinning (right top and bottom) using different solvents. (left top) N-methyl-2-pyrrolidone; (right top) dioxane; (left bottom) chloroform : methanol 3:1; (right bottom) isopranol. Scale bar is in all cases 10 μm. Membrane casting conditions: 20% w/w 75:25, water nonsolvent, 20°C. Electrospinning conditions: 10% 75:25 PLGA, flow rate 1.5 ml/h, fabrication time 15 min, voltage 10 kV, spinning gap 10 cm. All scaffolds were prepared in the tissue engineering group laboratories at the University of Bath.

can be classed as a single piece or particulate. Single-piece scaffolds such as sponges and blocks, and constructs made from fibers, tubes, or aggregated microparticles, can be used to fill large defects and are suitable for critical size bone defects, or soft tissues such as the liver. Injectable polymers, which form porous blocks *in situ*, can be used to fill less defined defects or to deliver cells and growth factors to a tissue. Films and porous sheets are used as wound dressings, for skin or surgical flaps to contain packing in a defect such as a tumor site. They are also classified as single-piece scaffolds, although several sheets are used to cover large areas of burns. Sheets are the most successful structure that has been developed commercially; MySkin®, Dermagraft®, and Epicel® are three wound dressing products.

Particulate scaffolds consist of many discrete particles that are loaded with cells, growth factors or drugs, or a combination of these either in the same or different particles. A review by Silva et al. (2007) [4] claims particulate materials can provide a "single system of … structural support, cell support, and controlled release…." The review draws on examples from drug delivery in which microcarriers have long been exploited, and from the field of tissue engineering that have developed such single systems, for example, from alginate-β-tricalcium phosphate (TCP) and PLGA with growth factor IGF-1. Here we are at the crossover between tissue engineering and drug delivery; a very exciting area for future research.

16.4 Scaffold Surface Modification

16.4.1 Improving Cell Culture on Polymers

Many polymeric biomaterials used in tissue engineering and regenerative medicine possess excellent biocompatibility, degradability, and mechanical properties but do not offer a cytocompatible environment. For the formation or regeneration of tissues, cells must be able to adhere to the scaffold, spread, migrate, proliferate and, if necessary, differentiate. For this sequence of events to occur, the cell must initially be able to recognize adhesive points on the polymer surface and this is where many manmade polymers are less efficacious than their natural counterparts. For example, the most widely used polymers in tissue engineering, Poly(L-lactic acid) (PLLA) and Poly(lactide-*co*-glycolic) acid (PLGA), lack recognition sites. Even if adhesion does occur, for subsequent spreading and migration, forces must be generated across extending cellular processes. Many polymers do not have cell recognition sites and without sufficient adhesive contacts between the material surface and the cytoskeleton, this process can not occur, and the cells will remain rounded [65]. Extracellular receptors that mediate cell-cell and cell-extracellular matrix (ECM) interactions are a number of proteins called integrins, cadherins, immunoglobulins, selectins, and proteoglycans. Extracellular matrix proteins are collagen, fibronectin, vitronectin, laminin, and proteoglycans. Many polymers used for tissue engineering scaffolds are hydrophobic and, hence, the infiltration of cells into the porous network is limited. Water contact angle of 65–95° for a typical hydrophobic surface. Charged surfaces demonstrate increased cell adhesiveness either due to direct interaction with charged groups on the cell surface or by promoting the adsorption of serum from the culture medium.

16.4.2 Adsorption of Serum Proteins

Cell culture flasks and plates are made from a modified form of polystyrene-termed tissue culture polystyrene (TCPS). Polystyrene itself is hydrophobic and, therefore, a poor substrate for cell adhesion and growth unless it is precoated with a cell-adhesive substrate. The TCPS on the other hand has been treated so that it has accessible oxygen-containing groups on its surface, rendering it more hydrophilic and greatly enhancing cell attachment. The mechanism of cell attachment to TCPS relies on the fact that most cell culture medium contains fetal bovine or other animal serum. Adsorption of extracellular matrix proteins to the TCPS from the serum component of the culture medium provides an adhesive surface for cell attachment. Specifically, vitronectin adsorption appears to be the mediator for cell adhesion to TCPS, while both vitronectin and fibronectin mediate adhesion to Primaria™, a polystyrene surface modified with oxygen- and nitrogen-containing groups [66,67].

For hydrophobic polymer scaffolds, the adhesion, spreading, and proliferation of cells can similarly be improved by incorporating chemical groups that increase hydrophilicity, resulting in serum protein adsorption, and there are a number of methods available to achieve this, depending on the polymer.

16.4.3 Methods of Scaffold Modification

Polymers can be modified during the polymerization process by selection of the initiator and monomer as described in Section 16.2 or after scaffold fabrication. Here the focus is on postscaffold fabrication modification. Methods include

- Plasma: etching, grafting, deposition
- Ion beam: entrapment of atoms
- Surface entrapment: swelling and collapse
- Direct ligation
- Ionic interactions
- Chemical treatment: generation of functional groups
- Coating: standard adsorption of biomolecules
- Photografting [68]
- Mineralization

16.4.3.1 Plasma Treatment

Macroporous scaffolds were fabricated using a 70/30 blend of PLLA and PLGA and then treated with an anhydrous ammonia plasma. Water contact angle was reduced from 70.1° to 15.9°, indicating a large increase in the scaffold wettability, and a cell seeding efficiency of 99% was achieved [69]. Plasma treatment alters the surface of a polymer scaffold without affecting the bulk properties.

16.4.3.2 Ozone

Oxidation of polymers with ozone can result in the generation of peroxides or hydroperoxides on the surface, and these reactive species can be used for the immobilization of natural or functionalized biomolecules. As with plasma treatment, this is a surface phenomenon and the bulk properties of the polymer are unaffected by the process. For example, the peroxide species are capable of initiating the polymerization of vinyl monomers and this has been utilized to great effect in altering the cellular response to polymers, particularly PLLA. Ho et al. grafted acryl succinimide to PLLA scaffolds following ozonolysis, generating an amine-reactive surface to which they coupled Arg-Gly-Asp-Ser (RGDS). The peptide-modified polymer was shown to greatly enhance the adhesion, proliferation, and mineralization of osteoblastic-like cells and, *in vivo*, a modified porous scaffold was shown to be biocompatible and supportive of cell ingrowth [70]. Grafting of type 1 atelocollagen on to ozone-treated PLLA membranes has also been demonstrated [71].

16.4.3.3 Hydrolysis and Aminolysis

While polyester biomaterials have excellent physical characteristics, are biocompatible and biodegradable, a major drawback is their hydrophobicity and lack of cellular binding sites, making them unsuitable for the growth of a number of cell types. However, there are various strategies for the surface modification of polyesters in order to render them cell adhesive with perhaps the simplest being chemical hydrolysis. In aqueous environments, polyesters undergo bulk hydrolysis as the rate of diffusion of the hydrolyzing moiety, in this case water, into the polymer is much greater than the rate of hydrolysis. Surface hydrolysis occurs when the opposite is true that is, the rate of hydrolysis is much more rapid than the rate of diffusion.

Surface hydrolysis of a poly(glycolic acid) fiber mesh with sodium hydroxide [72]. Fiber diameter decreased with increasing hydrolysis time, but bulk properties of the fibers were largely unchanged suggesting that modification was confined to the accessible surface groups of the polymer. The resultant increase in carboxylic acid and hydroxyl groups on the fiber surfaces improved the wettability of the PGA and enhanced cell seeding. The number of vascular smooth muscle cells that adhered to hydrolyzed scaffolds was more than double that for control scaffolds, and this was shown to result from increased adsorption of serum proteins from the culture medium. Surface hydrolysis results from the treatment with, for example, NaOH [73] and NaOCl (Figure 16.10). Aminolysis can also be used to functionalize polyester-like polyurethanes with amino groups [74]. The generation of carboxylic acid or amine functionality makes the polyesters amenable to covalent attachment of bioactive molecules.

16.4.3.4 Electrostatic Self-Assembly

Macroporous PLA scaffolds were treated with poly(ethylenimine) (PEI) to generate amine-modified, positively charged surfaces. This surface was then used to successively deposit alternate layers of gelatin, a cell-adhesive,

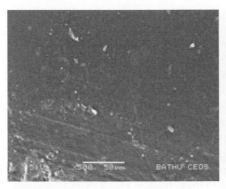
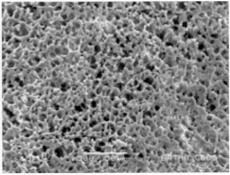

FIGURE 16.10
The effect of sodium hypochlorite on membrane morphology: outer surface of 5% PVA-PLGA hollow fiber treated with sodium hypochlorite (right) and untreated (left) NaOCl has removed the skin and both the outer and inner surfaces appear highly porous.

negatively charged derivative of collagen, and PEI using electrostatic self-assembly [75]. To determine the effect of this surface modification on cell adhesion and function, primary chondrocytes were seeded on the scaffolds and examined after a period of seven days. In comparison to untreated scaffolds, cell viability was over 70% higher on the PEI/gelatin surfaces, while the total intracellular protein content was nearly 60% higher, suggesting an increased rate of proliferation. When cell morphology and infiltration in the scaffolds were examined, it was found that the chondrocytes on the PEI/gelatin scaffolds were well spread and within the pores, while the cells on untreated PLA remained rounded and at the scaffold periphery.

16.4.3.5 Surface Entrapment

Surface entrapment of biomolecules onto a polymer scaffold is an effective, noncovalent means of improving its cytocompatibility by increasing hydrophilicity or incorporating adhesion sites. First described by Desai and Hubbell, this technique involves swelling the polymer surface in a solvent that is also compatible with the biomolecule of interest [76,77]. The biomolecule partially diffuses into the swollen surface layer of the polymer and, when the solvent is replaced by a nonsolvent, the surface collapses, entrapping the biomolecule within it.

PLGA was immersed in a solution of chitosan in acetone and, on immersion in water, the chitosan became entrapped on the scaffold. This modified matrix was implanted into a rabbit bone defect model and compared with unmodified scaffolds over a period of 2 to 12 weeks. Histological analysis demonstrated that the chitosan-modified scaffolds were more biocompatible, as evidenced by reduced inflammation and an increased bone formation compared to the controls [78].

16.4.3.6 Adsorption of Bioactive Molecules

The hydrophobic nature of many synthetic polymers means that adsorption of biomolecules that, by their very nature are hydrophilic, is not effective. However, this may be overcome by modifying the biomolecule with a moiety that is compatible with the scaffold surface. For example, Fahmy and colleagues modified the egg white protein avidin with a hydrophobic alkyl chain via the conjugation of palmitic acid. When a macroporous PLGA scaffold was incubated with the avidin-fatty acid conjugate, hydrophobic interactions between the polymer and the palmitic acid resulted in adsorption of avidin on the scaffold surface. This protein coating was still present after incubation in aqueous buffer at 37°C for three weeks and is readily applicable to a variety of polymers and bioactive coatings [79].

16.4.3.7 Covalent Attachment of Bioactive Molecules

Scaffolds are mainly modified by covalent attachment using short, bioactive molecules; proteins, polysaccharides, and, perhaps most frequently, peptide fragments such as the RGD, integrin-binding peptide motif, and the GRDYS. Amino groups were introduced into PLLA by treatment with 1,6-hexanediamine and gelatin, chitosan, and collagen then immobilized by cross-linking with glutaraldehyde [80]. Human umbilical vein endothelial cells exhibited increased adhesion, proliferation, and von Willebrand factor secretion on all of the surfaces tested in comparison to untreated PLLA. Poly(caprolactone) has been modified with RGD, IKVAV, and YIGSR peptides, attached following aminolysis using 1,6-hexanediamine [81]. Sulphydryl-terminated oligopeptides containing the YIGSR and IKVAV sequences were covalently attached using the sulpho-(N-maleimidomethyl)cyclohexane-1-carboxylate cross-linking agent [82], which enhanced the adhesion and neurite outgrowth of chick dorsal root ganglion neurons in comparison to unmodified scaffolds. Following aminolysis (or photografting of methacrylic acid) to PLLA, collagen was immobilized to the functionalized surface or layers of chondroitin sulphate and collagen were deposited via electrostatic interactions [83].

16.5 Bioreactors for Polymeric Scaffolds

Bioreactors provide an *in vivo*-like environment for the cells as well as improving nutrient delivery and waste product removal [84]. Common configurations are the packed bed [85,86] and more recently bioreactors that provide a blood vessel-like nutrient supply based on hollow fiber membranes [48,87,88]. In any system containing cells, the environment is a dynamic one with cells metabolizing, proliferating, and differentiating to varying degrees as a function of time. In the case of adherent cells, the transport phenomena are affected as the cells proliferate and lay down extracellular matrix, while is in many cases, as with polyesters, the scaffold degrades. Cell expansion and matrix deposition is only problematic with poorly designed scaffold-bioreactor systems when the scaffold becomes blocked over time or the bioreactor is unable to provide suitable media supply and waste removal within a suitable range of fluid dynamic properties. An understanding of the transport phenomena and its affects on construct development is vital to a viable and consistent product, as well as allowing better design of the bioreactor and scaffold in order to produce a more efficient tissue engineering system.

Bioreactor and scaffold selection should be combined and it is dependent on the end point of the culture; expansion of a cell population or development of a mature tissue construct. Mikos's group at Rice University, Houston, Texas have published a number of papers using a packed bed. The perfusion reactor with fibrous blocks was first reported in the paper by Bancroft et al. [86]. The rotating bioreactor, first developed by NASA for cartilage tissue engineering in space [89] is used with cell suspensions or with cell and microparticles to form aggregates and a number of designs are commercially available. Examples in bone tissue engineering include those carried out at Oxford for bone tissue engineering [90–92]. Chaudhuri et al. have utilized an expanded bed, first applied to cartilage engineering and then to bone engineering. The system allows maximum use of bioreactor volume while the particles are kept just suspended by the up-flow of the media.

Another bioreactor design is the membrane bioreactor. Membranes can be flat sheets, spiral wound flat sheets, hollow fibers, or tubes [93]. In biomedical applications, hollow fiber membranes as well as flat sheet and spiral-wound membranes, made from a vast number of polymers, have been successfully used for blood oxygenation and haemodialysis [94]. Membranes are growing in popularity in the field of tissue engineering. They are routinely used for oxygenation of cultures such as in the rotating wall bioreactor [95], stirred tanks [96], and as Transwell™ supports. Membranes as scaffolds are most widely used in liver and pancreas applications but have also been used in cartilage and bone tissue engineering. De Bartolo has published extensive work on the use of membranes for hepatocyte culture on a number of polymer membranes including polypropylene [97], a PEEK-polyurathane blend [98,99], polyether-sulphone [100], and has utilized both flat sheet and hollow fiber membranes. A comprehensive review on technologies for treating insulin diabetes [101] details the use of the flat sheet, hollow fiber, and spiral-wound membranes; a range of materials were highlighted for membrane fabrication for islet support, namely Thomapor® (polyamide), Cuprophane® HDF (cellulose), Amicon® XM-50 (polyvinyl chloride acrylic copolymer) and PEEK™ (polyetherether-ketone hollow fibers). Flat, hollow, and spiral membranes have been used to culture hepatocytes [94]. The PLLA membranes with surface modification have been used for chondrocyte culture [102] and were fabricated by using a tertiary system and temperature-induced phase separation (TIPS) [103].

(a)   (b)

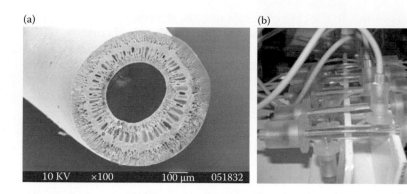

FIGURE 16.11
PLGA hollow fiber bioreactor. (a) PLGA hollow fiber bioreactor (HFB) operation in a 37°C, 5% CO_2 incubator (photo courtesy of B. Ainsworth) (b) SEM of the cross-section of a typical PLGA hollow fibre (Reproduced from Ellis, M. J., and Chaudhuri, J. B., *Biotechnol. Bioeng.* 96, 177–87, 2007.)

The PLGA hollow fiber bioreactor in Figure 16.11 was initially applied to bone, the lumen of the hollow fibers were designed to replicate the haversian canal system *in vitro* then *in vivo* to allow angiogenesis and innervation [48]. It has since been applied to expansion of human bone marrow stromal cells (HMSC) expansion [49] and liver tissue engineering. Other examples of polymer membrane scaffolds include a study on repairing segmental diaphyseal bone defects larger than the critical size in sheep tibia using PLA membrane tubes was reported as successful [47]. Flat sheet and spiral-wound membrane scaffolds are reported for the pancreas [101] and the early papers for mammalian- and human-hybridoma cell cultures [104,105]. Drug delivery using PLA hollow fiber membranes, commonly for contraceptives, has been previously reported [106,107] and highlight the scope to have a construct that is both a scaffold and a drug delivery device.

16.6 Conclusions

Polymer scaffolds for regenerative medicine are hugely versatile and there is an opportunity to tailor their properties at every stage of fabrication and use, and will have a dual function of being both a scaffold and for controlled drug release. Using a judicious choice of metal and ligand combination, it is possible to tailor the properties of the polymer, at the molecular level, for the desired application. There is a vast range of scaffold fabrication techniques and within these techniques the conditions can be set to produce the desired architecture. Careful selection of the bioreactor to house the polymer scaffold will complete the tissue engineering system.

References

1. Middleton, J. C., and Tipton, A. J. 2000. Synthetic biodegradable polymers as orthopedic devices. *Biomaterials* 21:2335–46.
2. Cheung, H. Y., Lau, K. T., Lu, T. P., and Hui, D. 2007. A critical review on polymer-based bio-engineered materials for scaffold development. *Composites Part B-Engineering* 38 (3):291–300.
3. Dawson, J. I., and Oreffo, R. O. C. 2008. Bridging the regeneration gap: Stem cells, biomaterials and clinical translation in bone tissue engineering. *Archives of Biochemistry and Biophysics* 473 (2):124–31.
4. Silva, G. A., Coutinho, O. P., Ducheyne, P., and Reis, R. L. 2007. Materials in particulate form for tissue engineering. 2. Applications in bone. *Journal of Tissue Engineering and Regenerative Medicine* 1 (2):97–109.
5. Weigel, T., Schinkel, G., and Lendlein, A. 2006. Design and preparation of polymeric scaffolds for tissue engineering. *Expert Review of Medical Devices* 3 (6):835–51.
6. Stevens, M. M. 2008. Biomaterials for bone tissue engineering. *Materials Today* 11 (5):18–25.
7. Stevens, M. M., and George, J. H. 2005. Exploring and engineering the cell surface interface. *Science* 310 (5751):1135–38.
8. Bilodeau, K., and Mantovani, D. 2006. Bioreactors for tissue engineering: Focus on mechanical constraints. A comparative review. *Tissue Engineering* 12 (8):2367–83.
9. Kumar, S., Wittmann, C., and Heinzle, E. 2004. Minibioreactors. *Biotechnology Letters* 26 (1):1–10.
10. Portner, R., Nagel-Heyer, S., Goepfert, C., Adamietz, P., and Meenen, N. M. 2005. Bioreactor design for tissue engineering. *Journal of Bioscience and Bioengineering* 100 (3):235–45.
11. Wang, D. L., Liu, W. S., Han, B. Q., and Xu, R. A. 2005. The bioreactor: A powerful tool for large-scale culture of animal cells. *Current Pharmaceutical Biotechnology* 6 (5):397–403.
12. Mushipe, M. T., Chen, X., Jennings, D., and Li, G. 2006. Cells seeded on MBG scaffold survive impaction grafting technique: Potential application of cell-seeded biomaterials for revision arthroplasty. *Journal of Orthopaedic Research* 24 (3):501–7.
13. Nich, C., and Sedel, L. 2006. Bone substitution in revision hip replacement. *International Orthopaedics* 30 (6):525–31.
14. Warnke, P. H. 2004. Growth and transplantation of a custom vascularised bone graft in a man. *Lancet* 364 (9436):766–70.
15. Salem, A. K., Rose, F., Oreffo, R. O. C., Yang, X. B., Davies, M. C., Mitchell, J. R., Roberts, C. J., et al. 2003. Porous polymer and cell composites that self-assemble *in situ*. *Advanced Materials* 15 (3):210–13.
16. Salgado, A. J., Coutinho, O. P., and Reis, R. L. 2004. Bone tissue engineering: State of the art and future trends. *Macromolecular Bioscience* 4 (8):743–65.
17. Albertsson, A. C., and Varma, I. K. 2003. Recent developments in ring opening polymerization of lactones for biomedical applications. *Biomacromolecules* 4 (6):1466–86.

18. O'Keefe, B. J., Hillmyer, M. A., and Tolman, W. B. 2001. Polymerization of lactide and related cyclic esters by discrete metal complexes. *Journal of the Chemical Society-Dalton Transactions* 15:2215–24.

19. Platel, R. H., Hodgson, L. M., and Williams, C. K. 2008. Biocompatible initiators for lactide polymerization. *Polymer Reviews* 48 (1):11–63.

20. Kricheldorf, H. R., Boettcher, C., and Tonnes, K. U. 1992. Polylactones. 23. Polymerization of racemic and meso D,L-lactide with various organotin catalysts stereochemical aspects. *Polymer* 33 (13):2817–24.

21. Dove, A. P., Gibson, V. C., Marshall, E. L., Rzepa, H. S., White, A. J. P., and Williams, D. J. 2006. Synthetic, structural, mechanistic, and computational studies on single-site beta-diketiminate tin(II) initiators for the polymerization of rac-lactide. *Journal of the American Chemical Society* 128 (30):9834–43.

22. Marshall, E. L., Gibson, V. C., and Rzepa, H. S. 2005. A computational analysis of the ring-opening polymerization of rac-lactide initiated by single-site beta-diketiminate metal complexes: Defining the mechanistic pathway and the origin of stereocontrol. *Journal of the American Chemical Society* 127 (16):6048–51.

23. Hormnirun, P., Marshall, E. L., Gibson, V. C., White, A. J. P., and Williams, D. J. 2004. Remarkable stereocontrol in the polymerization of racemic lactide using aluminum initiators supported by tetradentate aminophenoxide ligands. *Journal of the American Chemical Society* 126 (9):2688–89.

24. Kowalski, A., Duda, A., and Penczek, S. 2000. Mechanism of cyclic ester polymerization initiated with tin(II) octoate. 2. Macromolecules fitted with tin(II) alkoxide species observed directly in MALDI-TOF spectra. *Macromolecules* 33 (3):689–95.

25. Ovitt, T. M., and Coates, G. W. 2002. Stereochemistry of lactide polymerization with chiral catalysts: New opportunities for stereocontrol using polymer exchange mechanisms. *Journal of the American Chemical Society* 124 (7):1316–26.

26. Storey, R. F., and Sherman, J. W. 2002. Kinetics and mechanism of the stannous octoate-catalyzed bulk polymerization of epsilon-caprolactone. *Macromolecules* 35 (5):1504–12.

27. Qin, J. J., Gu, J., and Chung, T. S. 2001. Effect of wet and dry-jet wet spinning on the sheer-induced orientation during the formation of ultrafiltration hollow fiber membranes. *Journal of Membrane Science* 182 (1–2):57–75.

28. Chamberlain, B. M., Cheng, M., Moore, D. R., Ovitt, T. M., Lobkovsky, E. B., and Coates, G. W. 2001. Polymerization of lactide with zinc and magnesium beta-diiminate complexes: Stereocontrol and mechanism. *Journal of the American Chemical Society* 123 (14):3229–38.

29. Thakur, K. A. M., Kean, R. T., Hall, E. S., Kolstad, J. J., Lindgren, T. A., Doscotch, M. A., Siepmann, J. I., and Munson, E. J. 1997. High-resolution C-13 and H-1 solution NMR study of poly(lactide). *Macromolecules* 30 (8):2422–28.

30. Amgoune, A., Thomas, C. M., Ilinca, S., Roisnel, T., and Carpentier, J. F. 2006. Highly active, productive, and syndiospecific yttrium initiators for the polymerization of racemic beta-butyrolactone. *Angewandte Chemie-International Edition* 45 (17):2782–84.

31. Chmura, A. J., Davidson, M. G., Frankis, C. J., Jones, M. D., and Lunn, M. D. 2008. Highly active and stereoselective zirconium and hafnium alkoxide initiators for solvent-free ring-opening polymerization of rac-lactide. *Chemical Communications* (11):1293–95.

32. Chmura, A. J., Davidson, M. G., Jones, M. D., Lunn, M. D., Mahon, M. F., Johnson, A. F., Khunkamchoo, P., Roberts, S. L., and Wong, S. S. F. 2006. Group 4 complexes with aminebisphenolate ligands and their application for the ring opening polymerization of cyclic esters. *Macromolecules* 39 (21):7250–57.
33. Ma, H. Y., Spaniol, T. P., and Okuda, J. 2003. Rare earth metal complexes supported by 1,omega-dithiaalkanediyl-bridged, bis(phenolato) ligands: synthesis, characterization and ring-opening polymerization catalysis of L-lactide. *Dalton Transactions* (24):4770–80.
34. Zhong, Z. Y., Dijkstra, P. J., and Feijen, J. 2002. [(salen)Al]-mediated, controlled and stereoselective ring-opening polymerization of lactide in solution and without solvent: Synthesis of highly isotactic polylactide stereocopolymers from racemic D,L-lactide. *Angewandte Chemie-International Edition* 41 (23):4510–13.
35. Zhong, Z. Y., Dijkstra, P. J., and Feijen, J. 2003. Controlled and stereoselective polymerization of lactide: Kinetics, selectivity, and microstructures. *Journal of the American Chemical Society* 125 (37):11291–98.
36. Chisholm, M. H., Gallucci, J. C., and Phomphrai, K. 2004. Well-defined calcium initiators for lactide polymerization. *Inorganic Chemistry* 43 (21):6717–25.
37. Cheng, M., Attygalle, A. B., Lobkovsky, E. B., and Coates, G. W. 1999. Single-site catalysts for ring-opening polymerization: Synthesis of heterotactic poly(lactic acid) from rac-lactide. *Journal of the American Chemical Society* 121 (49):11583–84.
38. Coulembier, O., Lohmeijer, B. G. G., Dove, A. P., Pratt, R. C., Mespouille, L., Culkin, D. A., Benight, S. J., Dubois, P., Waymouth, R. M., and Hedrick, J. L. 2006. Alcohol adducts of N-heterocyclic carbenes: Latent catalysts for the thermally-controlled living polymerization of cyclic esters. *Macromolecules* 39 (17):5617–28.
39. Dove, A. P., Li, H. B., Pratt, R. C., Lohmeijer, B. G. G., Culkin, D. A., Waymouth, R. M., and Hedrick, J. L. 2006. Stereoselective polymerization of rac- and meso-lactide catalyzed by sterically encumbered N-heterocyclic carbenes. *Chemical Communications* (27):2881–83.
40. Jones, M. D., Davidson, M. G., Keir, C. G., Wooles, A. J., Mahon, M. F., and Apperley, D. C. 2008. *Dalton Transactions* 3655–57.
41. Yang, S., Leong, K.-F., Du, Z., and Chua, C.-K. 2001. Review: The design of scaffolds for use in tissue engineering. Part I. Traditional factors. *Tissue Engineering* 7 (6):679–89.
42. Yang, S., Leong, K.-F., Du, Z., and Chau, C.-K. 2002. Review: The design of scaffolds for use in tissue engineering. Part II. Rapid prototyping techniques. *Tissue Engineering* 8 (1):1–11.
43. Hutmacher, D. 2000. Scaffolds in tissue engineering bone and cartilage. *Biomaterials* 21:2529–43.
44. Calvert, J. W., Marra, K. G., Cook, L., Kumta, P. N., DiMilla, P. A., and Weiss, L. E. 2000. Characterization of osteoblast-like behavior of cultured bone marrow stromal cells on various polymer surfaces. *Journal of Biomedical Materials Research* 52 (2):279–84.
45. Klee, D., Ademovic, Z., Bosserhoff, A., Hoecker, H., Maziolis, G., and Erli, H. J. 2003. Surface modification of poly(vinylidenefluoride) to improve the osteoblast adhesion. *Biomaterials* 24 (21):3663–70.
46. El-Amin, S. F., Lu, H. H., Khan, Y., Burems, J., Mitchell, J., Tuan, R. S., and Laurencin, C. T. 2003. Extracellular matrix production by human osteoblasts cultured on biodegradable polymers applicable for tissue engineering. *Biomaterials* 24:1213–21.

47. Gugala, Z., and Gogolewski, S. 1999. Regeneration of segmental diaphyseal defects in sheep tibiae using resorbable polymeric membranes: A preliminary study. *Journal of Orthopaedic Trauma* 13 (3):187–95.

48. Ellis, M. J., and Chaudhuri, J. B. 2007. Poly(lactic-*co*-glycolic acid) hollow fibre membranes for use as a tissue engineering scaffold. *Biotechnology and Bioengineering* 96:177–87.

49. Morgan, S. M., Tilley, S., Perera, S., Ellis, M., Chaudhuri, J. B., and Oreffo, R. O. C. 2007. Expansion of human bone marrow stromal cells on poly-(DL-lactide-co-glycolide) (PDLLGA) hollow fibres designed for use in skeletal tissue engineering. *Biomaterials* 28 (35):5332–43.

50. Williamson, M. R., and Coombes, A. G. A. 2004. Gravity spinning of poly-caprolactone fibres for applications in tissue engineering. *Biomaterials* 25 (3):459–65.

51. Borden, M., El-Amin, S. F., Attawia, M., and Laurencin, C. T. 2003. Structural and human cellular assessment of a novel microsphere-based tissue engineered scaffold for bone repair. *Biomaterials* 24:597–609.

52. Park, H. C., Kim, Y. P., Kim, H. Y., and Kang, Y. S. 1999. Membrane formation by water vapor induced phase inversion. *Journal of Membrane Science* 156 (2):169–78.

53. Xu, Z. L., and Qusay, F. A. 2004. Polyethersulfone (PES) hollow fiber ultrafiltration membranes prepared by PES/non-solvent/NMP solution. *Journal of Membrane Science* 233 (1–2):101–11.

54. Mulder, M. 1996. *Basic principles of membrane technology,* 363. 2nd ed. Netherlands: Kluwer Academic Publishers Group.

55. Li, D. F., Chung, T. S., Ren, J. Z., and Wang, R. 2004. Thickness dependence of macrovoid evolution in wet phase-inversion asymmetric membranes. *Industrial & Engineering Chemistry Research* 43 (6):1553–56.

56. Ciapetti, G., Ambrosio, L., Savarino, L., Granchi, D., Cenni, E., Baldini, N., Pagani, S., Guizzardi, S., Causa, F., and Giunti, A. 2003. Osteoblast growth and function in porous poly epsilon-caprolactone matrices for bone repair: A preliminary study. *Biomaterials* 24 (21):3815-24.

57. Terai, H., Hannouche, D., Ochoa, E., Yamano, Y., and Vacanti, J. P. 2002. *In vitro* engineering of bone using a rotational oxygen-permeable bioreactor system. *Materials Science & Engineering C-Biomimetic and Supramolecular Systems* 20 (1–2):3–8.

58. Yoshimoto, H., Shin, Y. M., Terai, H., and Vacanti, J. P. 2003. A biodegradable nanofiber scaffold by electrospinning and its potential for bone tissue engineering. *Biomaterials* 24:2077–82.

59. Goldstein, A. S., Juarez, T. M., Helmke, C. D., Gustin, M. C., and Mikos, A. G. 2001. Effect of convection on osteoblastic cell growth and function in biodegradable polymer foam scaffolds. *Biomaterials* 22:1279–88.

60. Vehof, J., Fisher, J., Dean, D., Waerden, J.-P., Spauwen, P., Mikos, A., and Jansen, J. 2002. Bone formation in transforming growth factor Beta-1-coated porous poly(propylene fumerate) scaffolds. *Journal of Biomedical Materials Research* 60 (2):241–51.

61. Ginty, P. J., Howard, D., Upton, C. E., Barry, J. J. A., Rose, F., Shakesheff, K. M., and Howdle, S. M. 2008. A supercritical CO_2 injection system for the production of polymer/mammalian cell composites. *Journal of Supercritical Fluids* 43 (3):535–41.

62. Whitaker, M. J., Quirk, R. A., Howdle, S. M., and Shakesheff, K. M. 2001. Growth factor release from tissue engineering scaffolds. *Journal of Pharmacy and Pharmacology* 53 (11):1427–37.
63. Leong, K. F., Cheah, C. M., and Chua, C. K. 2003. Solid freeform fabrication of three-dimensional scaffolds for engineering replacement tissues and organs. *Biomaterials* 24 (13):2363–78.
64. Hadlock, T., Sundback, C., Hunter, D., Cheney, M., and Vacanti, J. P. 2000. A polymer foam conduit seeded with Schwann cells promotes guided peripheral nerve regeneration. *Tissue Engineering* 6 (2):119–27.
65. Gumbiner, B. M. 1996. Cell adhesion: The molecular basis of tissue architecture and morphogenesis. *Cell* 84 (3):345–57.
66. Steele, J. G., Dalton, B. A., Johnson, G., and Underwood, P. A. 1993. Polystyrene chemistry affects vitronectin activity: An explanation for cell attachment to tissue-culture polystyrene but not to unmodified polystyrene. *Journal of Biomedical Materials Research* 27 (7):927–40.
67. Steele, J. G., Dalton, B. A., Johnson, G., and Underwood, P. A. 1995. Adsorption of fibronectin and vitronectin onto primaria™ and tissue-culture polystyrene and relationship to the mechanism of initial attachment of human vein endothelial-cells and Bhk-21 fibroblasts. *Biomaterials* 16 (14):1057–67.
68. Ma, Z. W., Gao, C. Y., Juan, J., Ji, J., Gong, Y. H., and Shen, J. C. 2002. Surface modification of poly-L-lactide by photografting of hydrophilic polymers towards improving its hydrophilicity. *Journal of Applied Polymer Science* 85 (10):2163–71.
69. Yang, J., Shi, G. X., Bei, J. Z., Wang, S. G., Cao, Y. L., Shang, Q. X., Yang, G. G., and Wang, W. J. 2002. Fabrication and surface modification of macroporous poly(L-lactic acid) and poly(L-lactic-*co*-glycolic acid) (70/30) cell scaffolds for human skin fibroblast cell culture. *Journal of Biomedical Materials Research* 62 (3):438–46.
70. Ho, M. H., Lee, J. J., Fan, S. C., Wang, D. M., Hou, L. T., Hsieh, H. J., and Lai, J. Y. 2007. Efficient modification on PLLA by ozone treatment for biomedical applications. *Macromolecular Bioscience* 7 (4):467–74.
71. Suh, H., Hwang, Y. S., Lee, J. E., Han, C. D., and Park, J. C. 2001. Behavior of osteoblasts on a type I atelocollagen grafted ozone oxidized poly L-lactic acid membrane. *Biomaterials* 22 (3):219–30.
72. Gao, J. M., Niklason, L., and Langer, R. 1998. Surface hydrolysis of poly(glycolic acid) meshes increases the seeding density of vascular smooth muscle cells. *Journal of Biomedical Materials Research* 42 (3):417–24.
73. Shi, S., and Gronthos, S. 2003. Perivascular niche of postnatal mesenchymal stem cells in human bone marrow and dental pulp. *Journal of Bone and Mineral Research* 18 (4):696–704.
74. Zhu, Y. B., Gao, C. Y., He, T., and Shen, J. C. 2004. Endothelium regeneration on luminal surface of polyurethane vascular scaffold modified with diamine and covalently grafted with gelatin. *Biomaterials* 25 (3):423–30.
75. Zhu, H. G., Ji, J., and Shen, J. C. 2004. Biomacromolecules electrostatic self-assembly on 3-dimensional tissue engineering scaffold. *Biomacromolecules* 5 (5):1933–39.
76. Desai, N. P., and Hubbell, J. A. 1991. Solution technique to incorporate polyethylene oxide and other water-soluble polymers into surfaces of polymeric biomaterials. *Biomaterials* 12 (2):144–53.

77. Desai, N. P., and Hubbell, J. A. 1992. Surface physical interpenetrating networks of poly(ethylene-terephthalate) and poly(ethylene oxide) with biomedical applications. *Macromolecules* 25 (1):226–32.
78. Cai, K. Y., Yao, K. D., Yang, Z. M., Qu, Y. L., and Li, X. Q. 2007. Histological study of surface modified three dimensional poly(D,L-lactic acid) scaffolds with chitosan *in vivo*. *Journal of Materials Science-Materials in Medicine* 18 (10):2017–24.
79. Das, D. B. 2007. Multiscale simulation of nutrient transport in hollow fibre membrane bioreactor for growing bone tissue: Sub-cellular scale and beyond. *Chemical Engineering Science* 62 (13):3627–39.
80. Zhu, Y., Gao, C., Liu, X., He, T., and Shen, J. 2004. Immobilization of biomacro-molecules onto aminolyzed poly(L-lactic acid) toward acceleration of endothelium regeneration. *Tissue Engineering* 10 (1–2):53–61.
81. Santiago, L. Y., Nowak, R. W., Rubin, J. P., and Marra, K. G. 2006. Peptide-surface modification of poly(caprolactone) with laminin-derived sequences for adipose-derived stem cell applications. *Biomaterials* 27 (15):2962–69.
82. Yu, T. T., and Shoichet, M. S. 2005. Guided cell adhesion and outgrowth in peptide-modified channels for neural tissue engineering. *Biomaterials* 26 (13):1507–14.
83. Zhu, Y., Gao, C., Liu, Y., and Shen, J. 2004. Endothelial cell functions *in vitro* cultured on poly(L-lactic acid) membranes modified with different methods. *Journal of Biomedical Materials Research Part A* 69 (3):436–43.
84. Ellis, M., and Chaudhuri, J. 2004. In *Bioreactor design for clinical scale constructs*. Liverpool: Interdisciplinary Research Collaboration in Tissue Engineering.
85. Gomes, M. E., Holtorf, H. L., Reis, R. L., and Mikos, A. G. 2006. Influence of the porosity of starch-based fiber mesh scaffolds on the proliferation and osteogenic differentiation of bone marrow stromal cells cultured in a flow perfusion bioreactor. *Tissue Engineering* 12 (4):801–9.
86. Bancroft, G. N., Sikavitsas, V. I., and Mikos, A. G. 2003. Design of a flow perfusion bioreactor system for bone tissue-engineering applications. *Tissue Engineering* 9 (3):549–54.
87. Abdullah, N. S., Das, D. B., Ye, H., and Cui, Z. F. 2006. 3D bone tissue growth in hollow fibre membrane bioreactor: Implications of various process parameters on tissue nutrition. *International Journal of Artificial Organs* 29 (9):841–51.
88. Tischer, T., Schieker, M., Stengele, M., Pautke, C., Neth, P., Jochum, M., Mutschler, W., and Milz, S. 2004. 3D-culturing of human osteoblastic cells with vessel-like nutrient supply. *Zeitschrift Fur Orthopadie Und Ihre Grenzgebiete* 142 (3):344–49.
89. Freed, L. E., Langer, R., Martin, I., Pellis, N. R., and VunjakNovakovic, G. 1997. Tissue engineering of cartilage in space. *Proceedings of the National Academy of Sciences of the United States of America* 94 (25):13885–90.
90. Song, K. D., Liu, T. Q., Li, X. Q., Cui, Z. F., Ge, D., Sun, X. Y., and Ma, X. H. 2004. Three-dimensional fabrication of engineered bone in rotating wall vessel bioreactor. *Progress in Biochemistry and Biophysics* 31 (11):996–1005.
91. Song, K. D., Liu, T. Q., Li, X. Q., Cui, Z. F., Sun, X. Y., and Ma, X. H. 2007. Three-dimensional expansion: In suspension culture of SD rat's osteoblasts in a rotating wall vessel bioreactor. *Biomedical and Environmental Sciences* 20 (2):91–98.
92. Song, K. D., Yang, Z. M., Liu, T. Q., Zhi, W., Li, X. Q., Deng, L., Cui, Z. F., and Ma, X. H. 2006. Fabrication and detection of tissue-engineered bones with bio-derived scaffolds in a rotating bioreactor. *Biotechnology and Applied Biochemistry* 45:65–74.

93. Coulson, J. M., Richardson, J. F., Bankhiurst, J. R., and Harker, J. H. 1996. *Coulson & Richardson's chemical engineering: Particle technology and separation processes,* Vol. 2, 979, 4th ed. Oxford: Butterworth-Heinemann Ltd.

94. Drioli, E., and De Bartolo, L. 2006. Membrane bioreactor for cell tissues and organoids. *Artificial Organs* 30 (10):793–802.

95. Granet, C., Laroche, N., Vico, L., Alexandre, C., and Lafage-Proust, M. H. 1998. Rotating-wall vessels, promising bioreactors for osteoblastic cell culture: Comparison with other 3D conditions. *Medical and Biological Engineering and Computing* 36 (4):513–19.

96. Zhang, Z. Y., Szita, N., Boccazzi, P., Sinskey, A. J., and Jensen, K. F. 2006. A well-mixed, polymer-based microbioreactor with integrated optical measurements. *Biotechnology and Bioengineering* 93 (2):286–96.

97. De Bartolo, L., Catapano, G., Della Volpe, C., and Drioli, E. 1999. The effect of surface roughness of microporous membranes on the kinetics of oxygen consumption and ammonia elimination by adherent hepatocytes. *Journal of Biomaterials Science-Polymer Edition* 10 (6):641–55.

98. De Bartolo, L., Morelli, S., Gallo, M. C., Campana, C., Statti, G., Rende, M., Salerno, S., and Drioli, E. 2005. Effect of isoliquiritigenin on viability and differentiated functions of human hepatocytes maintained on PEEK-WC-polyurethane membranes. *Biomaterials* 26 (33):6625–34.

99. De Bartolo, L., Morelli, S., Rende, M., Gordano, A., and Drioli, E. 2004. New modified polyetheretherketone membrane for liver cell culture in biohybrid systems: adhesion and specific functions of isolated hepatocytes. *Biomaterials* 25 (17):3621–29.

100. De Bartolo, L., Morelli, S., Giorno, L., Carnpana, C., Rende, M., Salerno, S., Maida, S., and Drioli, E. 2006. Polyethersulfone membrane biohybrid system using pig hepatocytes: Effect of diclofenac on cell biotransformation and synthetic functions. *Journal of Membrane Science* 278 (1–2):133–43.

101. Silva, A. I., de Matos, A. N., and Brons, I. G., Mateus, M. 2006. An overview on the development of a bio-artificial pancreas as a treatment of insulin-dependent diabetes mellitus. *Medicinal Research Reviews* 26 (2):181–222.

102. Ma, Z. W., Gao, C. Y., Gong, Y. H., and Shen, J. C. 2003. Chondrocyte behaviors on poly-L-lactic acid (PLLA) membranes containing hydroxyl, amide or carboxyl groups. *Biomaterials* 24 (21):3725–30.

103. Hua, F. J., Kim, G. E., Lee, J. D., Son, Y. K., and Lee, D. S. 2002. Macroporous poly(L-lactide) scaffold 1. Preparation of a macroporous scaffold by liquid–liquid phase separation of a PLLA-dioxane-water system. *Journal of Biomedical Materials Research* 63 (2):161–67.

104. Brotherton, J. D., and Chau, P. C. 1990. Modeling analysis of an intercalated-spiral alternate-dead-ended hollow fiber bioreactor for mammalian-cell cultures. *Biotechnology and Bioengineering* 35 (4):375–94.

105. Brotherton, J. D., and Chau, P. C. 1995. Protein-free human-human hybridoma cultures in an intercalated-spiral alternate-dead-ended hollow-fiber bioreactor. *Biotechnology and Bioengineering* 47 (3):384–400.

106. Vandewitte, P., Esselbrugge, H., Peters, A. M. P., Dijkstra, P. J., Feijen, J., Groenewegen, R. J. J., Smid, J., et al. 1993. Formation of Porous Membranes for Drug Delivery Systems. *Journal of Controlled Release* 24 (1–3):61–78.

107. Eenink, M. J. D., and Feijen, J. 1987. Biodegradable hollow fibres for the controlled release of hormones. *Journal of Controlled Release* 6:225–47.

Index

A

AAm, *see* Acrylamide (AAm)
Acetylcholine esterase, 167, 174–175
Acidic fibroblast growth factor, 442
Acrylamide (AAm), 10, 220
  acrylamide-alginate hydrogel, 190
2-Acrylamido-2-methylpropane
    sulfonic acid, 27–28
Acrylate-derivatized alumoxane, 72
Acrylate monomer, 162
Acrylic esters, 194
Acrylic resin—poly(ethyl
    methacrylate), 61
Acute myeloid leukemia, 94, 96, 375, 389
Adenovirus, 25
Adhesion, 72–73, 87, 119, 123, 381, 389,
    407–408, 412, 414–416, 418,
    421–423, 439–440, 443, 446, 474,
    482–486
  adhesive bonding, 482
Adsorption isotherm, 306
Affi-gel, 156
Affinity adsorbent, 371, 375, 377
Affinity ligand, 94, 156, 364,
    369–370, 377
AFM, *see* Atomic force microscopy
    (AFM)
Agarose
  macroporous cryogel, 145–147
  monolith, 160, 172, 291
  sponge, 146–147
AGE, *see* Allyl glycidyl ether (AGE)
Aggregation, 30, 268, 294, 389
Agmatine-based mimetic of RGD
    peptide (RGDm), 389
Aldehyde-modified
    carboxymethylcellulose, 139
Alginate
  alginate-β-tricalcium phosphate
    (TCP), 482
  alginate-poly-L-ornithine, 443
  based anisotropic capillary
    hydrogel, 442–443

cryogel, 86, 147–148
  grade, 190
  sponge, 135–136, 442–443
Al(III) initiator, 473
Alkaline lysate, 170, 355
Alkaline lysis, 337
Alkaline phosphatase, 72, 421
Allyl glycidyl ether (AGE), 26, 34, 42,
    147, 271
Amberlite, 88, 94, 278
  Amberlite IRA-PVA cryogel, 278
  Amberlite IRA-401, 278
  Amberlite XAD-4, 94
Ammonium persulfate (APS), 33–35
Amperometric biosensor, 174–175
Ampicillin-resistant *Escherichia coli,* 368
Amylopectin, 141
Anchorage-dependent animal cell, 350
Anchorage-dependent cell line, 396
Anchorage-independent cell, 388
Angiogenesis, 385, 428, 438, 440–441,
    443, 475, 480, 488
Anion exchange, 88, 170, 278, 284, 298,
    303–307, 313, 315, 323, 338,
    343–346, 348–349, 356, 369
Anisotropic capillary hydrogel,
    442–443
Anisotropic macroporous solids, 68–69
Anisotropy, 249, 429
Antiapoptotic, 389
Antibacterial activity, 87
Antibacterial properties, 90
Anti-β-galactosidase, 172
Antibiotics, 23, 411
Antibodies, 373
  label, 75, 373, 376
Anticancer agent, 390
Anti-CD34-antibodies, 375–76, 378
Antigen drug, 301
Antigenicity, 295
Antisense, 301
Aorta, 88
Apoptosis, 388, 390, 392
APS, *see* Ammonium persulfate (APS)